Dictionary of Biology

D0760172

The Wordsworth
Dictionary
of Biology

–

Edited by Peter M. B. Walker

Wordsworth Reference

First published 1990 as *Chambers Biology Dictionary*
by W&R Chambers Ltd, Edinburgh.

This edition published 1995 by Wordsworth Editions Ltd,
Cumberland House, Crib Street, Ware, Hertfordshire SG12 9ET.

Copyright © W&R Chambers Ltd 1989.

All rights reserved. This publication may not be
reproduced, stored in a retrieval system,
or transmitted, in any form or by any means, electronic,
mechanical, photocopying, recording or otherwise,
without the prior permission of the publishers.

ISBN 1-85326-352-4

Printed and bound in Denmark by Nørhaven.

The paper in this book is produced from pure wood
pulp, without the use of chlorine or any other substance
harmful to the environment. The energy used in its
production consists almost entirely of hydroelectricity
and heat generated from waste materials, thereby
conserving fossil fuels and contributing little to the
greenhouse effect.

Contents

Preface

Arrangement

The entries in this dictionary are strictly alphabetical with single letter entries occurring at the beginning of each letter. Greek letters will be found under the nearest anglicized equivalent, as 'psi' under 'p'.

Numbers in chemical nomenclature: the convention is for any leading numerals to be ignored, so that '2,4,5,-T' will be found at the position determined by 'T'; in other areas the entry occurs at the position determined by the number spelt out.

Italic and Bold

Italic is used for:

    (1) alternative forms of, or alternative names for, the headword. Some entries list synonyms at the beginning of the entry. In others synonyms may be added in italics at the end of an entry after 'also';

    (2) terms derived from the headword, often after 'adj.' or 'pl.';

    (3) variables in mathematical formulae;

    (4) generic and specific names in binomial classification in the Biological Sciences;

    (5) for emphasis.

Bold is used for:

    cross-references, either after 'see', 'cf.' etc. or in the body of the entry. Such a cross-reference indicates that there is a headword elsewhere which amplifies the original entry.

Subject categories

Animal Behaviour
Biochemistry
Cell Biology
Ecology
Ethology
Forestry
Genetics

Immunology
Microscopy
Molecular Biology
Radiology
Statistics
Zoology

Abbreviations used in the dictionary

The following are the more common abbreviations used in this dictionary. Many others (including contractions, prefixes and symbols) occur in their alphabetical position in the text, especially at the beginning of each letter of the alphabet.

abbrev(s).	abbreviation(s)	*K*	kelvin(s)
adj(s).	adjective(s)	*L.*	Latin
ant.	antonym	*lb*	pound(s)
approx.	approximately	*m*	metre(s)
bp	boiling point	*min.*	minute(s)
c.	century	*MKS(A)*	metre-kilogram(me)-
°C	degree(s) Celsius		second(-ampere)
	(centigrade)	*mp*	melting point
ca	circa	*n.*	noun
cf.	compare	*N.*	North
CGS	centimetre-gram(me)-	*pl(s).*	plural(s)
	second	*r.a.m.*	relative atomic mass
CNS	central nervous	*rel.d.*	relative density
	system	*S.*	South
colloq.	colloquially	*S., sec.*	second(s)
conc.	concentrated	*SI*	Système International
CRO	cathode-ray		(d'Unités)
	oscilloscope	*sing.*	singular
CRT	cathode-ray tube	*sp*	species (*sing.*)
E.	East	*spp*	species (*pl.*)
esp.	especially	*syn.*	synonym
°F	degree(s) Fahrenheit	*TN*	trade (proprietary)
ft	foot, feet		name
g	acceleration due	*UK*	United Kingdom
	to gravity	*US*	United States
	gram(me)(s)		(of America)
Gk.	Greek	*USSR*	Union of Soviet
h., hr	hour(s)		Socialist Republics
in	inch(es)	*W.*	West
kg	kilogram(me)(s)	*y., yr*	year(s)
km	kilometre(s)	*yd*	yard(s)

Greek alphabet

The letters of the Greek alphabet, frequently used in technical terms, are given here for purposes of convenient reference. The roman letters refer to the dictionary letter at which any headwords beginning with Greek letters will be found.

A	α	alpha	a		N	ν	nu	n
B	β	beta	b		Ξ	ξ	xi	x
Γ	γ	gamma	g		O	o	omicron	o
Δ	δ	delta	d		Π	π	pi	p
E	ε	epsilon	e		P	ϱ	rho	r
Z	ζ	zeta	z		Σ	$\sigma\ \varsigma$	sigma	s
H	η	eta	e		T	τ	tau	t
Θ	$\theta\ \vartheta$	theta	t		Y	υ	upsilon	u
I	ι	iota	i		Φ	ϕ	phi	p
K	$\varkappa$	kappa	k		X	χ	chi	k
Λ	λ	lambda	l		Ψ	ψ	psi	p
M	μ	mu	m		Ω	ω	omega	o

Acknowledgements and Contributors

Extensive use has been made of the entries in *Chambers Science and Technology Dictionary* and I would like to thank all those who contributed to the individual sections of that dictionary. In addition the following have revised entries and contributed longer articles used in the panels:

Dr P. Fantes, M.A.

Dr J. Ham

Dr A. Maddy, M.Sc.

Professor A. Miller, F.R.S.E.

Professor J. M. Mitchison, F.R.S.

A. J. Tulett, B.Sc.

P.M.B.W.

The **Calvin cycle** on p. 44 is a modified form of Fig. 7.1 in *Photosynthesis: Metabolism, Control and Physiology*, D. W. Lawlor; published by Longman Scientific & Technical/John Wiley and Sons, Inc., 1987.

A

a- Prefix signifying *on*. Also shortened form of *ab-, ad-, an-, ap-.*

ABA Abbrev. for *ABscisic Acid.*

abambulacral Pertaining to that part of the surface of an Echinoderm lacking tube feet.

abdominal limbs Segmented abdominal appendages in most Crustacea which are used for swimming, setting up currents of water for feeding and/or respiration, or carrying eggs and young. In Diplopoda, segmented ambulatory appendages on the abdomen.

abdominal pores Apertures leading from the coelom to the exterior in certain Fish and in Cyclostomata.

abdominal reflex Contraction of the abdominal wall muscles when the skin over the side of the abdomen is stimulated.

abductor Any muscle that draws a limb or part away from the median axis by contraction, e.g. the abductor pollicis, which moves the thumb outward.

abiogenesis *Spontaneous generation*. The development of living organisms from non-living matter; either the spontaneous generation of yeasts, bacteria etc. believed in before Pasteur, or the gradual process postulated for the early Precambrian in modern theories of the origin of life.

abiotic Pertaining to non-living things.

ABO blood group substances See panel.

ABO blood group system See panel.

abomasum In ruminant Mammals, the fourth or true stomach. Also *reed, rennet.*

aboral Opposite to, leading away from, or distant from, the mouth. See **abambulacral**.

abortifacient Anything which causes artificial abortion; a drug which does this.

abortion (1) Expulsion of the foetus from the uterus during the first 3 months of pregnancy. Abortion may be spontaneous or induced. (2) Termination of the development of an organ.

abscess A localized collection of pus in infected tissue, usually confined within a capsule.

abscisic acid *Dormin*. A sesquiterpenoid plant growth substance, $(C_{15}H_{20}O_4)$ with a variety of reported effects, e.g. inhibiting growth, causing stomatal closure and promoting senescence, abscission and dormancy. Abbrev. *ABA.*

abscission The organized shedding of parts of a plant by means of an abscission layer.

absolute threshold The minimal intensity of a physical stimulus required to produce a response.

absorbed dose The energy absorbed by the patient from the decay of a radionuclide given for diagnostic or therapeutic purposes. The unit is a gray (*GY*). 1 GY = 1 Joule/kg.

absorption Used in immunology to describe the use of reagents to remove unwanted antibodies or antigens from a mixture.

abundance See **relative abundance, frequency.**

abyssal Refers to the ocean floor environment between ca 4000–6000 m. See **littoral, bathyal.**

abyssopelagic Refers to the region of deep water which excludes the ocean floor; floating in the ocean depths.

acantho- Prefix from Gk. *akantha*, spine, thorn.

Acanthocephala A phylum of elongate worms with rounded body and a protrusible proboscis, furnished with recurved hooks; there is no mouth or alimentary canal; the

---

ABO blood group system and substances

The most important of the antigens of human red blood cells for blood transfusion serology. Humans belong to one of four groups: A, B, AB and O. The red cells of each group carry respectively the A antigen, the B antigen, both A and B antigens, or neither. Natural antibodies (resulting from immunization by bacteria in the gut) are present in the blood against the blood group antigen which is absent from the red cells. Thus persons of group A have anti-B, of group B have anti-A, of group O have anti-A and anti-B, and group AB have neither. Before blood transfusion the blood must be cross-matched to ensure that red cells of one group are not given to a person possessing antibodies against them.

Blood group substances are large glycopeptides with oligosaccharide side chains bearing ABO antigenic determinants identical to those of the erythrocytes of the same individual, present in mucous secretions of persons who possess the secretor gene.

young stages are parasitic in various Crustaceans, the adults in Fish and aquatic Birds and Mammals. Thorny-headed worms.

acanthozooid In Cestoda, the proscolex, or head portion, of a bladder worm. Cf. **cystozooid**.

Acarina An order of small *Arachnida*, with globular, undivided body. The immature stages (hexapod larvae) have 6 legs. A large worldwide group, occupying all types of habitat, and of great economic importance. Many are ectoparasitic. Mites and ticks.

acarophily, acarophytism A symbiotic association between plants and mites.

acaulescent Having a short stem.

acauline, acaulose Stemless or nearly so.

accessorius A muscle which supplements the action of another muscle; in Vertebrates, the eleventh cranial nerve or spinal accessory.

accessory bud A bud additional to a normal axillary bud.

accessory cell (1) A **subsidiary cell** in a stomatal complex. (2) A cell other than a lymphocyte which takes part in an immune reaction, e.g. in *antigen presentation* and/or by modulating the function of the lymphocyte. Usually **macrophages** or **dendritic cells**.

accessory chromosome See **sex chromosome**.

accessory glands Glands of varied structure and function in connection with genitalia, esp. of Arthropoda.

accessory pigments Pigments found in chloroplasts and blue-green algae which transfer their absorbed energy to *chlorophyll a* during photosynthesis. They include *chlorophylls b, c* and *d*, the *carotenoids* and the *phycobilins*.

accessory pulsatory organs In some Insects and Molluscs, saclike contractile organs, pulsating independent hearts, and variously situated on the course of the circulatory system. Also *accessory hearts*.

Accipitriformes Order comprising the birds of prey. Hooked bill with fleshy cere, long talons with opposable hind toe, predaceous. Eagles, harriers, vultures, buzzards.

acclimation See **acclimatization**.

acclimatization, acclimation Reversible alteration of an organism's or individual's tolerance for environmental stress or conditions, e.g. for temperature conditions. The physiological adjustment usually takes days or weeks. See **hardening**.

accommodation For Piaget, one of the two main biological forces responsible for *cognitive* development; it refers to the process of modifying pre-existing cognitive organisations in order to absorb information that no longer fits into existing *schemata*. See **assimilation**.

accretion External addition of new matter; growth by such addition.

acellular Not partitioned into cells. Sometimes used for *unicellular* but for *multinucleate* or **coenocytic**.

acentric Having no centromere, applied to chromosomes and chromosome segments.

acentrous Having a persistent notochord with no vertebral centra, as in the Cyclostomata.

acephalous Showing no appreciable degree of cephalization; lacking a head-region, as Pelecypoda.

acervulus A dense cushionlike mass of conidiophores and conidia formed by some fungi. adj. *acervulate*.

acetabular bone In Crocodiles, the separate lower end of the ilium; in some Mammals, as *Galeopithecus*, an additional bone lying between the ilium and pubis and frequently fused with one or the other.

acetabulum In Platyhelminthes, Hirudinea, and Cephalopoda, a circular muscular sucker; in Insects, a thoracic aperture for insertion of a leg; in Vertebrates, a facet or socket of the pelvic girdle with which articulates the pelvic fin or head of the femur; in ruminant Mammals, one of the cotyledons of the placenta.

acetic fermentation The fermentation of dilute ethanol solutions by oxidation in presence of bacteria, esp. *Bacterium aceti*. Acetic acid is formed.

achene A dry indehiscent, 1-seeded fruit, formed from a single carpel, and with the seed distinct from the fruit wall. Also called *achaenocarp, akene*.

Achilles tendon In Mammals, the united tendon of the soleus and gastrocnemius muscles; the hamstring.

achlorhydria Absence of hydrochloric acid from gastric juice.

achondroplasia Dwarfism characterized by shortness of arms and legs, with a normal body and head. adj. *achondroplastic*.

achroglobin A colourless respiratory pigment occurring in some Mollusca and some Urochorda.

acicular (1) Needle-shaped. (2) The needle-like habit of crystals.

aciculum A stout internal chaeta in the parapodium of Polychaeta, acting as a muscle attachment.

acidosis A condition where the hydrogen ion concentration of blood and body tissues is increased, (normal range 36–43 nmol.l). *Respiratory acidosis* is caused by retention of carbon dioxide by the lungs; *metabolic acidosis* by retention of non-volatile acids (renal failure, diabetic ketosis) or loss of base (severe diarrhoea).

acid rain Rain that is unnaturally acid (pH 3 to 5.5) as a result of pollution of the atmosphere with oxides of nitrogen and sulphur from the burning of coal and oil.

acid soil complex Combination of aluminium and/or manganese toxicity with calcium deficiency which affects a relatively calcicole (or noncalcifuge) plant growing on an acid soil; preventable horticulturally by liming. Cf. **lime-induced chlorosis**.

aciniform Berry-shaped; e.g. in Spiders, the *aciniform* glands producing silk and leading to the median and posterior spinnerets.

acinostele A protostele in which the xylem is star-shaped in cross section with phloem between the arms as in the roots of most seed plants and the stems of *Lycopodium*.

acne Inflammation of a sebaceous gland. Pimples in adolescents are commonly due to infection with the acne bacillus.

acoelomate, acoelomatous Without a true coelom.

acoelomate triploblastica Animals with three embryonic cell layers but no coelom. They consist of the Platyhelminthes, Nematoda and some minor phyla, i.e. all the helminth phyla.

acoelous Lacking a gut cavity.

acontia In Anthozoa, free threads, loaded with nematocysts, arising from the mesenterics or the mesenteric filaments, and capable of being discharged via the mouth or via special pores.

acotyledonous Lacking cotyledons.

acquired behaviour Refers to a variety of processes which underlie attempts to adapt to environmental change; definitions are various and reflect the interest of different disciplines (e.g. psychology, physiology); most definitions assume long term changes in the central nervous system and exclude short term behaviour changes due to maturation, fatigue, sensory adaptation or habituation.

acquired character A modification of an organ during the lifetime of an individual due to use or disuse, and not inherited. For the *inheritance* of acquired characters, see **Lamarckism**.

acquired immunity Immunity resulting from exposure to foreign substances or microbes. Cf. **natural immunity**.

acquired immunodeficiency syndrome See panel.

acquired variation Any departure from normal structure or behaviour, in response to environmental conditions, which becomes evident as an individual develops.

Acrasiomycetes *Dictyosteliomycetes*. Cellular slime moulds. Class of slime moulds (Myxomycota) feeding phagotrophically as *myxamoebae* which aggregate to form a migratory **pseudoplasmodium** which eventually develops into a fruiting body, liberating most of the cells as spores. E.g. *Dictyostelium*.

acriflavine A deep orange, crystalline substance possessing antiseptic (bacteriostatic and bactericidal) properties; used in wound dressings.

acro- Prefix from Gk. *akros*, topmost, farthest, terminal.

acrocarp A moss in which the main axis is terminated by the development of reproductive organs. Any subsequent growth must be sympodial. Most are erect in habit. Cf. **pleurocarp**.

acrocentric Having centromere at the end, applied to chromosomes; a rod-shaped chromosome.

acrodont Said of teeth which are fixed by their bases to the summit of the ridge of the jaw.

acromion In higher Vertebrates, a ventral process of the spine of the scapula. adj. *acromial*.

acron In Insects, the embryonic, presegmental region of the head.

acropetal Towards the apex, away from the base.

acropodium That part of the pentadactyl limb of land Vertebrates which comprises the digits and includes the phalanges.

acrosome Structure forming the tip of a mature spermatozoon. adj. *acrosomal*.

acrotrophic In Insects, said of ovarioles in which nutritive cells occur at the apex.

acrylamide gel electrophoresis A gel for the electrophoretic separation of proteins and RNA according to their molecular weight. The monomer can be cast in the form of sheets or cylinders by polymerization *in situ* to give a clear gel. See **sodium dodecyl sulphate**.

ACTH *Adrenocorticotrophic hormone, corticotrophin.* A protein hormone of the anterior pituitary gland controlling many secretory processes of the adrenal cortex. Used medically to stimulate cortisol production as an anti-inflammatory measure.

actin A globular protein (G actin) which can polymerize into long fibres (F actin). Originally discovered in muscle in association with myosin, it is now known to be widely distributed at sites of cellular movement.

actinal In Echinodermata, see **ambulacral**; in Anthozoa, pertaining to the crown, including the mouth and tentacles; star-shaped.

actinic radiation Ultraviolet waves, which have enhanced biological effect by inducing chemical change; basis of the science of photochemistry.

actinin A protein frequently associated with actin, serving as a terminus for actin filaments. It was first found in the Z bands of striated muscle.

actino- Prefix from Gk. *aktis*, ray.

actinobiology The study of the effects of

acquired immunodeficiency syndrome

AIDS. Severe immunodeficiency due to infection by a retrovirus (HIV I) which infects predominantly **helper T lymphocytes** (T4 cells), leading to failure of **cell-mediated immunity**. It also infects macrophages and brain cells. The virus has a surface protein gp120 (a glycoprotein of 120 kd) which binds to the **CD4** receptors on the T4 cells, allowing the virus to infect these cells and to integrate a DNA copy of its genome into that of its host cell by **reverse transcription**. The infected cell then makes many copies of the virus and more gp120, some of which migrates to the cell's surface and there reacts with healthy T4 cells to form **syncytia** of ineffective T4 cells. Free gp120 may also circulate in the blood and lymph where it can bind to healthy T4 cells making them, like the infected cells, susceptible to attack by the body's immune system. Thus the processes available for the elimination of a foreign viral antigen have the paradoxical effect of also killing the very cells needed for the process to occur.

The course of the disease is spread over 6 or so years, marked after the first year by a continuous decline in the T4 cell population. In the early stages the patient may be more or less symptom-free, although the lymph nodes become chronically swollen after the first year. During this period a rare cancer, *Kaposi's sarcoma*, and a number of lymphomas are often found. After about 4 years immune dysfunction becomes increasingly obvious with the absence of **delayed type hypersensitivity** and then the onset of **opportunistic infections**. In the terminal stages of the disease patients may suffer a gradual loss of mental ability.

The disease is transmitted by sexual intercourse or by inoculation of blood, e.g. *via* infected needles or by blood transfusion with contaminated blood. The latter can be avoided in countries which can afford the rigorous screening needed, but is a continued hazard elsewhere. High risk groups are homosexuals and drug addicts but the infection now appears widespread among heterosexuals in central Africa, where a second, related, virus HIV 2 has been identified. This virus is endemic in parts of West Africa but does not appear to have caused an epidemic of an AIDS-like disease.

At present there is no cure and little hope of recovery, despite world-wide scientific and medical research. One approach is to try and disrupt the stage unique to retroviruses, reverse transcription, and here there has been some success with the use of 3'-azido-2',3'-dideoxythymidine (AZT), which in its triphosphate form, competitively binds to reverse transcriptase. In blind clinical trials, this compound was found both to prolong life and to improve its quality, but there are severe side effects at high doses and it does not provide a cure. Other related compounds with different toxicities are being tested in combined therapies. *Vaccination*, so successful with other viruses, may eventually prevent the disease but the AIDS virus presents special difficulties including its specificity towards the immune system. A preferred method is to use as antigen a part of the gp120 molecule attached to an adjuvant, but gp120 exists in many strain-specific variants and antigens containing the stable anti-CD4 sequence must produce antibodies with a specificity very like that of CD4 itself. Consequently, any secondary immune response against these antibodies will also be against T4 cells. It is, moreover, going to be very difficult to test a vaccine even in a population at risk. Its members may very well choose not to put themselves at risk despite vaccination and it would be several years before it could be shown that any HIV positive individuals are disease free.

radiation upon living organisms.
actinoid Star-shaped.
actinomorphic *Star-shaped*. Organisms which are radially symmetrical; divisible into two similar parts by any one of several longitudinal planes passing though the

active transport

The movement of a molecule across a cell membrane against its concentration or electrochemical gradient is known as active transport. Such a movement is an energy-requiring process and so active transport can only proceed in a metabolically active cell, ceasing when the energy supply in the form of ATP is restricted. That ATP is used is shown by the association between hydrolysis of ATP and solute flux when the sodium/potassium pump of *plasma membranes* is active. This establishes the ion asymmetry across the membrane by pumping sodium out of the cell and potassium in.

Solute movement which is directly linked to ATP hydrolysis is known as *primary transport* in contrast to *secondary transport*. In the latter the solute also moves against its concentration gradient, but its movement depends upon a gradient of a second solute itself established by a mechanism requiring the hydrolysis of ATP. Thus the entry of many sugars and amino acids into animal cells depends upon the initial establishment of a sodium ion gradient in the opposite direction from that which the sugar is to travel. The sugars or amino acids are taken along with the sodium ions as the latter move passively down their own concentration gradient.

Active transport systems consist of catalytic proteins, with many of the properties of enzymes which are stategically located in the membrane so as to direct the solute flux. In primary transport these proteins can be recognized by an ATPase activity which is solute dependent. Thus the sodium/potassium pump requires both sodium and potassium and the analogous calcium pump calcium ions. It is known that ATP hydrolysis is associated with the phosphorylation of pump proteins but the precise mechanism is not understood. Cf. **facilitated diffusion**.

centre. Includes starfish, sea urchins and plants in which the stamens are helically arranged rather than whorled.

Actinomycetales An order of Bacteria producing a fine mycelium and sometimes arthrospores and conidia. Some members are pathogenic in animals and plants. Some produce antibiotics, e.g. **streptomycin**.

Actinopterygii Subclass of Gnathostomata in which the basal elements of the paired fins do not project outside the body wall, the fin webs being supported by rays alone. Ganoid scales are diagnostic of the group and the skeleton is fully ossified in most members. The most widespread modern group of fishes, including cod, herring, etc.

activated Term applied to lymphocytes or macrophages which have undergone differentiation from a resting state, and have acquired new capacities such as the ability to secrete **lymphokines**, or in the case of macrophages, increased ability to kill and digest microbes.

activation A step in the fertilization process triggered by the incorporation of the spermatozoon into the egg cytoplasm, by which the secondary oocyte is stimulated to complete its division and becomes haploid.

activator (1) Of an enzyme, a small molecule which binds to it and increases its activ-

ity. (2) Of DNA transcription, a protein which, by binding to a specific sequence, increases the production of a gene product. (3) Any agency bringing about *activation*.

active chromatin See **transcriptionally active chromatin**.

active space The area surrounding an animal within which it can communicate.

active transport See panel.

activity Attribute of an amount of radionuclide. Describes the rate at which transformations occur. The unit is a **becquerel** (*Bq*).

Aculeata Stinging hymenoptera, e.g. bees, ants and some wasps.

aculeate Bearing prickles, or covered with needlelike outgrowths.

acuminate Having a long point bounded by hollow curves; usually descriptive of a leaf-apex. dim. *acuminulate*.

acute Bearing a sharp and rather abrupt point; said usually of a leaf-tip.

acute Said of a disease which rapidly develops to a crisis. Cf. **chronic**.

acute phase substances Proteins which appear in the blood in increased amounts shortly after the onset of infections or tissue damage. They are made in the liver and include **C-Reactive protein**, fibrinogen, proteolytic enzyme inhibitors, transferrin. The stimulus is interleukin-1 (IL-1) released by macrophages. These proteins probably serve to counteract some of the effects of tissue damage.

acyl-CoA Coenzyme A conjugated by a thioester bond to an acyl group, e.g. acetyl-CoA, succinyl-CoA. These compounds are intermediates in the transfer of the acyl groups, e.g. the formation of citric acid by the interaction of acetyl-AcA with oxalo-acetic acid.

acylic Having the parts of the flower arranged in spirals, not in whorls.

ad- Prefix signifying to, at, from L. *ad*.

adambulacral In Echinodermata, adjacent to the ambulacral areas.

Adam's apple In Primates, a ridge on the anterior or ventral surface of the neck, caused by the protuberance of the thyroid cartilage of the larynx.

adaptation *Evolutionary adaptation*, adjustment to environmental demands through the long term process of natural selection acting on the genotype; *sensory adaptation*, a short term change in the response of a sensory system as a consequence of repeated or protracted stimulation; *adaptation* (child psychology), a term used by Jean Piaget to describe the developmental process underlying the child's growing awareness and interactions with the physical and social world. The process of **assimilation**, **accommodation**, and **equilibration** are fundamental to this concept of psychological adaptation.

adaptation of the eye The sensitivity adjustment effected after considerable exposure to light (*light adapted*), or darkness (*dark adapted*).

adaptive radiation Evolutionary diversification of species from a common ancestral stock, filling available ecological niches. Also *divergent adaptation*.

adaptor hypothesis The prediction that some molecule would be needed to adapt the 4 base genetic code to the 20 amino-acid product. **tRNA** fulfils the prediction.

adaxial That surface of a leaf, petal etc. that during early development faced towards the axis (and usually, therefore, the upper surface of an expanded leaf). Cf. **abaxial**.

addict Someone physically dependent on a drug and who will experience withdrawal effects if the drug is discontinued.

additive genetic variance That part of the genetic variance of a *quantitative character* that is transmitted and so causes resemblance between relatives.

adductor A muscle that draws a limb or part inwards, or towards another part; e.g. *adductor mandibulae* in *Amphibia* is a muscle which assists in closing the jaws.

adelphous Said of an androecium in which the stamens are partly or wholly united by

their filaments.

adendritic Without dendrites.

adenine *6-aminopurine*, one of the five bases in nucleic acids in which it pairs with thymine in DNA and uracil in RNA.

Structure:

See **genetic code**.

adenitis Inflammation of a gland.

adeno- Prefix from Gk. *aden*, gland.

adenoid Lymphoid tissue in nasopharynx of children which may become enlarged as a result of repeated upper respiratory tract infection. Also *glandlike*.

adenopathy Disease or disorder of glandular tissue. The term is usually used in reference to lymphatic gland enlargement.

adenosine triphosphate, ATP The triphosphate of the nucleotide adenosine (adenine + ribose). As the predominant high-energy phosphate compound of all living organisms it has a pivotal role in cell energetics, mediating, by interconversion with the diphosphate (ADP), the energy transfer between *exergonic* and *endergonic* metabolic reactions. See **electron transfer chain, respiration.**

adenyl cyclase An enzyme which catalyses the formation of cyclic adenylic acid from ATP.

adhesion Abnormal union of two parts which have been inflamed; a band of fibrous tissue which joins such parts.

adhesion plaque A specialized region of the plasma membrane where stress fibres terminate.

adipo- Prefix from L. *adeps*, fat.

adipose tissue A form of connective tissue consisting of vesicular cells filled with fat and collected into lobules.

adjuvants In general remedies which assist others, more particularly substances which increase the immunogenicity of antigens when administered with them. One action is to provide a depot from which the antigen is released slowly, and another is to activate macrophages in the neighbourhood so as to ensure more effective antigen presentation. Some absorb the antigen onto mineral particles, such as aluminium hydroxide, others use water-in-oil emulsions with or without macrophage stimulatory substances such as muramyl dipeptide.

adlacrimal The lacrimal bone of Reptiles, so called to indicate that it is not homologous with the lacrimal bone of Mammals.

adnate Joined to another organ of a differant kind, as when the stamens are fused to the petals. Cf. **connate**.

adoral Adjacent to the mouth.

adrectal Adjacent to the rectum.

adrenal Adjacent to the kidney; pertaining to the adrenal gland.

adrenal cortex Outer portion of the adrenal gland which produces glucocorticoids.

adrenal gland See **suprarenal body**.

adrenaline See **catecholamine**. syn. *epinephrine*.

adrenal medulla The inner region of the adrenal gland which produces catecholamines.

adrenergic Pertaining to or causing stimulation of the sympathetic nervous system; applied to sympathetic nerves which act by releasing an adrenaline-like substance from their nerve endings.

adventitia Accidental or inessential structures; the superficial layers of the wall of a blood-vessel. adj. *adventitious*.

adventitious Applied to a plant part developed out of the usual order or in an unusual position. An *adventitious bud* is any bud except an **axillary** bud; it gives rise to an *adventitious branch*. An ·*adventitious root* develops from some part of a plant other than a pre-existing root.

adventive Not permanently established in a given habitat or area.

advertisement A conspicuous display that can involve coloration, posture or sound, and that serves to convey some information about the sender, e.g. age, sex, status, motivation.

aedeagus In male Insects the intromittent organ.

aegithognathous Of Birds, having a type of palate in which the maxillopalatines do not meet the vomer or each other, and in which the vomer is broad and truncate anteriorly; the palatines and pterygoids articulate with the basisphenoid rostrum.

aerenchyma Tissue with particularly well-developed, air-filled, intercellular spaces. Characteristic of the cortex of roots and stems of hydrophytes, where it probably facilitates gas exchange between the roots and the leaves. Cf. **pneumatophore**.

aerobe An organism which can live and grow only in the presence of free oxygen; an organism which uses aerobic respiration.

aerobic respiration See **respiration**.

aestival Occurring in summer, or characteristic of summer.

aestivation Prolonged summer torpor, as in some Insects. Cf. **hibernation**.

aetiology, etiology The medical study of the causation of disease.

afebrile Without fever.

affective behaviour Refers to a wide range of behaviour in which the emotional aspects of social interactions are salient and often fundamental, e.g. mother-infant interactions.

affective disorders A group of disorders whose primary characteristic is a disturbance of mood; feelings of elation or sadness become intense and unrealistic.

afferent Carrying towards, as blood vessels carrying nervous impulses to the central nervous system. Cf. **efferent**.

afferent arc The sensory or receptive part of a reflex arc, including the adjustor neurone(s).

affinity Measure of the strength of interaction or binding between antigen and antibody or between a receptor and its ligand.

aflagellar Lacking flagella.

aflatoxins Group of secondary metabolites produced by *Aspergillus flatus* and *A. parasiticus* which commonly grow on stored food esp. peanuts, rice and cotton seed. Some are highly toxic to cattle and are suspected of causing liver cancer in Africa.

afterbirth The placenta and membranes expelled from the uterus after delivery of the foetus. See **decidua**.

after-ripening The poorly-understood chemical and/or physical changes which must occur inside the dry seeds of some plants after shedding or harvesting, if germination is to take place after the seeds are moistened.

agamic See **parthenogenetic**.

agamogenesis Asexual reproduction.

agamogony See **schizogony**.

agamont See **schizont**.

agamospermy Reproduction by seed formed without sexual fusion. See **apomixis**.

agarics The *Agaricales*, an order of the Hymenomycetes containing the mushrooms and toadstools; ca 3000 species. The spores are borne on the surface of gills or in the lining of pores.

agarose gel electrophoresis Standard method for fractionating DNA fragments produced by restriction endonuclease digestion. Fragments migrate through the gel matrix under the influence of an electric field, at a rate which is a function of size, the smallest fragments moving the furthest. Commonly, a flat slab gel is cast between glass plates with a number of rectangular wells formed into one edge and with the DNA fragments initially placed in the wells. See **pulsed-field gel electrophoresis**.

age distribution The relative frequency, in an animal or plant population, of individuals of different ages. Generally expressed as a polygon or age pyramid, the number or percentage of individuals in successive age classes being shown by the relative width of horizontal bars.

agenesia, agenesis. Imperfect development (or failure to develop) of any part of the body.

ageotropic Not reacting to gravity. See **tropism**.

agglutination (1) Clumping of particles such as bacteria or red cells when linked by antibodies binding antigenic determinants present on the particles. (2) More generally, the formation of clumps by some Protozoa and spermatozoa.

agglutinin Specifically a constituent of the blood plasma of one individual which causes agglutination by reacting with a specific receptor in the red corpuscles in the blood of another individual.

aggregate species Abbrev. *agg.* A group of 2 or more closely similar species denoted, for convenience, by a single shared name, e.g. the blackberry, *Rubus fruticosus* agg.

aggregation A type of animal and plant dispersion in which individuals are closer to each other than they would be if they were randomly dispersed. Also *contagious distribution.*

aggressive behaviour Various and often non-overlapping meanings; most often refers to intentionally delivered noxious stimuli to a **conspecific**, but it can refer to any behaviour that intimidates or damages the conspecific, e.g. scent marking. See **agonistic behaviour**.

aggressive mimicry Resemblance to a harmless species in order to facilitate attack.

aglossate, aglossal Lacking a tongue. n. *aglossia*, congenital absence of the tongue.

Agnatha A superclass of eel-shaped chordates without jaws or pelvic fins. Lampreys and Hagfishes.

agnosia Loss of ability to recognize the nature of an object through the senses of the body.

agonistic behaviour A broad class of behaviour patterns, including all types of attack, threat, appeasement and flight, between members of the same species in response to a conflict between aggression and fear. Behaviour often alternates between attack and escape, e.g. across a territory boundary.

agoraphobia A phobia characterized by a fear of open spaces.

agraphia Loss of power to express thought in writing, as a result of a lesion in the brain.

agrestal Growing in cultivated ground, but not itself cultivated, e.g. a weed.

agroforestry Form of land use in which herbaceous crops and tree crops co-exist in an integrated scheme of farming. See **taungya**.

AIDS Abbrev. for *Acquired ImmunoDeficiency Syndrome.*

air bladder In Fish, an air-containing sac developed as a diverticulum of the gut, with which it may retain connection by the pneumatic duct in later life; usually it has a hydrostatic function, but in some cases it may be respiratory or auditory, or assist in phonation.

air chamber An air-filled cavity, e.g. towards the upper surface of the thallus of some liverworts, opening externally by a pore and containing photosynthetic cells, or in some hydrophytes.

air layering Horticultural method of vegetative propagation of esp. shrubby house plants, in which an aerial shoot is induced to form roots, by wounding and packing the wound with e.g. sphagnum moss, while still attached to the plant. It is severed and potted up after rooting.

air monitor Radiation (e.g. γ-ray) measuring instrument used for monitoring contamination or dose rate in air.

air plant See **epiphyte**.

air sinuses In Mammals, cavities connected with the nasal chambers and extending into the bones of the skull, esp. the maxillae and frontals.

air space Air-filled *intercellular spaces*, esp. large ones.

akaryote A cell lacking a nucleus, or one in which the nucleoplasm is not aggregated to form a nucleus.

akinete A non-motile thick-walled resting spore containing food reserves, formed without division by the direct modification of a vegetative cell in some Cyanophyceae.

Ala Symbol for **alanine**.

ala Any flat, wing-like process or projection, esp. of bone. adj. *alar, alary.*

alalia See **aphonia**.

alanine *2-aminopropanoic acid.* The L-isomer is a non-polar amino acid and constituent of proteins. Symbol Ala, short form A. R-group: CH_3— . See **amino acid**.

alary muscles In Insects, pairs of striated muscles arising from the terga and spread out fanwise over the surface of the dorsal diaphragm. Also *aliform muscles.*

ala spuria See **bastard wing**.

alate (1) Winged; applied to stems when decurrent leaves are present. (2) Having a broad lip (esp. of shells). (3) In Porifera, a type of triradiate spicule with unequal angles.

albinism The state of being an **albino**.

albino An abnormal plant lacking chlorophyll or normal plant pigments. An animal deficient in pigment in hair, skin, eyes etc. adj. *albinotic.*

albumen White of egg containing a number of soluble proteins, mainly ovalbumin. adj. *albuminous.*

albumin A general term for proteins soluble in water as distinct from saline. Specific albumins are designated by their sources, e.g. egg albumin from egg white, serum albumin from blood serum.

albuminous Endospermic.

albuminous cell Specialized parenchyma cell in gymnosperm phloem, associated with a sieve cell but not originating from the same precursor cell. Cf. **companion cell.**

alburnum Sapwood.

alcoholic fermentation A form of anaerobic **respiration** in which a sugar is converted to alcohol and carbon dioxide. Alcoholic fermentation by yeasts is important in baking, brewing and wine making.

alcoholism Disease produced by addiction to alcohol, manifesting itself in a variety of psychotic disorders, e.g. hallucinosis, delirium tremens.

aldohexoses The most important group of **monosaccharides.** including **glucose** and **galactose.** All have a formula which can be expressed: $OH.CH_2.(CH.OH)_4.CHO$.

aldolase The enzyme which catalyses the cleavage of fructose-1-6-diphosphate into glyceraldehyde-3-phosphate and dihydroxyacetone.

aldoses A group of monosaccharides with an aldehydic constitution, e.g. glucose. See **aldohexose.**

aldosterone A potent mineralocorticoid secreted by *zona glomerulosa* of the adrenal cortex which promotes the retention of sodium ions and water.

aldrin A chloro-derivative of naphthalene used as a contact insecticide, incorporated in plastics to make cables resistant to termites; persistent toxicity; formerly used in agriculture, esp. against wireworm.

alecithal Of ova, having little or no yolk.

aleurone Reserve protein occurring in seed granules, usually in the outermost layer, *aleurone layer*, of the endosperm, e.g. in cereals and other grasses.

alexia Word blindness; loss of the ability to interpret written language due to a lesion in the brain.

alexin Obsolete term for complement used by Bordet who first described complement.

algae Prokaryotic and eukaryotic photosynthetic organisms with *chlorophyll a* and other photosynthetic pigments, releasing O_2. Plant body unicellular, colonial, filamentous, siphoneous or parenchymatous, never with roots, stems or leaves. Sex organs unicellular or multicellular with all cells fertile (except Charales). The zygote does not develop into a multicellular embryo within the female sex organ. Includes Cyanophyceae and Prochlorophyceae (both prokaryotic), and Rhodophyceae, Cryptophyceae, Dinophyceae, Heterokontophyta, Haptophyceae, Euglenophyceae and Chlorophyta (all eukaryotic). Not a natural group, but the word is useful in many contexts.

algology The study of algae.

alien Believed on good evidence to have been introduced by man and now more or less naturalized.

aliform muscles See **alary muscles.**

Alismatidae A subclass or superorder of the monocotyledons. Aquatic and semi-aquatic herbs, typically with perianth segments free, sometimes differentiated into sepals and petals, mostly apocarpous, pollen trinucleate, mostly lacking endosperm in the mature seeds. Contains ca 500 spp. in 14 families.

alisphenoid A winglike cartilage bone of the Vertebrate skull, forming part of the lateral wall of the cranial cavity, just in front of the foramen lacerum; one of a pair of dorsal bars of cartilage in the developing Vertebrate skull, lying in front of the basal plate, parallel to the trabeculae; one of the sphenolateral cartilages.

alkaline phosphatase Enzyme commonly conjugated with antibodies for use in **indirect immunoassay.** It catalyses a reaction which deposits dye at the site of the bound antibody. Abbrev. *AP.*

alkalinity Of for example, a lake, $[HCO_3_-] + 2[CO_3{_2}_-] + [OH^-] - [H^+]$. (Square brackets indicate molar concentrations.)

alkaloids Natural organic bases found in plants; characterized by their specific physiological action and toxicity; used by many plants as a defence against herbivores, particularly insects. Alkaloids may be related to various organic bases, the most important ones being pyridine, quinoline, *iso*quinoline, pyrrole and other more complicated derivatives. Most alkaloids are crystalline solids, others are volatile liquids, and some are gums. They contain nitrogen as part of a ring, and have the general properties of amines.

alkylating drug Cytotoxic drug which acts by damaging DNA and interfering with cell replication. *Cyclophosphamide, chlorambucil, busulphan* and *mustine* are common examples.

allantois In the embryos of higher Vertebrates, a saclike diverticulum of the posterior part of the alimentary canal, having respiratory, nutritive or excretory functions. It develops to form one of the embryonic membranes. adj. *allantoic.*

allele Abbrev. for *allelomorph.* Any one of the alternative forms of a specified gene. Different alleles usually have different effects on the phenotype. Any gene may have several different alleles, called *multiple alleles.* Genes are *allelic* if they occupy the same *locus.* adj. *allelic.*

allelopathy Adverse influence exerted by one strain of plant over another of the same species by the production of a chemical inhibitor, often a terpenoid or phenolic.

allergen Antigenic substances which provoke an allergic response (see **allergy**). Commonly used to describe those which cause immediate type **hypersensitivity** such as pollens or insect venoms.

allergic Showing altered responsiveness to an antigen as the result of previous contact with that antigen. Responsiveness is usually increased, but can be decreased.

allergy The reaction of the body to a substance to which it has become sensitive, characterized by oedema, inflammation and destruction of tissue.

allo-antigen Antigen carried by an individual which is capable of eliciting an immune response in genetically different individuals but not in the individual bearing it.

alliaceous Looking or smelling like an onion.

allo- (1) Prefix from Gk. *allos*, other. (2) Thus used to describe a gene product, tissue etc. from a different individual of the same species, e.g. *allograft*.

allogamy Fertilization involving pollen and ovules from: (1) different flowers (whether on the same plant or not), including *geitonogamy* and *xenogamy*; (2) genetically distinct individuals of the same species (i.e. from another **genet.**). See **cross-fertilization, cross-pollination**. Cf. **autogamy**.

allograft Homograft between individuals differing in respect of one or more **histocompatibility antigens**.

allomeric Having the same crystalline form but a different chemical composition.

allometry The relationship between the growth rates of different parts of an organism.

allomone A chemical signal produced by one species of animal which influences the behaviour of members of another to the advantage of the signaller. Cf. **pheromone**.

allopatric Said of two species or populations not growing in the same geographical area; unable to interbreed by reason of distance or geographical barrier. Cf. **sympatric**.

allopatric speciation The accumulation of genetic differences in an isolated population leading to the evolution of a new species.

allopolyploid A polyploid of hybrid origin containing sets of chromosomes from two or more different species, often self-fertile but not interbreeding with the parental species. Allopolyploidy is important in speciation and in the evolution of some crop plants, e.g. wheat, *brassica*. See **amphidiploid**. Cf. **autopolyploid**.

allosteric protein Protein which alters its 3-dimensional conformation as a result of the binding of a smaller molecule, often leading to altered activity, e.g. of an enzyme.

allosteric site An enzymic site, distinct from the **active site**, which by binding molecules other than the substrate induces a *conformational change* which effects the enzyme's activity.

allotetraploid See **amphidiploid**.

allotropous flower A flower in which the nectar is accessible to every insect visitor.

allotype (1) Commonly used to describe identifiable differences between immunoglobulin molecules that are inherited as alleles of a single genetic locus. They are due to single amino acid substitutions in light or heavy chains, and are useful in population studies. (2) More rarely, in taxonomy, used to describe an additional type-specimen of the opposite sex to the original type specimen. (3) An animal or plant fossil selected as a species or subspecies, illustrating morphological details not shown in the holotype.

allozymes Different forms of an enzyme specified by *allelic genes*.

alopecia Baldness.

alpha-chain Heavy chain of **IgA**. Alpha-chain disease is a rare disease in which the intestine is infiltrated by lymphoma which makes alpha chains but no light chains, owing to a deletion involving the site required to link the two.

alpha diversity See **diversity**.

alphafetoprotein A plasma protein made by the foetus, but not by adults unless they have primary liver cancer or some other tumours in which foetal genes are expressed. Alphafetoprotein escapes into the maternal blood during pregnancy, and an abnormally high concentration at certain stages has been found to indicate that normal closure of the spinal canal of the foetus is incomplete (spina bifida).

alpha helix Important element of protein structure formed when a polypeptide chain turns regularly about itself to form a rigid cylinder stabilized by hydrogen bonding.

alpine Of high mountains.

alternate Leaves, branches etc. placed singly on the parent axis, i.e. not in pairs (opposite), not whorled. See **alternate host**.

alternate host One of the two (rarely more) hosts of a parasite which has the different stages of its life cycle in unrelated hosts. Cf. **alternative host**.

alternating cleavage See **spiral cleavage**.

alternation of generations The regular alternation of two (rarely, three) types of individual in the life history of an animal or plant; typically in plants a diploid **sporophyte** and a haploid **gametophyte** or in animals a sexually- and an asexually-produced form. They may be morphologically similar (isomorphic) or different (heteromorphic).

alternative host One of two or more possible hosts for a given stage in the life cycle of a parasite, particularly when it is not the commonest or the most important economically. Cf. **alternate host**.

alternative pathway of complement activation See **complement**.

altrices Birds whose young are hatched in a

---

amino acids

Aminoalkanoic acids, generally with the amino group on the carbon adjacent to the carboxyl group. Some 20 of these are the building blocks of proteins. All except glycine are chiral, and usually have the L- or S-configuration. Some amino acids can be synthesized in an animal body; these vary somewhat from species to species. Those which cannot be synthesized are called *essential amino acids*. For man, these are arginine, histidine, isoleucine, leucine, lysine, methionine, phenylalanine, threonine, tryptophan and valine.

All amino acids, except proline, terminate in the group on the right and it is this group which determines that they are L-isomers. Individual side chains are attached at the R position and shown under each amino acid.

$$\begin{array}{c} COOH \\ \uparrow \\ H_2N \blacktriangleright C \blacktriangleleft H \\ \downarrow \\ R \end{array}$$

---

very immature condition, generally blind, naked, or with down feathers only, unable to leave the nest, fed by the parents, e.g. the Perching Birds, Passeriformes.

altruism Broad class of animal behaviour in which an individual benefits another at the risk of its own life or expense.

alula See **alia spuria**.

alveolar, alveolate Having pits over the surface, and resembling honeycomb.

alveolus A small pit or depression on the surface of an organ; the cavity of a gland; a small cavity of the lungs; in higher Vertebrates, the tooth socket in the jaw bone; in Echinodermata, part of Aristotle's lantern, one of five pairs of grooved ossicles which grasp the teeth; in Gastropoda, the glandular end-portion of the tubules of the digestive gland, secreting enzymes.

ambergris A greyish white fatty substance with a strong but agreeable odour, obtained from the intestines of diseased sperm whales; sometimes found floating on the surface of the sea. It is used in perfumery as a fixative; on suitable treatment it yields ambreic acid.

amber mutation A base change in a coding sequence of DNA, which gives the stop codon UAG, resulting in a shortened gene product.

ambi- L. form of Gk. *amphi-*.

ambisexual, ambosexual Pertaining to both sexes; activated by both male and female hormones.

amblyopia Dimness of vision, from the action of noxious agents on the optic nerve or retina.

ambrosia Certain Fungi which are cultivated for food by some Beetles (see **ambrosia beetle**), the pollen of flowers collected by social Bees and used in the feeding of the larvae.

ambrosia beetle Beetles of the family Scotylidae which cultivate the fungus *Monilia candida* in galleries in wood to feed their lar-

vae and themselves.

ambulatory Having the power of walking; used for walking.

ameiosis Nonpairing of the chromosomes in synapsis (meiosis).

amelification The formation of enamel.

ameloblast A columnar cell forming part of a layer immediately covering the surface of the dentine, and secreting the enamel prisms in the teeth of higher Vertebrates.

amenorrhoea, amenorrhea Absence or suppression of menstruation.

amentia Mental deficiency; failure of the mind to develop normally, whether due to inborn defect, or to injury or disease.

amentum A catkin. adj. *amentiform*.

ametabolic Having no obvious metamorphosis.

amino acids See panel.

aminoglycosides Group of antibiotics particularly effective against **Gram-negative bacteria**. Common members of this group are *gentamycin, neomycin, streptomycin* and *tobramycin*.

amino group Essential component of the **amino acids**.

amitosis, amitotic division Direct division of the nucleus by constriction, without the formation of a spindle and chromosomes; direct nuclear division, occurring in the meganuclei of the *Ciliophora*. Cf. **mitosis, meiosis**.

ammonification, ammonization The release of ammonia from amino acids (and ultimately protein) in decaying organic matter by soil bacteria. A step in the mineralization of nitrogen. Commonly followed, except in waterlogged and/or acid soils, by **nitrification**.

amnesia *Loss of memory*. Common in dissociation states of hysteria. In a concussed patient *retrograde amnesia* is loss of memory of events immediately preceding the concussion.

amniocentesis Method for diagnosing

foetal abnormalities in which foetal cells removed from amniotic fluid at about the 16th week of gestation are cultured and used in diagnostic assays, including **probing** with DNA sequences to detect disease-associated alleles.

amnion In Insects, the inner cell envelope covering and arising from the edge of the germ band; in higher Vertebrates, one of the embryonic membranes, the inner fold of blastoderm covering the embryo, formed of ectoderm internally and somatic mesoderm externally.

Amniota Those higher Vertebrates which possess an amnion during development, i.e. Reptiles, Birds, and Mammals. adj. *amniote*.

amniotic cavity In *Amniota*, the space between the embryo and the amnion.

amniotic fluid Liquid filling the amniotic cavity.

amniotic folds Protrusions round the periphery of the blastoderm which give rise to the amnion and the chorion.

Amoebida An order of Sarcodina, the members of which extrude lobose pseudopodia and generally lack a skeleton, or have only a simple shell; their ectoplasm is never vacuolated.

amoebocyte A metazoan cell having some of the characteristics of an amoeboid cell, esp. as regards form and locomotion; a leucocyte.

amoeboid Of a cell, having no fixed form, creeping, and putting out pseudopodia.

amoeboid movement Locomotion of an individual cell by means of pseudopodia.

amphetamine The sulphate is used as a drug for its vasomotor, respiratory and stimulant effects. Popularly called '*purple hearts*'.

amphiaster During cell division by meiosis or mitosis, the two asters and the spindle connecting them.

Amphibia Class of semi-aquatic chordates with larvae possessing gills and anamniotic eggs. Frogs, Toads, Salamanders.

amphibious Adapted for both terrestrial and aquatic life.

amphiblastic Of ova, showing complete but unequal segmentation.

amphibolic Capable of being turned backwards or forwards, as the 4th toe of Owls.

amphicondylous, amphicondylar Having two occipital condyles.

amphicribral bundle A vascular bundle in which a central strand of xylem is surrounded by phloem.

amphidiploid An **allopolyploid** containing the diploid set of chromosomes from each of two species.

amphimixis True sexual reproduction, with the fusion of two gametes to form a zygote. Cf. **apomixis**.

Amphineura A class of bilaterally symmetrical Mollusca in which the foot, if present, is broad and flat, the mantle is undivided, and the shell is absent or composed of eight valves. Coat-of-Mail Shells etc.

amphiont See **zygote**.

amphiphloic Having phloem on both sides of the xylem, e.g. solenostele.

amphiplatyan Having both ends flat, as in certain types of vertebral centrum.

amphipneustic Possessing both gills and lungs; in dipterous larvae, having the prothoracic and posterior abdominal spiracles only functional.

Amphipoda An order of Malacostrica in which the carapace is absent, the eyes are sessile, and the uropods styliform; the body is laterally compressed. They show great variety of habitat, being found on the shore, in the surface waters of the sea, in fresh water, and in the soil of tropical forests. Some are parasitic. Whale Lice, Sandhoppers, Skeleton Shrimps etc.

amphipodous Having both ambulatory and natatory appendages.

amphirhinal Having two external nares.

amphistomatal, amphistomatic Leaf etc. having stomata on both surfaces. Cf. **epistomatal**.

amphistomous Having a sucker at each end of the body; as leeches.

amphithecium Outer layer(s) of a developing sporophyte of a Bryophyte giving rise to capsule wall. Cf. **endothecium**.

amphitrichous Having a flagellum at each end of the cell.

amphitropous Said of an ovule bent like a 'V' and attached, to its stalk, near the middle of its concave side.

ampholines Mixtures of aliphatic amino acids with a range of iso-electric points which are used to establish the pH gradients used in **iso-electric focusing**.

ampicillin A semisynthetic penicillin with a broad range of activity against those bacteria causing bronchitis, pneumonia, gonorrhoea, certain forms of meningitis, enteritis, biliary and urinary tract infections.

amplexicaul Said of a sessile leaf with its base clasping the stem horizontally.

amplexus See **copulation**.

amplification The process by which multiple copies of genes or DNA sequences are formed.

ampulla Any small membranous vesicle; in Vertebrates, the dilation housing the sensory epithelium at one end of a semicircular canal of the ear; in Mammals, part of a dilated tubule in the mammary gland; in Fish, the terminal vesicle of a neuromast organ; in Echinodermata, the internal expansion of the axial sinus below the madreporite. adj. *ampullary*.

amyelinate Of nerve fibres, nonmedullated, lacking a myelin sheath.

amygdala A lobe of the cerebellum; one of the palatal tonsils.

amyl-, amylo- Prefixes meaning starch.

amylase Enzyme which hydrolyses the internal 1,4-glycosidic bonds of starch. The amylase found in human saliva is known as *ptyalin*.

amyloid An insoluble fibrillary material deposited around blood vessels. This can arise in different ways, but one form which occurs with myelomas making light chains or in very prolonged infections with raised immunoglobulin levels, results from aggregates of free light chains.

amylolytic Starch-digesting.

amylopectin A polymer, α-1$\rightarrow$4 linked with α-1$\rightarrow$6 branches, of glucose. A constituent of starch.

amylose The sol constituent of starch paste; linear polymer of glucose units having α-1$\rightarrow$4 glucosidic bonds. Cf. **amylopectin**.

amylum Starch.

an Prefix from Gk. *an*, not. See **ap-**.

ana Prefix from Gk. *ana*, up, anew.

anabiosis A temporary state of reduced metabolism in which metabolic activity is absent or undetectable. See **cryptobiosis**.

anabolism The chemical changes proceeding in living organisms with the formation of complex substances from simpler ones. adj. *anabolic*.

anabolite A substance participating in anabolism.

anaemia, anemia Diminution of the amount of total circulating haemoglobin in the blood.

anaerobe An organism which can grow in the absence or near absence of oxygen. Facultative anaerobes can utilize free oxygen; obligate anaerobes are poisoned by it. adj. *anaerobic*.

anaerobic Living in the absence of oxygen. *Anaerobic respiration* is the liberation of energy which does not require the presence of oxygen.

anaerobic respiration See **respiration**.

anaerobiosis Existence in the absence of oxygen. adj. *anaerobic*.

anaesthesia, anesthesia Correctly, loss of feeling but often applied to the technique of pain relief for surgical procedures.

anal cerci In Insects, sensory appendages of one of the posterior abdominal somites, generally the 11th, retained throughout life.

anal character In psychoanalytic theory, an adult personality derived from unresolved conflicts (**fixation**) during the anal stage of psychosexual development. It is characterized by a **reaction-formation** against impulsiveness, resulting in personality traits of an extremely controlled and compulsive type

(e.g. extreme tidiness, parsimony). See **anal phase**.

analgesia Loss of sensibility to pain.

analogous organs Organs which are similar in appearance and/or function but which are neither equivalent morphologically nor of common evolutionary origin, e.g. foliage leaves and cladodes; the wings of birds and insects. Cf. **homologous organs**.

analogy Likeness in function but not in evolutionary origin, e.g. tendrils, which may be modified leaves, branches, inflorescences; the wings of Birds and of Insects. adj. *analogous*. Cf. **homology**.

anal phase, stage In psychoanalytic theory, the second phase of *psychosexual development*, during which the focus of pleasure is on activities related to retaining and expelling the faeces; occurs in the second year of life. See **anal character**.

analplerotic Reactions which replenish deficiencies of metabolic intermediates, e.g. the formation of oxaloacetate by the carboxylation of pyruvate.

anal suture In the posterior wings of some Insects, a line of folding, separating the anal area of the wing from the main area.

analysis of variance The partition of the total variation in a set of observations into components corresponding to differences between and within subclassifications of the data, used as a method of comparing subclassification means.

anamnesis The recollection of past things; the patient's recollections of symptoms and past illnesses.

anamnestic Antibody response following antigenic stimulation in which antibodies occur reactive with an antigen previously encountered, but different from that which elicited the response.

Anamniota Vertebrates without an amnion during development, i.e. Amphibia and Fish.

anamniotic Lacking an amnion during development. Also *anamniote*.

anamorph An asexual or imperfect stage of a fungus, esp. Deuteromycotina. Cf. **teliomorph**.

anaphase The stage in mitotic or meiotic nuclear division when the chromosomes or *half-chromosomes* move away from the equatorial plate to the poles of the spindle; more rarely, all stages of mitosis leading up to the formation of the chromosomes.

anaphylatoxin Peptides released from **complement** components C3 and C5 during complement activation which act on mast cells to release histamine etc. as in anaphylaxis.

anaphylaxis, anaphylactic shock An acute immediate hypersensitivity reaction following administration of an antigen to a subject resulting from combination of the antigen with IgE on mast cells or **basophils** which causes these cells to release histamine and other vasoactive agents. An acute fall in blood pressure may be so severe as to be fatal. Other symptoms include bronchospasm, laryngeal oedema and urticaria.

anaplasia Loss of the differentiation of a cell associated with proliferative activity; a characteristic of a malignant tumour.

anapophysis In higher Vertebrates, a small process just below the postzygapophysis which strengthens the articulation of the lumber vertebrae.

anapsid Having the skull completely roofed over, i.e. having no dorsal foramina other than the nares, the orbits, and the parietal foramen.

Anapsida A subclass of the reptiles containing the oldest known forms, characterized by the temporal region of the skull lacking the fenestrae. Turtles.

anarthrous Without distinct joints.

anatomy (1) The study of the form and structure of animals and plants; it includes the study of minute structures, and thus includes **histology**. (2) Dissection of an organized body in order to display its physical structure.

anatropous Said of an inverted ovule, so that the micropyle is next to the stalk.

anaxial Asymmetrical.

anconeal Pertaining to, or situated near, the elbow.

anconeus An extensor muscle of the arm attached in the region of the elbow.

andro- Prefix from Gk. *aner*, gen. *andros*, man, male.

androconia In certain male Lepidoptera, scent scales serving to disseminate the **pheromones** which serve the purpose of sexual attraction.

androcyte Cells in an antheridium which will metamorphose to form antherozoids.

androdioecious Said of a species, having some individuals with male flowers only and others hermaphrodite flowers only. Cf. **dioecious**.

androecium The male part of a flower, consisting of one or more stamens. Cf. **gynoecium**.

androgen General term for a group of male sex hormones, which e.g. stimulate the growth of male secondary sex characteristics. Cf. **oestrogen**.

androgenesis (1) Development from a male cell. (2) Development of an egg after entry of male germ cell without the participation of the egg nucleus.

androgynous Bearing staminate and pistillate flowers on distinct parts of the same inflorescence; having the male and female organs on or in the same branch of the thallus.

andromonoecious A species in which all

of the plants bear both male and hermaphrodite flowers. Cf. **monoecious**.

androphore An elongation of the receptacle of the flower between the corolla and the stamens.

androsporangium Sporangium in which androspores are produced.

androspore In heterosporous plants, same as **microspore**. Cf. **gynospore**.

anecdysis The intermoult period in Arthropoda.

anemochorous Dispersed by wind.

anemophily Pollination by means of wind. Dispersal of spores by wind. adj. *anemophilous*.

anemotaxis Orientation to an odour source based upon wind direction.

anencephaly Neural tube defect in which skull and cerebral hemispheres fail to develop. adj. *anencephalic*.

anergy Absence of ability to give the expected allergic responses, esp. delayed-type hypersensitivity. Occurs when the lymphocytes or monocytes needed are absent or suppressed.

aneuploid A cell or individual with missing or extra chromosomes (or parts of chromosomes in a *segmental aneuploid*). Cf. **euploid**.

aneurysm Pathological dilatation, fusiform or saccular, of an artery.

angio- Prefix from Gk. *angion*, denoting a case or vessel.

angioblast An embryonic mesodermal cell from which the vessels and early blood cells are derived.

angiography The study of the cardiovascular system by means of radio-opaque media.

angiology The study or scientific account of the anatomy of blood and lymph vascular systems.

angiosperms, flowering plants See panel.

Anguilliformes Order of Osteichthyes with pelvic fins and girdle absent or reduced; elongate forms. Eels.

angular divergence The angle subtended at the mid-line of an apical meristem of a shoot by the mid-point of two successive leaf primordia. This varies between species but, where the phyllotaxis is spiral, it is commonly the Fibonacci angle 137.5°.

anima, animus Term used in Jungian psychology to denote the unconscious feminine component in men and the unconscious masculine component in women.

animal charcoal The carbon residue obtained from carbonization of organic matter such as blood, flesh etc.

animal electricity A term used to denote the ability possessed by certain animals (e.g. electric eel) of giving powerful electric

angiosperms, flowering plants

The group, often classified as a class, Angiospermae, or as a division, Anthophyta or Magnoliophyta, which contains those seed plants in which the ovules are enclosed within carpels (in contrast to the unenclosed ovules of the **Gymnosperms),** the pollen germinating on a stigma and pollen tubes growing to the ovule(s) in the ovary; there is characteristically double fertilization. The carpels and stamens are usually borne in flowers, which are often more or less showy and/or fragrant and thus attractive to insect, bird or bat pollinators or, alternatively, more or less inconspicuous and wind pollinated. The xylem usually has vessels and the phloem sieve tubes and companion cells. There are about 220 000 species in two classes *Monocotyledones* and *Dicotyledones*.

Angiosperms dominate most terrestrial habitats, the most obvious exceptions being the boreal coniferous forests, many freshwater habitats and a few intertidal marine habitats e.g. eel-grass.

The earliest undisputed angiosperm fossils (pollen and leaves) date from the Lower Cretaceous; angiosperms may have originated in, and certainly underwent rapid diversification in this period. Many of the mesozoic groups of Gymnosperma have been suggested as the immediate ancestors of the angiosperms but the latter's evolutionary origins are not clear. The angiosperm's success in replacing the dominant gymnosperms of the early Mesozoic has been ascribed to the protection offered to the ovules by the ovary, to the potential of stigmatic isolating mechanisms and of insect pollination to increase the rate of speciation, to faster growth and maturation and to the more efficient transport afforded by the more advanced phloem and by esp. the xylem vessels.

With the major exception of the conifers, important for soft-wood timber, most plants of economic importance are angiosperms and almost all agriculture and horticulture is based on them. They also provide food like the grains and pulses (e.g. Gramineae, Leguminosae), leaf and root vegetables (Cruciferae, Umbelliferae, Solanaceae), fruit (Rosaceae, *Citrus*), oil (Palamae, Cruciferae) and sugar (Gramineae, Chenopodiaceae), beverages, herbs (Labiatae, Umbelliferae) and spices, fodder (Gramineae), constructional materials such as timber (dicotyledenous trees), bamboo (Gramineae) and rattans (Palmae), fibres (Malvaceae, flax), many important drug plants (*Digitalis, Vinca, Atropa*) and much of the world's fuel.

Although perhaps 2000–5000 species are important in these ways, fewer than 200 are of major importance in world trade and fewer than 20 provide the bulk of the world's food. Many more species are cultivated as ornamental and amenity plants.

shocks.

animal field In developing blastulae, a region distinguished by the character of the contained yolk granules, and representing the first rudiment of the germ band.

animal pole In the developing ovum, the apex of the upper hemisphere, which contains little or no yolk; in the blastula, the corresponding region, wherein the micromeres lie.

animism Attributing feelings and intentions to non-living things. In Piagetian theory children's thinking is characterized by animism in the years two to six.

aniso- Prefix from Gk. *an*, not, *isos*, equal.

anisocercal Having the lobes of the tail-fin unequal.

anisogamete A gamete differing from the other conjugant in form or size. adj. *anisogamous*.

anisogamy Sexual fusion of gametes that differ in size but not necessarily in form. See **heterogamy, isogamy, oögamy.**

anisopleural Bilaterally asymmetrical.

anisotropic Of ova, having a definite polarity, in relation to the primary axis passing from the animal pole to the vegetable pole. n. *anisotropy*.

ankylosis, anchylosis The fusion of two or more skeletal parts, esp. bones.

anlage See **primordium.**

anneal To reform the duplex structure of a

nucleic acid.

Annelida A phylum of metameric Metazoa, in which the perivisceral cavity is coelomic, and there is only one somite in front of the mouth; typically there is a definite cuticle and chitinous setae arising from pits of the skin; the central nervous system consists of a pair of preoral ganglia connected by commissures to a postoral ventral ganglionated chain; if a larva occurs it is a trochophore. Earthworms, Ragworms, Leeches.

annual A plant that flowers and dies within a period of one year from germination. Cf. **ephemeral, biennial, perennial**.

annual ring A growth ring formed over a year.

annulus Any ring-shaped structure. (1) The fourth digit of a pentadactyl forelimb. (2) In Arthropoda, subdivision of a joint forming jointlets. (3) In Hirudinea, a transverse ring subdividing a somite externally. (4) A membranous frill present on the stipe of some agarics. (5) A patch or a crest of cells with thickened walls occurring in the wall of the sporangium of ferns, and bringing about dehiscence by setting up a strain as they dry. (6) A zone of cells beneath the operculum of the sporangium of a moss, which break down and assist in the liberation of the operculum. adj. *annular, annulate*.

anodontia Absence of teeth.

anoestrus In Mammals, a resting stage of the oestrus cycle occurring between successive heat periods.

anomaly Any departure from the strict characteristics of the type.

anomeristic Of metameric animals, having an indefinite number of somites.

anorexia Loss of appetite.

anorexia nervosa Chronic failure to eat due to fear of gaining weight or to emotional disturbance; results in malnutrition, semi-starvation and sometimes death.

anosmia Loss, partially or completely, of the sense of smell.

anoxia, anoxaemia, anoxemia Deficiency of oxygen in the blood; any condition of insufficient oxygen supply to the tissues; any condition which retards oxidation processes in the tissues and cells.

anoxybiosis Life in absence of oxygen.

Anseriformes An order of Birds with webbed feet. The members of this order are unusual in the possession of an evaginable penis; they are all aquatic forms, living on the animals found living in the mud at the bottom of shallow waters and in marshes; some are powerful fliers. Geese, Ducks, Screamers, Swans.

antagonism A relationship between different organisms in which one partly or completely inhibits the growth of, or kills, a second, esp. when due to a toxic metabolite.

See **antibiotic, allelopathy**.

ante- Prefix from L. *ante*, before.

antebrachium The region between the brachium and the carpus in land Vertebrates; the fore-arm.

antecubital In front of the elbow.

antenna In Arthropoda, one of a pair of anterior appendages, normally many-jointed and of sensory function; in Angler Fish the elongate first dorsal fin-ray, which bears terminally a skinny flap, used by the fish to attract prey. pl. *antennae*. adjs. *antennary, antennal*.

antennal glands The principal organs of Crustacea. They open at the bases of the osmoregulatory appendages from which they take their name. Also called *maxillary glands*.

antennule A small antenna; in some Arthropoda (as the Crustacea) which possess two pairs of antennae, one of the first pair.

antepetalous, antipetalous Inserted opposite to the petals.

anteposition Situation opposite, and not alternate to, another plant member.

anterior (1) The side of a flower next to the bract, or facing the bract. (2) That end of a motile organism which goes first during locomotion. (3) In animals in which cephalization has occurred, nearer the front or cephalad end of the longitudinal axis. (4) In human anatomy, ventral.

antero- Prefix from *anterior*, former.

anterograde amnesia Loss of memory for events after injury to the brain or mental trauma; with little effect on information acquired previously.

antesepalous, antisepalous Inserted opposite to the sepals.

anther Fertile part of a stamen, usually containing 4 sporangia, and producing pollen.

anther culture See **plant cell culture**.

antheridiophore, antheridial.receptacle A specialized branch of a thallus bearing antheridia.

antherozoid A motile male gamete, spermatozoid or sperm.

anthesis The opening of a flower bud; by extension, the duration of life of any one flower, from the opening of the bud to the setting of fruit.

anthocyanins A large group of water-soluble, flavonoid, glycoside pigments in cell vacuoles responsible for the red, purple and blue colours of flowers, fruit and leaves in most flowering plants. Cf. **betalains**.

anthogenesis A form of parthenogenesis in which both males and females are produced by asexual forms, as in some Aphidae.

anthophilous Flower-loving; feeding on flowers.

anthophore An elongation of the floral receptacle between the calyx and corolla.

Anthophyta (1) Usually Angiospermae. See **angiosperms**. (2) Rarely Spermatophyta.

Anthozoa A class of Cnidaria in which alternation of generations does not occur, the medusoid phase being entirely suppressed; the polyps may be solitary or colonial; the gonads are of endodermal origin. Also *Actinozoa*.

anthracnose One of a number of plant diseases characterized by black, usually sunken, lesions; mostly caused by one of the fungi of the Melanconiales.

anthrax An acute infective disease caused by the anthrax bacillus, communicable from animals to man in whom it causes cutaneous malignant pustules and lung, intestinal and nervous system infection. Also *woolsorter's disease*, notifiable disease in animals for which vaccines are available.

anthropogenic Resulting from or influenced by man's activities.

anthropoid Resembling Man; pertaining to, or having the characteristics of the Anthropoidea.

anthropomorph A conventional design of the human figure; resembling a human in form or in attributes.

anthropophyte A plant introduced incidentally in the course of cultivation.

anti- Prefix from Gk. *anti*, against.

anti-auxin Compound that in low concentrations will directly interfere with **auxin** action, e.g. tri-chlorobenzoic acid.

antibiosis A state of mutual antagonism. Cf. **symbiosis**.

antibiotic resistance The property of micro-organisms or cells, which can survive high concentrations of a normally lethal agent. Normally acquired by the selection of a rare resistant mutant in the presence of low concentrations of the agent, but can be added by **genetic manipulation**.

antibody Immunoglobulin with combining site able to combine specifically with antigenic determinants on an antigen. See **immunoglobulin**.

antical The upper surface of a thallus, stem or leaf.

anticlinal Perpendicular to the nearest surface. If a cell divides anticlinally the daughter cells will be separated by an anticlinal wall. Cf. **periclinal**.

anticodon The sequence of three bases on **tRNA** which binds to the codon of **mRNA**. The complement of the coding triplet.

antidromic Contrary to normal direction, e.g. applied to nerve cells, when the impulse is conducted along the axon towards the cell body.

antigen A substance which has determinant groups which can interact with specific receptors on lymphocytes or on antibodies. The term is often used to include substances which can stimulate an immune response, although these are more correctly termed *immunogens*.

antigenic determinant A small part of the antigen which has a structure complementary to the recognition site on a T cell receptor or an antibody. Most antigens are large molecules with several different antigenic determinants, which interact with lymphocytes with differing specific recognition sites.

antigenic variation Many viruses, bacteria and protozoa can develop new antigenic determinants as a result of genetic mutation and selection during multiplication in their hosts. If the variation involves the antigenic component which stimulates protective immunity, the variant can cause infection in subjects who would otherwise be immune to that microbe.

antiglobulin Term used to describe antibodies against immunoglobulins which can be used to detect the presence of immunoglobulins bound to the surface of cells or microbes, and so to detect the binding of antibodies to them.

antihistamine A substance or drug which inhibits the actions of histamine by blocking its site of action.

anti-idiotype Antibody which recognizes the combining site of an antibody against an antigenic determinant on an antigen. The combining site of the anti-idiotype may thus resemble the shape of the determinant on the original antigen. If anti-idiotypes are used in turn to elicit antibodies against them, it is possible to obtain anti-anti-idiotypes which recognize the determinant on the original antigen, even though this was never administered. In principle such a method could be used to immunize against antigens which in themselves are dangerous or impractical.

anti-lymphocytic serum Serum containing antibodies reactive with surface antigens on lymphocytes and capable of killing or otherwise suppressing their capabilities. It is used, usually as Ig concentrate, as an immunosuppressive agent to prevent tissue graft rejection or in severe autoimmune disease.

antimetabolite Drugs used in treatment of cancer which are incorporated into new nuclear material and prevent normal cell division. Common examples are methotrexate, cytoarabinose and fluorouracil.

antimutagen A compound which inhibits the action of a mutagen.

anti-nuclear factor Auto-antibody reactive with nucleic acids (DNA or RNA) present in the blood of subjects with **systemic lupus erythematosus** (SLE) and some other autoimmune conditions. Used diagnostically.

antiperistaltic Said of waves of contraction

passing from anus to mouth, along the alimentary canal. Cf. **peristaltic**. n. *antiperistalsis*.

antipetalous See **antepetalous**.

antipodal cells Wall-less cells, usually three, typically haploid, derived by mitotic division of the megaspore, lying in the embryo sac at the end remote from the micropyle.

antipyretic Counteracting fever; a remedy for fever.

antisepsis The inhibition of growth, or the destruction of bacteria in the field of operation by chemical agent; the principle of antiseptic treatment.

antithetic alternation of generations (1) See **antithetic theory of alternation**. (2) Sometimes the same as *heteromorphic alternation of generations*.

antithetic theory of alternation The hypothesis that the sporophyte is a novel phase in the life cycle resulting from the postponement of meiosis. Cf. **homologous theory of alternation**.

antitoxin Antibody capable of neutralizing toxins which are made usually by microbes, e.g. tetanus antitoxin, diphtheria antitoxin. Used in treatment when the main damaging agent is the toxin. Produced by immunizing horses or other animals, but liable to cause **serum sickness**. Nowadays antibodies from pre-immunized humans are used when possible, and in future human monoclonal antibodies may become available.

antitranspirant Substance which reduces transpiration (and, usually, photosynthesis) e.g. by causing stomatal closure or by forming a more or less impermeable surface film; of some use horticulturally when transplanting.

antivivisectionists Those who oppose experiments on live animals.

antorbital In front of the orbit. In Vertebrates, a small bone in the nasal region.

antrorse Directed or bent forward.

antrum A sinus, as the maxillary sinus in Vertebrates; a cavity, as the **antrum of Highmore**. pl. *antra*.

Anura See **Salientia**.

anural, anurous Without a tail; pertaining to *Anura*.

anus The opening of the alimentary canal by which indigestible residues are voided, generally posterior. adj. *anal*.

anxiety An unpleasant psychological state most commonly identified with intense fear; the situational determinants are not always apparent.

anxiolytic Drug used for relieving anxiety states.

aorta In Arthropoda, Mollusca, and most Vertebrates, the principal arterial vessel(s) by which the oxygenated blood leaves the

heart and passes to the body; in Amphibians, the principal artery by which blood passes to the posterior part of the body, formed by the union of the systemic arteries; in Fish (*ventral aorta*), the vessel by which the blood passes from the heart to the gills, and (*dorsal aorta*) the vessel by which the blood passes from the gills to the body. adj. *aortic*.

aortic arches In Vertebrates, a series of pairs of vessels arising from the *ventral aorta*.

AP See **alkaline phosphatase**.

ap- Another form of *an-*.

apatetic coloration See **cryptic coloration**.

aperturate Said of pollen grains having one or more apertures, that is, areas of the wall where the exine is thinner or absent and through which the pollen tube may emerge. Cf. **colpus, pore**.

apetaly Absence of petals. adj. *apetalous*.

apex The end of an organ or plant part remote from its point of attachment or origin, e.g. root tip or shoot tip. The top or pointed end of anything. adj. *apical*.

apgar score A scoring system for assessing a baby's condition at birth of which a value of 0, 1 or 2 is given to each of five signs: colour, heart rate, muscle tone, breathing effort and response to stimulation. A score of 10 indicates the baby is in excellent condition.

aphagia Inability to swallow or feed.

aphasia Loss of, or defect in, language function due to a lesion in certain association areas of the brain. May be unable to comprehend speech or written word (receptive) or be defective in writing or speech (expressive).

apheliotropic Turning away from the sun.

aphids Insects of the family Aphididae (order Hemiptera). Reproduction is either sexual or parthenogenetic, oviparous or viviparous, giving rise to a complex life cycle. Greenfly.

aphonia Loss of voice.

aphotic zone The zone of the sea below about 1500 metres which is essentially dark. Cf. **photic zone**.

aphototropic (1) Usually *not phototropic*. (2) Less commonly and confusingly, *negatively phototropic*.

apical body See **acrosome**.

apical cells In some Invertebrates, e.g. the Limpet (*Patella*), during cleavage of the ovum, a quartette of small cells at the apex of the egg.

apical dome The usually dome-shaped part of an apical meristem distal to the most recently formed leaf primordium.

apical dominance The influence of a terminal bud in inhibiting or controlling the growth of buds or lateral branches on the shoot below it, ceasing if the terminal bud is destroyed.

apical growth (1) The elongation of tubular cell or hypha by continued growth at the apex only (the normal pattern for root hairs, pollen tubes and fungal hyphae). (2) The condition in which the only transverse divisions in a filament of cells take place in the apical cell.

apical meristem A group of meristematic cells at the tip of a thallus, stem or root, which divide to produce the precursor of the cells of the thallus, or of the primary tissues of root or shoot. There may or may not be a distinct apical cell.

apical placentation The condition in which the ovule or ovules is/are inserted at the top of the ovary.

apical plate In various pelagic larval forms, such as trochophores, tornariae, echinoplutei and larvae of some Crinoidea, an aggregation of columnar ectoderm cells at the apical pole, usually bearing cilia.

apical sense organ In Ctenophora, an elaborate sensory structure formed of small otoliths united into a morula, supported on 4 pillars of fused cilia and covered by a roof of fused cilia.

apiculate Ending in a short, sharp point.

aplacental Without a placenta.

aplanetic Non-motile or lacking a motile stage.

aplanogamete A non-motile gamete.

aplanospore A non-motile spore, e.g. autospore, hypnospore. Cf. **zoospore**.

aplasia Defective structural development.

apneustic Possessing no organs specialized for respiration; in some aquatic insect larvae, having no functional spiracles, respiration taking place through the general body surface or by means of gills.

apneustic centre In the higher Vertebrates, that part of the brain which controls the inflation of the lungs.

apnoea Cessation of breathing. Recognized to occur in sleep in obese and other persons. *Sleep apnoea syndrome.*

apo- Prefix from Gk. *apo*, away.

apocarpous Said of a gynoecium consisting of two or more free (i.e. not fused) carpels. Cf. **syncarpous**.

Apoda (1) An order of Amphibians having a cylindrical snakelike body without limbs, reduced eyes and an anterior sensory tentacle; burrowing forms, living near water and feeding chiefly upon earthworms. Caecilians.

apodal, apodous Without feet; without locomotor appendages.

apodeme In Arthropoda, an ingrowth of the cuticle forming an internal skeleton and serving for the insertion of muscles; in Insects, more particularly, an internal lateral chitinous process of the thorax.

apodous larva A type of insect larva in which the trunk appendages are completely suppressed; formed in some Coleoptera, Diptera, Hymenoptera and Lepidoptera.

apogamy The development of a sporophyte directly from a cell of the gametophyte without fusion of gametes so that the resulting sporophyte has the same chromosome number as the parent gametophyte. Cf. **apospory**, **apomixis**. adj. *apogamous*.

apomixis (1) *Agamospermy*, reproduction by seeds formed without sexual fusion. (2) Any form of asexual reproduction, including vegetative propagation.

apophysis In Vertebrates, a process from a bone, usually for muscle attachment; in Insects, a ventral chitinous ingrowth of the thorax for muscle insertion.

apoplast That part of the plant body which is external to the living protoplasts, i.e. the cell walls, the intercellular spaces and the lumina of dead cells such as xylem vessels and tracheids or in some contexts, the water filled parts of this space. Cf. **symplast**.

apoprotein The protein component of a conjugated protein. The *globin* of haemoglobin.

aporogamy The entrance of the pollen tube into the ovule by a path other than through the micropyle.

aposematic coloration Warning coloration, often yellow and black as in some stinging insects.

apospory The development of a gametophyte directly from a sporophyte cell without meiosis and the formation of spores. The resulting gametophyte has the same chromosome number as the parent sporophyte. Cf. **apogamy**, **apomixis**.

apostrophe The position assumed by chloroplasts in bright light, when they lie against the radial walls of the cells of the palisade layer of the mesophyll.

apothecium An open ascocarp, often cup- or saucer-shaped.

appeasement behaviour Submissive behaviour which inhibits attack by a conspecific, often by minimizing threat signals or by mimicking sexual or infantile behaviours, e.g. crouching or sexual invitation.

appendage In plants, a general term for any external outgrowth which does not appear essential to growth or reproduction of the plant. In animals, a projection of the trunk, as the parapodia and tentacles of Polychaeta, sensory tentacle of Apoda, fins of Fish, and limbs of land Vertebrates; in Arthropoda, almost exclusively one of the paired, metamerically arranged, jointed structures with sensory, masticatory or locomotor function, but used for the wings of Insecta.

appendix An outgrowth.

appendix vermiformis , In some Mammals, the distal rudiment of the caecum of the intestine, which in Man is a narrow, blind tube of gut, from 25–250 mm in length.

appetitive behaviour A term used to refer to the active exploratory phase that precedes the presumed goal of a behaviour sequence. Traditionally the appetitive phase is said to lead to **consummatory behaviour**.

applied psychology That part of psychology which puts its knowledge to work in practical situations, e.g. in vocational guidance and assessment, education and industry.

apposition The addition of new material to a cell wall at the surface next to the plasmalemma. Cf. **intussusception**.

appressed Flattened, and pressed close to, but not united with, another organ.

appressorium A flattened outgrowth which attaches a parasite to its host; esp. a modified hypha, closely applied to the host epidermis. A narrow infection hypha or penetration tube is pushed into the cell or space below the attachment.

apterism The condition of winglessness, either primitive or secondary, found in many Insects. adj. *apterous*.

Apterygota A subclass of small, primitively wingless Insects showing litle metamorphosis. Bristletails and Springtails.

aptitude A specific ability or capacity to learn. *General aptitude* refers to the capacity for acquiring knowledge in a wide range of areas, as opposed to a *specific aptitude*, such as the ability to acquire musical skills. *Aptitude testing* is an attempt to measure individual differences in potential for learning, as opposed to *achievement testing*, which measures present levels of competence in a given area.

apyrexia Absence of fever.

aquatic See **Hydrophyte**.

aqueduct A channel or passage filled with or conveying fluid; in higher Vertebrates, the reduced primitive ventricle of the midbrain.

aqueductus Sylvii In Vertebrates, the ventricle of the midbrain; the iter.

aqueductus vestibuli In Craniata, a narrow tube arising from the auditory sac and opening on the dorsal surface of the head, as in some fishes, or ending blindly. The endolymphatic duct.

aqueous humour In Vertebrates, the watery fluid filling the space between the lens and the cornea of the eye.

aqueous tissue Water-storage tissue, made up of large, thin-walled, hyaline cells.

aquiculture Augmentation of aquatic animals of economic importance by direct methods; cultivation of the resources of sea and inland waters as distinct from exploitation.

Arachnida A class of mainly terrestrial Arthropoda in which the head and thorax are continuous (**prosoma**). The head bears pedipalps and chelicerae but no antennae. There are four pairs of ambulatory legs. Spiders, Harvest-men, Mites, Ticks, Scorpions etc.

arachnidium In Spiders, the spinnerets and silk glands.

arachnoid *Cobweblike*. Formed of entangled hairs or fibres; pertaining to or resembling the Arachnida; one of the 3 membranes which envelop the brain and spinal cord of Vertebrates, lying between the dura mater and the pia mater.

Araneae An order of Arachnida in which the prosoma is joined to the apparently unsegmented opisthosoma by a waist. Spinnerets and several kinds of spinning glands occur. The pedipalps are modified in the male for the transmission of sperm. Spiders.

araneous Cobweblike.

arboretum An area devoted to the cultivation of trees and other woody plants.

arbuscule (1) A dwarf tree or shrub of treelike habit. (2) A much-branched haustorium formed within the host cells by some endophytic fungi. See **vesicular arbuscular mycorrhiza**.

arch A curved or arch-shaped skeletal structure supporting, covering or enclosing an organ or organs, as *haemal arch*, *neural arch*, *zygomatic arch*.

archaeo- Prefix from Gk. *archaios*, ancient.

archaeostomatous Having a persistent blastopore, which gives rise to the mouth.

arche- Prefix from Gk. *archē*, beginning.

archecentra In Vertebrates, centra formed by the enlargement of the bases of the arched elements which grow around the notochord outside its primary sheath. Cf. **chordacentra**. adj. *archecentrous, arcicentrous, arcocentrous*.

archegonial chamber A small cavity at the micropylar end of the female gametophyte of some gymnosperms, e.g. cycads, into which the spermatozoids are liberated to swim to the archegonia.

Archegoniatae In some classifications, one of the main groups within the plant kingdom, including the Bryophyta and Pteridophyta. Characterized by the presence of the archegonium as the female organ, and by the regular alternation of gametophyte and sporophyte in the life cycle.

archegoniophore, archegonial receptacle A specialized branch of a thallus bearing archegonia.

archegonium A sessile or stalked organ, bounded by a multicellular wall, and flask-shaped in general outline. It consists of a chimney-like neck containing an axial series of neck-canal cells, and a swollen venter below, containing a single egg and a ventral-canal cell. The archegonium is the female organ of Bryophyta and Pteridophyta, and, in a slightly simplified form, of most Gymnospermae.

archencephalon In Vertebrates, the primitive forebrain; the cerebrum.

archenteron Cavity in the gastrula, enclosed by endoderm. It opens to the exterior at the blastopore.

archesporium The tissue in a sporangium that gives rise to the spore mother cells, including the region of the nucellus giving rise to the megaspore mother cells.

archetype A notion associated with the psychology of Carl Jung; it refers to an emotionally laden image assumed to be present in the unconscious mind of all human beings throughout history; an aspect of the *collective unconscious*.

archi- Prefix from Gk. *archi-*, first, chief.

Archiannelida A class of Annelida, of small size and marine habit, which usually lack setae and parapodia and have part of the epidermis ciliated; the nervous system retains a close connection with the epidermis; they resemble the Polychaeta in many of their characteristics.

archiblastic Exhibiting total and equal segmentation; pertaining to the protoplasm of the egg; pertaining to an **archiblastula**.

archiblastula A regular spherical blastula, having cells of approximately equal size.

archicoel See **blastocoel**.

archinephric In Vertebrates, pertaining to the archinephros (see **pronephros**); in Invertebrates, pertaining to the larval kidney or **archinephridium**.

archinephridium In Invertebrates, the larval excretory organ, usually a solenocyte.

archipallium In Vertebrates, that part of the cerebral hemispheres not included in the olfactory lobes and corpora striata, and comprising the hippocampus and the olfactory tracts and associated olfactory matter; that part of the pallium excluding the neopallium.

architype A primitive type from which others may be derived.

Archosauria Subclass of diapsid reptiles which were the dominant forms in the Mesozoic. Only surviving group are the Crocodiles and Alligators.

aricentrous See **archecentra**.

arcuate Bent like a bow.

area monitoring The survey and measurement of types of ionizing radiation and dose levels in an area in which radiation hazards are present or suspected.

area opaca In developing tetrapods, a whitish peripheral zone of blastoderm, in contact with the yolk.

area pellucida In developing tetrapods, a central clear zone of blastoderm, not in direct contact with the yolk.

area vasculosa In developing tetrapods, part of the extraembryonic blastoderm, in which the blood vessels develop.

Arecaceae Same as **Palmae**.

Arecidae Subclass or superorder of monocotyledons. Trees, shrubs, terrestrial herbs and a few free-floating aquatics, mostly with broad, petiolate leaves often net-veined, inflorescence of usually numerous small flowers generally subtended by a spathe and often aggregated into a spadix. Contains ca. 6400 spp. in 5 families including Palmae, Araceae, Pandanaceae and Lemnaceae.

arenaceous, arenicolous (1) Plants growing best in sandy soil. (2) Animals occurring in sand. (3) Composed of sand or similar particles, as the shells of some kinds of Radiolaria.

areola (1) One of the spaces between the cells and fibres in certain kinds of connective tissue. (2) In the Vertebrate eye, that part of the iris bordering the pupil. (3) In Mammals, the dark-coloured area surrounding the nipple. pl. *areolae*.

areolar, areolate (1) Divided into small areas or patches. (2) Pitted. (3) Pertaining to an areola.

areolar tissue A type of connective tissue consisting of cells separated by a mucin matrix in which are embedded bundles of white and yellow fibres.

areole (1) A small area delimited in some way, esp. (a) an island into which a reticulated and veined leaf is divided by the veins, and (b) an area demarcated by the network of cracks in a lichen thallus. (2) A cushion, representing a condensed lateral shoot from which spines, branches and flowers arise in cacti.

Arg Symbol for **arginine**.

argentate Of silvery appearance.

argillicolous Living on a clayey soil.

arginine *2-amino-5-guanidopentanoic acid.* The L-isomer is a 'basic' amino acid. Symbol Arg, with short form R.

Side chain:
$$\begin{array}{c} HN \\ \diagdown \\ \diagup \\ H_2N \end{array} C-NH-CH_2-CH_2-CH_2-$$

See **amino acid**.

arid zone A zone of latitude 15°–30°N and S in which the rainfall is so low that only desert and semi-desert vegetation occurs, and irrigation is necessary if crops are to be grown.

aril An outgrowth on a seed, formed from the stalk or from near the micropyle. It may be spongy or fleshy, or may be a tuft of hairs.

Aristotle's lantern In Echinoidea, the framework of muscles and ossicles supporting the teeth, and enclosing the lower part of the oesophagus.

arm In Echinodermata, a prolongation of the

body in the direction of a radius; in Cephalopoda, one of the tentacles surrounding the mouth; in bipedal Mammals, one of the upper limbs.

armed Protected by prickles, thorns, spines, barbs etc.

arousal A general psychophysiological concept referring to the effect of various nonspecific stimulation or motivational factors on a number of physiological variables, e.g. heart rate, skin resistance. It is used to describe differences in responsiveness to general stimulation, usually along a continuum from drowsiness to alertness, for example.

array A set of values for a particular variate.

arrectores pilorum In Mammals, unstriated muscles attached to the hair follicles, which cause the hair to stand on end by their contraction.

arrhenotoky Parthenogenetic production of males.

arrhythmia Abnormal rhythm of the heart beat.

artefact, artifact Any apparent structure which does not represent part of the actual specimen, but is due to faulty preparation. Particularly a microscope image which has no counterpart in reality.

arterial system That part of the vascular system which carries the blood from the heart to the body.

arteriole A small artery.

artery One of the vessels of the vascular system, that conveys the blood from the heart to the body. adj. *arterial*.

arthritic Pertaining to the joints; situated near a joint.

arthrodia A joint.

arthrodial membranes In Arthropoda, flexible membranes connecting adjacent body sclerites and adjacent limb joints, and occurring also at the articulation of the appendages.

Arthrophyta Division of the plant kingdom, the horsetails and allies, here treated as the class Sphenopsida.

Arthropoda A phylum of metameric animals, having jointed appendages (some of which are specialized for mastication) and a well-developed head; there is usually a hard chitinous exoskeleton; the coelom is restricted, the perivisceral cavity being haemocoelic. Centipedes, Millipedes, Insects, Crabs, Lobsters, Shrimps, Spiders, Scorpions, Mites, Ticks etc.

arthrospore Spore resulting from hyphal fragmentation.

Arthus reaction A type III **allergic reaction** named after the person who first described it.

articular(e) Pertaining to, or situated at, or near, a joint. In Vertebrates, a small cartilage

at the angle of the mandible, derived from the Meckelian, and articulating with the quadrate forming the lower half of the jaw hinge. pl. *articularia*.

articulated Jointed or segmented; divided into portions that may easily be separated.

articulation The movable or immovable connection between 2 or more bones.

artificial classification A classification based on one or a few arbitrarily chosen characters, and giving no attention to the natural relationships of the organism; the old grouping of plants into trees, shrubs, and herbs was an *artificial classification*.

artificial community A plant community kept in existence by artificial means,e.g. a garden habitat or a cloche.

artiodactyl Possessing an even number of digits.

Artiodactyla The one order of the Mammalia containing the 'even-toed' hooved 'Ungulates', i.e. those with a **paraxonic foot**. Includes Pigs, Peccaries, Hippopotami, Camels, Llamas, Giraffes, Sheep, Buffaloes, Oxen, Deer, Gazelles and Antelopes.

arundinaceous Reedlike and thin.

ascertainment In human genetics, the way by which families come to the notice of the investigator. The method of ascertainment may lead to biassed data.

Aschelminthes Phylum of invertebrate animals which have in common the possession of a pseudocoelom and an unsegmented elongate body with terminal anus and a nonmuscular gut.

ascidium A pitcher-shaped leaf or part of a leaf.

ascites Accumulation of fluid in the peritoneal cavity.

ascocarp, ascoma The fruiting body of the Ascomycotina consisting of a sterile wall more or less enclosing the asci. See **apothecium, perithecium**.

ascolichen One of the majority of *lichens* in which the fungal constituent is an ascomycete.

ascomycete Fungus of the Ascomycotina.

Ascomycotina, Ascomycetes Subdivision or class of those Eumycota or true fungi in which the sexual spores are formed in **asci** usually within ascocarps. No motile stages. Usually mycelial with hyphae with simple septa; some are yeasts. Asexual reproduction is by conidia. They include the Hemiascomycetes, Plectomycetes, Pyrenomycetes and Discomycetes.

ascorbic acid See **vitamin C**.

ascospore Spore, typically, uninucleate and haploid, formed within an ascus.

ascus Specialized, usually more or less cylindrical cell within which (usually 8) ascospores are formed following fusion of 2 heterokaryotic nuclei in the ascomycete re-

---

assimilation

(1) In Piagetian theory of child behaviour, one of the two main biological forces responsible for cognitive development; it refers to the ability to absorb new information into pre-existing cognitive organizations (*schemata*). See accommodation.

(2) The metabolic processes, mostly anabolic, by which the predominantly inorganic substances, taken up by plants, are converted into the constituents of the plant body. It includes photosynthesis.

(3) In animals the conversion of food material into protoplasm, after it has been ingested, digested and absorbed.

(4) Resemblance of an animal to its surroundings, not only by coloration but also by configuration.

---

production. pl. *asci.*

asepalous Devoid of sepals.

aseptate Not divided into segments or cells by septa.

asexual Without sex; lacking functional sexual organs.

asexual reproduction Any form of reproduction not depending on a sexual process or on a modified sexual process.

Asn Symbol for **asparagine.**

Asp Symbol for **aspartic acid.**

asparagine The monoamide of aspartic acid. Symbol Asn, short form N.

Side chain:
$$\underset{O}{\overset{H_2N}{\diagdown}} C - CH_2 -$$

See **amino acid.**

aspartic acid   *2-aminobutanedioic acid.* The L-isomer is a polar amino acid and constituent of proteins. Symbol Asp, short form D.

Side chain:
$$\underset{O}{\overset{HO}{\diagdown}} C - CH_2 -$$

See **amino acid.**

aspect (1) Degree of exposure to sun, wind etc. of a plant habitat. (2) Effect of seasonal changes on the appearance of vegetation.

aspergillosis A disease of the lungs caused by the fungus *Aspergillus fumigatus.* May cause allergic reaction in the bronchioles to give asthma or may infect old cavities to give a ball-like growth or *aspergilloma.* In severely immune compromised patients may spread beyond the lung.

Aspergillus A form genus of Deuteromycotina. Includes parasites (causing e.g. *aspergillosis*), saprophytes, food-spoilage organisms (see **aflatoxins**). Used to prepare soy sauce and industrial enzymes.

aspermia Complete absence of spermatozoa.

asphyxia Suffocation which is due to lack of inspired oxygen.

asplanchnic Having no gut.

assimilation See panel.

assimilatory quotient Same as **photosynthetic quotient.**

association (1) A set of plants or animals usually occupying a particular area, consisting of a definite population of species, having a characteristic appearance and habitat, and stable in its duration. (2) In certain Sporozoa, adherence of individuals without fusion of nuclei.

associative learning Refers to learning through the formation of associations between ideas or events based on their cooccurrence in past experience. The term originated in philosophy, but it is now most often used as a synonym for learning through both *classical* and *operant conditioning* procedures.

assortative mating Non-random mating caused by e.g. pollinating insects, which may cause preferential inbreeding or outbreeding.

astelic Not having a stele.

aster A group of radiating fibrils formed of microtubules surrounding the centrosome, seen immediately prior to and during cell division, and more prominent in animal than plant nuclei.

Asteraceae Same as **Compositae.**

Asteridae Subclass or superorder of dicotyledons. Some trees and shrubs but mostly herbs, mostly sympetalous, stamens as many as corolla lobes or fewer, mostly with 2 fused carpels or with a pseudomonomerous ovary. Contains ca 56 000 spp. in 43 families including Solanaceae, Scrophulariaceae, Labiatae, Verbenaceae, Rubiaceae and Compositae.

Asteroidea A class of Echinodermata, having a dorsoventrally flattened body of pentagonal or stellate form; the arms merge into the disk; the tube feet possess ampullae and lie in grooves on the lower surface of the arms; the anus and madreporite are aboral, and there is a well-developed skeleton; free-living carnivorous forms. Starfish.

asthma A chronic disease characterized by difficulty in breathing, accompanied by wheezing and difficulty in expelling air from the lungs. This is due to constriction of the bronchi and their blocking by viscid mucous secretions. In some cases (so-called extrinsic asthma) the condition is due to inhaled or ingested allergens and the main symptoms are caused by histamine and other mediators released from mast cells in a type I allergic reaction.

astomatous (1) Lacking stomata. (2) Without a mouth.

astrocyte A much branched, star-shaped neuroglia cell.

astrosclereide A sclereide with radiating branches ending in points.

asulam Translocated herbicide used widely to control bracken in grasslands, upland pastures and forest plantations.

asymmetry An irregular form. The condition of an organism in which no plane can be found which will divide the body into two similar halves, as in Snails. adj. *asymmetric*.

asynapsis Absence of pairing of chromosomes at meiosis.

atactostele Stele characteristic of the stems of monocotyledons, consisting of many vascular bundles apparently scattered throughout the ground tissue.

atavism The appearance in an individual of characteristics believed to be those of its distant ancestors.

ataxia, ataxy Incoordination of muscles, leading to irregular and uncontrolled movements; due to lesions in the nervous system.

ataxia telangiectasia Human clinical syndrome in which spontaneous chromosome rearrangements occur at a high rate, preferentially involving non-homologous chromosomes.

atlas The first cervical vertebra.

atmometer Apparatus, like a potometer, but designed to measure water loss from a wet, non-living surface, e.g. a porous pot.

atokous Having no offspring; sterile.

atopy A constitutional or hereditary tendency to develop high levels of IgE and immediate hypersensitivity to allergens, esp. those which are absorbed across the respiratory mucosa.

ATPase An enzyme which converts ATP to ADP. In this process the free energy change of the exergonic hydrolysis is used to drive an endergonic reaction, e.g. muscle myosin possesses ATPase activity and the ATP breakdown is coupled to the movement of the *myosin* fibres relative to the *actin*.

atresia Disappearance by degeneration; as the follicles found in the Mammalian ovary. adjs. *atresic, atretic*.

atriopore The opening by which the atrial

cavity communicates with the exterior.

atrium In Platyhelminthes, a space into which open the ducts from the male and female genital organs; in pulmonate Mollusca, a cavity into which the vagina and the penis open and which itself opens to the exterior; in Protochordata, the cavity surrounding the respiratory part of the pharynx; in Vertebrates, the anterior part of the nasal tract; in Reptiles and Birds, the cavity connecting the bronchus with the lung chambers; in the developing Vertebrate heart, the division between the sinus venosus and the ventricle, which will later give rise to the auricles.

atrophy Degeneration, i.e. diminution in size, complexity or function, through disuse.

atropus See **orthotropous**.

attachment theory A wide variety of research efforts which centres around the proposal, put forward by Bowlby, that the emotional bond between human infants and their caretakers has an evolutionary history and that various attachment behaviours, e.g. smiling, crying, have physical and psychological functions of enhancing survival. The theory includes perspectives from ethology, psychoanalysis and developmental psychology.

attention *Selective attention*. That aspect of perception which implies a readiness to respond to a particular stimulus or aspects of it.

attenuated vaccine Live bacterial or virus vaccine in which the microbes have been selected or otherwise treated in such a way as to greatly diminish their capacity to cause disease but still to retain their ability to evoke protective immunity, e.g. *poliomyelitis, measles* and *yellow fever* vaccines.

attenuation Lessening of the capacity of a pathogen to cause disease.

attitude An inferred disposition to feel, think and act in certain ways which is used to explain the variation between individuals in their response to similar situations; attitudes are assumed to represent the effects of past experience on behaviour through their effects on the cognitive and emotional structuring of perception.

attitude scale Standard procedure for measuring attitudes.

attribution theories Theories concerned with how individuals explain everyday events by attributing the cause of a person's behaviour to either situational or personal factors, or some combination of each.

auditory, aural Pertaining to the sense of hearing or to the apparatus which subserves that sense; the 8th cranial nerve of Vertebrates, supplying the ear.

auditory ossicles Three small bones, the *incus, malleus* and *stapes*, bridging the tympanic cavity of the middle ear in mammals.

aural See **auditory**.

auricle (1) A chamber of the heart connecting the afferent blood vessels with the ventricle. (2) The external ear of Vertebrates. (3) Any lobed appendage resembling the external ear. Also called *auricula*. (4) A small ear-shaped lobe at the base of a leaf or other organ.

auricularia In Holothuria and Asteroidea, a pelagic ciliated larva, having the cilia arranged in a single band, produced into a number of short processes.

auriculoventricular Pertaining to, or connecting, the auricle and ventricle of the heart; e.g. the *auriculoventricular connection*, a bundle of muscle fibres which transmits the wave of contraction from the auricle to the ventricle, in higher Vertebrates.

Australasian region One of the primary faunal regions into which the land surface of the globe is divided; includes Australia, New Guinea, Tasmania, New Zealand and the islands south and east of Wallace's line.

aut- See **auto-**.

autecology The study of the ecology of any individual species. Cf. **synecology**.

authoritarian personality A term introduced into social psychology in a famous post war study of antisemitism; it refers to a cluster of social attitudes and personality attributes that are hostile and rigid. See **F-scale**.

autism *Infantile autism. A childhood psychosis* originating in infancy (before 30 months), characterized by a lack of responsiveness in social relationships, language abnormality, and a need for constant environmental input, or *sameness*; stereotypic motor habits, overactivity and epilepsy are often associated with it.

autoallogamy The condition of a species in which some individual plants are capable of self-pollination and others of cross-pollination.

auto-antibody Antibody which reacts specifically with an antigen present on normal constituents of the body of the individual in whom the antibody was made. B lymphocytes able to make auto-antibodies are present in healthy individuals, but are suppressed or not normally stimulated. In auto-immune disease the regulation mechanisms break down and auto-antibodies are present in the blood, which may or may not be responsible for the disease process.

auto-, aut- Prefixes from Gk. *autos*, self.

autocatalysis Reaction or disintegration of a cell or tissue, due to the influence of one of its own products.

autochthonous In an aquatic community, said of food material produced within the community; more generally, indigenous, inherited, hereditary (e.g. *autochthonous*

species, *autochthonous* characteristics).

autocidal control Method of insect pest control by release of sterile or genetically altered individuals into the wild population.

autodiploid A totally homozygous diploid cell or a plant derived from such a cell, resulting from chromosome doubling in a haploid cell. See **plant cell culture**.

autoecious, autoxenous Completing its life cycle in a single host species. Cf. **heteroecious**.

auto-erotism A condition where sensual pleasure is sought and gratified in one's own person, without the aid of an external love object; e.g. masturbation, thumb sucking. See **narcissism**.

autogamy Fertilization involving pollen and ovules from (1) the same flower or sometimes (2) (more widely) the same plant or genetically identical individuals (same **genet** or **clone**). Similarly for animals, where also the fusion of sister-cells or of 2 sister-nuclei. See **self-pollination**, **self-fertilization**. Cf. **allogamy**.

autogenic (1) Resulting from processes internal to the system. Cf. **allogenic**. (2) See **autonomic**.

autograft A mass of tissue, or an organ, moved from one region to another within the same organism.

auto-immunity A condition in which T or B lymphocytes capable of recognizing 'self' constituents are present and activated so as to cause damage to cells by cell-mediated immunity or to release auto-antibodies, and so to cause *auto-immune diseases*. There are many such conditions, the clinical manifestations of which depend upon the nature of the cells damaged and whether circulating **immune complexes** are formed.

autolysis The breakdown of living matter caused by the action of enzymes produced in the cells concerned; self-digestion. adj. *autolytic*. See **lysosomes**.

automatism An automatic act done without the full co-operation of the personality, which may even be totally unaware of its existence. Commonly seen in hysterical states, such as fugues and somnambulism, but may also be a local condition as in automatic writing.

autonomic, autonomous Independent; self-regulating; spontaneous.

autonomic movement Movement in parts of organisms maintained by an internal stimulus, e.g. the beating of flagella, cyclosis, chromosome movements and circumnutation. Cf. **paratonic movement**.

autonomic nervous system In Vertebrates, a system of motor nerve fibres supplying the smooth muscles and glands of the body. See **parasympathetic nervous system**, **sympathetic nervous system**.

autoplasma In tissue culture, a medium

prepared with plasma from the same animal from which the tissue was taken. Cf. **heteroplasma, homoplasma**. adj. *autoplastic*.

autoplastic transplantation Reinsertion of a transplant or graft from a particular individual in the same individual. Cf. **heteroplastic, homoplastic, zenoplastic**.

autopodium In Vertebrates, the hand or foot.

autopolyploid A **polyploid** containing 3 or more **basic chromosome sets** all from the same species.

autoradiography See panel.

autoshaping The classical conditioning of an *operant response* that is not reinforced by instrumental conditioning.

autosome A chromosome that is not one of the *sex-determining* chromosomes.

autospasy The casting of a limb or part of the body when it is pulled by some outside agent, as when the Slow-worm casts its tail.

autospore A non-motile spore, one of many formed within the parent algal cell and having all the characteristics of the parent in miniature before it is set free. Characteristic of some Chlorococcales.

autostyly In Craniata, in Dipnoi and all tetrapods, a type of jaw suspension in which the hyoid arch is broken up and the hyomandibular attached to the skull. adj. *autostylic*.

autotetraploid A polyploid containing 4 similar sets of chromosomes all from the same species.

autotomy Voluntary separation of a part of the body (e.g. limb, tail), as in certain Worms, Arthropods and Lizards.

autotransplantation See **autoplastic transplantation**.

autotrophic Able to elaborate all its chemical constituents from simple, inorganic compounds (esp. all its carbon compounds from CO_2). Cf. **heterotrophic**.

autotrophic bacteria Bacteria which obtain their energy from light and inorganic compounds, and which are able to utilize carbon dioxide in assimilation.

autumn wood See **latewood**.

auxanometer A device for recording the elongation of a plant stem, leaf etc. traditionally by means of a lever and smoked drum.

auxin A plant growth substance, *indole-3-acetic acid*, (IAA), or any of a number of natural or artificial substances with similar effects. Auxins promote root initiation, cell elongation, xylem differentiation and may be involved in apical dominance and tropism. At high concentrations some synthetic auxins are used as herbicides.

auxocyte Any cell in which meiosis has begun; an androcyte, sporocyte, spermatocyte or oöcyte, during the period of growth.

auxotonic Of muscle contraction, of or against increasing force.

auxotroph A variant organism requiring the addition of special nutrients or growth factors before they will grow, e.g. aminoacid requiring bacteria. Cf. **prototroph**.

available Said of that part of e.g. water or mineral nutrient in the soil or fertilizer, which can be drawn upon by a plant. Cf. **unavailable**.

avascular Not having blood vessels.

average General term for *mean, median,* and *mode*.

aversive, aversion therapy A form of *behaviour therapy* in which the undesirable response is paired with an aversive stimulus.

aversive stimulus A stimulus that an animal will attempt to avoid or escape from; it can be used experimentally to *punish* or *negatively reinforce* a response.

Aves A class of Chordata adapted for aerial life. The forelimbs are modified as wings, the sternum and pectoral girdle are modified to serve as origins for the wing muscles, and the pelvic girdle and hind limbs to support the entire weight of the body on the ground. The body is covered with feathers and there are no teeth. Respiratory and vascular systems are modified for homoiothermy. Birds.

avian leucosis *Lymphoid leukosis*. Occurs naturally only in the chicken, caused by members of the leukosis/sarcoma group of avian retroviruses. The virus is transmitted efficiently through the embryo, and infected birds become depressed prior to death. There are few typical clinical symptoms, but diffuse or nodular tumours are found in most organs. Involvement of the *bursa* is considered **pathognomonic** and the disease is controlled by eradication. The virus is important in cancer research.

avidin A protein which binds very strongly to biotin. It can be labelled by fluorescence or by attachment of enzymes, and is used to reveal antibodies to which biotin has been conjugated.

avidity A measure of the strength of binding between an antigen and an antibody. Since antigens are liable to have several combining sites the avidity is an average and less precise than affinity.

avitaminosis The condition of being deprived of vitamins; any deficiency disease caused by lack of vitamins.

Avogadro number, constant The number of atoms in 12 g of the pure isotope ^{12}C; i.e. the reciprocal of the *atomic mass unit* in grams. It is also by definition the number of molecules (or atoms, ions, electrons) in a **mole** of any substance and has the value $6.022\,52 \times 10^{23}\ mol^{-1}$. Symbol N_A or L.

awn (1) A long bristle borne on the glumes and/or lemmas of some grasses, e.g. barley.

autoradiography

Originally used to show the distribution of radioactive molecules in cells and tissues after injecting the organism with, or growing the cells in a medium containing a radioactive precursor. It is now widely used to show the distribution of radiolabelled molecules separated on the basis of size, charge etc. Photographic film or emulsion is exposed after applying it to the section or fixed cell, or to filter paper *blotted* from the separating medium; the distribution of developed grains is viewed directly or under the microscope. Similar procedures exist for fluorescent and other labels.

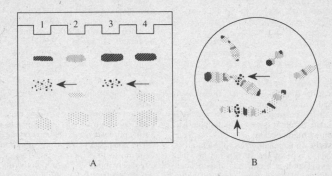

A   B

The diagram shows in (A) the effect of superimposing the developed photographic film on a pattern of bands separated after gel electrophoresis. A ^{32}P labelled polynucleotide has hybridized to the arrowed bands in channels 1 and 3 and the high-energy β-particles, emitted from the polynucleotide, have exposed the film. See **Southern blot**. In (B) a high-resolution photographic emulsion has been poured over a stained chromosome preparation, which is then exposed and developed. The arrowed regions show where the radioactivity is localized under the microscope. In this case ^{3}H which emits low-energy particles is used to provide the precise localization of the radioactive *probe*.

(2) A similar structure on another organ.

axenic culture A culture of a single species in the absence of all others; pure culture.

axial Relating to the axis of a plant or organ; longitudinal.

axial skeleton The skeleton of the head and trunk; in Vertebrates, the cranium and vertebral column, as opposed to the appendicular skeleton.

axiate pattern The morphological differentiation of the parts of an organism, with reference to a given axis.

axil The upper hollow where the adaxial surface of leaf or bract attaches to a stem.

axile Coinciding with the longitudinal axis.

axilemma In medullated nerve fibres, the whole of the medullary sheath.

axile placentation That in which the placentas are in the angles formed where the septa along the central axis of an ovary meet the 2 or more locules, e.g. tomato.

axillary Situated in or arising from an axil;

esp. of buds, shoots, flowers, inflorescences etc.

axillary air sac In Birds, one of the paired air sacs, lying in the axillary position It communicates with the median interclavicular air sac.

axis (1) A central line of symmetry of an organ or organism. (2) A stem or root. (3) A rachis. (4) In higher Vertebrates, the second cervical vertebra.

axon The process of a typical nerve cell or neuron which transmits an impulse or action potential away from the cell body.

axoneme The central core of a **cilium** consisting of **microtubules** and associated proteins.

3′-azido-2′,3′-dideoxythymidine Abbrev. *AZT.* An analogue of thymidine in which N_3 is substuted for OH in the 3′-position of the sugar. Its triphosphate competitively binds to reverse transcriptase causing chain termination. It prolongs the life but does not cure patients with AIDS. See **acquired immuno-deficiency syndrome**.

azonal soil Immature soil. Cf. **zonal soil**.

Azotobacter Free-living genus of bacteria in soil and water which are able to fix free nitrogen in the presence of carbohydrates.

azygomatous Lacking a zygomatic arch.

azygos An unpaired structure. adj. *azygous*.

azygospore A structure resembling a zygospore in morphology, but not resulting from a previous sexual union of gametes or of **gametangia**.

B

Babesia A genus of protozoal parasites which occur in the erythrocytes of mammals.

baccate Resembling a berry.

Bacillaceae A family of bacteria included in the order *Eubacteriales*. Many are able to produce highly resistant endospores; large Gram-positive rods; aerobic or anaerobic; includes many pathological species, e.g. Bacillus anthracis (anthrax). See **clostridium**.

Bacillariophyceae *Diatomophyceae*. The diatoms, a class of eukaryotic algae in the division Heterokontophyta. The cell wall or *frustule* contains silica. Mostly unicellular, some colonies and chains of cells. Mostly phototrophic (sometimes auxotrophic); some are heterotrophs. Ubiquitous; fresh water and marine, planktonic, benthic and epiphytic, in soils. Fossil deposits of frustules constitute diatomaceous earth or kieselguhr. Two orders: Centrales and Pennales.

bacilluria Presence of bacilli in the urine.

bacillus (1) A rod-shaped member of the Bacteria. (2) Genus in the family *Bacillaceae*. pl. *bacilli*.

backcross The mating of an individual to one of its parents or parental strains. In *Mendelian genetics* a mating of a *heterozygote* to the *recessive homozygote*, producing a 1:1 ratio in the progeny.

background radiation Radiation coming from sources other than that being observed.

back-mutation See **reversion**.

bacteria See panel.

bactericide A substance which destroys bacteria. adj. *bactericidal*. Cf. **bacteriostat**.

bacteriocin A toxin produced by one class of bacteria which kills another, usually related class.

bacteriology The scientific study of bacteria.

bacteriophage A virus which infects bacteria. syn. *phage*.

bacteriostat A substance which inhibits growth but does not kill bacteria.

bacteriotoxin A toxin destructive to bacteria.

bacteriotropin A substance, usually of blood serum, which renders bacteria more subject to the action of antitoxin or more readily phagocytable, e.g. opsonin.

bacteroid The enlarged, X- or Y-shaped, nitrogen-fixing form of a *Rhizobium* bacterium within a legume root nodule.

Bacteroidaceae A family of pleomorphic bacteria included in the order *Eubacteriales*. Usually Gram-negative rods; anaerobic; inhabit the intestine and mucous membrane of higher Vertebrates.

bagassosis Respiratory disease similar to farmer's lung occurring in persons who inhale dust from mouldy sugar cane. Due to Type III hypersensitivity reaction induced by thermophilic mould spores.

BAL *British anti-lewisite*, dithioglycerol. Antidote for poisoning by **lewisite** and other poisons, particularly arsenic and mercury.

balance Equilibrium of the body; governed from the cerebellum, in response to stimuli from the eyes and extremities and esp. from the **semicircular canals** of the ears.

balance theories Theories concerned with how an individual's attitudes and perceptions of other people, events or objects, relate to each other.

Balbiani rings Very large *puffs* at specific sites on Dipteran (fruit-fly) salivary gland chromosomes. Occur when RNA, coding for secretory proteins, is being transcribed at these sites.

baleen In certain Whales, horny plates arising from the mucous membrane of the palate and acting as a food strainer.

ballistospore A spore that is violently projected, e.g. the basidiospore of the basidiomycete fungi.

bands See panel.

banding techniques See panel.

Bang's bacillus *Brucella abortus*; the cause of contagious abortion in animals and of undulant fever in man.

BAP See **6-benzylaminopurine**.

baragnosis Loss of the ability to judge differences between the weights of objects.

barb Any hooked, bristlelike structure, such as a hooked or doubly-hooked hair on a plant or one of the lateral processes of the rachis of the feather which form the vane.

barbate Bearded; bearing tufts of long hairs.

barbel In some Fish, a finger-shaped tactile or chemosensitive appendage arising from one of the jaws.

barbule In Birds, one of the processes borne on the barbs of a feather, by which the barbs are bound together.

bark A non-technical term applied to all the tissues outside the cambium, i.e. the corky and other material which can be peeled from a woody stem.

barophil Used of organisms that grow and metabolize as well (or better) at increased pressures then at atmospheric pressure.

Barr body The densely-staining **heterochromatin** of the inactive X chromosome. See **X-inactivation**.

Bartholin's duct An excretory duct of the sublingual gland.

Bartholin's glands In some female Mammals, glands (corresponding with Cowper's glands in the male) lying on either side of the

bacteria

A large group of unicellular or multicellular organisms, lacking chlorophyll, multiplying rapidly by simple fission, some species developing a highly resistant resting ('spore') phase. They are the principle members of the *prokaryotes*, whose genetic material is not packaged into chromosomes contained in a nucleus. Instead they have a single circle of DNA free in the cytoplasm, which reproduces by a 'rolling circle' mechanism. They lack mitochondria, oxidative phosphorylation occurring across the plasma membrane, and they are bounded by a cell wall situated outside and separated from the plasma membrane by a periplasmic space.

The taxonomy of bacteria is difficult, not only because of the obvious problems of size and rapid mutation but also because a phylogenetic classification, so powerful in Botany and Zoology, is largely unprofitable in these microorganisms. Instead morphological, chemical and pathogenic characteristics are used. In particular bacteria can be divided into those which stain with Gram's method and have a simpler kind of cell wall and those which do not. Colony colour and shape on agar containing a defined nutrient or growth in the presence of antibiotics and other chemicals are often used to screen and identify cultures.

In shape they can be spherical, rodlike, spiral or filamentous. Some are motile by means of flagella and they occur in every natural habitat often in large numbers, as much as 10^8 per gram. Since Pasteur's time they have been actively studied and classified in relation to disease but they are also important in the production of chemicals, enzymes and antibiotics. This commercial production of bacterial products has become increasingly important as **genetic manipulation** has allowed the expression of genes coding for substances like human insulin and growth hormones.

Bacteria can be infected with viruses called *bacteriophage* or *phage* which commonly occur in two states in their host, depending on the strain of host and phage. In the first or *lysogenic* state the genetic material of the phage after penetrating its host either becomes incorporated into the bacterial DNA or remains as a separate *episome* in the bacterial cytoplasm. In either case the phage DNA replicates in step with its host's DNA and it does not destroy the host. In the second or *lytic* state the infective DNA multiplies rapidly and covers itself with a protein coat, using the host's synthetic machinery, before rupturing the cell wall and dispersing into the medium. External stimuli can convert the lysogenic to the lytic state and episomes can also exist in the cytoplasm as multiple copies called *plasmids*. Lysogenic phage or plasmids can confer new properties on their hosts such as the ability to produce a toxin or resistance to further attack by related lytic phage. In the laboratory, phage can be used to transfer DNA sequences from a previous host to a new one, a process called *transduction*.

In an analogous process, *transformation*, pure DNA can be ingested and recombined with the host DNA to alter its genome. Finally, in a 'sexual' process DNA can be transferred from a + to a − strain in a few bacterial types. All these methods rely on selection by the research worker for the easy demonstration of gene transfer and it is common to have a marker for, say, antibiotic resistance joined to the DNA sequence needed to allow the easy destruction of unaltered cells.

upper end of the vagina.

basal area A measure of the extent of trees in an area, being the total cross-sectional area of the trunks.

basal body (1) A cylindrical structure found at the base of cilia composed of 9 sets of triplet **microtubules** which serves as a centre for the growth of microtubules in culture. In flagellate or ciliate *Protozoa*, zoospores or spermatozoids, it occurs as a small,

bands and banding techniques

Metaphase chromosomes, when stained with a variety of **banding techniques** exhibit a pattern of transverse bands of varying width characteristic of specific chromosomes. The bands appear to be made by distinctive interactions between specific types of DNA sequence and proteins. Factors such as base and sequence composition, e.g. blocks of short repeats, are involved. **Polytene** chromosomes also exhibit characteristic bands, with or without staining. They are much more numerous than those of metaphase chromosomes and the more stretched state of the polytene chromosomes would be expected to show a greater number of bands. The two kinds of banding patterns may therefore be related. As shown in the diagram, the combination of size, shape and banding pattern gives each chromosome a unique appearance, and it is this which has made *eukaryotic* gene mapping possible.

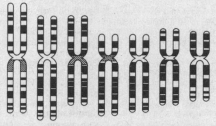

A stylized diagram of banded chromosomes.

Banding techniques are methods of treating chromosomes to produce patterns of bands characteristic of an individual chromosome, as an aid to recognition. All involve staining fixed **metaphase** chromosomes and fall into 5 main groups.

(1) *Q-banding*, staining with quinacrine mustard or 33258 Hoechst.

(2) *G-banding*, removal of some protein followed by staining with Giemsa.

(3) *R-banding*, heat or alkali treatment followed by Giemsa or acridine orange staining (gives reverse pattern of G banding).

(4) *T-banding*, variant of R-banding which mainly stains the **telomeric** regions.

(5) *C-banding*, treatment which mainly stains **constitutive heterochromatin**.

In mammals, G- and Q-banding give generally similar patterns, but in plants Q- and C-bands appear to be more closely related.

deeply staining granule at the base of the locomotory *organelle*. Also *basal granule*, *blepharoplast*. (2) Part of a **thallus** fixed to the substrate by rhizoids. Also *basal corpuscle*.

basal ganglia A localized concentration of grey matter deep in the cerebral hemispheres and the mid brain. They are concerned with the regulation of movement and are often referred to as the extra-pyramidal system. Disease of the basal ganglia gives *Parkinsonism* and *chorea*.

basal lamina A thin sheet of extracellular matrix underlying epithelia. It contains, in addition to collagen and other proteins, the distinctive glycoprotein laminin.

basal placentation Placentation in which the plant's ovules are attached to the bottom of the locule in an ovary.

basal plates In the developing Vertebrate skull, a plate of cartilage formed by the fusion of the parachordals and the trabeculae; in Crinoidea, certain plates situated at or near the top of the stalk; in Echinoidea, certain plates forming part of the apical disk.

base The end of an organ or plant part nearest to its point of attachment or origin.

basement membrane A membrane lying between an epithelium and the underlying connective tissue.

basi- Prefix from Gk. *basis*, base.

basic chromosome set, number The haploid set of chromosomes as found in the gametes. Because species may have evolved by polyploidy, aneuploidy or chromosome rearrangement, it may also be possible to infer a basic chromosome set for the ancestor. See **chromosome complement**.

basicity The number of hydrogen ions of an acid which can be neutralized by a base.

basic number See **chromosome complement**.

basiconic In Insects, said of certain subconical and immobile sensilla arising from the general surface of the cuticle.

basidiocarp The fruiting body of the Basidiomycotina.

basidioma Same as **basidiocarp**.

Basidiomycotina, Basidiomycetes Subdivision or class of those Eumycota or true fungi in which the sexual spores are formed on a basidium. No motile stages. Usually mycelial with septate hyphae. Sexual reproduction typically involves the fusion of a fruiting body in which the basidia develop. Includes the Tiliomycetes comprising the rusts (Uredinales) and smuts (Ustilaginales), the Hymenomycetes and the Gasteromycetes.

basidiospore Spore, typically uninucleate and haploid, formed at the end of a sterigma on a basidium.

basidium Specialized, usually more or less club-shaped, cell on which (typically 4) basidiospores are formed following the fusion of 2 heterokaryotic nuclei and meiosis, in the reproduction of the Basidiomycetes.

basifixed Said of an anther which is attached by its base to the filament.

basifugal Away from the base, towards the apex.

basilar Situated near, pertaining to, or growing from the base.

basilar membrane In Mammals, a flat membrane, part of the partition of the cochlea, containing the collection of auditory nerves in the inner ear which translate mechanical vibrations of differing frequencies into nerve impulses, which are passed to the brain.

basipetal Towards the base, away from the apex.

basiphil Having a marked affinity for basic dyes. Also *basophil(e)*.

basis cranii In Craniata, the floor of the cranium which is formed from the basal

plate of the embryo.

basophil White blood cell with an affinity for basic stains whose granules contain vasoactive amines and has an important role in infection.

basophilia An increase of basophil cells in the blood.

basophil leucocyte Cell present in the blood with properties similar to **mast cells** which binds IgE and can release histamine and other mediators on contact with specific antigen. So called because its granules bind basic dyes.

bast Phloem.

bastard wing In Birds, quill feathers, usually three in number, borne on the thumb or first digit of the wing. Also *ala spuria*, *alula*.

batch culture A culture initiated by the inoculation of cells into a finite volume of fresh medium and terminated at a single harvest after the cells have grown. Cf. **continuous culture**.

Batesian mimicry Convergent resemblance between two animals, advantageous in some way to one of them.

batho-, bathy- Prefixes from Gk. *bathys*, deep, used esp. with relation to sea-depths.

bathophilous Adapted to an aquatic life at great depths.

bathyal Refers to the ocean-floor environment between ca 200 and 4000 m. The three zones of increasing depth are **littoral**, **abyssal** and **bathyal**. There are numerous definitions of the depth range of these zones.

bathybic Relating to, or existing in, the deep sea, e.g. plankton floating well below the surface.

bathylimnetic Living in the depths of lakes and marshes.

bathymetric See **abyssopelagic**.

bathysmal Same as abyssal, below bathyal.

batrachian Relating to the Salientia (i.e. Frogs and Toads).

Bayesian Proceeding from a definition of probability as a measurement of belief; considering statistical inference as a process of re-evaluating such probabilities on the basis of empirical observation.

B cell See **B lymphocyte**.

BCG *Bacille Calmette Guérin*. A living attenuated strain of *Mycobacterium tuberculosis* used as a vaccine to protect against tuberculosis. Induces sensitivity to *purified protein derivative* (see **tuberculin**). Developed by Calmette and Guérin in France in 1909 and first used there in 1921.

B-chromosomes Accessory non-essential chromosomes present in variable numbers in addition to the normal *A-chromosomes*. They are usually small and heterochromatic.

beak See **rostrum**

bearded Having an awn; bearing long hairs

like that of a beard.

beetle A member of the insectan order of Coleoptera.

Beggiatoales Chemosynthetic sulphur-oxidizing bacteria, some of which resemble filamentous algae, and which may in fact be more closely related to them than to the true bacteria. They contain intracellular Sulphur granules.

behaviourism An approach to psychology that considers only observable behaviour as appropriate subject matter for study, and which views with distrust explanations which refer to non-observable mental events, not directly available for objective verification (e.g. consciousness, imagery).

behaviour therapy *Behaviour modification*. A general approach to psychological treatment which holds that behaviour disorders are the result of maladaptive learning and are best remedied by re-education based on the principles of learning theory. The focus is on behaviour, rather than on a hypothetical unconscious process.

belemnoid Dart-shaped.

Bellini's ducts In the kidney of Vertebrates, ducts formed by the union of the primary collecting tubules and opening into the base of the ureter at the pelvis of the kidney.

belt transect A strip of ground marked between two parallel lines so that its vegetation may be recorded and studied. See **transect**, **quadrat**.

Bence-Jones protein A protein present in the urine of some persons with **myelomas**, which precipitates on heating to 60°C but redissolves at 80°C. It consists of light chain dimers.

benthon, benthos Collectively, the sedentary animal and plant life living on the sea or lake bottom. Cf. **nekton, plankton**. adj. *benthic*.

benzodiazepines Class of drug used as an **anxiolytic** or **hypnotic**. *Diazepam* (Valium) is commonly used for relieving anxiety and *nitrazepam* (Mogadon) for inducing hypnosis, although the hazards of addiction with these drugs is being increasingly recognized.

6-benzylaminopurine *6-benzyladenine, PAB*. A synthetic **cytokinin**.

Bergmann's law In warm-blooded animals and within a species southern forms are smaller than northern forms.

beri-beri A disease causing peripheral nerve lesions and/or heart failure due to a deficiency of the vitamin B_1 (thiamine).

berry (1) A fleshy fruit, without a stony layer, usually containing many seeds.(2) The eggs of Lobster, Crayfish, and other macruran Crustacea. (3) Part of the bill in Swans.

betacyanins Red pigments of the betalain type, e.g. the red pigment of the beetroot.

beta diversity See **diversity**.

betalains A group of nitrogen-containing pigments functionally replacing other pigments including anthocyanins.

beta-microglobulin A protein which forms part of the structure of class I major **histocompatibility antigens**, but is present in small amounts in blood, urine and seminal plasma.

beta-oxidation The oxidative degradation of the fatty acid chains of lipids into 2-carbon fragments by the cleavage of the penultimate C-C bonds.

beta-pleated sheet Important element of protein structure resulting from hydrogen bonding between parallel polypeptide chains.

betaxanthins Yellow pigments of the **betalain** type.

bhang See **cannabis**.

bicarpellary Said of an ovary consisting of two carpels.

biceps A muscle with two insertions. adj. *bicipital*.

bicipital groove A groove between the greater and lesser tuberosities of the humerus in Mammals.

bicollateral bundle A vascular bundle with two strands of phloem adjacent to the single strand of xylem; one placed centrifugally, the other centripetally. Cf. **collateral bundle**.

bicuspid, bicuspidate Having two cusps, as the premolar teeth of some Mammals.

bicuspid valve The valve in the left auriculoventricular aperture in the Mammalian heart. Also *mitral valve*.

biennial A plant that flowers and dies between its first and second years from germination and which does not flower in its first year. Cf. **annual, perennial**.

bifacial leaf A dorsiventral leaf; typically with palisade mesophyll towards the upper surface and spongy mesophyll below. The commonest sort. Ant. *unifacial leaf*.

bifid Divided halfway down into two lobes; forked.

bifurcate Twice-forked; forked. v. *bifurcate*. n. *bifurcation*.

bigeneric hybrid A hybrid resulting from a cross between individuals from two different genera, e.g. *Triticale*, a hybrid between wheat (*Triticum*) and rye (*Secale*).

bilabiate With two lips.

bilateral Having, or pertaining to, two sides.

bilateral cleavage The type of cleavage of the zygote formed in *Chordata*.

bilateral symmetry The condition when an organism is divisible into similar halves by one plane only. Cf. **radial symmetry**.

bile duct The duct formed by the junction of the hepatic duct and the cystic duct, leading into the intestine.

bile salts Breakdown products of cholesterol which accumulate in bile and are responsible for the detergent activity of bile.

bilocular Consisting of two loculi or chambers.

bimanous Having the distal part of the two forelimbs modified as hands, as in some Primates.

bimastic Having two nipples.

binary fission Division of the nucleus into two daughter nuclei, followed by similar division of the cell-body.

binaural Listening with two ears, the result of which is a sense of directivity of the arrival of a sound wave. Said of a stereophonic system with two channels (matched) applying sound to a pair of ears separately, e.g. by earphones. The effect arises from relative phase delay between wavefronts at each ear.

binomial (binominal) nomenclature The system (introduced by Linnaeus) of denoting an organism by 2 Latin words, the first name of the genus, the second the specific epithet. The 2 words constitute the name of the species, e.g. *Homo sapiens; Bellis perennis*. See **species**.

binomial distribution The probability distribution of the total number of outcomes of a particular kind in a predetermined number of trials, the probability of the outcome being constant at each trial, and the different trials being statistically independent.

binovular twins Twins resulting from the fertilization of 2 separate ova.

binucleate phase Same as **dikaryophase**.

bio-assay (1) The quantitative determination of a substance by measuring its biological effect on e.g. growth, that is the use of an organism to test the environment. (2) Determination of the power of a drug or of a biological product by testing its effect on an animal of standard size.

biochemistry The chemistry of living things; physiological chemistry.

bioclimatology The study of the effects of climate on living organisms.

biocoenosis The association of animals and plants together, esp. in relation to any given feeding area. adj. *coenotic*.

biodegradation The breaking-down of substances by bacteria.

bio-electricity Electricity of organic origin.

bio-engineering Provision of artificial means (electronic, electrical etc.) to assist defective body functions, e.g. hearing aids, limbs for thalidomide victims etc. Cf. **genetic engineering**.

biofeedback Refers to procedures whereby subjects are given information about physiological functions that are not normally available to conscious experience (e.g. heart rate, blood pressure etc.) with the object of gaining some conscious control over them.

biogas Gas, mostly methane and carbon di-

oxide, produced in suitable equipment by bacterial fermentation of organic matter (see biomass) and used as fuel.

biogenesis The formation of living organisms from their ancestors and of organelles from their precursors.

biogeographic regions Regions of the world containing recognizably distinct and characteristic endemic fauna or flora.

biological clock An internal, physiological timekeeping system underlying e.g. circadian rhythms and photoperiodism.

biological constraint A general term in learning theory that refers to the fact that certain behaviours are more easily learned by some organisms than by others, and conversely, that some behaviours are not easily learned by some organisms.

biological containment Alteration of the genetic constitution of an organism so as to minimize its ability to grow in a non-laboratory environment.

biological control Use of an organism to control a disease, pest or weed.

biological form Same as *physiological race*.

biological half-life Time interval required for half of a quantity of radioactive material absorbed by a living organism to be eliminated naturally.

biological magnification Process whereby the concentration of a pollutant, within living tissues, increases at each link in a food chain.

biological oxygen demand An indication of the amount of oxygen needed to oxidize fully, by biological means, reducing material in a water sample. It is expressed as the concentration of oxygen gas in parts per million chemically equivalent to the reducing agents in the water. Abbrev. *BOD*.

biological race *Physiological race*. A race occurring within a taxonomic species; distinguished from the rest of the species by slight or no morphological differences, but by evident differences of habitat, food-preference or occupation which inhibit interbreeding.

biological shield Screen made of material highly absorbent to radiation used as a protection.

biological warfare The use of bacteriological (biological) agents and toxins as weapons. Cf. **chemical warfare**.

bioluminescence The production of light by living organisms, as Glow worms, some deep-sea fish, some bacteria, some fungi.

biomass (1) The total dry mass of an animal or plant population. (2) Organic matter (mostly from plants) harvested as a source of energy (by burning or **biogas** production) and/or as a chemical feedstock.

biomass, pyramid of A diagrammatic

representation of the biomass of a series of organisms, organized according to *trophic layers* to form a pyramid.

biome The largest land community region recognized by ecologists, e.g. tundra, savanna, grassland, desert, temperate and tropical forest.

biometeorology The study of the effects of atmospheric conditions on living things.

biometrical genetics Same as **quantitative genetics**.

biometry Statistical methods applied to biological problems.

bionics The various phenomena and functions which characterize biological systems with particular reference to electronic systems.

biophysics The physics of vital processes; study of biological phenomena in terms of physical principles.

biopsy Diagnostic examination of tissue (e.g. tumour) removed from the living body.

biosphere That part of the earth (upwards at least to a height of 10 000 m, and downwards to the depths of the ocean, and a few hundred metres below the land surface) and the atmosphere surrounding it, which is able to support life. A term that may be extended theoretically to other planets.

biosynthesis The synthesis of complex molecules using enzymes and biological structures like ribosomes and chromosomes either within or without the cell.

biosystematics The study of relationships with reference to the laws of classification of organisms; *taxonomy*.

biota The fauna and flora of a given region.

biotechnology The use of organisms or their components in industrial or commercial processes, which can be aided by the techniques of **genetic manipulation** in developing e.g. novel plants for agriculture or industry. See **plant cell culture, plant genetic manipulation**.

biotic Relating to life.

biotic barrier Biotic limitations affecting dispersal and/or survival of animals and plants.

biotic climax A community that is maintained in a stable condition because of some biotic factor, e.g. grazing. See **climax**.

biotic factor The activities of any organisms that determine which plants grow where.

biotin See **vitamin B**.

biotinylation Labelling a probe with conjugated biotin, whose high affinity for avidin or anti-biotin antibodies is exploited to mark the spot to which the probe binds by indirect immunoassay.

biotope A small habitat in a large community, e.g. a cattle dropping on a grass prairie, whose several short seral stages

comprise a microsere.

biotroph A parasite which feeds off the living cells of its host and therefore needs their continued functioning, e.g. rust and smut fungi. Cf. **necrotroph**. See **obligate parasite**.

biotype A group of individuals within a species with identical, or almost identical, genetic constitution.

biparous Having given birth to 2 young.

bipedal Using only 2 limbs for walking.

bipinnate Said of a compound pinnate leaf with its main segments pinnately divided.

bipolar Having two poles; having an axon at each end, as some nerve cells.

bipolar disorder Affective disorder characterized by shifts from one emotional extreme (euphoria, intense activity) to another (depressive episodes). Formerly *manic-depressive psychosis*.

bipolar germination Germination of a spore by the formation of 2 germ tubes, one from each end.

biradial symmetry The condition in which part of the body shows radial, part bilateral symmetry; as in some Ctenophora.

biramous Having two branches; forked, as some Crustacean limbs. Cf. **uniramous**.

birth mark See **naevus**.

biseriate In 2 whorls, cycles, rows or series.

biserrate Of a leaf margin, having a series of sawlike teeth, which are themselves serrated.

bisexual Possessing both male and female sexual organs. See **hermaphrodite**.

bisexuality (1) Having the physical or psychological attributes of both sexes. (2) Being sexually attracted to both sexes.

bisporangiate Said of a strobilus which consists of megasporophylls and microsporophylls, with megasporangia and microsporangia.

bivalent One of the pairs of homologous chromosomes present during meiosis.

bivalve Having the shell in the form of two plates, as in Bivalvia.

Bivalvia Class of Mollusca with the body usually enclosed by paired shell valves joined by a hinge and closed by adductor muscles. Ctenidia or gills are used for filter feeding and there are inhalent and exhalent siphons. Mostly sessile aquatic forms. Mussels, Clams, Oysters, Scallops. Also called *Pelecypoda*.

bivoltine having two broods in a year. Cf. **univoltine**.

bladder (1) Any membranous sac containing gas or fluid; esp. the urinary sac of Mammals. (2) A small hollow spherical trap for catching and digesting small animals on the bladder-wort, *Utricularia*.

bladderworm See **cysticercus**.

blade The flattened part of a leaf (i.e. the

lamina), sepal, petal or thallus.

blanket bog Bog vegetation forming an extensive and continuous layer of peat over flat and undulating landscape, deriving its mineral nutrients largely from rainfall, and found in cold wet parts of the world.

blasto-, -blast Prefix and suffix from Gk. *blastos*, bud.

blastema Anlage; a mass of undifferentiated tissue; the protoplasmic part of an egg as distinguished from the yolk.

blastochyle Fluid in the blastocoel.

blastocoel The cavity formed within a segmenting ovum; cavity within a blastula; primary body cavity; segmentation cavity. Also *archicoel*.

blastocyst *Germinal vesicle.* In Mammalian development, a structure resulting from the cleavage of the ovum; it consists of an outer hollow sphere and an inner solid mass of cells.

blastoderm In eggs with much yolk, the disk of cells formed on top of the yolk by cleavage.

blastodisc In a developing ovum, the germinal area.

blastomere One of the cells formed during the early stages of cleavage of the ovum.

blastopore The aperture by which the cavity of the gastrula retains communication with the exterior.

blastospore A spore produced by budding.

blastula A hollow sphere, the wall of which is composed of a single layer of cells, produced as a result of the cleavage of an ovum. Also *blastodermic vesicle, blastosphere*.

blastulation A form of cleavage resulting in the production of a blastula.

bleb A small vesicle containing clear fluid.

bleeding The exudation of xylem sap, phloem sap or latex from wounds. See **root pressure**.

blephar-, blepharo- Prefixes from Gk. *blepharon*, eyelid.

blepharism Spasm of the eyelids.

blepharoplast See **basal body** (1).

blight The common name for a number of plant diseases characterized by the rapid infection and death of the leaves or the whole plant, e.g. potato late blight (caused by the fungus *Phytophthora infestans*).

blind spot In Vertebrates, an area of the retina where there are no visual cells (due to the exit of the optic nerve) and over which no external image is perceived.

blister A thin-walled circumscribed swelling in the skin containing clear or blood-stained serum; caused by irritation.

blocking antibody Antibody which combines preferentially with an antigen so as to prevent it from combining with IgE on mast cells, and thereby prevents type I allergic reactions. Blocking antibodies are usually IgG, and the aim of hyposensitization treatment of acute allergies is to stimulate blocking antibody production.

blood count The number of red or white corpuscles in the blood.

blood flukes Trematodes of the genus *Schistosoma*, parasitic on man and domestic animals *via* various species of water snail as intermediate host. They can attack the liver, spleen, intestines and urinary system and occasionally the brain, causing *bilharziasis, schistosomiasis*.

blood islands In developing Vertebrates, isolated syncytial accumulations of reddish mesoderm cells containing primitive erythroblasts, which give rise respectively to the walls of the blood vessels and to the red corpuscles.

blood pressure The pressure of blood in the arteries, usually measured by sphygmomanometry. The maximum occurs in *systole* and the minimum in *diastole*. Normal young adults will have a blood pressure of approximately 120/80 mmHg.

blood substitutes Plasma, albumin and dextran can be used to substitute volume for loss of blood. Newer substances are being investigated which may be able to transport oxygen.

blood sugar The level of glucose in the blood, normally between 3.2–5.2 mmol/l in the fasting state.

bloom (1) A covering of waxy material occurring on the surface of some leaves and fruits, resulting in a whitish cast. (2) A visible, often seasonal occurrence of very large numbers of algae in the plankton of fresh water or sea. See **red tide**.

Bloom's syndrome Human clinical syndrome in which frequent chromosomal rearrangements involving homologous chromosomes occur.

blotting, blot The method by which biological molecules are transferred from, usually, a gel to a membrane filter. In the former they can be separated physically by e.g. size, and in the latter they can be tested for the presence of specific sequences by radioactive or similar **probes**. See **Southern blot, immunoblot**.

blubber In marine Mammals, a thick fatty layer of the dermis.

blue-green algae, blue-green bacteria See **Cyanophyceae**.

blunt-ended DNA DNA cleaved straight across the double stranded molecule, without forming any single stranded ends. An effect of some **restriction enzymes**.

B lymphocyte See panel.

B-memory cell See panel.

body cavity The perivisceral space, or cavity, in which the viscera lie; a vague term, sometimes used incorrectly to mean **coelom**.

B lymphocyte

Lymphocytes derived from precursors in the bone marrow (or in birds the *Bursa of Fabricius*, a tissue budding off from the hind gut) which do not undergo differentiation in the thymus. They make immunoglobulins, which are present at the cell surface and act as specific receptors for antigens and, when stimulated, B lymphocytes manufacture and secrete large amounts of their characteristic immunoglobulin into the circulation. This constitutes the antibody response.

Stimulation by **thymus-dependent antigens** requires cooperation of helper T lymphocytes, which release B cell growth and differentiation factors, whereas stimulation by thymus-independent antigens does not need such cooperation.

Some **mitogens** such as *pokeweed mitogen* can stimulate B cells irrespective of the antigen specificity of their Ig receptors. B lymphocytes are produced continuously throughout life. Most die soon unless stimulated by antigen. Following such stimulation they either differentiate towards antibody secretion or become **B-memory cells**.

A **B-memory cell** is a resting B cell which is derived from a B cell which has been stimulated by a specific antigen in a **germinal centre** so as to multiply without going on to secrete antibody. B-memory cells live and recirculate through the blood between **lymphoid tissues** for many weeks. When they encounter the appropriate antigen again, with T cell help, they rapidly differentiate into antibody secreting cells (or into more B-memory cells). This is the basis of a secondary or booster antibody response. One of the aims of prophylactic immunization is to elicit B-memory cells.

---

body cell Generally, a somatic cell; more particularly, the cell that divides to give the two sperm cells in the gymnosperm pollen tube. Cf. **germ cells**.

body-section radiography See **tomography**.

body wall The wall of the perivisceral cavity, comprising the skin and muscle layers.

bog Wetland vegetation forming an acid peat which is mainly composed of dead individuals of the moss, *Sphagnum*. See **blanket bog, raised bog, valley bog**.

bole The trunk of a tree.

boll A capsule, esp. of cotton.

bolting Premature flowering and seed production, esp. of a biennial crop plant during its first year; 'running to seed'.

bone A variety of connective tissue in which the matrix is impregnated with salts of lime, chiefly phosphate and carbonate.

bone tolerance dose The dose of ionizing radiation which can safely be given in treatment without bone damage.

book gill See **gill book**.

book lung See **lung book**.

booster response The heightened response to readministration of an antigen, or to a microbial infection, which occurs when prior contact with the antigen has elicited **B-memory cells** and **helper T lymphocytes**.

booted Of an animal with feet which are

protected by horny scales.

bordered pit A **pit** in which the secondary wall overarches the pit membrane, markedly narrowing the cavity towards the cell lining. Characteristic of tracheids and vessel elements. Cf. **simple pit**.

boreal Of the North. The *boreal zone* is the geographical region where short summers and long cold winters occur. The *boreal period* in northern Europe extended from 7500 to 5500 BC and had warm summers and cold winters. *Boreal forests* occur in the boreal zone and boreal period.

bosset In deer, the rudiment of the antlers in the first year.

botry-, botryo- Prefixes from Gk. *botrys*, a bunch of grapes; shaped thus.

botryose, botryoid, botrytic Branched; like a bunch of grapes. See **racemose**. adj. *botryoidal*.

bottom yeast Brewer's yeast, *Saccharomyces cerevisiae*, which accumulates at the bottom of the medium during fermentation and used in brewing lagers. Cf. **top yeast**.

botulism Severe and often fatal poisoning due to eating bottled and canned food contaminated by the anaerobic bacterium (*Clostridium botulinum*), which secretes a potent neurotoxin, *botulin*. Causes many cases of food poisoning in man but all types of animals can be affected by eating contaminated food. Fish farms have outbreaks of *C. bo-tulinum* Type E, which can grow in temperatures as low as 5°C.

boundary layer, unstirred layer Surface layer of gas or liquid across which molecular movement is diffusion limited. This has a significant effect on the uptake of CO_2 by leaves or of some solutes by cells.

bound water Water held by matric forces. Cf. **matric potential**.

bouquet stage See **pachytene**.

bouyant density The density of molecules, particles or viruses as determined by flotation in a suitable liquid. A gradient of CsCl will separate DNA according to its base composition, different DNA molecules banding at discrete positions.

Bowman's capsule In the Vertebrate kidney, the dilated commencement of an uriniferous tubule.

Bowman's glands In some Vertebrates, serous glands of the mucous membranes of the olfactory organs.

boxplot The graphical representation of the frequency distribution of a set of values by means of a strip showing the relative positions of the smallest and largest values, the median and the quartiles of a set of values.

braccate Of Birds, having feathered legs or feet.

brachiate, brachiferous Branched; having widely spreading branches; bearing arms.

brachi-, brachio- Prefixes from L. *brachium*, arm.

Brachiopoda A phylum of solitary non-metameric Metazoa, with a well-developed coelom; sessile marine forms, with a lophophore in the form of a double vertical spiral, and usually with a bivalve shell. Brachiopods range from early geological periods up to the present time; they occur in all seas, often at great depths.

brachium The proximal region of the forelimb in land Vertebrates; a tract of nerve-fibres in the brain; more generally, any armlike structure, as the rays of Starfishes. adj. *brachial*.

brachy- Prefix from Gk. *brachus*, short.

brachycerous Having short antennae, as some Diptera.

brachydactyly, brachydactylia Abnormal shortness of fingers or toes.

brachyodont Said of Mammals having low-crowned grinding teeth in which the bases of the infoldings of the enamel are exposed; used also of the teeth. Also *brachydons*. Cf. **hypsodont**.

brachypterism In Insects, the condition of having wings reduced in length. adj. *brachypterous*.

brachysclereid *Stone cell*. A more or less isodiametric cell which has a thick lignified

wall, e.g. in the flesh of the pear fruit. See **sclereid**.

brachyurous, brachyural Said of decapodan Crustacea, in which the abdomen is reduced and bent forward underneath the laterally expanded cephalothorax by which it is completely hidden.

bracken poisoning A disease occurring in cattle and horses due to the ingestion of bracken. In cattle, the main symptoms are multiple haemorrhages and high fever, associated with bone marrow damage and later tumours of the gut. In horses, nervous symptoms are shown and the disease is essentially an induced thiamin deficiency.

bracket fungus Basidiomycotina which have the fruiting body projecting as a rounded bracket from the side of a tree trunk or stump.

brackish Salty, but not as salty as sea water. Brackish water occurs in estuaries, creeks and deep wells.

bract A leaf, often modified or reduced, which subtends a flower or an inflorescence.

bracteate Having bracts.

bracteole A small leaf-like organ occurring along the length of a flower stalk between the true subtending bract and the calyx.

bract scale The structure in conifers that subtends the *ovuliferous scale* and may be more or less fused to it.

brady- Prefix from Gk. *bradys*, slow.

bradycardia Slowness of the beating of the heart.

brain A term used loosely to describe the principal ganglionic mass of the central nervous system; in Invertebrates, the pre-oral ganglia; in Vertebrates, the expanded and specialized region at the anterior end of the spinal cord, developed from the three primary cerebral vesicles of the embryo.

brain stem In Vertebrates, regions of the brain conforming to the organization of the spinal cord, as distinct from such suprasegmental structures as the cerebral cortex and the cerebellum.

brain stimulation Technique for studying the neurophysiological basis of some behaviour patterns thought to be under central nervous system (CNS) control; involves stimulation of the CNS by electrical or chemical means, sometimes producing behaviour that appears *motivated*.

branch gap A region of parenchyma in the vascular cylinder of the stem, located above the level where the *branch traces* bend out towards the branch. Cf. **leaf gap**.

branchia In aquatic animals, a respiratory organ consisting of a series of lamellar or filamentous outgrowths; a gill. adj. *branchial*.

branchial arch In Vertebrates, one of a series of bony or cartilaginous structures lying in the pharyngeal wall posterior to the hyoid arch; it prevents the gill-slits from collapsing.

branchial basket (1) In Cyclostomata and cartilaginous Fish, the skeletal framework supporting the gills. (2) In the larvae of certain Dragonflies (Anisoptera), an elaborate modification of the rectum associated with respiration.

branchial chamber In Urochordata, cavity of the pharynx.

branchial clefts See **gill slits**.

branchial heart In Vertebrates, a heart such as that of the Cyclostomata, in which all the blood entering the heart is deoxygenated and passes thence directly to the respiratory organs; in Cephalopoda, special muscular dilations which pump blood through the capillaries of the ctenidia.

branchial rays Branches of the hyoid and branchial arches which support the gills and gill-septa.

Branchiopoda Subclass of Crustacea, the members of which are distinguished by the possession of numerous pairs of flattened, leaf-like, lobed swimming feet which also serve as respiratory organs; mainly freshwater forms including the Fairy Shrimps, Brine Shrimps, Tadpole Shrimps, Clam Shrimps, Water Fleas.

branchiostegal Pertaining to the gill-covers.

branchiostegal membrane In Fish, the lower part of the opercular fold below the operculum. Also *branchiostege*.

brand fungi See **Ustilaginales**.

brand spore The thick-walled resting spore of the brand fungi; it is black or brown, and forms sooty masses.

Brassica A genus of the Cruciferae which includes cabbage, broccoli, kale, rape, turnip, swede, mustard.

Brassicaceae Same as **Cruciferae**.

Braun Blanquet system Method of classifying vegetation in the European school of phytosociology, first enunciated by Braun Blanquet in 1921. With the advent of computers, its use has declined.

breaking The development of streaks and stripes in flowers owing to virus infection in e.g. Rembrandt tulips.

breaking of the meres The sudden development of large masses of blue-green algae (Cyanophyceae) in small bodies of fresh water.

breast bone In higher Vertebrates, the sternum.

breathing An activity of many animals, resulting in the rapid movement of the environment (water or air) over a respiratory surface. Now usually referred to the more general concept of **respiration**.

breathing root See **pneumatophore**.

brevi- Prefix from L. *brevis*, short.

bright-field illumination The common method of illumination in microscopy in which the specimen appears more or less dark on a bright background. Cf. **phase-contrast microscopy**, **interference microscopy**, **dark-ground illumination**.

broad spectrum *Wide spectrum*. Said of drugs effective against a wide range of micro-organisms.

Bromeliaceae Family of ca 2500 spp. of monocotyledonous flowering plants (superorder Commelinidae). Terrestrial and esp. epiphytic herbs (including tank epiphytes and atmospheric plants) from tropical and subtropical America. Many are **CAM** plants. The flowers often have showy bracts and are bird- or insect-pollinated. It includes the pineapple (the only major CAM crop plant) and some plants grown for fibre.

bronch-, broncho- Prefixes from Gk. *bronchos*, windpipe.

bronchia The branches of the bronchi. adj. *bronchial*.

bronchiole One of the terminal subdivisions of the bronchia.

bronchus One of two branches into which the trachea divides in higher Vertebrates and which lead to the lungs. pl. *bronchi*. adj. *bronchial*.

brood A set of offspring produced at the same birth or from the same batch of eggs.

brown algae The *Phaeophyceae*.

brown forest soil, brown earths A dark brown and friable soil with no visible layering, which is well aerated with only a thin litter layer, **mull** humus and pH of say 5–7, formed under deciduous forests at humid, temperate latitudes. In Britain, a common soil type with good potential for agriculture. Cf. **brown podzolic soil**.

brown podzolic soil An acid soil, usually formed from a brown forest soil in areas of high rainfall, with a pale layer from which elements and particles have been leached to a deeper zone where iron is often precipitated to form an impenetrable layer.

brown rot (1) Fungal diseases of plum and other fruit trees, infecting shoots and fruit. (2) The fungal decay of timber in which celluloses are preferentially attacked. Cf. **white rot**.

Brucellaceae A family of obligate parasites belonging to the order *Eubacteriales*. Gram-negative cocci or rods; aerobic or facultatively anaerobic; many pathogenic species, e.g. *Pasteurella pestis* (plague), *Pasteurella multocida* (fowl cholera, swine plague, haemorrhagic septicaemia), *Brucella abortus* (undulant fever in Man, contagious abortion in cattle, goats and pigs).

bruise Rupture of blood vessels in a tissue, with extravasation of blood, as a result of a blow which does not lacerate the tissue.

bruit A sound or murmur due to vascular blood flow, heard over heart, blood vessels and vascularized organs.

brush border See **microvillus**.

Bryophyta *Bryophytes*. (1) Division of the plant kingdom containing ca 25 000 spp. of small, rootless, thalloid or leafy, non-vascular plants; includes the liverworts (*Hepaticopsida*), the hornworts (*Anthoceropsida*) and the mosses (*Bryopsida*), having alternation of generations in which the gametophyte is the dominant generation, the sex organs are archegonia and antheridia and the sporophyte is more or less parasitic on the gametophyte. (2) In some confusing usages, the mosses alone. See **Bryopsida**.

Bryopsida *Musci*. The mosses. Class of those Bryophyta, ca 15 000 spp., which have a leafy (not thalloid) gametophyte with the leaves not strictly in 2 or 3 ranks, multicellular rhizoids and, in most, a capsule (sporophyte) with both a columella and a lid (operculum). Includes *Sphagnum*. See **acrocarp** and **pleurocarp**.

Bryozoa See **Ectoprocta**.

bubo An inflamed and swollen lymphatic gland, esp. in the groin.

bubonic plague A form of plague in which there is great swelling of a lymphatic gland, esp. those in the groin. See **plague**.

buccal Pertaining to, or situated in or on, the cheek or the mouth.

buccal cavity The cavity within the mouth opening but prior to the commencement of the pharynx.

buccal glands Glands opening into the buccal cavity in terrestrial Craniata; the most important are the salivary glands.

bucco- Prefix from L. *bucca*, cheek.

buccopharyngeal respiration *Buccal respiration*. Breathing by means of the moist vascular lining of the mouth cavity or diverticula thereof, as in some Amphibians and certain Fish which have become adapted to existence on land.

bud (1) An unexpanded shoot consisting of a short rudimentary stem bearing immature and primordial leaves and/or flowers. At least in extant spermatophytes, buds are expected at shoot tips (terminal or apical buds) and in leaf axils (axillary buds); other buds are accessory or adventitious. (2) An outgrowth of a parent organism that becomes detached and develops into a new individual, esp. from some yeasts and other fungi.

budding (1) A method of asexual reproduction by growth and specialization. This is followed by the separation by constriction of a part of the parent, as found in some yeasts. (2) Bud grafting, used esp. for the propagation of fruit trees and some woody ornamentals including roses, in which a bud, together

with more or less of the underlying stem, from the scion variety is grafted into a suitable root-stock.

bud scale A simplified leaf or stipule on the outside of a bud, forming part of a covering which protects the contents of the bud.

bud sport A shoot, branch, inflorescence or flower differing markedly from the rest of the plant with the differences persisting in vegetatively propagated offspring; due to nuclear or cytoplasmic mutation when the sport will often be chimeric, or in horticulture to a change in the structure of a pre-existing chimera.

build-up of radiation Increased radiation intensity in an absorber over what would be expected on a simple exponential absorption model. It results from scattering in the surface layers and increases with increasing width of the radiation beam.

bulb Any bulb-shaped structure, such as the organ of storage and perennation, usually underground, consisting of a short stem bearing a number of overlapping swollen fleshy leaf bases and/or scale leaves, with or without a *tunic*, the whole enclosing next year's bud, e.g. onion. Cf. **corm**, **rhizome**. adj. *bulbar*.

bulbiferous Having, on the stem, bulbs or bulbils in place of ordinary buds.

bulbil (1) Small bulb or tuber developing above ground from an axillary bud or in an inflorescence and functioning in asexual reproduction. (2) A contractile dilatation of an artery. (3) Any small bulblike structure.

bulbus arteriosus In many Vertebrates, a strongly muscular region following the conus arteriosus.

bulbus oculi The eyeball of Vertebrates.

bulimia An abnormal increase in the appetite, often part of the symptoms of **anorexia nervosa**.

bulla (1) A blister or bleb. A circumscribed elevation above the skin, containing clear fluid; larger than a **vesicle**. (2) In Vertebrates, with a flask-shaped tympanum, the spherical part of that bone which usually forms a protrusion from the surface of the skull.

bullate (1) Having a blistered or puckered surface. (2) Bubblelike. (3) Bearing one or more small hemispherical outgrowths.

bulliform cell An enlarged epidermal cell present, with other similar cells, in longitudinal rows in the leaves of some grasses and alleged to be *motor* cells causing the rolling and unrolling of leaves in response to changes of water status.

bundle See **vascular bundle**.

bundle cap Strand of sclerenchyma or parenchyma adjacent to the xylem and/or phloem sides of a vascular bundle.

bundle end The much simplified termination of a small vascular bundle in the meso-phyll of a leaf.

bundle sheath A sheath of one or more layers of parenchymatous or sclerenchymatous cells, surrounding a vascular bundle in a leaf.

bunion An enlarged deformed joint of the big toe where it joins the foot, as a result of pressure of tight-fitting shoes or boots, with overlying **bursitis**.

bunodont Said of mammalian teeth in which the cusps remain separate and rounded. Also *bunoid*. Cf. **lophodont**, **selenodont**.

bunt Disease of cereals caused by a *smut* fungus in which the grain of infected plants is transformed into a mass of spores.

Burkitt lymphoma A malignant tumour of B cells, esp. affecting the jaw and the gut, common in children in hot humid regions of Africa but not confined to these regions. **Epstein-Barr virus** is present and may be responsible for malignant transformation occurring in a B cell population subject to constant antigenic stimulation. Associated with a specific chromosomal rearrangement affecting chromosome 8q24.

burr A fruit covered with hooks to aid in dispersal by animals.

bursa Any saclike cavity; more particularly, in Vertebrates, a sac of connective tissue containing a viscid, lubricating fluid, and interposed at points of friction between skin and bone and between muscle, ligament, and bone.

bursa copulatrix A special genital pouch

of various animals acting generally as a female copulatory organ.

bursa inguinalis The cavity of the scrotal sac in Mammals.

bursa of Fabricius A sac-like structure arising as a diverticulum from the cloaca of young birds, composed of primary follicles containing B lymphocyte precursors. The bursa is the only source of these cells in birds and removal of the bursa at hatching (or by certain viral infections) results in a severe B cell deficiency.

bursa omentalis In Mammals, a sac formed by the epiploon or great omentum.

bursicon In Insects, a hormone produced by neurosecretory cells of the brain and released by neurochaemal organs in the thoracic and abdominal ganglia. It effects many post-ecdysal processes such as cuticular tanning.

bursiform Resembling a bag or pouch.

butterfly flower A flower pollinated by butterflies.

buttress root Form of *prop root* which thickens unevenly to produce a flat, apparently supporting, structure something like a buttress.

byssinosis Respiratory disease among workers in the vegetable fibre industry, characterized by chest tightness on returning to work after a period of absence. Due to sensitization by substances present in the fibre dust.

byssus In Bivalvia, a tuft of strong filaments secreted by a gland in a pit (*byssus pit*) in the foot and used for attachment. adj. *byssogenous, byssal*.

C

C Symbol for **complement**. The component proteins for complement are designated C1–C9′ and in activated form as C1⁻–C5⁻.

C1⁻-inhibitor An inhibitor of activated esterase formed from complement, C1. It also inhibits some other esterases activated during blood clotting. Normally present in blood but congenitally absent or inactive in some persons, who are liable to attacks of angioneurotic oedema due to vasoactive peptides released by the esterase.

C3 Method of photosynthesis in which CO_2 is fixed directly by **ribulose 1,5-bisphosphate carboxylase oxygenase** into the 3-carbon compound, 3-phosphoglyceric acid, which is subsequently converted to sugars etc. by the **Calvin cycle**. Most plants are C3 plants. Cf. **C4**, **CAM**.

C3a, C5a Peptide fragments which are split off from **complement** proteins, C3 and C5, respectively during conversion to their enzymically active forms. The peptides are chemotactic for leucocytes and cause local increase in vascular permeability. They act as *anaphylatoxins* causing histamine release from local mast cells.

C3b receptors Receptors present on cell membranes which can bind C3b, the activated form of complement C3, or its breakdown products, C3bi or C3d. The receptors are designated CR1, CR2 or CR3. The most important, CR1 is present on mononuclear phagocytes, granulocytes, B lymphocytes and some other cells. The receptor enables immune complexes or microbes which have bound complement to become attached to the cells and increases their ingestion. It is also probably involved in modulating the activation of B lymphocytes by immune complexes.

C4 *C4 photosynthetic pathway*. See panel.

C4 pathway evolution C4 photosynthesis has evolved apparently independently in a number of angiosperm families (e.g. Gramineae, Euphorbiaceae, Chenopodiaceae). In some sorts of C4 plant, the amino-acids, aspartate (C4) and alanine (C3), are shuttled between the cell types in place of malate and pyruvate, being formed from and forming oxaloacetate and pyruvate by transamination in the two cell types. See panel.

Cactaceae The Cactus family, ca 2000 spp. of dicotyledonous flowering plants (superorder Caryophillidae). Most are leafless, spiny, stem-succulent, CAM plants of the semi-deserts of the New World. Of little economic importance other than as ornamentals, but some edible fruit.

caducibranchiate Animals possessing gills at one period of the life cycle only, as in the young of some Caudata.

caducous Falling off at an early stage. Cf. **deciduous**.

caecum Any blind diverticulum or pouch, esp. one arising from the alimentary canal.

caenogenesis A phenomenon whereby features which are adaptations to the needs of the young stages develop early and disappear in the adult stage. adj. *caenogenetic*.

Caenorhabditis elegans A small nematode worm which is ideally suited for genetic, molecular and cellular studies of eukaryotic development.

caesious, caesius Bearing a bluish-grey waxy covering (bloom).

caespitose, cespitose Of plants growing from the root in tufts, as with many grasses. dim. *caespitulose*.

caffeine A weak central nervous stimulant found in coffee and tea.

Calamitales Order of extinct, mainly Carboniferous Sphenopsida. Sporophytes were mostly large trees with substantial hollow trunks having secondary xylem, and whorled or opposite branches. Co-dominant with the Lepidodendrales in the Carboniferous swamps.

calamus The proximal hollow part of the scapus of a feather; quill. pl. *calami*.

calcaneum In some Vertebrates, the fibulare or large tarsal bone forming the heel; more generally, the heel itself; in Birds, a process of the metatarsus.

calcar In Insects, a tibial spine; in Amphibians, the prehallux; in Birds, a spur of the leg, or more occasionally, of the wing; in Bats, a bony or cartilaginous process of the calcaneum supporting the interfemoral part of the patagium.

calcarate Bearing one or more spurs.

calcareous Containing, or coated with, calcium carbonate (lime).

calcicole *Calciphile*. Plants found on or confined to soils containing free calcium carbonate. Cf. **calcifuge**.

calciferol See **vitamin D**.

calciferous, calcigerous Producing or containing calcium salts.

calcification The accumulation of calcium carbonate on or in cell walls or e.g. in diseased or dead tissue such as the walls of arteries.

calcifuge *Calciphobe*. Plants not normally found on, or intolerant of, soils containing free calcium carbonate. Cf. **calcicole**.

calcigerous glands In some Oligochaeta, a pair of oesophageal glands producing a limy secretion to control the acid/base balance of the body; in some Amphibians, the glands of Swammerdam, calcareous concre-

C₄, C₄ photosynthetic pathway

A form of **photosynthesis** in which the first products of CO_2 fixation are C_4 acids (acids with 4 carbon atoms) rather than the phosphoglyeric acid (3 carbons) of the commoner C_3 plants. Typically, CO_2 is fixed in the mesophyll cells of a leaf with **Krantz anatomy** by **PEP carboxylase** into the 4-carbon oxaloacetic acid which is then reduced to malic acid. This is then transported, probably through the **plasmodesmata**, to the bundle sheath cells where it is de-carboxylated. The CO_2 thus released is refixed, by **ribulose 1,5-bisphosphate carboxylase oxygenase** in the **Calvin cycle,** into sugars in the normal way while the 3-carbon pyruvic acid remaining is returned to the mesophyll cells. The whole cycle acts as a pumping mechanism for CO_2, promoting CO_2 uptake and, by concentrating CO_2 in the bundle sheath cells, reducing photo-respiration.

C_4 metabolism uses more energy (as ATP from light reactions) than C_3 but most C_4 plants live in hot sunny places where there is plenty of energy avail-able in the form of light. Under such conditions, C_4 plants typically photosyn-thesize more rapidly than C_3 plants and C_4 crop plants like maize sorghum, sugar beet and some tropical fodder grasses, can outyield C_3 crops. See **C₄ pathway evolution.**

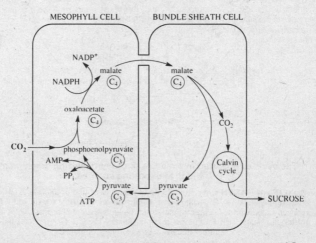

The metabolism of carbon (the Hatch-Slack pathway) in one type of C₄ plant.

AMP = adenosine monophosphate; ATP = adenosine trisphosphate; $NADP^+$ oxidized-, NADPH = reduced- nicotinamide adenine dinucleotide phosphate; PP_i = inorganic pyrophosphate. The C_3, C_4 in circles indicate the numbers of carbon atoms in the compounds.

tions lying on either side of the vertebrae, close to the exit points of the spinal nerves.

calcium phosphate precipitation Tech-nique for introducing foreign DNA or chro-mosomes into cells: co-precipitation with calcium phosphate facilitates the uptake of DNA or chromosomes.

callose A polymer of glucose, linked through $\beta-1\rightarrow3$, occurring e.g. as a cell wall constituent, esp. in the sieve areas of sieve elements.

callous, callose Hardened, usually thick-ened and often like horn in appearance.

callus (1) A tissue consisting of large, thin-

Calvin cycle, photosynthetic carbon reduction cycle

Abbrev. *PCR*. The cyclical sequence of reactions in which carbon dioxide is fixed by **ribulose bisphospate carboxylase oxygenase** and reduced to produce e.g. sugars, in all photosynthetic plants and algae and most other autotrophic organisms.

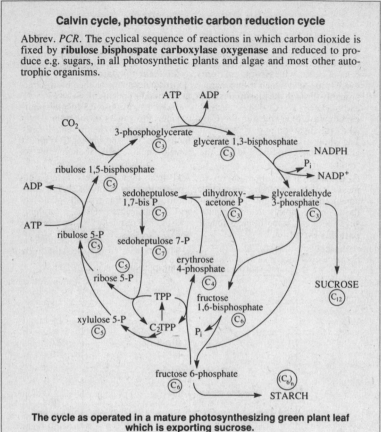

The cycle as operated in a mature photosynthesizing green plant leaf which is exporting sucrose.

The ATP and NADPH come from the light reactions of photosynthesis. ADP = adenosine diphosphate; ATP = adenosine triphosphate; $NADP^+$ = oxidized-, NADPH reduced-nicotinamide adenine dinucleotide phosphate; P = phosphate; P_i = inorganic (ortho)phosphate; TPP = thiamine pyrophosphate. The $C_3...C_{12}$ in circles indicate the number of carbon atoms in the compounds.

---

walled parenchymatous cells developing as a result of injury, as in wound healing and grafting, or in **plant cell culture**. (2) An accumulation of *callose*. (3) Hard basal projection at the base of the floret or spikelet of some grasses. (4) Newly formed bony tissue between the broken ends of a fractured bone.

calmodulin A calcium binding protein which is virtually ubiquitous in eukaryotic cells. It plays a central role in controlling the calcium levels in cytoplasm.

Calvin cycle See panel.

calcycle Syn. for *calyx* (1–3).

calypter One or two small lobes at the pos-

terior margin of the base of the wing in some Diptera.

calyptra The layer of cells, developed from part of the archegonium wall that protects the developing sporophyte in mosses and liverworts.

calyptrate Of Diptera, possessing **calypters**.

calyptrogen The layer of meristematic cells that gives rise to the root cap.

calyptron In calyptrate Diptera, the enlarged squama which covers the haltere. Also called *calypter*.

calyx (1) Outer whorl of the perianth, often

green and protective, composed of free or fused sepals. (2) A pouch of an oviduct, in which eggs may be stored. (3) In some Hydrozoa, the cuplike exoskeletal structure surrounding a hydroid. (4) In Crinoidea, the body as distinct from the stalk and arms. (5) In some mammals, part of the pelvis of the kidney.

calyx tube The tube formed by fused sepals.

CAM Abbrev. for *crassulacean acid metabolism*.

cambial initial One of the permanently meristematic cells of a cambium.

cambium A layer of meristematic cells, lying parallel to the surface of a stem or root, which undergo **periclinal** divisions to give secondary tissues. See **vascular cambium**, **cork cambium**.

cAMP See **cyclic adenosine monophosphate**.

campaniform Dome-shaped; as *campaniform sensilla* of certain Insects, which are mechanoreceptors, occurring widely on the body.

campanulate Bell-shaped.

campylotropous Said of an ovule which is so curved that the micropyle and the stalk are approximately at right angles and the stalk appears to be attached to the side.

canal An elongated intercellular space containing air, water or a secretory product such as resin or oil.

canal cell One of the short-lived cells present in the central cavity of the neck of the archegonium.

canaliculate Marked longitudinally by a channel or groove.

canaliculus Any small channel; in the liver, an intercellular bile channel; in bone, one of the ramified passages uniting the lacunae; in nerve cells, a fine channel penetrating the cytoplasm of the cell body. Adj. *canalicular*.

cancellous, cancelled Having a spongy structure, with obvious interstices.

cancer Any *malignant neoplasm*. An uncontrolled growth of cells which exhibits invasiveness and remote growth.

canine Pertaining to, or resembling, a dog; in Mammals, a pointed tooth with single cusp, adapted for tearing, and occurring between the incisors and premolars; pertaining to a canine tooth; pertaining to a ridge or groove on the surface of the maxillary.

canker A plant disease characterized by well defined necrotic lesions of a main root, stem or branch in which the tissues outside the xylem disintegrate.

cannabis *Indian hemp*. The plant *Cannabis sativa* or *Cannabis indica*. The dried flowers, exuded resin and leaves are used to produce the drug hashish, marihuana or bhang. Cannabis is widely recognized as a drug of addiction.

cannon bone In the more advanced Artiodactyla, the characteristic bone formed by the fusion of the two metapodials in the limb, associated with the reduction of the number of toes to two.

canopy The leaves, stems and branches of a plant or an area of vegetation, considered as a whole.

canopy cover Percentage of the ground occupied by the vertical projection of all the individuals of one plant species. The sum of such percentages for all plant species gives the total canopy cover.

cap A modified base added to the 5′ ends of eukaryotic messenger RNA molecules.

capillary (1) Of very small diameter; slender, hairlike. (2) Any thin-walled vessel of small diameter, forming part of a network, which aids rapid exchange of substances between the contained fluid and the surrounding tissues, as *bile capillaries, blood capillaries, lymph capillaries*.

capillary soil water Water held between the particles of the soil by capillarity.

capitate Having an enlarged tip, as *capitate antennae*.

capitellum An enlargement or boss at the end of a bone, for articulation with another bone; more particularly, the smaller of the two articular surfaces on the distal end of the mammalian humerus, for articulation with the radius; the distal knoblike extremity of the haltere in Diptera.

capitulum (1) An inflorescence on which the sessile flowers or florets are crowded on the surface of the enlarged apex of the **peduncle**, the whole group being surrounded, and covered in the bud by an envelope of bracts forming an **involucre**; the whole inflorescence superficially appearing to be one flower, as in the daisy, *Bellis*. See **Compositae**. (2) A terminal expansion, as that of some shaft bones, tentacles or hairs.

capping The phenomenon whereby proteins at cell membranes are caused to accumulate in clusters and then to move to one end of the cell when crosslinked, e.g. by antibody against them. Thus antibodies can cause proteins to disappear from the cell surface (though they may be reformed if antibody is removed).

caprification The fertilization of the flowers of fig trees by the agency of Fig Insects, a family of Chalcids (Agaonidae); the process of hanging caprifigs in the female trees.

capsomere Proteins which form regular structures on the surface of a virus.

capsular polysaccharides Those present as constituents of bacterial capsules which often resist engulfment by granulocytes. Antibodies against these polysaccharides can coat the bacteria so that they become susceptible to phagocytosis, and are usually protective.

capsule (1) That part of the sporophyte of a bryophyte which contains the spores. (2) A fruit, dry when mature, composed of more than one carpel, which splits at maturity to release the seeds. (3) A coating of mucilaginous material outside the wall of a bacterial cell. (4) A fibrous or membranous covering enclosing an abdominal structure or *viscus*, e.g. the kidney. The name is applied also to certain areas in the brain which are formed by nerve fibres. (5) A soluble case of gelatine or similar substance in which a medicine may be enclosed.

caput An abrupt swelling at the distal end of a structure. pl. *capita*. adj. *capitate*.

carapace An exoskeleton shield covering part or all of the dorsal surface of an animal; as the bony dorsal shield of a tortoise, the chitinous dorsal shield of some Crustacea.

carbamyl phosphate The phosphate ester of carbamic acid which is an intermediate in the biosynthesis of urea and pyrimidines.

carbohydrates A group of compounds represented by the general formula $C_x(H_2O)_y$. Substances found in plants and animals, e.g. sugars, starch, cellulose. The carbohydrates also comprise other compounds of a different general formula but closely related to the above substances, e.g. rhamnose, $C_6H_{12}O_5$. The carbohydrates can be divided into **monosaccharides**, **oligosaccharides**, **polysaccharides**. The carbohydrate element in diet supplies energy, provided by the oxidation of the constituent elements.

carbon cycle The biological circulation of carbon from the atmosphere into living organisms and, after their death, back again. See **carbon dioxide**, **photosynthesis**.

carbon dating *Radiocarbon dating*. Atmospheric carbon dioxide contains a constant proportion of radioactive ^{14}C, formed by cosmic radiation. Living organisms absorb this isotope in the same proportion. After death it decays with a half-life 5.57×10^3 years. The proportion of ^{12}C to the residual ^{14}C indicates the period elapsed since death.

carbon film technique One of the methods used in electron microscopy to provide a supporting film for the specimen. The film is prepared by subliming carbon in a vacuum and is itself supported on a metal grid.

carbon fixation The synthesis of organic compounds from carbon dioxide, most notably in photosynthesis.

carbonic anhydrase An enzyme in blood corpuscles catalysing the decomposition of carbonic acid. It is essential for the effective transport of carbon dioxide from the tissues to the lungs.

carbon replica technique A method used in electron microscopy for making a surface replica of a specimen. It is coated with a structureless carbon film, and the film and specimen are subsequently removed by dissolving in an appropriate solvent.

carboxydismutase Same as *ribulose bisphosphate carboxylase*.

carboxylase An enzyme which by the mediation of biotin, its **prosthetic group**, carboxylates its substrate as pyruvate carboxylase converts the monocarboxylic acid, pyruvate into *dicarboxylic oxaloacetate* or as ribulose bisphosphate carboxylase, PEP carboxylase, incorporates carbon dioxide into its substrate.

carbuncle Circumscribed staphylococcal infection of the subcutaneous tissues.

carcinogenesis The production and development of cancer.

carcinoma A disorderly growth of epithelial cells which invade adjacent tissue and spread via lymphatics and blood vessels to other parts of the body. See **sarcoma**.

cardiac muscle The contractile tissue forming the wall of the heart of Vertebrates.

cardiac valve A valve at the point of junction of the fore- and mid-intestine in many insects. Also *oesophageal valve*.

cardinal In Bivalvia and Brachiopoda, pertaining to the hinge; more generally, primary, principal, as the *cardinal sinuses* or *veins*, being the principal channel for the return of blood in the lower vertebrates.

cardioblast A mesodermal cell in an embryo, destined to take part in the formation of the heart.

cardiolipin Phospholipid hapten purified from beef heart which is the active antigen in the **Wassermann reaction** and other serological tests for syphilis.

cardo The hinge of the bivalve shell. pl. *cardines*.

carina A median dorsal plate of the exoskeleton of some Cirripedia; a ridge of bone resembling the keel of a boat, as that of the sternum of flying Birds.

carinate Shaped like a keel; having a projection like a keel. Also *tropeic*.

carious *Cariose*. Appearing as if decayed.

carnassial In terrestial Carnivora, large flesh-cutting teeth from the first lower molar and the last upper premolars.

Carnivora An order of primarily carnivorous Mammals, terrestrial or aquatic; usually with three pairs of incisors in each jaw and large prominent canines; the last upper premolar and the first lower molar frequently modified as carnassial teeth; collar bone reduced or absent; four or five unguiculate digits on each limb. Cats, Lions, Tigers, Panthers, Dogs, Wolves, Jackals, Bears, Raccoons, Skunks, Seals, Sea-Lions and Walruses.

---

carrier

(1) In human genetics particularly, a *heterozygote* for a recessive disorder.

(2) A non-radioactive compound added to a *tracer* quantity of the same compound, which is radio-labelled, to assist in its recovery after some chemical or biological process or precipitation.

(3) A molecule or molecular system which brings about the transport of a solute across a cell membrane either by **active transport**, or **facilitated diffusion**. See **pump**.

(4) An organism harbouring a parasite but showing no symptoms of disease, esp. if it acts as a source of infection.

(5) A more or less inert material used either as a diluent or vehicle for the active ingredient of e.g. a fungicide or as a support e.g. for cells in a bioreactor.

---

carnivorous Flesh eating.
carnivorous plant *Insectivorous plant*. One of ca 400 species belonging to several unrelated families, mostly growing on substrates poor in mineral nutrients, which trap and digest insects and other small animals.
carotenes Red to yellow **carotenoids**, unsaturated tetraterpene hydrocarbons ($C_{40}H_{56}$). Accessory and photoprotective pigments in chloroplasts and in chromoplasts in some fruits and in carrot roots.
carotenoids Red, orange and yellow *terpenoids*, based on tetraterpenes (C_{40}), including the *carotenes* (simple hydrocarbons) and the *xanthophylls* (oxygenated hydrocarbons) in chloroplasts and chromoplasts in all the plant and algal groups as **accessory pigments** or as pigments protecting the major photosynthetic pigments against photo-oxidation. See **sporopollenin**.
carotid arteries In Vertebrates, the principal arteries carrying blood forward to the head region.
carpal *Carpale*. One of the bones composing the **carpus** in Vertebrates. pl. *carpals, carpalia*.
carp-, carpo-, -carp, -carpous Prefixes and suffixes from Gk. *karpos*, fruit.
carpel A female organ in a flower, bearing and enclosing one or more ovules, and forming singly or with others the **gynaecium**. Typically like a leaf, folded longitudinally so the edges come together, and bearing one or more ovules on a placenta along the line of the junction. Comprises ovary, style (usually) and stigma.
carpellate Having functional carpels, female.
carpus In land Vertebrates, the basal podial region of the fore limb; the wrist.
carr Fen vegetation with a conspicuous component of tree species. See **fen**.
carrier See panel.
carrying capacity Maximum number of individuals of a species which can live on an area of land, usually calculated from food requirements.
cartilage A form of connective tissue in which the cells are embedded in a stiff matrix of **chondrin**.
caruncle (1) An outgrowth from the neighbourhood of the micropyle of a seed. The seed is said to be *carunculate*. (2) Any small fleshy outgrowth. (3) In some *Polychaeta*, a fleshy dorsal sense organ. (4) In some *Acarina*, a tarsal sucker. (5) In embryo chicks, a horny knob at the tip of the beak.
cary-, caryo- See **kary-, karyo-**.
Caryophyllaceae Family of ca 2000 spp. of dicotyledonous flowering plants (superorder Caryophyllidae). Mostly herbs, mostly temperate. Typically with opposite leaves, flowers with five free petals, and a superior ovary with free-central placentation. Of little economic importance other than as ornamentals, e.g. pinks and carnations (Dianthus).
Caryophyllidae Subclass or superorder of dicotyledons. Mostly herbs, most have trinucleate pollen (binucleate is commoner in flowering plants), most have betalains rather than anthocyanins (except Caryophyllaceae) and/or free-central or basal placentation. Contains ca 11 000 spp. in 14 families including Caryophyllaceae, Chenopodiaceae and Cactaceae. (Corresponds approximately to older group Centrospermae.)
caryopsis A dry, indehiscent, one-seeded fruit, characteristic of the grasses, with the ovary wall (pericarp) and seed coat (testa) united, e.g. a grain of wheat.
caseation The process of becoming cheese-like, e.g. in tissue infected with tubercle bacillus the cells break down into an amorphous cheese-like mass.
casein The principal albuminous constituent of milk, of which it is present as a calcium salt. Transformed into insoluble paracasein

(cheese) by enzymes. Casein is a raw material for thermoplastic materials used for insulators, handles, buttons, artificial fibres and bristles etc. Also used in adhesives, nerve tonics and for priming artists' canvases.

caseous Cheeselike; having undergone **caseation**.

casparian strip, casparian band A band running round the cell in which apparently the whole thickness of the primary wall is impregnated with suberin and/or lignin making it impermeable to water and solutes. Typical of the root *endodermis* where it occurs in all radial and transverse walls, preventing the movement in the **apoplast** of water and solutes between the cortex and the stele.

caste In some social Insects, one of the types of polymorphic individuals composing the community.

castration anxiety In psychoanalytic theory, the male child's fear that his penis will be cut off as punishment for his sexual desire for the mother.

casual species An introduced plant which occurs but is not established in places where it is not cultivated.

cata- See **kata-**.

catabolism A metabolic process of breaking down complex molecules into simpler ones and releasing energy.

catadromous See **katadromous**.

catalase An enzyme which catalyses the oxidation of many substrates by hydrogen peroxide. In the absence of a suitable substrate it destroys hydrogen peroxide converting it to water.

cataphyll A non-foliage leaf inserted low on a shoot, e.g. a scale on a rhizome or a bud scale. Cf. **hypsophyll** .

catch muscle A set of smooth muscle fibres which form part of the adductor muscle in bivalve Molluscs, and are capable of keeping the valves closed by means of a sustained tonus; any set of smooth muscle fibres associated with striated muscle fibres for a similar purpose. Also called *arrest muscle*.

catecholamines A series of compounds derived from dihydroxyphenylalanine (DOPA), dopamine, noradrenaline (norepinephrine) and adrenaline (epinephrine). They function as *neurotransmitters*, adrenaline also acting as a hormone.

catenation The arrangement of chromosomes in chains or in rings.

caterpillar A type of eruciform larva, found in Lepidoptera, which typically possesses abdominal locomotor appendages (prolegs).

cathexis A charge of mental energy attached to any particular idea or object.

catkin An inflorescence with the flowers sessile on a common axis and typically pendulous, unisexual and wind-pollinated. A common inflorescence of deciduous, north temperate trees.

cauda The tail, or region behind the anus; any tail-like appendage; the posterior part of an organ, as the *cauda equina*, a bundle of parallel nerves at the posterior end of the spinal cord in Vertebrates. adj. *caudal, caudate*.

caudad Situated near, facing towards, or passing to, the tail region.

Caudata See **Urodela**.

caudate Bearing a tail-like appendage.

caudex A trunk or stock.

caul- Prefix or suffix meaning stem.

caul In the higher Vertebrates, the amnion; more generally any enclosing membrane.

caulescent Having a stalk or a stem.

cauliflory Production of flowers on trunks, branches and old stems of woody plants rather than near the ends of smaller twigs. Occurs in some trees of tropical forest, e.g. cocoa.

cauline (1) Pertaining to a stem. (2) Leaves borne on an obvious stem, not at the extreme base, and well above soil level, i.e. not *radical*.

cavernosus, cavernous Honeycombed; hollow; containing cavities, e.g. *corpora cavernosa*.

cavitation The sudden development of a gas bubble in a previously sap-filled xylem *conduit* as a result of excessive tension. See **embolism**.

cavum A hollow or cavity; a division of the concha.

C-banding See **banding techniques**.

CD *Cluster of Differentiation*. Term approved by international agreement to designate molecules on leucocytes which are recognized by various different specific antibodies (usually monoclonal). In some cases the molecules are known and identify subsets of lymphocytes or stages of differentiation, e.g. CD1 identifies T cells in the thymus cortex, CD4 and CD8 identify helper and cytotoxic subsets of T cells.

cDNA cloning Procedure by which DNA complementary to mRNA is inserted into a **vector** and propagated. Because mRNA has no **introns**, such cDNA clones can be made to produce a normal polypeptide product.

cDNA, complementary DNA A DNA sequence complementary to any RNA. Formed naturally in the life cycle of RNA viruses by **reverse transcriptase**, which is widely used in the laboratory to make DNA complements of mRNA.

-cele Suffix from Gk. *kele*, tumour, hernia.

cell (1) The unit, consisting of nucleus and cytoplasm, of which plants and animals are composed; in the former surrounded by a non-living wall. (2) The whole bacterium or yeast, the *bacterial cell*. (2) One of the spaces into which the wing of an Insect is divided by the veins. adj. *cellular*.

cell cavity See **lumen**.

cell cycle See panel.

cell division The formation of two daughter cells from one parent cell. See **amitosis, mitosis** and **meiosis**.

cell enlargement, extension The growth in volume or length of a plant cell, produced from a meristem as it matures, involving *vacuolation* and the synthesis of protoplasmic and wall materials.

cell-free Applied to biological phenomena like **translation** and **transcription** which can be made to occur in the laboratory in the absence of cells.

cell fusion The merging of cells by fusion of their plasma membranes resulting in a bi- or multi-nucleate complex. Fusion can be induced by various agents like polyethylene glycol and is the crucial step in the formation of **hybridomas** during **monoclonal antibody** production.

cell genetics The study of genetics, particularly the location of genes on chromosomes, by means of cells grown in culture.

cell line Strictly a cell culture derived from a single progenitor cell and with therefore a homogeneous genetic constitution.

cell lineage The developmental history of individual cells of an embryo during cell division following fertilization.

cell-mediated immunity Specific immunity which depends on the presence of activated **T lymphocytes** acting as cytotoxic cells and/or releasing **lymphokines** which activate monocytes and macrophages. Cell mediated immunity is responsible for protecting against intracellular microbes, but also for the rejection of **allografts** and for **delayed type hypersensitivity reactions**.

cellobiose, celliose $C_{12}H_{22}O_{11}$, a disaccharide, obtained by complete hydrolysis of cellulose. $G–\beta 1\rightarrow 4–G$ (G=glucose). It is the repeating unit for cellulose.

cell plate A thin partition, bounded by a membrane, growing centrifugally by the coalescence of vesicles across the equatorial plane of the telophase spindle to effect the division of the cytoplasm and to become the basis of the middle lamella of the new wall. Characteristic of cell division in larger plants and many algae. See **cleavage, phragmoplast**.

cell transformation See **transformation** (2).

cellular slime moulds See **Acrasiomycetes**.

cellulase An enzyme or mixture of enzymes capable of catalysing the hydrolysis of cellulose to cellobiose or glucose.

cellulose See panel.

cell wall See panel.

cement In Mammalian teeth, a layer resembling bone, covering the dentine beyond the enamel.

censer mechanism A means of seed liberation in which the seeds are shaken out of the fruits as the stem of the plant sways in the wind (e.g. in the poppy).

censor, censorship In psychoanalytic theory, a powerful unconscious inhibitory mechanism in the mind, which prevents anything painful to the conscious aims of the individual from emerging into consciousness. It is responsible for the distortion, displacement and condensation present in dreams.

centi- Prefix meaning one-hundredth.

centiMorgan Measure of the distance between the loci of two genes on the same chromosome, obtained from the *cross-over frequency*. 1 *centiMorgan* =1% *crossing-over*; 1 *Morgan* =100 *centiMorgans* when summed over short distances between intervening loci.

centipedes See **Chilopoda**.

central cylinder See **stele**.

central dogma The postulate that genetic information resides in the nucleic acid and passes to the protein sequence, but cannot flow from protein to nucleic acid.

central nervous system The main ganglia of the nervous system with their associated nerve cords, consisting usually of a brain or cerebral ganglia and a dorsal or ventral nerve cord which may be double, together with associated ganglia. Abbrev. *CNS*.

centric Said of a diatom which is radially symmetrical.

centric leaves Cylindrical leaves, with the palisade tissues arranged uniformly around the periphery of the cylinder, e.g. Onion.

centrifugal (1) A process, starting at the centre and developing or moving towards the outside. (2) Of nerves, see **efferent**.

centrifuge Rotating machine which uses centrifugal force to separate molecules from solution, particles and solids from liquids, and immiscible liquids from each other. It depends on differences in the relative densities of the substances to be separated. Used widely in science where forces up to 500 000 g may be obtained in an *ultracentrifuge*. In industry they are used in sugar and cream production, separating water from fuel and swarf from cutting oil etc.

centriole A structure virtually identical with the basal body of cilia, which serves as the centre for the polymerization of the microtubules of specialized fibrillar structures, including sperm flagelli and the mitotic spindle. Centrioles are normally arranged in pairs at right angles to each other.

cell cycle

The period between one cell division and the next. Most work has been done on growing cells or on the non-growing but dividing cells of early embryos. Synthesis of most of the major components (protein and RNA) is continuous during the cycle, except in some cells during mitosis. DNA synthesis however is periodic except in fast-growing bacteria. The period of DNA synthesis is called the S period (or *phase*), see the diagram, and is usually preceded by a G1 period and followed by a G2 period which lasts until mitosis. A minority of other processes are also periodic like mitosis itself and cell divsion, the synthesis of nuclear histones and some enzyme activities. Non-dividing cells are often said to be in the G0 period.

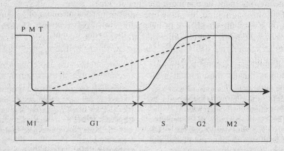

The synthetic events in the cell cycle of a growing eukaryotic cell.

M1, M2 = first and second mitoses, S = period of DNA synthesis, G1 and G2 = the periods before and after DNA synthesis, P = prophase, M = metaphase, T = telophase. (—) DNA synthesis. (- - -) RNA and protein synthesis.

There is no general model of how the cell cycle is controlled and why cell division occurs at a particular time. Two important *control points* are before mitosis and at the initiation of DNA synthesis near the G1/S boundary. This latter point has been called *start*, and is a branch point at which the cell either goes through the cell cycle or differentiates. In yeast, bacteria and some other cells, there is evidence that the cell has to grow to a critical size before it can pass a control point but this cannot apply to early embryos. In other cases, time is more important than size since the cell may have to go through a fixed time sequence of events before it passes a critical point.

Genetic tools for studying the cell cycle have become important in recent years, especially in yeast and bacteria. In particular, cell division cycle (*cdc*) mutants have been isolated which are **temperature-sensitive conditional mutants.** At the *restrictive temperature*, mutant cells are blocked at specific points in the cell cycle, providing new insights into what is happening in molecular terms at the control points. There appears at present to be a complex regulatory network but this is a rapidly developing research field. Recent work has shown that there is homology between mitotic control genes and their products in widely different cell types such as fission yeast and mammalian cells.

centripetal (1) A process starting at the outside and developing or moving towards the centre. (2) Of nerves, see **afferent**.

centrolecithal Having the yolk in the centre.

centromere Also *primary constriction*. In mitotic metaphase chromosomes it consists of a narrow region in which **chromatids** are joined. It has flanking **kinetochores** to which the microtubules of the spindle attach at mitosis.

centrosome Minute, self-duplicating struc-

cellulose

Occurs in **microfibrils** in the walls of most plant cells and since the secondary walls in the wood of trees commonly contain 50–60% cellulose, it is the most abundant organic molecular species on earth.

It has the formula $(C_6H_{10}O_5)n$, a complex polyose with many thousands of residues. The chief source of cellulose is wood, cotton and other fibrous materials (e.g. flax, hemp, nettle etc.). Pure cellulose is obtained by removing all incrustations of lignin resins and other inorganic and organic matter by treatment with alkali, acids, sodium sulphite etc. Cellulose is soluble in cuprammonium hydroxide (**Schweitzer's reagent**), ammoniacal copper carbonate, a solution of zinc oxide in conc. hydrochloric acid.

It is the raw material for the manufacture of paper, rayon, cellulose lacquers, films. Cellulose can undergo many chemical transformations, e.g. strong acids transform it into **amyloid**; it can be hydrolysed and oxidized (*cellulose hydrates, hydrocelluloses, oxycelluloses*) and esterified (*cellulose acetates; cellulose nitrates; benzyl-cellulose; cellulose xanthate*). It contains the β–1-4 glucoside linkage. See **cellobiose**.

ture near the interphase nucleus, from which the fibres of the spindle radiate at mitosis. In animal cells it contains a **centriole**.

centrum The basal portion of a vertebra which partly or entirely replaces the notochord, and from which arise the neural and haemal arches, transverse processes etc.

cepaceous Smelling or tasting like onion or garlic.

cephalad Situated near, facing towards, or passing to, the head region.

cephal-, cephalo- Prefixes from Gk. *kephalē*, head.

cephalic Pertaining to, or situated on or in, the head region.

cephalization The specialization of the anterior end of a bilaterally symmetrical animal as the site of the mouth, the principal sense organs and the principal ganglia of the central nervous system; the formation of the head.

Cephalochordata A subphylum of the Chordata having a persistent notochord, metameric muscles and gonads, a pharynx having a large number of gill-slits which are enclosed in an atrial cavity, and lacking paired fins, jaws, brain, and skeletal structures of bone and cartilage; marine sand living forms Lancelets.

Cephalopoda A class of bilaterally symmetrical Mollusca in which the anterior part of the foot is modified into arms or tentacles, while the posterior part forms a funnel leading out from the mantle cavity, the mantle is undivided, and the shell is a single internal plate, or an external spiral structure, or absent. Squids, Octopods and Pearly Nautilus.

cephalosporins Group of antibiotics derived from the fungus *Cephalosporium*,

effective against a broad spectrum of organisms (*wide-spectrum antibiotic*). Common examples are *cefuroxime* and *cefotaxime*.

cephalothorax In some Crustacea, a region of the body formed by the fusion of the head and thorax.

ceramic filter A deep filter, usually fingerlike, with fine pores in which small particles or bacteria become trapped. Also called *Pasteur filter*. Now largely superseded by **membrane filters**.

cerat- Same as *kerat-*.

cercal Pertaining to the tail.

cercaria The final larval stage of Trematoda which develops directly into the adult; usually characterized by the possession of a round or oval body, bearing eye spots and a sucker, and a propelling tail.

cercus In some Arthropoda, a multiarticulate sensory appendage at the end of the abdomen.

cere In Birds, the soft skin covering the base of the upper beak.

cerebellar fossa See **cerebral fossa**.

cerebellum A dorsal thickening of the hindbrain in Vertebrates. adj. *cerebellar*. Also *epencephalon*.

cerebral Pertaining to the brain; pertaining to the **cerebrum**.

cerebral flexure The bend which develops between the axis of the forebrain and that of the hindbrain, in adult Craniata.

cerebral fossa In Mammals, a concavity in the cranium corresponding to the cerebrum.

cerebral hemispheres See **cerebrum**.

cerebr-, cerebro- Prefixes from L. *cerebrum*, brain.

cerebroside The simplest glycolipid, consisting of N-acyl sphingosine with either a glucose or a a galactose residue.

cell wall

A structure external to the **plasmalemma** of a plant cell, secreted by the cell and enclosing it. It is typically tough, sometimes rigid, and protective or skeletal in function but with relatively little effect on solute influx or efflux, which are mainly controlled by the plasmalemma. The properties of cell walls largely determine the shape of cells and, hence, the morphology of plants.

In vascular plants the cell wall of an undifferentiated, parenchyma, cell typically consists of microfibrils of cellulose (say 40% of the dry mass) embedded in a matrix of other substances, rather like a glass-fibre-reinforced plastic. The microfibrils are flattened threads, 5–8 nm wide and relatively long. The matrix includes polysaccharides such as hemicelluloses and pectins together with some protein, especially in growing primary walls. The wall contains much water and is porous, thus allowing free passage to water and to solutes less than about 4 nm in diameter. If, as nearly always is the case, the concentration of solutes in the cell is greater than that in the medium, the cell being *hyperosmotic* to its surroundings, then the wall restrains the tendency of the cell to expand by the osmotic uptake of water. Equilibrium occurs when the excess hydrostatic pressure, the *turgor pressure*, generated by the stretching of the wall, is equal to the difference of osmotic pressure inside and outside the cell. The turgor pressure of a turgid cell might be 5–20 bars.

Cell walls are typically flexible but turgor gives them rigidity and enables non-woody plants to stand erect. Cf. the **hydrostatic skeleton** of some soft-bodied animals. The cell walls of vascular plants have thin areas of **pits** and are traversed by **plasmadesmata**. The walls of some sorts of cells are impreganted with lignin which more or less obliterates the pores and makes the wall incompressible and rigid. The walls of xylem tracheids and vessels, which conduct liquid under tension and must, therefore, resist implosion, are lignified as are those of the secondary xylem or wood cells generally and of most sclerenchyma cells. Woody stems are therefore stiff regardless of water content. Other cell walls contain suberin, cutin, sporopollenin or silica.

The arrangement of the *microfibrils* appears to relate to the function of the cell and to its manner of growth. See **multinet growth**. Microfibrils are approximately longitudinal in many fibres. Growth of the wall may be controlled by the matrix.

In algae and fungi, cell walls contain other components but there is often a microfibrillar element e.g. of cellulose or xylan in some algae or chitin in some fungi, embedded in a matrix of other polysaccharides. Some walls are impregnated with silica as in diatoms or calcium carbonate as in some red and green algae and may be intrinsically rigid. In some algae, especially motile unicells, the walls are turgor-resisting devices and cell volume may be controlled by the cells being iso-osmotic or having contractile vacuoles as in marine and freshwater forms, respectively.

The cell walls of *prokaryotes, bacteria* and *blue-green algae* are generally turgor-resisting devices but are constructed differently. See **cell envelope**.

cerebrospinal Pertaining to the brain and spinal cord.

cerebrum A pair of hollow vesicles or hemispheres forming the anterior and largest part of the brain of Vertebrates.

ceriferous Wax-bearing, wax-producing.

ceroma Syn. for *cere*.

cerous See *cere*.

ceruminous glands Modified sweat glands occurring in the external auditory meatus of Mammals and producing a waxy secretion.

cervical ganglia Two pairs of sympathetic ganglia, anterior and posterior, situated in the neck of Craniata.

cervical smear The taking of a small sample of cells from the uterine cervix for the detection of cancer or the pre-cancerous stage.

cervicum In Insects, the neck or flexible intersegmental region between the head and the prothorax; in higher Vertebrates, the neck or narrow flexible region between the head and the trunk. adj. *cervical*. Pertaining to the neck or to the cervix uteri.

cervine Dark-tawny.

cervix uteri Neck of the uterus, situated partly above and partly in the vagina.

Cestoda A class of Platyhelminthes, all the members of which are endoparasites; there is a tough cuticle; the alimentary canal is lacking; hooks and suckers for attachment occur at what is considered to be the anterior extremity. Tapeworms.

Cetacea An order of large aquatic carnivorous Mammals; the forelimbs are fin-like, and hindlimbs lacking; there is a horizontal flattened tailfin; the skin is thick with little hair, there are two inguinal mammae flanking the vulva, and the neck is very short. Whales, Dolphins and Porpoises.

CFA Abbrev. for *Complete Freund's Adjuvant*.

chaeta In Invertebrates, a chitinous bristle, embedded in and secreted by an ectodermal pit.

chaetiferous Also *chaetigerous, chaetophorous*. Bearing bristles.

Chaetognatha A phylum of hermaphrodite Coelomata, having the body divided into three distinct regions: head, trunk and tail; the head bears two groups of sickle-shaped setae; small transparent forms, of carnivorous habit, occur in the surface waters of the sea. Arrowworms.

chaetoplankton Planktonic organisms bearing bristle-like outgrowths from the cells the function of which is taken to be to increase drag and reduce sedimentation rate.

Chaetopoda A group of Annelida, the members of which are distinguished by the possession of conspicuous setae; it includes the **Polychaeta** and the **Oligochaeta**.

chain terminator See **stop codon**.

chalaza (1) The basal portion of the nucellus of an ovule. (2) One of two spirally twisted spindlelike cords of dense albumen which connect the yolk to the shell membrane in a bird's egg.

chalazogamy The process of entry of the pollen tube through the chalaza of the ovule. Cf. **porogamy**.

chalice A flask-shaped gland consisting of a single cell, esp. numerous in the epithelia of mucous membranes.

chalk gland A secreting organ, present in some leaves around which a deposit of calcium carbonate accumulates as in many species of *Saxifraga*.

chamaephyte Woody or herbaceous plant with perennating buds within 25 mm above the soil surface. Includes *cushion plants*. See **Raunkaier system**.

character In *Mendelian genetics*, the abnor-

mality or variant caused by a gene. In *quantitative genetics*, whatever is measured for study, e.g. weight or yield.

Charadriiformes Order of wading and swimming birds found on sea coasts and inland waters. Most feed on animal life. Waders, Gulls and Auks.

Charales *Stoneworts*. Small order in the Charophyceae of macroscopic fresh and brackish water algae with a distinct axis; anchored by rhizoids to the bottom, and with whorled branches. One very large cell runs the length of each internode. Oögamous.

Charophyceae Class of the green algae (Chlorophyta) characterized by motile cells (if produced) scaly or naked and asymmetric, with 2 flagella inserted laterally or subapically and associated and with a **multilayered structure**; no phycoplast; haplontic, the zygote being a resting stage. Includes unicellular, sarcinoid, filamentous and parenchymatous sorts. Predominantly freshwater. Includes *Klebsormidium*, *Spirogyra*, the **desmids**, *Coleochaeta* and *Chara*. (In earlier classification only the Charales were placed in the Charophyceae.)

chasmocleistogamous Producing both chasmogamous and cleistogamous flowers.

chasmogamous Said of flowers which open normally to expose the reproductive organs. Cf. **cleistogamous**.

cheek In Trilobita, the pleural portion of the head; in Mammals, the side of the face below the eye, the fleshy lateral wall of the buccal cavity.

cheilitis Inflammation of the lip.

chela In Arthropoda, any chelate appendage. adjs. *cheliferous, cheliform*.

chelate Of Arthropoda, having the penultimate joint of an appendage enlarged and modified so that it can be opposed to the distal joint like the blades of a pair of scissors to form a prehensile organ. Cf. **subchelate**.

chelating agent A chemical agent which combines with unwanted metal ions. Used to treat heavy metal poisoning, thus sodium calcium EDTA is a chelating agent promoting the excretion of lead.

chelicerae In Arachnida, a pair of pre-oral appendages, which are usually chelate.

Chelicerata Arthropod sub-phylum comprising animals with two major body regions, an anterior prosoma and a posterior opisthosoma and with the foremost appendages bearing chelae. Spiders, Scorpions, Horseshoe Crabs.

Chelonia An order of Reptiles in which the body is encased in a horny capsule consisting of a dorsal carapace and a ventral plastron, the jaws are provided with horny beaks in place of teeth, and the lower temporal arcade alone is present. Tortoises and Turtles.

cheluviation Leaching of iron and aluminium oxides from soils after the formation of soluble complexes with esp. polyphenols from fresh litter of conifers or heath plants. See **podsol**.

chemautotroph An organism which derives energy from the oxidation of inorganic compounds for the assimilation of simple materials, e.g. carbon dioxide and ammonia.

chemical potential μ. (1) A measure of the (Gibbs) free energy associated with a given uncharged chemical species under given conditions and hence of its relative ability to perform work. Any non-ionized substance including water moving by diffusion or osmosis will tend to move spontaneously down it own chemical potential gradient.

For the j th component:

$$\mu_j = \mu^*_j + RT.\ln(a_j) + P\overline{V}_j$$

where μ^*_j = the chemical potential of j in some arbitrary standard state, R = gas constant, T = absolute temperature, $\ln$ = natural logarithm, a = activity ($\cong$ concentration), P = hydrostatic pressure, V = partial molar volume. The PV term is important for water but not for solutes. Cf. **electrochemical potential**. (2) Same as *electrochemical potential*.

chemiosmosis The mechanism, discovered by P. Mitchell, underlying the formation of ATP synthesis by oxidative phosphorylation. The energy for ATP synthesis is derived from electrochemical gradients across the inner membrane of the mitochondrion. A similar mechanism operates during **photophosphorylation**. See **oxidative phosphorylation**.

chemokinesis Random movement of cells such as leucocytes stimulated by substances in the environment.

chemonasty A plant movement provoked, but not orientated by a chemical stimulus. See **nasty**.

chemoreceptor A sensory nerve ending, receiving chemical stimuli.

chemostat A culture vessel in which steady state growth is maintained by appropriate rates of harvest and the addition of the ingredients of the medium.

chemosynthesis The use, as by some bacterium, of energy derived from chemical reactions (e.g. oxidation of sulphur or of ammonia) in the synthesis from inorganic molecules of their organic requirements. Cf. **photosynthesis**.

chemosynthetic autotroph See **chemautotroph**.

chemotaxis Stimulation of movement by a cell or organism towards or away from substances producing a concentration gradient in the environment, e.g. the C5a comple-

ment peptide and leukotriene cause granulocytes to move into sites where they are released.

chemotaxomony The use of chemical evidence (of both primary and secondary metabolism) in taxonomy.

chemotherapy Treatment of disease by chemical compounds selectively directed against invading organisms or abnormal cells.

chemotropism A differential growth movement or curvature of part of an organ in a direction related to the concentration gradient of a chemical.

Chenopodiaceae Family of ca 1500 spp. of dicotyledonous flowering plants (superorder Caryophyllidae). Mostly herbs, temperate and subtropical, mostly in saline habitats. Flowers inconspicuous and wind pollinated. Some are C_4 plants. Includes sugar beet, mangelwurzels, beetroot, leaf beets (all species of *Beta*), spinach and quinoa.

chernozem Grassland soil formed in subhumid cool to temperate areas, with humus at the surface with a blackish layer of mineral soil just beneath which grades downward to a lighter layer where lime has accumulated. Occurs under tall-grass communities such as Russian Steppes, North American Prairies and Argentinian Pampas.

chiasma (1) Point of contact between **chromatids** visible during meiosis and involved in **crossing over**. pl. *chiasmata*. (2) A structure in the central nervous system, formed by the crossing over of the fibres from the right side to the left side and vice versa.

childhood psychosis A group of childhood disorders characterized by disturbed social relationships, speech impairment, and bizarre motor behaviour. Three sub-groups are recognized: infant **autism**, late onset psychosis, e.g. **childhood schizophrenia**, and disorders due to degeneration of the central nervous system.

childhood schizophrenia A childhood disorder which manifests itself after a period of normal development, often in adolescence, when the child begins to show severe disturbances in social adjustment and reality contact. As a diagnostic category it is not distinguished from adult **schizophrenia**.

Chilognatha See **Diplopoda**.

Chilopoda Class of Arthropoda having the trunk composed of numerous somites each bearing one pair of legs; the head bears a pair of uniflagellate antennae, a pair of mandibles, and two pairs of maxillae; the first body somite bears a pair of poisonclaws; the genital opening is posterior; active carnivorous forms, some of considerable size and dangerous to Man;

some are phosphorescent. Centipedes.

chimera (1) In animals or plants, an individual exhibiting two or more different genotypes in patches derived from two or more different embryos which have become fused, naturally or artificially, at an early stage to make a single embryo. Botanically the name is not applied to ordinary grafted plants where the stock differs from the scion but rather where cells of one sort have come to form a layer over a core of cells of the other throughout the shoot. See **periclinal chimera**, **mericlinal chimera**, **sectorial chimera**. Cf. **mosaic**. (2) A DNA molecule with sequences from more than one organism. adj. *chimeric*.

Chiroptera An order of aerial Mammals having the forelimbs specially modified for flight; mainly insectivorous or frugivorous nocturnal forms. Bats.

chiropterophilous Pollinated by bats.

chi-squared distribution The distribution of many quadratic forms in statistics, often encountered as the distribution of the sample variance and of a statistic measuring the agreement of a set of empirically observed frequencies with theoretically derived frequencies. The central chi-squared distribution is indexed by one parameter, the degrees of freedom.

chitin A nitrogenous polysaccharide with the formula $(C_8H_{13}N_5)_n$ occurring as skeletal material in many Invertebrates particularly the cuticle of Arthropoda.

Chlamydobacteriales Chemosynthetic bacteria-like organisms, probably closely related to the true bacteria: filamentous: characterized by the deposition of ferric hydroxide in or on their sheaths. Occur in fresh water, particularly moorland bogs, where their oxidation of iron (II) to iron (III) results in the deposition of **bog iron ore**.

chlamydospore A hyphal cell that becomes thick-walled, separates from the parent mycelium and functions as a spore.

chloragen, chloragogen cells In Oligochaeta, yellowish flattened cells occurring on the outside of the alimentary canal, and concerned with nitrogenous excretion.

Chlorella Microscopic unicellular green algae (Chlorophyceae, Chlorococcales) reproducing by autospores, no sexual reproduction; easily grown in laboratory culture and used in biochemical studies.

chlorenchyma Tissue composed of cells containing chloroplasts, e.g. leaf mesophyll.

Chlorococcales An order of the Chlorophyceae; the members are unicellular, but may form colonies of uninucleate, or multinucleate, cells, which never divide vegetatively; asexual reproduction by zoospores or autospores, and sexual reproduction by biflagellate gametes.

chlorocruorin A green respiratory pigment of certain *Polychaeta*. Conjugated protein containing a prosthetic group similar to, but not identical with, reduced haematin.

Chlorophyceae Class of the green algae (Chlorophyta) characterized by: motile cells (if produced) radially symmetrical with 2, 4 or many flagella inserted apically and with four, cruciate flagellar roots; with a phycoplast; haplontic, the zygote being a resting stage. Predominantly fresh water. Includes flagellate, coccoid motile and non-motile colonial (**coenobia**), sarcinoid, filamentous and parenchymatous sorts. Examples include *Chlamydomonas*, *Volvox*, *Chlorella*, *Oedogonium*. See **Chlorococcales**.

chlorophylls Green pigments (variously substituted porphyrin rings with magnesium) involved in photosynthesis. Chlorophyll a is the primary photosynthetic pigment in all those organisms that release oxygen, i.e. all plants and all algae including the blue-green algae. The other chlorophylls are *accessory pigments*; chlorophyll b in vascular plants, bryophytes and green algae; chlorophyll c in brown algae, diatoms, chrysophytes etc.

Chlorophyta The green algae, a division of eukaryotic algae characterized by: chlorophyll a and b; irregularly stacked thylakoids; no chloroplastal endoplasmic reticulum; mitochondrial cristae; flagella not heterokont; storing starch in the chloroplasts. Contains the classes Chlorophyceae, Ulvophyceae and Charophyceae, and is obviously ancestral to land plants.

chloroplast See panel.

chloroplast DNA The *chloroplast genome*, which is organized like that of prokaryotes and codes for some but not all chloroplast proteins. See **ribulose 1,5-bisphosphate carboxylase oxygenase**. Genetic recombination within the chloroplast genome has been reported during sexual reproduction of some algae (e.g. *Chlamydomonas*) and in **cybrids**. In higher plants, inheritance of chloroplast characters is usually maternal and recombination has not been reported even in the few with biparental inheritance.

chloroplast ER This is an extension of the **endoplasmic reticulum** which encloses the chloroplast to make a total of four membranes round it and occurs in some groups of eukaryotic algae, e.g. the Heterokontophyta. In the Dinophyceae and Euglenophyceae the chloroplasts are surrounded by a total of three membranes.

chlorosis Deficiency of chlorophyll in a normally green part of a plant so that it appears yellow-green, yellow or white, as a result of mineral deficiency, inadequate light or infection.

choana A funnel-shaped aperture; pl. *choanae*, the internal nares of Vertebrates.

chloroplast

A *plastid*, one or more in a cell, containing the membranes, pigments and enzymes necessary for photosynthesis in a eukaryotic alga (of any colour) or green plant.

A chloroplast in a leaf cell of a typical (C_3) vascular plant is an oblate ellipsoid, say 5×2 μm bounded by an envelope of two lipoprotein membranes which enclose an aqueous *stroma* traversed by **thylakoids**. The stroma contains enzymes, DNA and, sometimes, granules of starch. The thylakoids (interconnected, flattened, membrane-bound sacs) are organized into **grana** and **stroma lamellae**. The thylakoid membranes contain the **photosynthetic pigments**. The stroma contains the enzymes for the **Calvin cycle** and other processes.

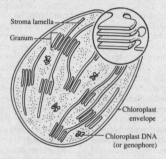

Stroma lamella
Granum
Chloroplast envelope
Chloroplast DNA (or genophore)

Diagrammatic vertical section through a chloroplast of a vascular plant.

Inset shows the connections between a stroma lamella and three discs of a granum.

During photosynthesis, energy from light is used to pump protons from the stroma into the lumen of the thylakoid. This generates both an electrical potential difference (lumen positive) and a pH difference (lumen acid) across the membrane. The resultant *proton motive force* drives protons back into the stroma through a coupling factor in the membrane, the movement being coupled to the synthesis of ATP from ADP and P_i. See **photophosphorylation, chemiosmotic hypothesis**. NADP is also reduced. The reduced NADP and ATP are used to drive reactions in the stroma, esp. the Calvin cycle in which CO_2 is reduced to sugars.

Chloroplasts of *green algae* vary much in size, shape and number per cell. They may also contain **pyrenoids** but are otherwise similar to those of vascular plants. In other algae the thylakoids may lie free in the stroma (red algae) or be associated in pairs (Cryptophyceae) or threes (several groups including the Heterokontophyta).

Prokaryotic photosynthetic organisms do not have chloroplasts. The pigments may be associated with flattened or tubular thylakoids or vesicles usually lying free in the cytoplasm in which the functions of the stroma occur.

Chloroplasts may represent photosynthetic prokaryotes that, in the remote past, became symbiotic within the cell of a eukaryotic host, the **endosymbiotic hypothesis**.

Choanichthyes See **Sarcopterygii**.

choanocyte In Porifera, a flagellate cell, in which a collar surrounds the base of the flagellum.

choice point The position in a T-maze or other type of maze or in any apparatus involved in discrimination training, when an animal can make only one of two or more alternative responses.

cholera An acute bacterial infection by

Vibrio cholerae in Eastern countries; characterized by severe vomiting and diarrhoea, drying of the tissues and painful cramps; spread by infected food and water.

choline *Ethylol-trimethyl-ammonia hydrate*, $OH.CH_2.CH_2.NMe_3.OH$, a strong base, present in the bile, brain, yolk of egg etc., combined with fatty acids or with glyceryl-phosphoric acid (**lecithin**). It is concerned in regulating the deposition of fat in the liver, and its acetyl ester is an important neurotransmitter. See **acetyl choline**.

chomophyte A plant growing on rock ledges littered with detritus, or in fissures and crevices where root hold is obtainable.

chondral Pertaining to cartilage.

chondr-, chondrio-, chondro- Prefixes from Gk. *chondros*, cartilage, grain.

Chondricthyes Class of cartilaginous Fishes in which the skeleton may be calcified but not ossified. Teeth not fused to the jaw but serially replaced; fertilization internal. Rays, Dogfish, Sharks and Chimeras. Cf. **Osteichthyes**.

chondrification Strictly, the formation of chondrin; hence, the development of cartilage. Also *chondrogenesis*.

chondrin A firm, elastic, translucent, bluish-white substance having a gelatinous nature, which forms the ground-substance of cartilage.

chondroblast A cartilage cell which secretes the chondrin matrix.

chondroclast A cartilage cell which destroys the cartilage matrix.

chondrocranium The primary cranium of Craniata, formed by the fusion of the parachordals, auditory capsules and trabeculae.

chondrosamine $C_6H_{13}O_5N$, or *2-aminogalactose*, is the basis of *chondroitin*, which is the substance of cartilage and similar body tissues.

chondroskeleton The cartilaginous part of the Vertebrate skeleton.

chorda (1) Any stringlike structure, e.g. the *chordae tendineae*, tendinous chords attaching the valves of the heart. (2) The **notochord**.

chordacentra Vertebral centra formed from the notochordal sheaths. Cf. **archecentra**. adj. *chordacentrous*.

Chordata A phylum of the Metazoa containing those animals possessing a notochord.

chordotonal organs In Insects, sense organs consisting of bundles of scolophores, sensitive to pressure, vibrations and sound.

choria A neurological disorder, characterized by spasmic, jerky and involuntary movements, particularly of the face, tongue, hands and arms.

chorion In higher Vertebrates, one of the foetal membranes, being the outer layer of the amniotic fold; in Insects, the hardened eggshell lying outside the vitelline membrane.

chorionic villus sampling Method for diagnosing human foetal abnormalities in the 6th to 10th week of gestation, in which small pieces of foetally-derived chorionic villi are removed through the cervix; chromosomes in the tissue are examined for abnormalities, and DNA may be extracted and **probed** for disease-associated alleles. Although the method can be used early in pregnancy, there is an associated risk to the foetus, higher than with **amniocentesis**. Abbrev. *CVS*.

choroid The vascular tunic of the Vertebrate eye, lying between the retina and the sclera. adj. *choroidal*.

choroid plexus In higher Craniata, the thickened, vascularized regions of the *pia mater* which is in immediate contact with the thin epithelial roofs of the diencephalon and medulla.

chromatic adaptation Differences in amount or proportion of photosynthetic pigments in response to the amount and colour of light. It can be (1) phenotypic as in many Cyanophyceae or (2) constitutive as in the distribution of littoral algae (red lowest, green at top, brown between).

chromatid One of the two, thread-like structures joined at the **centromere**, which constitute a single **metaphase** chromosome, each containing a single, double-helical DNA molecule with associated protein.

chromatin Network of more or less de-condensed DNA, and associated proteins and RNA, in the interphase nucleus forming higher-ordered structures in the nucleus of **eukaryotes**.

chromatin bead See **nucleosome**.

chromatography See panel.

chromatophore A cell containing pigment granules which may change its shape and colour effect on nervous or hormonal stimulation.

chrom-, chromo-, chromat-, chromato- Prefixes from Gk. *chrôma, chrômatos*, colour.

chromoblast An embryonic cell which will develop into a chromatophore.

chromocentre Mass of localized interphase **chromatin**; usually refers to the fused, centromeric regions of dipteran salivary gland **polytene** chromosomes.

chromomere One of the characteristic granules of compacted chromatin in serial array on a metaphase chromosome.

chromonema Complete single thread of chromatin.

chromophil, chromophilic Staining heavily in certain microscopical techniques.

chromophobe, chromophobic Resisting stains, or staining with difficulty, in certain microscopical techniques.

---

chromatography

A method of separating (often complex) mixtures.

Adsorption chromatography depends on using solid adsorbents which have specific affinities for the adsolved substances. The mixture is introduced onto a column of the adsorbent, e.g. alumina, and the components eluted with a solvent or series of solvents and detected by physical or chemical methods.

Partition chromatography applies the principle of **countercurrent distribution** to columns and involves the use of two immiscible solvent systems: one solvent system, the *stationary phase*, is supported on a suitable medium in a column and the mixture introduced in this system at the top of the column; the components are eluted by the other system, the *mobile phase*. See **paper chromatography**.

---

chromoplast (1) A **plastid** containing a pigment, esp. a yellow or orange plastid containing carotenoids. (2) A photosynthetic plastid, now usually called a chloroplast, of a non-green alga.

chromosomal aberration Any visible abnormality in chromosome number or structure, including **trisomy** and **translocations**.

chromosomal chimera That in which there are differences in number or morphology of the chromosomes.

chromosomally-enriched DNA library DNA library made from DNA enriched for one specific chromosome; **flow-sorted** chromosomes are a possible source.

chromosome See panel.

chromosome arm See panel.

chromosome complement The set of chromosomes characteristic of the nuclei of any one species of plant or animal, i.e. having the *basic number* of chromosomes. **Chromosome set** refers to the haploid set.

chromosome cores Non-histone protein network left when **histones**, DNA and RNA are removed from mammalian metaphase chromosomes.

chromosome mapping Assigning **genes** or **bands** to specific regions of the haploid chromosome complement. Each chromosome is numbered in order of size, short arms being designated p and long arms q. The bands are numbered consecutively outwards along each arm with the regions within a band also numbered. Thus 11p2.3 refers to the third region of band 2 on the short arm of chromosome 11. The ends are designated *ter*, as in 3pter, the end of the short arm of chromosome 3.

chromosome-mediated gene transfer Transfer of genetic material into a cell by introducing foreign chromosomes. Techniques include **calcium phosphate precipitation** and **electroporation**.

chromosome set The whole of the chro-

mosomes present in the nucleus of a gamete, usually consisting of one each of the several kinds that may be present.

chromosome sorting The ability to sort chromosomes by DNA content or size. See **fluorescence activated cell sorter**.

chronic granulomatous disease Inherited disease of male children characterized by recurrent abscesses and granuloma formation. The neutrophil leucocytes are deficient in a cytochrome component necessary for generating active oxygen free radicals for intracellular killing of microbes, even though these are ingested normally.

chrysalis The pupa of some Insects, esp. Lepidoptera; the pupa-case.

Chrysophyceae The golden-brown algae, a class of eukaryotic algae, in the division Heterokontophyta, ca 800 spp. Naked, scaly (silica) or walled. Flagellated unicellular and colonial, amoeboid, coccoid, palmelloid, simple and branched filamentous and a few thalloid types. Fresh-water mostly. Prototrophs; osmotrophic and phagotrophic heterotrophs.

Chrytridiomycetes Class of the Mastigomycotina with posteriorly uniflagellate zoospores and gametes. Body unicellular or mycelial. Mostly aquatic and saprophytic, or parasitic on algae, fungi or plants. Includes *Synchytrium endobioticum* causing potato wart disease, and *Blastocladia*.

chunking A process of reorganizing materials in memory which allows a number of items to be fitted into one larger unit, a form of *encoding*.

chyle In Vertebrates, lymph containing the results of the digestive processes, and having a milky appearance due to the presence of emulsified fats and oils.

chylification, chylifaction Formation of **chyle**.

chylomicron The plasma lipoprotein which has the lowest density and which transports dietary lipids from the intestine

chromosome

In eukaryotes the deeply-staining rod-like structures seen in the nucleus at cell division, made up of a continuous thread of DNA which with its associated proteins (mainly *histones*) forms higher order structures called **nucleosomes** and has special regions, **centromere** and **telomere**. See **sex determination**. Normally constant in number for any species, there are 22 pairs and 2 sex chromosomes in the human. In micro-organisms the DNA is not associated with histones and does not form visible condensed structures.

A *chromosome arm* is that part of a chromosome from the **centromere** to the end.

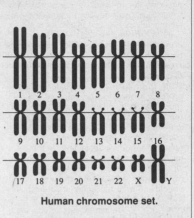

Human chromosome set.

to the liver and adipose tissue.

chyme In Vertebrates, the semifluid mass of partially digested food entering the small intestine from the stomach.

chymotrypsin A peptidase of the mammalian digestive system which is specific for peptide bonds adjacent to amino acids with aromatic or bulky hydrophobic side chains.

cicatrix The scar left after the healing of a wound; a scar which marks the previous attachment of an organ or structure, particularly in plants.

Ciconiiformes An order of Birds having a desmognathous palate and usually webbed feet; all are long-legged birds of aquatic habit, living mainly in marshes and nesting in colonies. They are powerful flyers and some migrate over long distances. Storks, Herons, Ibises, Spoonbills and Flamingoes.

cilia (1) Fine hair-like protrusions of the cell surface which beat in unison to create currents of liquid over the cell surface or propel the cell through the medium. Each cilium has a complex and characteristic internal structure built around 9 peripheral pairs and one central pair of microtubules. (2) In Mammals, the eyelashes; in Birds, the barbicels of a feather. adjs. *ciliated, ciliate.*

ciliary In general, pertaining to or resembling cilia; in Vertebrates, used of certain structures in connection with the eye, as the *ciliary ganglion, ciliary muscles, ciliary process.*

ciliate, ciliated (1) Having a fringe of long hairs on the margin. (2) Having flagella.

ciliograde Moving by the agency of cilia.

Ciliophora A class of Protozoa, comprising forms which always possess cilia at some

stage of the life cycle, and usually have a meganucleus.

ciliospore In Protozoa, a ciliated swarmspore. In Suctoria, a bud produced by asexual reproduction.

cilium Sing. of *cilia.* A flagellum esp. any one of the many short flagella of some protozoa.

cingulum Any girdle-shaped structure. In Annelida, the clitellum; in Rotifera, the outer post-oral ring of cilia; in Mammals, a tract of fibres connecting the hippocampal and callosal convolutions of the brain; also in Mammals, a ridge surrounding the base of the crown of a tooth and serving to protect the gums from the hard parts of food.

circadian rhythm A cyclical variation in the intensity of a metabolic or physiological process, or of some facet of behaviour, of ca 24 hours.

circinate Coiled, like a watch spring inwardly from base towards the apex. The leaf vernation in most ferns, some cycads and some seed ferns.

circulatory system A system of organs through which is maintained a constant flow of fluid, which facilitates the transport of materials between the different organs and parts of the body.

circumnutation The rotation of the tip of an elongating stem, so that it traces a helical curve in space.

cirrate, cirriferous Bearing cirri.

cirrhosis A disease of the liver in which there is increase of fibrous tissue and destruction of liver cells. Gk. *kirros*, orange-tawny.

Cirripedia Subclass of marine Crustacea,

generally of sessile habit when adult; the young are always free-swimming; the adult possesses an indistinctly segmented body which is partially hidden by a mantle containing calcareous shell plates; there are six pairs of biramous thoracic legs; attachment is by antennules; many species are parasitic. Barnacles.

cirrose Curly, like a waved hair. Consisting of diverging filaments.

cirrus In Protozoa, a stout conical vibratile process, formed by the union of cilia; in some Platyhelminthes, a copulatory organ formed by the protrusible end of the vas deferens; in Annelida, a filamentous tactile and respiratory appendage; in Cirripedia, a ramus of a thoracic appendage; in Insects, a hairlike structure on an appendage; in Crinoidea, a slender jointed filament arising from the stalk or from the centrodorsal ossicle and used for temporary attachment; in Fish, a barbel.

cisternum A compartment or vesicle, often flattened, formed within the cytoplasm by membranes of the **endoplasmic reticulum** or **Golgi apparatus**.

cistron A *gene* defined as a stretch of DNA specifying one *polypeptide*.

citric acid *2-hydoxypropane-1,2,3-tricarboxylic acid*. $C_6H_8O_7$, an important hydroxy-tricarboxylic acid, occurs in the free state in many fruits, esp. lemons, but is now prepared commercially, largely by fermentation with *Aspergillus*. Much used for flavouring effervescent drinks.

citric acid cycle See **tricarboxylic acid cycle**.

cladistics Method of classifying organisms into groups (taxa) based on 'recency of common descent' as judged by the possession of shared derived (i.e. not primitive) characteristics.

cladode (1) Strictly a **phylloclade** of one internode. (2) More commonly, any phylloclade.

cladogram A branching diagram (dendrogram) reflecting the relationships between groups of organisms determined by the methods of **cladistics**.

cladophyll Same as **phylloclade**.

clamp connection In some Basidiomycetes, a small connecting hypha across the septa between two adjacent cells of dikaryotic hypha. Formed as the two nuclei, lying one behind the other, in the terminal cell divide. Facilitates the maintenance of the dikaryon by allowing the second of the two daughter nuclei to pass the cell wall.

claspers In Insects, an outer pair of gonapophyses; in male Selachian Fish, the inner narrow lobe of the pelvic fin, used in copulation; more generally, any organ which is used by the sexes for clasping one another during copulation.

class In biometry, a group of organisms all falling within the same range, as indicated by the unit of measurement employed; in Zoology, one of the taxonomic groups into which a phylum is divided, ranking next above an order. In Botany the taxonomic rank below *division* and above *order*; the names end in -phyceae (algae), -mycetes (fungi) or -opsida (other plants).

class frequency The number of observations in a set of data in a particular class interval.

classical conditioning A learning procedure involving the repeated pairing of two stimuli, one of which (the unconditioned stimulus or *UCS*) already elicits a response (the unconditioned response), and the other (the conditioned stimulus or *CS*), which does not. After one or more pairing the CS comes to elicit a response very similar to that initially elicited only by the UCS, even if the CS is now presented on its own. This learned response is referred to as the conditioned response.

class interval A subset of the range of values of a variate.

clathrin A protein which is the major structural component of the proteinaceous layer of **coated pits** and **vesicles**. Clathrin takes the form of a triskelion and the association of triskelions into hexagons and pentagons forms the protein lattice of the coated pit and vesicle.

clavate adj. from *clava*; shaped like a club, e.g. *clavate* antennae.

clave A gradual swelling at the distal end of a structure, resembling a club.

clavicle In Vertebrates, the collar bone, an anterior bone of the pectoral girdle. adj. *clavicular*.

claw (1) A curved, sharp-pointed process at the distal extremity of a limb; a nail which tapers to a sharp point. (2) The narrow, elongated lower portion of a petal in some plants.

clearing agent In microscopical technique, a liquid reagent which has the property of rendering objects immersed in it transparent and so capable of being examined by transmitted light.

cleavage (1) Division of the cytoplasm by infurrowing of the plasmalemma. Cf. **cell plate**. (2) The series of mitotic divisions by which the fertilized ovum is transformed into a multicellular embryo.

cleavage-nucleus The nucleus of the fertilized ovum produced by the fusion of the male and female pronuclei; in parthenogenetic forms, the nucleus of the ovum.

cleidoic egg Egg of a terrestrial animal with a protective shell.

cleistocarp Same as **cleistothecium**.

cleistogamy The production of flowers, often inconspicuous, which do not open and in which self-pollination occurs. Cf. **chasmogamy**.

cleistothecium A more or less globose **ascocarp** with no specialized opening.

climacteric (1) A critical period of change in a living organism, e.g. the menopause. (2) The period of ripening of some fruit, e.g. apples, characterized by an increased rate of respiration.

climatic factor A condition such as average rainfall, temperature, and so on, which plays a controlling part in determining the features of a plant community and/or the distribution and abundance of animals. Cf. **edaphic factor**, **biotic factor**.

climax The end point in a succession of vegetation, when the community has reached an approximately steady state, in equilibrium with local conditions. See **monoclimax theory**, **polyclimax theory**, **sere**.

cline A quantitative gradation in the chacteristics of an animal or plant species across different parts of its range associated with changing ecological, geographic or other factors, e.g. **ecocline**, **geocline**.

clinical psychology A branch of psychology concerned with the application of research findings in the field of mental health.

clinostat See **klinostat**.

clitellum A special glandular region of the epidermis of *Oligochaeta* which secretes the cocoon and the albuminoid material which nourishes the embryo.

clitoris In female Mammals, a small mass of erectile tissue, homologous with the glans penis of the male, situated just anterior to the vaginal aperture.

cloaca Generally, a posterior invagination or chamber into which open the anus, the genital ducts, and the urinary ducts; in Urochordata, the median dorsal part of the atrium; in Holothuroidea, the wide posterior terminal part of the alimentary canal into which the respiratory trees open.

clock-driven behaviour Rhythmic behaviour under the influence of an endogenous clock, found throughout the animal kingdom and involving many of the annual, lunar and daily rhythms of behaviour found in animals. Its presence must be determined by isolation and other experimental procedures. See **zeitgeber**.

clonal selection Hypothesis proposed by F. M. Burnet which explains immune responses as due to selective stimulation by antigens of those lymphocytes bearing receptors capable of recognizing the antigens, from among a large population bearing a great variety of different receptors. The expansion of lymphocyte clones accounted for

specific antibody production and immunological memory.

clone (1) Organisms, cells or micro-organisms all derived from a single progenitor by asexual means. They have therefore an almost identical genotype. In plants it includes those derived by vegetative propagation such as grafting and taking cuttings. (2) Used loosely to describe the procedures by which a **vector** with an inserted DNA sequence makes multiple copies of itself and thus of the inserted sequence, i.e. *cloning*.

clonic phase The third phase in a *grand mal seizure* in which the muscles contract and relax rhythmically while the body jerks in violent spasms.

closed community A plant community which occupies the ground without leaving any closed spaces bare of vegetation.

closed mitosis Mitosis during which the nuclear envelope remains more or less intact, e.g. some algae and fungi. Ant. **open mitosis**.

closed vascular bundle A vascular bundle which does not include any cambium and which will not, therefore, form secondary tissues.

closing layer One of the alternating layers of compact and loose suberized tissue in the lenticels of some species of plant.

closing membrane Same as **pit membrane**.

clostridium An ovoid or spindle-shaped bacterium, specifically one of the anaerobic genus *Clostridium*, which contains several species pathogenic to man and animals: viz., *Cl. botulinum* (botulism), *Cl. chauvei* (blackleg), *Cl. tetani* (tetanus), *Cl. welchii* (gas gangrene).

clot The semisolid state of blood or of lymph when they coagulate. See **thrombosis**, **embolism**.

club moss In various usages, the Lycopsida, the orders Lycopodiales and Selaginellales, or just the Lycopodiaceae.

Clupeiformes An order of Osteichthyes, mostly planktonic feeders containing some economically important species. Herrings and Anchovies.

cluster In statistical sampling, a division of the population to be sampled into subsets, a sample of which will be taken.

cluster analysis (1) Hierarchical classification technique, often used to reveal patterns of similarity among species lists from many sites. (2) A statistical method of classifying observations into subsets, members of which satisfy some criterion of similarity.

cluster cup The popular name for an *aecidium*.

cnemidium In Birds, the lower part of the leg, bearing usually scales instead of feathers.

cnemis The shin or tibia.

Cnidaria A phylum of Metazoa comprising forms which are of aquatic habit; they show radial or biradial symmetry; possess a single cavity in the body, the enteron, which has a mouth but no anus; and generally have only two germinal layers, the ectoderm and the endoderm, from one of which the germ cells are always developed. Polyps, Corals, Sea anemones, Jelly fish and Hydra.

cnidoblast a thread-cell or stinging-cell, containing a **nematocyst**; characteristic of the Cnidaria. Also *cnida*.

CNS Abbrev. for *Central Nervous System*.

co-adaptation Correlated adaptation or change in two mutually dependent organisms.

coagulation The irreversible setting of protoplasm on exposure to heat, extreme pH or chemicals. Cf. **denaturation**.

coal ball A calcareous nodule, usually containing abundant petrified plant remains found in some seams of coal.

coalescent Grown together, esp. by union of walls.

coat (1) See **integument** (ovule). (2) See **testa** (seed coat).

coated pit A small surface invagination of the plasma membrane distinguished by a thick proteinaceous layer composed largely of *clathrin* on its cytoplasmic side. The pits invaginate to form vesicles within the cytoplasm. See **receptor mediated endocytosis**.

coated vesicle A cytoplasmic vesicle with a *clathrin* coat formed by the invagination of a coated pit.

coccoid Small, unicellular, walled, spherical and non-motile.

coccolith Small (say 2–10 μm) calcified scale covering the cells of coccolithophorids (flagellated unicellular algae of the Haptophyceae). Abundant as fossils in chalk.

coccus A spherical or near spherical bacterium with a diameter from 0.5–1.25 μm.

coccyx A bony structure in Primates and Amphibians, formed by the fusion of the caudal vertebrae; urostyle. pl. *coccyges*.

cochlea In Mammals, the complex spirally coiled part of the inner ear which translates mechanical vibrations into nerve impulses.

cochleate Spirally twisted, like the shell of a snail; *cochleariform*.

cocoon In Insects, a special envelope constructed by the larva for protection during the pupal stage; it consists either of silk or of extraneous matter bound together with silk.

coding capacity The number of different protein molecules which a DNA molecule could specify. Because of the presence of exons, reiterated sequences and other noncoding regions, it is not a very helpful concept to apply to whole chromosomes or the genomes of higher organisms.

coding sequence That part of a nucleic acid molecule which can be transcribed and translated into polypeptide using the genetic code.

codominant (1) Describes a pair of alleles which both show their effects in heterozygotes, e.g. many blood group genes. (2) One of two or more species, which together dominate a plant community.

codon A *triplet* of three consecutive bases in the DNA or in messenger RNA, which specifies (*codes for*) a particular *amino acid* for incorporation in a *polypeptide*. See **genetic code**.

coefficient of variation A dimensionless quantity measuring the relative dispersion of a set of observations, calculated as the ratio of the **standard deviation** to the mean of the data values (sometimes expressed as a percentage.

coele-, -coele Prefix and suffix from Gk. *koilia*, large cavity (of the belly).

Coelenterata See **Cnidaria**.

coeliac In Vertebrates, pertaining to the belly or abdomen.

coelom The secondary body cavity of animals, which is from its inception surrounded and separated from the primary body cavity by mesoderm. adj. *coelomic, coelomate*.

Coelomata A group of Metazoa, including all those animals which possess a **coelom** at some stage of their life-history.

coelomere In metameric animals, the portion of coelom contained within one somite.

coelomoduct A duct of mesodermal origin, opening at one end into the coelom, at the other end to the exterior.

coelomostome In Vertebrates, the ciliated funnel by which the nephrocoel opens into the splanchnocoel.

coelozoic Extracellular; living within one of the cavities of the body.

coenobium An algal *colony* in which the number and arrangement of cells are initially determined and which grows only by enlargement of the cells, e.g. *Volvox, Pediastrum*.

coenocyte, coenocytia A multinucleate cell. Multinucleate syncytial tissues formed by the division of the nucleus without division of the cell, as striated muscle fibres and the trophoblast of the placenta.

coenogamete A multinucleate gamete.

coenosarc In Hydrozoa, the tubular common stem uniting the individual polyps of a hydroid colony.

coenosteum In Corals and Hydrocorralinae, the common calcareous skeleton of the whole colony.

coenzyme Small molecules which are essential in stoichiometric amounts for the activity of some enzymes. Their loose association with enzymes distinguish them from prosthetic groups which fulfil a similar role but are tightly bound to the enzyme.

coenzyme A The coenzyme which acts as a carrier for acyl groups (A stands for acetylation). See **acyl-CoA**.

coenzyme Q See **ubiquinone**.

cognition, animal See **cognitive ethology**.

cognition, human Pertaining to the intellect and the mental processes involved in the obtaining and processing of knowledge and information; cognitive psychology includes studies of perception, memory, concept formation and problem solving.

cognitive dissonance In social psychology, the most influential of **balance theories**, originating with Festinger, that assumes that beliefs, attitudes and knowledge about related experiences must be consistent with one another. If they are not, the individual experiences an unpleasant state of dissonance and is motivated to reduce it by reinterpreting some aspect of their experience in a way that will maximize consistency (consonance).

cognitive ethology A branch of ethology introduced by Griffin, concerned with the issues of whether or not conscious awareness, and/or intention, should be taken into account in explanations of animal behaviour.

cognitive map A mental representation of physical space.

cognitive therapy An approach to therapy which holds that emotional disorders are caused primarily by irrational but habitual forms of thinking. It is related to behaviour therapy because it regards such patterns as forms of behaviour and attempts to help the individual discover and change inappropriate thought patterns.

coherent United, but so slightly that the coherent organs can be separated without very much tearing.

cohesion The union of plant members of the same kind, as when petals are joined in a sympetalous corolla.

cohesion mechanism Any mechanism in a plant, in particular one concerned with the dehiscence of a sporangium, which depends on the cohesive powers of water (i.e. that a mass of water resists disruption).

cohesion theory The generally accepted explanation of water movement through the xylem, that the water is drawn though the vessels or tracheids under tension, the columns of water being maintained by the cohesion of the water molecules and their adhesion to the walls.

cohesive end DNA in nature or when cut by many restriction enzymes often has short single-stranded sequences at their end. If the complementary sequence occurs elsewhere, the two ends will cohere. The basis of most

sequence insertion procedures. syn. *sticky end*.

cohort A taxonomic group ranking above a superorder.

colchicine $C_{22}H_{25}O_6N$, an alkaloid obtained from the root of the autumn crocus, *Colchicum autumnale*; pale yellow needles, mp 155°–157°C. Used to inhibit chromosome separation and so double the original number of chromosomes; much used in plant breeding for making *tetraploids*.

cold agglutinin Antibody against red cells which causes agglutination at temperatures below body temperature but not at 37°C. Often reactive with the I antigen of red cells and induced by infection with *Mycoplasma pneumoniae*.

cold-blooded Having a bodily temperature which is dependent on the environmental temperature; poikilothermal. Cf. **warm-blooded**.

Coleoptera An order of Insecta, having the fore-wings or elytra thickened and chitinized, meeting in a straight line; the hind-wings, if present, are membranous; the mouthparts are adapted for biting. Beetles.

coleoptile The sheath, probably a much modified first leaf, enclosing the epicotyl in the embryo of grasses and growing up, during germination, as far as the soil surface to protect the expanding leaves. It is a classical object for the study of auxin action and of phototropism.

coleorrhiza The sheath enclosing the radicle of the embryo of grasses, through which the radicle grows at germination.

collagen Family of fibrous proteins abundant in the extracellular matrix, tendons and bones of animals.

collar In plants, the junction between the stem and root of a plant, usually situated at soil level. In animals, the rim of a choanocyte; in Hemichorda, a collarlike ridge posterior to the proboscis; in Gastropoda with a spiral shell, the collarlike fleshy mantle edge protruding beyond the lip of the shell; more generally, any collarlike structure.

collar cell See **choanocyte**.

collateral (1) Running parallel or side by side. (2) Having a common ancestor several generations back.

collateral bud An accessory bud located to the side of an axillary bud.

collateral bundle A vascular bundle having phloem on one side only of the xylem, usually the abaxial side.

collecting cell A cell of the mesophyll of a leaf lying below and in contact with cells of the palisade from which it is presumed to collect photosynthetic products for transfer to the vascular tissue.

collective fruit A fruit derived from several flowers, as a mulberry.

collective unconscious Jung: those aspects of unconscious mental life that represent the accumulated experiences of the human species, in contrast with the unconscious life of an individual based on personal experience.

collenchyma A mechanical tissue, typical of leaf veins, petioles and the outer cortex of stems, of more or less elongated cells with unevenly thickened non-lignified primary walls.

colleterial glands One or two pairs of accessory reproductive glands, present in most female Insects. Their secretion forms the oötheca in Orthoptera, and a cement which fastens the eggs to the substratum in many other insects.

colliculus A small prominence, as on the surface of the optic lobe of the brain; a rounded process of the arytaenoid cartilage.

colloid From Gk. *kolla*, glue. Name originally given by Graham to amorphous solids, like gelatine and rubber, which spontaneously disperse in suitable solvents to form lyophilic sols. Contrasted with crystalloids on the one hand and with lyophobic sols on the other. The term currently denotes any colloidal system.

colon In Insects, the wide posterior part of the hind-gut; the large intestine of Vertebrates. adj. *colonic*.

colony (1) A collection of individuals living together and in some degree interdependent, as a *colony* of polyps, a *colony* of social Insects; strictly, the members of a colony are in organic connection with one another. (2) The vegetative form of many species of algae in which the sister cells are connected in a group to function as a unit. In many sorts (e.g. *Synura*), the colonies, of no fixed number of cells, grow by division and reproduce asexually by fragmentation. See **coenobium**. (3) A fungal mycelium grown, e.g. on an agar plate, from one spore. (4) A bacterial colony similarly initiated and grown.

colony stimulating factors Substances made by a number of cells which cause haemopoietic stem cells to proliferate and differentiate into mature forms, appearing as colonies in tissue culture. There are separate factors for granulocytes and macrophages, for eosinophyl leucocytes and for erythrocytes. Abbrev. *CSF*.

Colorado beetle A black-and-yellow striped beetle (*Leptinotarsa decemlineata*), which feeds upon potato leaves, causing great destruction.

colostrum A clear fluid accumulated in the mammary glands in the latter part of pregnancy. It precedes the flow of milk at the onset of lactation and is an important source of maternal antibodies for the infant.

colostrum corpuscles Large cells containing fat particles, which appear in the secretion of the mammary glands at the commencement of lactation.

colour blindness The lack of one or more of the spectral colour sensations of the eye. The commonest form, Daltonism, consists of an inability to distinguish between red and green. Even persons of normal sight may be colour blind to the indigo of the spectrum.

colour vision The ability of animals to discriminate light of different wavelengths. Depending on the animal there may be specialized receptors (cones in vertebrates) which are preferentially sensitive to two, three or more wavelengths. Trichromatic vision is most common.

colpus An elongated aperture in the wall of a pollen grain. Cf. **aperturate**.

columella In plants, a small column, esp. (1) The axial part of a root cap in some species, in which the cells are arranged in longitudinal files. (2) The sterile tissue in the centre of the sporangium of bryophytes and some fungi. (3) A radial rod in the wall of a spore or pollen grain. In Mammals, the central pillar of the cochlea; in lower Vertebrates, the auditory ossicle connecting the tympanum with the inner ear; in some lower Tetrapods, the epipterygoid; in spirally coiled gastropod shells, the central pillar; in the skeleton of some Corals, the central pillar. adj. *columellar*.

column The central portion of the flower of an orchid (probably an outgrowth of the receptacle of the flower), bearing the anther, or anthers and the stigmas. In Crinoidea, the stalk; in Vertebrates, a bundle of nerve fibres running longitudinally in the spinal cord; the edge of the nasal septum; more generally, any columnar structure, as the vertebral column.

columnar epithelium A variety of epithelium consisting of prismatic columnar cells set closely side by side on a basement membrane, generally in a single layer.

coma (1) A state of complete unconsciousness in which the patient is unable to respond to any external stimulation. (2) A tuft of hairs or leaves.

comatose Being in a state of coma.

comb In Ctenophora, a ctene; the framework of hexagonal wax cells produced by social Bees to shelter the young or for storing food.

comedo A blackhead. A collection of cells, sebum and bacteria, filling the dilated orifices of the sebaceous glands near hair follicles.

comfort behaviour Behaviours that have to do with body care, e.g. grooming, scratching, preening.

Commelinidae Subclass or superorder of

monocotyledons. Almost all are terrestrial herbs, often of moist places, with the perianth differentiated into sepals and petals, or reduced and not petaloid, syncarpous and usually with a starchy endosperm. Contains ca 19 000 spp. in 25 families including Cyperaceae, Gramineae, Bromeliaceae and Zingiberaceae.

commensalism An external, mutually beneficial partnership between 2 organisms (*commensals*); one partner may gain more than the other. adj. *commensal*.

comminuted Reduced to small fragments, e.g. *comminuted* fracture.

commissural bundle A small vascular bundle interconnecting larger bundles.

commissure (1) A juncture or suture. A line of junction between two organs or structures, esp. a surface by which carpels are united. (2) A bundle of nerve fibres connecting two nerve centres.

common bundle A vascular bundle belonging in part to a stem and in part to a leaf.

communication No generally accepted definition; different researchers use the term according to their interest in: (1) goal-directed behaviour used to influence other individuals; (2) species-specific behaviour adapted through evolution for a signalling function.

communication, non-verbal Communication by means other than language in its spoken or written form.

community Any group of animals or plants growing together under natural conditions and forming a recognizable sort of vegetation, e.g. oak wood, blanket bog, field or pond; such a community is not necessarily stable.

companion cell A parenchyma cell, with dense cytoplasm and a conspicuous nucleus, in the phloem of an angiosperm, adjacent to and originating from the same mother cell as a sieve tube member. See **sieve element**.

comparative psychology Refers to a field of animal study associated mostly with North American psychologists; ideally it stresses attention to the issues of development, motivation and causality, and relies on the method of cross-species comparison for making generalisations about the mechanisms underlying behaviour. Some comparative psychologists focus their studies on a single species for convenience, e.g. the white rat, and tend to focus on the area of learning.

compartment Region of an insect embryo within which all cells give rise to the same adult structure, e.g. leg, forewing.

compensation point *CO_2 compensation point*. The light intensity at which, under specified conditions, photosynthesis and respiration just balance so that there is no net

exchange of CO_2 nor O_2. For C_3 plants it is 50–70 ppm, for C_4 plants 0–10 ppm.

competition The struggle between organisms for the necessities of life (water, light, etc.).

competitive exclusion principle The ecological 'law' that two species cannot occupy the same ecological niche or utilize the same limiting resource. One species always outcompetes the other.

complanate Flattened, compressed.

complement (1) See **chromosome complement, chromosome set**. (2) See panel.

complementary Relationships between single strands of DNA and RNA are complementary to each other if their sequences are related by the *base-pairing rules*, thus ATCG is complementary to the sequence TAGC, and can pair with it by hydrogen bonding.

complementary genes Two *non-allelic* genes which must both be present for the manifestation of a particular character.

complementation The full or partial restoration of normal function when two recessive mutants, both deficient in that function, are combined in a double heterozygote. *Complementing* mutants are *non-allelic*, *non-complementing* are *allelic*.

complement deficiency Hereditary deficiencies of complement components are uncommon in man although strains of laboratory animals exist lacking C3, C4, C5 or C6. Absence of any single component is compatible with life but persons lacking the early acting components, esp. with diminished C3, are unusually liable to bacterial infection and often show signs of immune complex disease. The genes for C4, C2 and Factor B lie within the **major histocompatibility complex**.

complement fixation Term synonymous with activation but often applied to a system *in vitro* which detects complement by its capacity to cause lysis of red cells with antibody on the surface. Activation of complement by combination of antigen with antibody prior to adding the red cells diminishes the amount of complement available to lyse the red cells. This is a sensitive method for detecting the presence of antigen or antibody but has been superseded by other methods such as **ELISA**.

complete Freund's adjuvant See **Freund's adjuvant**.

complex A term introduced by Jung to denote an emotionally toned constellation of mental factors formed by the attachment of instinctive emotions to objects or experiences in the environment, and always containing elements unacceptable to the self. It may be recognized in consciousness, but is usually repressed and unrecognized.

---

complement

Name given to a system of proteins in the blood which act in series to produce a variety of biological effects. The system is triggered by antibodies when they become cross-linked through combination with antigens, irrespective of the antibody specificity.

The *classical pathway* consists of 9 separate components, of which C1 to C5 are enzymes (specific esterases) in an inactive form. Combination of C1 with aggregated immunoglobulin Fc activates it; it activates in turn the next component by splitting off small peptides; this in turn acts on the next component *seriatim*. This results in a mechanism which amplifies the effects at each stage, but the activations are transient and regulated by inhibitory substances also present in blood.

The most important products of the enzyme cascade are C3b and the peptides **C3a** and **C5a** which aid engulfment and killing of microbes by phagocytic cells and initiate inflammatory reactions. C5⁻ acts on C6 and this causes the remaining components C6 to C9 to arrange themselves into tubular structures which become inserted in the lipid layer of the outer cell membrane, causing the cell to leak its contents. Many cells with attached antibody and some microbes can be killed this way.

The *alternative pathway* of complement activation, which can be triggered by a variety of microbial polysaccharides and by some antigen complexes, does not involve C1, C2, C4 but is initiated by combination with **properdin** and two factors, D and B, which activate C3 to C3b. The biological effects of either pathway are similar.

See also **chromosome complement, chromosome set**.

---

complexity of DNA, RNA A measure, obtained from renaturation kinetics, of the number of copies of a given sequence in, e.g. a genome or the mRNA in a cell.

complex tissue A tissue made up of cells or elements of more than one kind.

complicate Folded together.

Compositae *Asteraceae*. The daisy family, ca 25 000 spp. of dicotyledonous flowering plants (superorder Asteridae). The largest family of dicotyledons. Mostly herbs and shrubs, cosmopolitan. The inflorescence is a head (*capitulum*) made up of many small individual florets or florets surrounded by an involucre of bracts, the whole resembling, and functioning biologically as a single flower. The florets have a gamopetalous corolla that is usually **tubular** or **ligulate**, and the ovary is inferior and develops into a one-seeded indehiscent, dry fruit. The modified calyx or pappus is composed of hairs, scales or bristles, and often develops as a feathery parachute aiding wind dispersal. Includes relatively few economic plants, such as sunflower (for oil), lettuce, endive, chicory and a number of ornamentals, e.g. chrysanthemum and various daisies. The insecticide pyrethrum comes from the heads of a species of *Tanacetum*.

compost (1) Rotted plant material and/or animal dung etc. used as a soil conditioner.

(2) A medium in which plants (esp. plants in pots) are grown, composed of one or more of sand, soil, grit, peat, perlite, vermiculite etc. with lime and fertilizers as necessary.

compound Consisting of several parts; a leaf made up of several distinct leaflets; an inflorescence of which the axis is branched etc. Cf. **simple**.

compound eyes Paired eyes consisting of many facets or ommatidia, in most adult Arthropoda.

compound reflex A combination of several reflexes to form a definite coordination, either simultaneous or successive.

compression wood The **reaction wood** of conifers, as formed on the underside of horizontal branches, characterized by denser structure, higher lignin content and microfibrils in shallower helices than normal wood.

compressor A muscle which by its contraction serves to compress some organ or structure.

compulsion An action which an individual may consider irrational but feels compelled to do.

concanavalin A A **lectin** derived from jack beans (*Canavalia ensiformis*) which binds to oligosaccharides present in the membrane glycoproteins on many cells. The lectin has 4 binding sites and so can cause cross-linking of the glycoproteins. It is a very effective polyclonal mitogen for T cells, causing them to secrete **lymphokines**.

concentric vascular bundle A bundle in which a strand of xylem is completely surrounded by a sheath of phloem (*amphicribral*) or vice versa (*amphivasal*).

conceptacle A flask-shaped cavity in a thallus, opening to the outside by a small pore, and containing reproductive structures, e.g. of *Fucus*.

concha In Vertebrates, the cavity of the outer ear; the outer or external ear; a shelf projecting inwards from the wall of the nasal cavity to increase the surface of the nasal epithelium.

conchiolin A horny substance forming the outer layer of the shell in Mollusca.

concolorate Having both sides the same colour.

concolor, concolorous Uniform in colour.

concrescence Union of originally distinct organs by the growth of the tissue beneath them.

concrete operations, stage, period In Piagetian theory, the period between ages 8 and 11 years, when a child acquires the ability to think logically, but only in very concrete terms; the ability to reason abstractly is still very limited. This limitation is reflected in their inability to solve particular types of problems, e.g. those involving the mental operation of *reversibility*. See **concrete thinking, conservation**.

concrete thinking A form of reasoning that is strongly tied to the immediate situation, or to very tangible and specific information, as opposed to abstract reasoning. See **concrete operations**.

condensed Said of an inflorescence with the flowers crowded together and nearly or quite sessile.

condensed chromatin See **heterochromatin**.

conditional lethal See **lethal**.

conditional probability The probability of occurrence of an event given the occurrence of another *conditioning* event.

conditional probability distribution The distribution of a random variable given the value of another (possibly associated) random variable or event.

conditioned reflex Reflex action by an animal to a previously neutral stimulus as the result of **classical conditioning**.

conduct disorders Childhood disorders involving antisocial behaviour.

conducting tissue Xylem and phloem in vascular plants; leptoids and hydroids in Bryophytes.

conduit Functional element of the conducting system of the xylem; either a whole

vessel or a single *tracheid*. Water moves within conduits though cell lumens and perforations in vessels and between conduits through pits (which may resist the spread of an **embolism**).

conduplicate Folded longitudinally about the midrib so that the two halves of the upper surface are brought together.

condyle A smooth rounded protuberance, at the end of a bone, which fits into a socket on an adjacent bone, as the *condyle* of the lower jaw, the occipital *condyles*. Adjs. *condylar, condyloid*.

cone (1) See **strobilus**. (2) Light sensitive structures in the retina of many Vertebrates which respond preferentially to particular wavelengths and thus provide the basis for colour vision. Cf. **rod**.

confabulation A tendency to fill in memory gaps with invented stories.

confervoid Consisting of delicate filaments.

confidence interval An interval so constructed as to have a prescribed probability of containing the true value of an unknown parameter.

conflict (1) *Animal behaviour*: refers to situations in which an individual animal appears motivated to engage in more than one activity, usually inferred from the simultaneous presence of causal factors that would normally impel the animal to behave in two or more incompatible ways. (2) *Human behaviour*: refers to situations in which the tendency to respond in a particular way is inhibited either because of (1) environmental factors (e.g. fear of punishment), or (2) because of internal inhibitions (e.g. conscience). The study of conflict is a vast and diverse area, and includes *social conflict*, which focuses on conflicts between individuals or between groups of individuals.

confocal microscope A form of light microscope in which an aperture in the illuminating system confines the illumination to a small spot on the specimen and a corresponding aperture in the imaging system (which may be the same aperture in reflecting and fluorescence devices) allows only light transmitted, reflected or emitted by the same spot to contribute to the image. The spots are made, by suitable mechanical or optical means, to scan the specimen as in a television raster. Compared to conventional microscopy, confocal techniques offer improved resolution (say $0.2\ \mu m$ in the x and y dimensions and $0.7\ \mu m$ in the z direction) and improved rejection of out-of-focus noise.

congeneric Belonging to the same genus.

congenic Applied to inbred cells of animals which have been bred to be genetically identical except in respect to a single gene locus. See **recombinant inbred strains**.

congenital Dating from birth or from before birth.

congenital deformity Malformation present at birth. Does not have to be a genetically determined defect, but may be due to environmental factors *in utero*, e.g. thalidomide.

conidiophore A simple or branched hypha bearing one or more conidia.

conidiosporangium A sporangium capable of direct germination, as well as producing zoospores.

conidium An asexual fungal spore. Produced exogenously from a hyphal tip, never within a sporangium. adj. *conidial*.

Coniferales The conifers. Order of ca 600 spp. of gymnosperms (class Conferopsida). Also fossils from the Jurassic. Trees (mostly) and shrubs, with simple, usually small, leaves. Reproductive organs in unisexual cones. Siphonogamous. Dominating large areas of the earth (though reduced since the Cretaceous) and including many important *soft wood* trees for timber and pulp. Includes pines, spruces, cypresses, monkey puzzles.

Coniferopsida Class of gymnosperms dating from the Carboniferous. Mostly substantial trees, with pycnoxylic wood, simple leaves, saccate pollen and flattened (platyspermic) seed. Includes Cordaitales (Carboniferous and Permian), Volztiales (Permian–Jurassic), and the Coniferales, Taxales (yews) and, in some classifications, the Ginkgoales.

coniferous Cone-bearing; relating to a cone-bearing plant.

conjugate division Simultaneous mitosis of a pair of associated nuclei in e.g. a dikaryotic cell.

conjugation (1) In prokaryotes, the process by which genetic information is transferred from one bacterium to another. (2) Sexual reproduction in some algae and fungi in which there is fusion of 2 non-flagellated gametes or protoplasts.

conjugation tube *Fertilization tube*. A tubular outgrowth of a cell through which one non-flagellated gamete moves to fuse with another during conjugation.

conjunctiva In Vertebrates, the modified epidermis of the front of the eye, covering the cornea externally and the inner side of the eyelid.

conjunctive tissue Secondary tissue occupying the space between the vascular bundles where the secondary xylem does not form a solid cylinder.

connate Of plant or animals parts which are firmly joined, particularly of like parts.

connecting thread See **plasmodesma**.

connective A bundle of nerve fibres uniting two nerve centres.

connective tissue A group of animal tissues fulfilling mechanical functions, developed from the mesoderm and possessing a large quantity of nonliving intercellular matrix, which usually contains fibres; as bone, cartilage, and areolar tissue.

connivent Converging and meeting at the tips.

consciousness Several related and general meanings: (1) alert and capable of action; (2) awareness of the environment, sentience; (3) awareness of a person's thoughts and feelings; (4) the ability of an individual to perceive their own mental life; (5) states of mind of which one is aware, and which organize and co-ordinate one's activities, as opposed to subconscious factors organizing and guiding behaviour.

consensus sequence A DNA sequence found with minor variations and similar function in widely divergent organisms.

conservation In behaviour it is a term in Piagetian theory, referring to the understanding that certain physical attributes such as quantity, volume and mass, do not change despite various transformations in their physical appearance. It is a knowledge gained between 8 and 11 years of age, during the periods of **concrete operations**. In ecology it refers to (1) the protection of natural ecosystems from the hand of man with the intention of preserving them as heritage or as a practical gene-bank and (2) the wise management of ecosystems, allowing exploitation at a level which does not impair the future capacity to produce.

consolidation of learning, memory A process that is thought to follow a learning trial or experience and that continues for some time after the learning event, during which the memory for the event becomes stable and durable, presumably a physiological process involving a structural change in the brain.

conspecific Relating to the same species. Often used as a noun.

constancy The percentage of sample plots in a plant community containing a particular species.

constant region The carboxy-terminal half of the light or the heavy chain of an immunoglobulin molecule. Termed constant because the amino acid sequence is the same in all molecules of the same class or subclass.

constitutive enzyme An enzyme that is formed under all conditions of growth. Cf. **inducible enzyme**.

constitutive heterochromatin Heterochromatin which is always condensed. **Satellite DNA** is found in these regions and coding sequences are apparently absent.

constriction Narrow, localized region in a chromosome, normally found at the centromere (*primary constriction*), and often also at other sites (*secondary constrictions*), including **nucleolar organizing region**.

constrictor A muscle which by its contraction constricts or compresses a structure or organ.

consumers In an ecosystem, the heterotrophic organisms, chiefly animals, which ingest either other organisms or particulate organic matter. Cf. **decomposers**, **producers**.

consummatory act, behaviour, phase Historically in animal behaviour studies, the end phase of goal oriented behaviour, typified by a series of responses directed at that goal, often of a stereotyped nature; it follows the **appetitive phase**, a goal seeking phase of behaviour. The rigid distinction between appetitive and consummatory phases of a behaviour sequence has been largely abandoned, although the terms are still used descriptively.

contact herbicide A herbicide which kills those plant parts that it comes into contact with, e.g. ioxynil, paraquat. Cf. **soil-acting herbicide**, **translocated herbicide**.

contact hypersensitivity Hypersensitivity reaction provoked by application to the skin of substances which act as **haptens** or as antigens. Usually due to prior sensitization by the chemical, and may be immediate or delayed type.

contact inhibition The inhibition of movement of some kinds of cells in tissue culture which occurs when they touch each other. See **transformation (2)**.

contact insecticide One which kills on contact with insect surface (body, legs etc.); used against sucking insects (e.g. aphids, mosquitoes) which are not affected by insecticides acting only through the alimentary system; e.g. pyrethrins, rotenone, DDT.

contagion The communication of disease by direct contact between persons, or between an infected object and a person. adj. *contagious*.

contagious distribution Pattern of distribution of plants and animals in which individuals occur closer together than would be expected on a random basis.

contiguity The closeness in time of two events which is sometimes regarded as the condition leading to association, esp. in **classical conditioning** procedures.

contingency table A table giving the frequency of observations cross-classified by variate values.

continuous culture A culture maintained at a steady state over a period, usually in a **chemostat**. Cf. **batch culture**.

continuous reinforcement *Schedules of reinforcement* in which every correct behaviour is reinforced.

continuous variation Variation of a character whose measurements do not fall into distinct classes, but take any value within certain limits.

continuum The pattern of overlapping populations in a large but definite community with component populations distributed along a gradient of e.g. altitude.

contorted Said of petals in a bud if each overlaps its neighbour on one side and is overlapped on the other. See **aestivation**.

contraceptive Any agent which prevents the fertilization of the ovum with a spermatozoon. See **oral contraceptives**.

contractile root A root, some part of which shortens (by a change in shape of the inner cortical cells) so as to pull e.g. a herbacious plant closer to the ground or a bulb or corm deeper into the soil.

contractile tissue A group of animal tissues which possess the property of contractility; more commonly spoken of as muscle.

contractile vacuole In some Protozoa, a cavity, filled with fluid, which periodically collapses and expels its contents into the surrounding medium, so ridding the animal of surplus fluid.

contractility The power of becoming reduced in length, exhibited by some cells and tissues, as muscle; the power of changing shape.

contracture Muscular contraction which persists after the stimulus which caused it has ceased.

contralateral Pertaining to the opposite side of the body. Cf. **ipsilateral**.

conus Any cone-shaped structure or organ.

conus arteriosus In some lower Vertebrates, a valvular region of the truncus arteriosus, adjacent to the heart.

conus medullaris The conical termination of the spinal cord.

convergence See **convergent evolution**.

convergent evolution The tendency of unrelated species to evolve similar structures, physiology or appearance due to the same selective pressures, e.g. the eyes of Vertebrates and Cephalopods or the succulent CAM plants of deserts.

convergent thinking A type of ability which is sampled in standard intelligence tests. Rational, logical thought which works within certain confines, as in solving a mathematical problem. See **creativity**, **divergent thinking**.

conversion disorder The loss or impairment of some motor or sensory function for which there is no known organic cause. Formerly *conversion hysteria*.

convolute (1) Coiled, folded or rolled, so that one half is covered by the other, as the cerebral lobes of the brain in higher Vertebrates. (2) Gastropod shells in which the outer whorls overlap the inner. n. *convolution*.

coomassie blue TN for dye used as a sensitive stain to locate proteins after their fractionation by **electrophoresis** or **isoelectric focusing** in a gel matrix.

co-operation A category of interaction between two species where each has a beneficial effect on the other, increasing the size or growth rate of the population, but, unlike mutualism, not a necessary relationship. Termed **protoco-operation** by some, since its basis is neither conscious nor intelligent, as in man.

Copepoda Subclass of Crustacea, mainly of small size. Some are parasitic, others planktonic where they form an important food source for pelagic Fish like herring.

coppice (1) A traditional form of woodland in which trees are cut to near ground level every 10–15 years and allowed to grow again from the **stool**, to produce poles for firewood, charcoal and fencing (esp. hurdle-making). (2) To cut such trees. In a *coppice-with-standard*, some trees are left to grow for several coppice cycles to form timber.

coprodaeum In Birds that part of the cloaca into which the anus opens.

coprolalia The utterance of filthy words by the insane.

coprophagous Dung-eating.

coprophilia Pleasure or gratification obtained from any dealing with faeces.

coprophilous, coprophilic Growing on or in dung.

coprozoic Living in dung, as some Protozoa.

copula A structure which bridges a gap or joins two other structures, as the series of unpaired cartilages which unite successive gill arches in lower Vertebrates.

copulation In Protozoa, a type of syngamy in which the gametes fuse completely; in higher animals, union in sexual intercourse.

copulation tube See **conjugation tube**.

copy number The number of genes or plasmid sequences per genome which a cell contains.

Coraciiformes An order of Birds, most of which are short-legged arboreal forms, nesting in holes and having nidicolous young. Mainly tropical and often brightly coloured. Kingfishers, Bee-eaters.

coracoid In Vertebrates, a paired posterior ventral bone of the pectoral girdle, or the cartilage which gives rise to it.

coral The massive calcareous skeleton formed by certain species of *Anthozoa* and some *Hydrozoa*; the colonies of polyps that form this skeleton. adjs. *coralline, coralloid, coralliferous, corallaceous, coralliform*.

corbicula The pollen basket of Bees, consisting of the dilated posterior tibia with its fringe of long hairs.

cordate Said of a leaf base which has the form of the indented end of a conventional heart.

coremium (1) A ropelike strand of anastomosing hyphae. (2) A tightly packed group of erect conidiophores, somewhat resembling a sheaf of corn.

coriaceous, corious Firm and tough, like leather in texture.

corium The dermis of Vertebrates.

cork *Phellem*. A tissue of dead cells with suberized cells which form a protective layer replacing the epidermis in older stems and roots of many seed plants. See **cork cambium, periderm**.

cork cambium *Phellogen*. The layer of meristematic cells lying a little inside the surface of an older root or stem and forming cork on its outer surface and phelloderm internally. See **periderm**.

corm Organ of perennation and vegetative propagation in e.g. *Crocus* consisting of a short, usually erect and tunicated, underground stem of one year's duration, next year's rising on top.

cormophyte In former systems of classification, a plant of which the body is differentiated into roots, stems and leaves.

corn (1) In Britain, wheat as well as other cereals; in US, maize. (2) Localized overgrowth of the horny layer of the skin due to local irritation, the overgrowth being accentuated at the centre.

cornea In Invertebrates, a transparent area of the cuticle covering the eye, or each facet of the eye; in Vertebrates, the transparent part of the outer coat of the eyeball in front of the eye. adj. *corneal*.

corneous Resembling horn in texture.

corniculate, cornute (1) Shaped like a horn. (2) Bearing a horn or hornlike outgrowth.

cornua Hornlike processes; as the posterior *cornua* of the hyoid. adj. *cornual, cornute*.

corolla Inner whorl of the **perianth** esp. if different from the outer and then often brightly coloured, composed of the petals which may be free or fused to one another.

corona (1) A trumpetlike outgrowth from the perianth, as in the daffodil. (2) A ring of small leafy undergrowths from the petals, as in campion. (3) A crown of small cells on the Oögonium of Charophyta. (4) In Echinoidea, the shell or test. (5) In Crinoidea, the disk and arms as opposed to the stalk. (6) In Rotifera, the discoidal anterior end of the body. (7) The head or upper surface of a structure or organ. adj. *coronal*.

corona radiata A layer of cylindrical cells surrounding the developing ovum in Mammals.

coronary circulation In Vertebrates, the

system of blood vessels (coronary arteries) which supply the muscle of the heart-wall with blood.

coronet (1) The junction of the skin of the pastern with the horn of the hoof of a horse. (2) The knob at the base of the antler in deer.

coronoid (1) In some Vertebrates, a membrane bone on the upper side of the lower jaw. (2) More generally, beak-shaped.

corpora allata In Insects, endocrine organs behind the brain which secrete **neotenin**, the juvenile hormone. In some species they are paired and laterally placed, but in others they fuse during development to form a single median structure, the *corpus allatum*.

corpora bigemina In Vertebrates, the optic lobes of the brain.

corpora cardiaca In Insects, paired neurohaemal organs lying behind the brain, and containing the nerve endings of the neurosecretory cells in the brain which produce several hormones including one involved in moulting. This hormone is released into the blood at the *corpora cardiaca*.

corpora cavernosa In Mammals, a pair of masses of erectile tissue in the penis.

corpora geniculata In the Vertebrate brain, paired protuberances lying below and behind the thalamus.

corpora lutea See **corpus luteum**.

corpora pedunculata In Insects, the *mushroom* or *stalked bodies*, which are the most conspicuous formations in the protocerebral lobes of the brain.

corpora quadrigemina The optic lobes of the Mammalian brain, which are transversely divided.

corpus Inner core of cells, dividing in several planes and distinct from the more superficial *tunica* in a shoot apical meristem, giving rise to the inner tissues of the shoot. See **tunica-corpus concept**.

corpus adiposum See **fat-body**.

corpus albicans See **corpus mamillare**.

corpus callosum In the brain of placental Mammals, a commissure connecting the cortical layers of the two lobes of the cerebrum.

corpuscle A cell which lies freely in a fluid or solid matrix and is not in continuous contact with other cells.

corpus luteum The endocrine structure developed in the ovary from a Graafian follicle after extrusion of the ovum, secreting progesterone; the yellow body. pl. *corpora lutea*.

corpus mamillare In the brains of higher Vertebrates, a protuberance on the floor of the hypothalamic region in which the fornix terminates.

corpus spongiosum In Mammals, one of the masses of erectile tissue composing the penis.

corpus striatum In the Vertebrate brain,

the basal ganglionic part of the wall of each cerebral hemisphere.

correlation *Mutual relationship*, e.g. the condition of balance existing between the growth and development of various organs of a plant.

correlation The tendency for variation in one variate to be accompanied by linear variation in another.

correlation coefficient A dimensionless quantity taking values in the range -1 to 1 measuring the degree of linear association between two variates. A value of -1 indicates a perfect negative linear relationship, 1 a perfect positive relationship.

cortex (1) In plants generally, the outer part of a thallus or an organ. (2) Specifically, the tissue (often collenchyma and parenchyma), in a stem or root, between the epidermis and the vascular tissue (i.e. from hypodermis to endodermis inclusive). (3) In animals, the superficial or outer layers of an organ. Cf. **medulla**. adj. *cortical*.

cortical microtubules Microtubules in the cytoplasm just below the plasmalemma in an interphase plant cell, commonly parallel to and perhaps controlling the shape of the developing cellulose microfibrils in the wall.

corticate Possessing or producing a cortex.

corticolous Living on the surface of bark.

cortisone A crystalline hormone isolated from the adrenal cortex.

Corti's organ In Mammals, the modified epithelium forming the auditory apparatus of the ear, in which nerve fibres terminate.

corymb A **racemose inflorescence** with the upper flower stalks shorter than the lower so that all the flowers are at approximately the same level.

Corynebacteriaceae A family of bacteria belonging to the order *Eubacteriales*. Grampositive rods; some pleomorphic species, mainly aerobes, occur in dairy products and the soil. Some pathogenic species, e.g. *Erysipelothrix rhusiopathiae* (swine erysipelas). *Corynebacterium diphtheriae* (diphtheria).

cosmine The dentine-like substance forming the outer layer of the cosmoid scales of Crossopterygii.

cosmoid scale In Crossopterygii, the characteristic type of scale consisting of an outer layer of cosmine, coated externally with vitrodentine, a middle bony vascular layer and an inner isopedine layer. Cf. **ganoid scale**.

costa In Vertebrates, a rib; in Insects, one of the primary veins of the wing; in Ctenophora, one of the meridional rows of ctenes; more generally, any riblike structure. adj. *costal, costate*.

cost-benefit analysis An assessment of the relative costs (in terms of the necessary investment of carbohydrate or nitrogen etc.) and benefits (in terms of enhanced photosynthesis, reduced losses to herbivores, increased probability of establishment of offspring etc.), and hence the likely selective advantage, of any observed or imagined morphological or physiological variation, such as hairier leaves, synthesis of novel toxin, larger seeds etc.

Cot curve A plot of concentration against time for the renaturation of DNA, gives a measure of the number of different sequences present.

coterminous Of similar distribution.

cotransport The **active transport** of a solute that is driven by a concentration gradient of some other solute, usually an ion, e.g. the entry of amino acids into animal cells depends upon a sodium ion gradient across the **plasma membrane**.

cotyledon *Seed leaf.* The first leaf or one of the first leaves of the embryo of a seed plant; typically one in monocotyledons, two in dicotyledons, two to many in gymnosperms. In non-endospermic seed, e.g. pea, the cotyledons may act as storage organs. See **hypogeal**, **epigeal**, **endosperm**.

cotyledonary placentation Having the villi in patches, as Ruminants.

cotyloid Cup-shaped; pertaining to the acetabular cavity. In Mammals, a small bone bounding part of the acetabular cavity.

cotype An additional type specimen, being a brother or sister of the same brood as the type specimen.

Coulter counter TN for a method of counting individual cells by pumping a suspension through an orifice and measuring the change in capacitance as each cell passes.

counter-conditioning A procedure for weakening a classically conditioned response by associating the stimuli that evoke it to a new response that is incompatible with it.

countershading A type of protective coloration in which animals are darker on their dorsal surface than on their ventral surface, thus ensuring that illumination from above renders them evenly coloured and inconspicuous.

counter-transference In psychoanalytic theory, the analyst's emotional response to the client, often involving personal and unconscious feeling projected onto the client. See **projection**.

counts The disintegrations that a radionuclide detector records.

coupling When two specified *non-allelic* genes are on the same chromosome, having come from the same parent, they are in *coupling*. Cf. **repulsion**.

coupling factors Proteins of the mitochondrial inner membrane, which are essential for the *coupling* of the passage of electrons along the electron transport chain

with the synthesis of ATP.

courtship behaviour Refers to a wide range of behaviours throughout the animal kingdom, often very conspicuous, leading to copulation and rearing of young.

cover The percentage of the ground surface covered by a plant species.

cover slip The thin slip of glass used for covering a specimen that is being observed under a microscope. Essential for all but the lowest magnifications.

coverts See **tectrices**.

Cowper's glands In male Mammals, paired glands whose ducts open into the urethra near the base of the penis.

coxa In Insects, the proximal joint of the leg. adj. *coxal*.

C₃ plant See C₃.

C₄ plant See C₄.

CR1, CR2, CR3 Cell surface receptors for C3b and its decay products.

cramp Painful spasm of muscle.

Crampton's muscle In Birds, a muscle of the eye which by its contraction decreases the diameter of the eyeball, and so aids the eye to focus objects near to it.

cranial flexures Flexures of the brain in relation to the main axis of the spinal cord, transitory in lower Vertebratres, permanent in higher Vertebrates. See **nuchal flexure**, **pontal flexure**, **primary flexure**.

Craniata A subphylum of the Chordata. Also *Vertebrata*.

craniosacral system See **parasympathetic nervous system**.

cranium That part of the skull which encloses and protects the brain; the brain case. adj. *cranial*.

crassualacean acid metabolism See panel.

Crassulaceae Family of ca 1500 spp. of dicotyledonous flowering plants (superorder Rosidae). Most are leaf-succulent perennial CAM plants; widespread mostly in warm dry temperate regions. Of little economic importance other than as ornamentals, e.g. *Sedum, Kalanchoe*.

C reactive protein A plasma protein normally present in low amounts but increased greatly by trauma or infection, i.e. an *acute phase protein*. It was originally identified by its ability to bind a carbohydrate from *Streptococcus pneumoniae* containing phosphoryl choline groups (C-carbohydrate), but it can bind to nucleic acids, to some lipoproteins and can activate **complement**. Its biological function is unknown.

creatine phosphate Phosphate ester of creatine which is a high energy compound capable of converting ADP to ATP. This capacity is exploited in the creatine phosphate of muscle as a short term source of energy during bursts of muscular activity.

crassulacean acid metabolism, CAM

Form of **photosynthesis** characteristic of desert and some other succulent plants in which CO_2 is taken up (stomata open) during the night and fixed (*via* **PEP carboxylase**) into malic acid from which it is released (stomata shut) during the day and then refixed by **ribulose 1,5-bisphosphate carboxylase oxygenase** in the normal way. Such CAM plants lose perhaps one tenth as much water in transpiration than a C_3 plant in fixing equivalent amounts of carbon, although the rate at which the carbon is fixed is less. Some plants are facultatively CAM, using C_3 when well watered and switching to CAM when water is scarce.

CAM has evolved apparently independently in several angiosperm families (e.g. Crassulaceae, Cactaceae, Bromeliaceae) and in a few ferns. A very few crop plants, e.g. pineapple and sisal, are CAM plants. A few submerged aquatic plants (without stomata) also operate a form of CAM, presumably taking advantage of the higher concentrations of dissolved CO_2 during the hours of darkness. See also C_3, C_4.

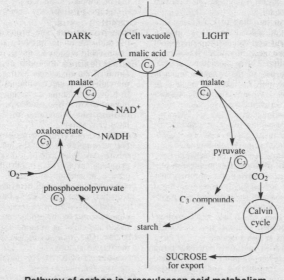

Pathway of carbon in crassulacean acid metabolism.

NAD^+ = oxidized-, NADH = reduced-nicotinamide adenine dinucleotide. The C_3,C_4 in circles indicate the number of carbon atoms in the compounds.

creativity An area of study which attempts to explore the ability to construct original and viable products, ideas etc., and to go beyond conventional developments.

cremaster In the pupae of Lepidoptera, an organ of attachment developed from the tenth abdominal somite; in Mammals, a muscle of the spermatic cord; in Metatheria, a muscle whose contraction causes the expression of milk from the mammary gland.

crenate Having rounded teeth; scalloped.

crepitation The explosive discharge of an acrid fluid by certain Beetles, which use this as a means of self-defence.

crepuscular Active at twilight or in the hours preceding dawn.

crest A ridge or elongate eminence, esp. on a bone.

cribellum In certain Spiders, a perforate oval plate, lying just in front of the anterior spinnerets, which produces a broad strip of silk composed of a number of threads.

cribriform Perforate, sievelike; as the *cribriform plate*, a perforate cartilaginous element of the developing Vertebrate skull, which later gives rise to the ectethmoid.

cribrose Pierced with many holes; resembling a sieve.

cricoid Ring-shaped; as one of the cartilages of the larynx.

Crinoidea A class of Echinodermata with branched arms; the oral surface directed upwards; attached for part of the whole of their life by a stalk which springs from the aboral apex; suckerless tube feet; open ambulacral grooves; no madreporite, spines or pedicellariae. Most members extinct.

crispate, crisped Having a frizzled appearance.

crissum In Birds, the region surrounding the cloaca or the feathers situated on that area. adj. *crissal*.

crista (1) An infolding of the inner membrane of a mitochondrion. (2) A ridge or ridgelike structure, such as the projection of the transverse crests of lophodont molars.

crista acustica (1) A chordotonal apparatus forming part of the tympanal organ in Tetigoniidae and Gryllidae. (2) A patch of sensory cells in the ampulla, utricle and saccule of the vertebrate ear.

cristate Bearing a crest.

crithidial Pertaining to, or resembling, the flagellate genus *Crithidia*; said of a stage in the life cycle of some Trypanosomes.

critical period See **sensitive period**.

critical point method Technique for preparing tissue or **metaphase** chromosomes for electron microscopy, by **freeze-drying** at the critical point of water. Preserves structural features relatively well.

crochet A hook which aids in locomotion, and is associated with the apex of the abdominal legs in Insect larvae.

Crocodilia An order of Reptiles having upper and lower temporal arcades, a hard palate, an immovable quadrate, loose abdominal ribs, socketed teeth; large powerful amphibious forms. Crocodiles, Alligators, Caimans, Gavials. Also *Loricata*.

croissant vitellogène In the developing oöcyte, a crescentic area surrounding the archoplasm, in which the mitochondria are grouped.

crop See **proventriculus** (2).

cross The mating together of individuals of two different breeds, varieties, strains or genotypes. The progeny are *cross-bred*.

cross-fertilization The fertilization of the female gametes of one individual by the male gametes of another individual. Also called *allogamy*.

cross matching Procedure used in select-ing blood for transfusion. The red cells to be transfused are mixed with the serum from the patient and if no agglutination occurs the red cells are suitable for transfusion, but an antiglobulin test may be necessary to detect *incomplete antibody*.

Crossopterygii A subclass of the class *Osteichthyes*, first known as fossils from the Middle Devonian period, and persisting to the present. They are of interest because the pectoral fins which are lobed and branched at their tips are attached to the girdle, an arrangement which could have led to the evolution of the tetrapod limb. Living forms include the *Coelocanths* and *Dipnoi* (lungfish).

cross-over See panel.

cross-over site See panel.

cross pollination The conveyance of pollen from an anther of one flower to the stigma of another, either on the same or on a different plant of the same, or related, species.

cross protection The protection offered by prior, systemic infection by one virus against infection by a second, related virus. Deliberate infection with a symptomless strain of tomato mosaic virus is used commercially to protect tomatoes from infection by other, more damaging strains. The phenomenon is also used experimentally to establish the relatedness of different isolates of viruses.

crown (1) A very short rootstock. (2) The part of a polyp bearing the mouth and tentacles; the distal part of a deer's horn; the grinding surface of a tooth; the disc and arms of a Crinoid; crest; head.

crown gall Disease of dicotyledons, esp. of fruit bushes and trees, caused by a soil bacterium *Agrobacterium tumefaciens* and characterized by the production of large, tumour-like galls: See **Ti plasmid**, **opine**.

crozier The young ascus when it is bent in the form of a hook.

cruciate, cruciform Having the form of, or arranged like, a cross.

Cruciferae Family of ca 3000 spp. of dicotyledonous flowering plants (superorder Dilleniidae). Mostly herbs, rarely shrubs, cosmopolitan. The flowers characteristically have four sepals, four petals and six stamens, all free, and superior ovary of 2 fused carpels. Includes the genus *Brassica*, and a number of minor vegetable crops, e.g. water-cress, and many ornamentals, e.g. wallflower.

cruor The coagulated blood of Vertebrates.

crural Pertaining to or resembling a leg. See **crus**.

crureus A leg muscle of higher Vertebrates.

crus The zeugopodium of the hind-limb in Vertebrates; the shank; any organ resem-

cross-over

An exchange of segments of homologous chromosomes during **meiosis** whereby linked genes become recombined (see diagram); also the product of such an exchange. The *cross-over frequency* is the proportion of gametes bearing a cross-over between two specified gene loci. It ranges from 0 for allelic genes to 50% for genes so far apart that there is always a cross-over between them. See also **chiasma, centiMorgan**.

The *cross-over site* is the place in the chromosome where breakage and reunion of DNA strands occur during **recombination**.

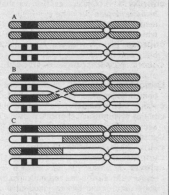

bling a leg or shank. pl. *crura*. adj. *crural*.

Crustacea A class of Arthropoda, mostly of aquatic habit and mode of respiration; the second and third somites bear antennae and the fourth a pair of mandibles. Shrimps, Prawns, Barnacles, Crabs, Lobsters etc.

crustose Forming a crust; esp. of lichens, having a crust-like thallus closely attached to, and virtually inseparable from the surface on which it is growing.

cryoglobulin, cryoprecipitate Precipitate which forms in serum at temperatures below about 10°C but goes into solution at body temperature. It is usually due to **rheumatoid factor** interacting with immunoglobulin, although constituents like heparin and fibrinogen may be included. Cryoglobulins occur in some proliferative B cell disorders and in **systemic lupus erythematosus**. Their presence may cause restricted blood flow and vessel spasm in cold extremities.

crypt A small cavity; a simple tubular gland.

cryptic coloration Protective resemblance to some part of the environment or camouflage, from simple **countershading** to more subtle mimicry of, e.g. leaves or twigs. Cf. **aposematic coloration**.

crypto- Prefix from Gk. *kryptos*, hidden.

cryptobiosis The state in which an animal's metabolic activities have come effectively, but reversibly, to a standstill. See **anabiosis**.

cryptogam In earlier systems of classification, a plant without flowers or cones in which the method of reproduction was not apparent, i.e. algae, fungi, bryophytes and pteridophytes.

Cryptophyceae Small class of eukaryotic algae. Biflagellate unicells with **periplast**; chloroplasts with chlorophyll a and c and phycobilins, thylakoids paired, chloroplast ER and nucleomorph present, reserve polysaccharide starch stored between chloroplast and chloroplast ER. Freshwater and marine.

cryptophyte (1) A member of the **Cryptophyceae**. (2) Herb with perennating buds below soil (or water) surface. Includes geophyte, heliophyte and hydrophyte. See **Raunkaier system**.

cryptorchid An animal in which one or both testes have not descended from the abdominal cavity to the scrotum within a reasonable time.

cryptozoic Living in dark places, as in holes, caves, or under stones and tree trunks.

crystal Crystalline inclusion in a plant cell, usually of calcium oxalate. Types include the **druse** and the **raphide**.

crystalline cone The outer refractive body of an ommatidium which acts as a light guide.

crystalline lens The transparent refractive body of the eye in Vertebrates, Cephalopoda etc. It is compressible by muscles and fo cuses images of objects emitting light onto the retina.

crystalline style In Bivalvia, a transparent rod-shaped mass secreted by a diverticulum of the intestine; composed of protein with an adsorbed amylolytic enzyme.

crystalloid A crystal of protein in e.g. a cell of a storage organ.

crystal sac A cell almost filled with crystals of calcium oxalate.

CSF Abbrev. for *Colony Stimulating Factor*.

ctene One of the comb-plates or locomotor organs of Ctenophora, consisting of a row of strong cilia of which the bases are fused.

ctenidium Generally, any comblike structure; in aquatic Invertebrates a type of gill consisting of a central axis bearing a row of filaments on either side; in Insects, a row of spines resembling a comb.

ctenoid Said of scales which have a comb-like free border.

Ctenophora A phylum of triploblastic animals showing biradial symmetry; they have a system of gastrovascular canals and typically eight meridional rows of swimming plates or ctenes, composed of fused cilia. Sea Acorns, Comb-Bearers.

cubical epithelium A form of columnar epithelium in which the cells are short.

cubital See **secondary**.

cubital remiges The primary quills connected with the ulna in Birds.

cucullate Hood-shaped.

culm The stem esp. the flowering stem, of grasses and sedges.

cultivar A subspecific rank used in classifying cultivated plants and indicated by the abbreviation cv. and/or by placing the name in single quotation marks; defined as an assemblage of cultivated plants which is clearly distinguished by any characters (morphological, physiological, cytological, chemical etc.), and which when reproduced (sexually or asexually, as appropriate) retains its distinguishing characters.

culture A micro-organism, tissue or organ growing in or on a medium or other support; to cultivate such in this way. See **plant tissue culture**.

cumulative distribution function A function giving the probability that a corresponding continuous random variable takes a value less than or equal to the argument of the function.

cumulative dose Integrated radiation dose resulting from repeated exposure.

cumulus The mass of cells surrounding the developing ovum in Mammals.

cumulus oöphorus See **zona granulosa**.

cuneate, cuneal, cuneiform Wedge-shaped.

cupula Any domelike structure, e.g. the apex of the lungs, the apex of the cochlea.

cupule One of a number of more or less cup-shaped organs, esp. the structure that encloses the fruits of oak, beech, chestnut, birch etc. e.g. the acorn cup (Cupuliferae).

curare South American native poison from the bark of species of *Strychnos* and *Chondodendron*.

curarine Paralysing toxic alkaloid ($C_{19}H_{26}ON_2$) extracted as *d*-tubocurarine chloride from crude curare; used in anaesthesia as a muscle relaxant.

cursorial Adapted for running.

cushion plant Plant with many densely crowded upright shoots not more than a few centimetres high, forming a cushion-like mass on the ground; typical of alpine and arctic floras. Also *chamaephyte*.

cusp A sharp-pointed prominence, as on teeth.

cutaneous Pertaining to the skin.

cuticle A nonliving layer secreted by and overlying the epidermis, e.g. the layer of **cutin** on the outside of some plant cell walls, especially the shoot epidermis where it forms a continuous layer which, with the **epicuticle**, has relatively low permeability to water and gases.

cuticular transpiration The loss of water vapour from a plant through the cuticle.

cuticulin The outermost layer of the insect epicuticle, consisting of lipoprotein.

cutin A mixture of fatty substances esp. of cross-linked polyesters based on mostly C_{16} and C_{18} aliphatic acids and hydroxyacids in the **cuticle** of plants.

cutinization The formation of cutin; the deposition of cutin in a cell wall to form a cuticle.

cutis The dermis or deeper layer of the Vertebrate skin.

cut-off posture In ethology, a term referring to postures that remove social stimuli (e.g. a potential mate, or opponent) from sight, and thus may serve to reduce the actor's arousal in a conflict situation.

cutting A piece of a plant, usually shoot, root or leaf, which is cut off and induced to form adventitious roots and/or buds as a means of vegetative propagation. See **rooting compound**.

Cuvierian ducts In lower Vertebrates, a pair of large venous trunks entering the heart from the sides.

cv Abbrev. for *cultivar*.

c-value paradox The paradox that some very similar animals and plants have unexpectedly large differences in the amount of their genomic DNA, e.g. amphibian genomes vary by over a 100-fold. Not simply an increase in the number of sequence copies per genome.

cyanogenesis The release from plant parts, usually after wounding, of hydrogen cyanide by cytoplasmic glycosidase action on a vacuolar glycoside containing, e.g. mandelonitrile. Occurs in leaves of cherry laurel (*Prunus laurocerasus*), seeds of bitter almonds and fronds of bracken. Possibly a deterrent to herbivores.

Cyanophyceae *Myxophyceae, Cyanobacteria*, Blue-green algae. Prokaryotic organisms with chlorophyll a and phycobilins. Reserve carbohydrate α, $1{\rightarrow}4$ glucan. Asexual reproduction by spores, division fragmentation or homogonia. Unicellular, colonial or filamentous. Sorts with **heterocysts** fix nitrogen. Planktonic sorts may have **gas-vacuoles**. Occur in fresh- and saltwater (planktonic and benthic), in soils and as nitrogen-fixing symbionts in *Azolla* and the roots of Cycads and some flowering plants and in some lichens. See **gliding**.

cyanosis Blueness of the skin and the mucous membranes due to insufficient oxygenation of the blood. May be peripheral due to poor circulation or central due to failure of oxygenation.

Cycadales The cycads. Order of gymnosperms (Cycadopsida), widespread in the Mesozoic, now ca 65 spp. in Central America, S. Africa, SE Asia and Australia. Stems stout, unbranched, manoxylic; leaves large, pinnate, with haplocheilic stomata. Dioecious. Reproductive organs in large cones (except female *Cycas*), Zooidogamous. Radiospermic. The pith of 2 spp. is a minor source of sago.

Cycadopsida Class of gymnosperms containing the superficially similar Cycadales and Cycadeoidales and a number of other orders. Probably not a natural group.

cybrid A cell, callus, plant etc. typically resulting from **protoplast fusion** and **protoplast culture**, possessing the nuclear genome of one plant with at least some part of the chloroplastal or mitochondrial genome of the other, as opposed to a *hybrid* in which some parts of both parental genomes are present. See **chloroplast, somatic hybridization**.

cycle A series of occurrences in which conditions at the end of the series are the same as they were at the beginning. Usually, but not invariably a cycle of events is recurrent. Typical of many biological events such as the **cell cycle**.

cyclic A flower having the parts arranged in whorls, rather than in spirals.

cyclic adenosine monophosphate A derivative of adenosine monophosphate in which the phosphate forms a ring involving the 3′ and 5′ hydroxyl groups of ribose. It is of major metabolic importance through its diverse effects on many enzymes. Abbrev. *cAMP*.

cyclo- Prefix from Gk. *kyklos*, circle.

cycloid Evenly curved; said of scales which have an evenly curved free border.

cyclophosphamide A potent alkylating agent used as an anticancer drug, but also as an immunosuppressive agent which acts particularly on B lymphocytes. Use is restricted by its toxicity to bone marrow and the bladder.

cyclosis The circulation of protoplasm within a cell.

cyclospondylous Showing partial calcification of cartilaginous vertebral centra in the form of concentric rings.

cyclosporin A A cyclic peptide used as an immunosuppressive agent which has a selective action on the generation of helper T cells, which do not become functional while the drug is present. It is useful in preventing graft rejection but, because it produces renal damage easily, blood levels need repeated monitoring.

Cyclostomata An order of the class Agnatha. Aquatic and gill-breathing, with a round suctorial mouth; buccal cavity contains a muscular tongue bearing horny teeth used to rasp the flesh from the prey; cartilaginous endoskeleton; no fins or limb girdles; slimy skin with no scales. Lampreys and Hagfish.

cyesis Pregnancy. See **pseudocyesis**.

cymose inflorescence, cyme An inflorescence in which the main stem and each subsequent branch ends in a flower, with any further development of the inflorescence coming from a lateral branch or lateral branches arising below the flower. Cf. **racemose inflorescence**.

cynopodous Having nonretractile claws, as dogs.

Cyperaceae The sedge family, ca 4000 spp. of monocotyledonous flowering plants (superorder Commelinidae). Mainly rhizomatous, perennial, grasslike herbs, cosmopolitan esp. in temperate and arctic regions, often in wet habitats. The aerial stems are typically solid, triangular in section and bear grass-like leaves in three ranks; the flowers are inconspicuous and wind-pollinated. The leaves and stems of some are used for making hats, baskets, mats and paper (papyrus) and for thatching. Includes the large genus *Carex* (1000 spp.).

cypress knee A vertical upgrowth from the roots of swamp cypress (*Taxidium*) apparently a **pneumatophore**.

Cypriniformes An order of Osteichthyes, almost entirely inhabiting fresh water, with over 3000 species. Characins, Loaches and Carp.

Cys Symbol for **cysteine**.

cyst A nonliving membrane enclosing a cell or cells; any bladderlike structure, as the gall bladder or the urinary bladder of Vertebrates; a sac containing the products of inflammation. adjs. *cystic, cystoid, cystiform*.

cysteine *2-amino-3-mercaptopropanoic acid*. The L-isomer is an 'acidic' amino acid found in proteins, often in its oxidized form, **cystine**. Symbol Cys, short form C.
R group: HS–CH₂— See **amino acid**.

cystic Pertaining to the gall bladder; pertaining to the urinary bladder.

cystic duct The duct from the gall bladder which meets the hepatic duct to form the common bile duct.

cysticercus Bladderworm; larval stage in many tapeworms, possessing a fluid-filled sac containing an invaginated scolex.

cysticolous Cyst-inhabiting.

cystidium A swollen, elongated, sterile hypha, occurring among the basidia of the hymenium of some Hymenomycetes, usually projecting beyond the surface of the hymenium.

cystine The dimer resulting from the oxidation of cysteine. The resulting disulphide bridge is an important structural element in proteins, as it often connects groups otherwise distant in the protein chain.

cystitis Inflammation of the bladder.

cystogenous Cyst-forming; cyst-secreting.

cystolith Mass of calcium carbonate within a plant cell, on a stalk-like projection from the cell wall.

cystozooid In Cestoda, the bladder or tail portion of a bladderworm. Cf. **acanthozooid**.

cytase A general term for an enzyme able to break down the β–$1 \to 4$ link of cellulose.

cytochimera See **chromosome chimera**.

cytochromes Proteins of the electron transfer chain which can carry electrons by virtue of their haem **prosthetic groups**. Cytochromes b, c1 and c have the same prosthetic group as haemoglobin. Cytochromes a and a3 have the related haem A and together form the terminal complex of the chain, cytochrome oxidase.

cytogenesis The formation and development of cells.

cytogenetic map See **chromosome mapping**.

cytogenetics Study of the chromosomal complement of cells, and of chromosomal abnormalities and their inheritance.

cytokinesis The contraction of an equatorial belt of cytoplasm which brings about the separation of two daughter cells during cell division of animal tissues. In plants the division of the cytoplasm as distinct from the nucleus. See **cell plate**, **cleavage**.

cytokinin Any of a group of plant **growth substances**, derivatives of adenine, e.g. zeatin, synthesized esp. in roots and promoting cell division and bud formation, delaying senescence and, sometimes, promoting flowering and breaking dormancy. Also the artificial analogues of the above.

cytology The study of the structure and functions of cells.

cytolysis Dissolution of cells.

cytophilic antibody Antibodies which bind to **Fc receptors** on the cell membrane.

cytoplasm That part of the cell outside the nucleus but inside the **cell wall** if it exists.

cytoplasmic inheritance Inheritance of traits coded for by the chloroplast or mitochondrial genomes, maternal because of the inheritance of chloroplasts and mitochondria through the egg rather than the sperm or male cell.

cytoplasmic male sterility Lack of functional pollen as a maternally inherited trait, resulting from a defective mitochondrial genome. See **male sterility**. Cf. **cytoplasmic inheritance**.

cytorrhysis Process in which a plant cell wall collapses inwardly, following water loss as a result of the exposure of the cell to a solution of a macromolecular solute to which the cell wall is impermeable, of higher osmotic pressure than that of the cell contents. Turgor will be zero or possibly negative. Cf. **plasmolysis**.

cytosine One of the five major bases found in nucleic acids. It pairs with guanine in both DNA and RNA.

Formula:

See **DNA**, **genetic code**.

cytoskeleton Structures composed of protein which serve as skeletal elements within the cell, e.g. **microtubules**, **microfilaments**.

cytosol The non-particulate components of the cytoplasm.

cytotaxis Rearrangement of cells as a result of stimulation.

cytotaxonomy The use of studies of chromosome number, morphology and behaviour in taxonomy.

cytotoxic Able to kill cells. Applies to cytotoxic T lymphocytes, to killer cells and **natural killer cells**, and also to damage mediated by **complement**.

cytotoxin A toxin having a destructive action on cells.

cytotrophoblast The inner layer of the trophoblast; layer of Langerhans.

D

dactyl A digit. adj. *dactylar*.

damping-off Collapse and death of seedlings around emergence, due usually to fungal attack by *Pythium* and *Fusarium* spp. when conditions are unfavourable for the seedlings.

dark ground illumination See panel.

dark reactions Those reactions in photosynthesis in which CO_2 is fixed and reduced. They depend on energy and reducing power from the **light reactions**. See **Calvin cycle**.

dart Any dartlike structure, e.g. in certain Snails, a small pointed calcareous rod which is used as an incentive to copulation; in certain Nematoda, a pointed weapon used to obtain entrance to the host.

Darwinian theory See **natural selection**.

dasypaedes Birds which when hatched have a complete covering of down. Cf. **altrices**.

daughter Offspring belonging to the first generation, whether male or female, as *daughter cell*, *daughter nucleus*.

day-neutral plant A plant in which flowering period is not sensitive to day length. Cf. **long-day plant**, **short-day plant**. See **photoperiodism**.

death In a cell or an organism, complete and permanent cessation of the characteristic activities of living matter.

decalcification The process of absorption of lime salts from bone.

Decapoda (1) An order of Malacostraca with 3 pairs of thoracic limbs modified as maxillipeds, and 5 as walking legs. Shrimps, Prawns, Crabs, Lobsters etc. (2) A suborder of Cephalopoda having 8 normal arms and 2 longer partially retractile arms; the suckers are pedunculate, there is a well-developed internal shell, and lateral fins are present; actively swimming forms, usually carnivorous. Squids and Cuttlefish.

decarboxylase An enzyme that catalyses the removal of CO_2 from its substrate.

decay The process of spontaneous transformation of a radionuclide.

decerebrate Lacking a cerebrum.

decerebrate tonus A state of reflex tonic contraction of certain skeletal muscles following upon the separation of the cerebral hemispheres from the lower centres.

decidua In Mammals, the modified mucous membrane lining the uterus at the point of contact with the placenta, which is torn away at parturition and then ejected; the afterbirth; the maternal part of the placenta.

deciduate Said of Mammals in which the maternal part of the placenta comes away at birth. Cf. **indeciduate**.

deciduous (1) Falling off, usually after a lengthy period of functioning. (2) Plants which shed leaves habitually before a cold period. Ant. *evergreen*.

decomposers In an ecosystem, heterotrophic organisms, chiefly bacteria and fungi, which break down the complex compounds of dead protoplasm, absorbing some of the products of decomposition, but also releasing simple substances usable by producers. Cf. **consumers**, **producers**.

decompound Two or more times *compound*.

decondensed chromatin See **euchromatin**.

deconjugation The separation of the paired chromosomes before the end of the prophase meiosis.

decorticated Deprived of bark; devoid of cortex.

decumbent Lying flat, usually with a turned-up tip.

decurrent (1) Running down, as when a leaf

---

dark ground illumination

A method for the microscopic examination of living material, e.g. micro-organisms, tissue culture cells, by scattered light. A special condenser with a circular stop illuminates the specimen with a numerical aperture larger than that collected by the objective (see the diagram). Only the scattered rays pass through the objective to reach the eye and the specimen appears luminous against a dark background. It can also be used to detect smoke and other particles too small to be resolved by the light microscope.

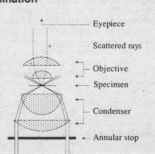

margin continues down the stem as a wing. (2) With several roughly equal branches, as in shrubs and in the crowns of some trees, esp. when old. Cf. **excurrent**.

decussate With leaves in opposite pairs, each pair being at right angles to the next. See **phyllotaxis**.

decussation Crossing over of nerve tracts with interchange of fibres.

dedifferentiation Changes in a differentiated tissue, leading to the reversion of cell types to a common indifferent form, such as the meristematic state.

deep therapy X-ray therapy of underlying tissues by **hard radiation** (usually produced at more than 180 kVp) passing through superficial layers.

defective virus Virus unable to replicate without a *helper*, e.g. a **plasmid** providing a replicative function.

defence mechanism In psychoanalytic theory, a collective term for a number of reactions that try to ward off or lessen anxiety, by a variety of means that seek to keep the source of anxiety out of consciousness, e.g. repression.

defibrillator Electrical apparatus to arrest ventricular **fibrillation**.

deficiency The absence, by loss or inactivation, of a gene or a part of a chromosome, that is normally present.

deficiency disease Any disease resulting from the deprivation of nutritional substances necessary to good health or growth, such as vitamins or essential minerals unavailable in the soil.

definite In plants (1) sympodial or (2) cymose.

definitive Final, complete; fully developed; defining or limiting.

deflexed Bent sharply downwards.

deforestation Permanent removal of forest.

degeneration Evolutionary retrogression; the process of returning from a higher or more complex state to a lower or simpler state.

degenerative disorders See **dementia**.

deglutition The act of swallowing.

dehiscence The spontaneous opening at maturity of a fruit, anther, sporangium or other reproductive body, permitting the release of the enclosed seeds, spores etc. adj. *dehiscent*. Cf. **indehiscent**. More generally the act of splitting open as in **diapedesis**.

dehydration Excessive loss of water from the tissues of the body.

dehydrogenase Enzymes which catalyse the oxidation of their substrate with the removal of hydrogen atoms.

de-individuation A term used by social psychologists to denote a loss of a sense of personal responsibility in conditions of relative anonymity, where a person cannot be identified as an individual but only as a member of a group. The individual engages in activities he or she would not normally do, presumably because of a weakening of internal controls (e.g. shame or guilt).

delamination The division of cells in a tissue, leading to the formation of layers.

delayed-type hypersensitivity Hypersensitivity state mediated by T lymphocytes. When the antigen is introduced locally, e.g. in the skin, a gradual local accumulation of T cells and monocytes results. The visible reaction is reddening and local swelling, increasing for 24–48 hrs. and then subsiding, sometimes leaving a small scar due to necrosis of blood vessels. Tuberculin testing is a good example. Frequently this and other types of hypersensitivity co-exist, and reactions are not clear cut.

deletion Same as **deficiency**.

deletion mutation That in which a base or bases are lost from the DNA. Cf. **base substitution**.

delinquency, delinquent Conduct disorder, usually against the law, but including a range of antisocial, deviant or immoral behaviour. In Britain, young people are considered *juvenile delinquents* if they are under 17 years of age with criminal convictions.

deltoid (1) Having the form of an equilateral triangle. (2) Any triangular structure, as the deltoid muscle of the shoulder.

delusions An irrational belief which an individual will defend with intensity, despite overwhelming evidence that it has no basis in reality; common among schizophrenic mental disorders.

deme A local population of interbreeding organisms.

dementia Refers to degeneration of various functions governed by the central nervous system, including motor reactions, memory and learning capacity, problem solving etc. These functions normally decline with age, but several dementia syndromes result from pathological organic deterioration of the brain. See **senile dementia**.

dementia praecox Obsolete term for **schizophrenia**.

demersal Found in deep water or on the sea bottom; as Fish eggs which sink to the bottom, and midwater and bottom-living Fish as opposed to surface Fish (e.g. Herring) and Shellfish. Cf. **pelagic**.

demifacet One of the two half facets formed when the articular surface for the reception of the capitular head of a rib is divided between the centra of two adjacent vertebrae.

Demospongiae A class of *Porifera* usually distinguished by the possession of a skeleton

---

dendritic cell

A cell that has branching processes. The term is used in immunology to describe two distinct kinds of cells which have different functions. (1) *Follicular dendritic cells*. Present in **germinal centres** where they are in intimate contact with dividing B cells. Antigen-antibody complexes become trapped at the surface of the dendritic processes and are retained there for long periods. They are involved in the generation of **B-memory cells**. (2) Cells with dendritic morphology which occur mainly in the thymus-dependent areas of lymph nodes and the spleen. Here they are often termed *interdigitating cells*.

Very similar cells occur as Langerhans cells in the skin. Dendritic cells of this kind are not phagocytic and they typically express Ia. They are very effective accessory cells in stimulating T lymphocytes.

---

composed of siliceous spicules, or of spongin, e.g. bath sponges.

demulcent Soothing; allaying irritation.

denaturation The destruction of the native conformation or state of a biological molecule by heat, extremes of pH, heavy metal ions, chaotropic agents etc., resulting in loss of biological activity. Specifically in DNA, the breakage of the hydrogen bonds maintaining the double-helical structure, a process which can be reversed by *renaturation* or annealing.

dendr-, dendro- Prefixes from Gk. *dendron*, tree; tree-like or branching.

dendrite A branch of a **dendron**.

dendritic cell See panel.

dendrochronology Science of reconstructing past climates from the information stored in tree trunks as annual radial increments of growth.

dendrogram A branching diagram after the style of a family tree reflecting similarities or affinities of some sort. See **cladogram, phenogram**.

dendrograph An instrument which is used to measure the periodical swelling and shrinkage of tree trunks.

dendroid (1) Tall, with an erect main trunk, as tree ferns. (2) Freely branched.

dendron The afferent or receptor process of a neuron. Often much branched. Cf. **axon**.

denervated Deprived of nerve supply.

denial Refusal to acknowledge some unpleasant feature of the external world or some painful aspect of one's own experiences or emotions (a *defence mechanism*).

denitrification See **nitrate-reducing bacteria**.

dens epistrophei See **odontoid process**.

density dependence The phenomenon whereby performance of organisms within a population is dependent on the extent of crowding.

density gradient centrifugation The separation of cells, cell organelles or macromolecules according to their density differences by centrifugation to their density equilibrium positions in density gradients established in appropriate solutions. A solution of a highly soluble salt, like CsCl, can be centrifuged to an equilibrium gradient of density in which DNA of different densities will separate. It is an important method of physically separating DNA molecules according to their base composition because DNA density is affected by base composition, as well as by the binding of small molecules and by DNA strandedness.

dental formula A formula used in describing the dentition of a Mammal to show the number and distribution of the different kinds of teeth in the jaws; thus a Bear has in the upper jaw 3 pairs of incisors, 1 pair of canines, 4 pairs of premolars, and 2 pairs of molars; and in the lower jaw 3 pairs of incisors, 1 pair of canines, 4 pairs of premolars, and 3 pairs of molars. This is expressed by the formula: $\dfrac{3143}{3142}$.

dentary In Vertebrates, a membrane bone of the lower jaw which usually bears teeth. In Mammals, it forms the entire lower jaw.

dentate Having a toothed margin.

denticles Any small toothlike structure; the placoid scales of Elasmobranchii.

dentine A hard calcareous substance, allied to bone, of which teeth and placoid scales are mainly composed. Ivory is dentine. adj. *dentinal*.

dentirostral Having a toothed or notched beak.

dentition The kind, arrangement and number of the teeth; the formation and growth of the teeth; a set of teeth, as the milk dentition.

denuded quadrat A square piece of ground, marked out permanently and cleared of all its vegetation, so that a study may be made of the manner in which the area is reoccupied by plants.

depersonalization A condition in which an individual experiences a range of feelings of unreality or remoteness in relation to the

desmosome

Strong intercellular junctions which bind cells together, either at discrete points at the surface, *spot desmosomes*, or as continuous bands around cells, *belt desmosomes*. The *hemidesmosome* is a similar structure binding an epithelial cell to the basal lamina.

The two membranes of associated cells remain separated in the desmosome by an intercellular space which is traversed by protein filaments that extend into the adjacent cytoplasm of both cells (see diagram).

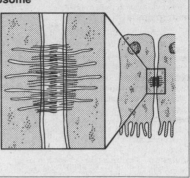

self, to the body or to other people, even extending to the feeling of being dead. Primarily a symptom in a range of neurotic syndromes.

depilate To remove the hair from.

depressant Lowering functional activity; a medicine which lowers functional activity of the body.

depressor A muscle which by its action lowers a part or organ; a reagent which, when introduced into a metabolic system, slows down the rate of metabolism.

derm See **dermis**.

dermal Relating to the epidermis or other superficial layer of a plant member or in animals to the skin; more strictly, pertaining to the dermis. Also *dermic*.

dermal branchiae See **papulae**.

dermatogen A **histogen** precursor of the epidermis.

Dermatophagoides pteronyssinus Mite often present in house dust. Antigens extracted from mites and their faeces are a common cause of allergy to house dust in W. European countries.

dermatophyte A parasitic fungus which causes a skin disease in animals or man, e.g. ring worm, athlete's foot.

dermis, derm The inner layers of the integument, lying below the epidermis and consisting of mesodermal connective tissue. adjs. *dermal, dermic*.

dermomuscular layer A sheet of muscular tissue underlying the skin in lower Metazoa; it consists of longitudinal and, usually, circular layers.

dertrotheca In Birds, the horny covering of the maxilla. Also *dertrum*.

descending Running from the anterior part of the body to the posterior part, or from the cephalic to the caudal region.

desensitization, systematic A form of **behaviour therapy**, used esp. in the treat-

ment of **phobias**, in which fear is reduced by exposing the individual to the feared object in the presence of a stimulus that inhibits the fear; usually some form of relaxation is involved. See **counter conditioning**.

desertification Formation of deserts from vegetated zones by the action of drought and/or increased populations of humans and herbivores.

desmids Green algae, two families of the Charophyceae. Unicellular, the cells usually symmetrical about a median constriction, often elaborate in shape, and moving by the secretion of mucilage. Characteristic of rather acid fresh water habitats.

desmognathous In Birds, said of a type of palate in which the vomers are small or wanting and the maxillopalatines meet in the middle line; the palatines and pterygoids articulate with the basisphenoid rostrum.

desmosome See panel.

desquamation The shedding of the surface layer of the skin.

determinate (1) Of a limited extent. (2) Sympodial. (3) Cymose.

determinate cleavage A type of cleavage in which each blastomere has a predetermined fate in the later embryo.

detorsion In Gastropoda, partial or complete reversal of torsion, manifested by the untwisting of the visceral nerve loop and the altered position of the ctenidium and anus.

detritovore Organism which eats detritus.

detritus Organic material formed from decomposing organisms.

Deuteromycotina *Deuteromycetes, Fungi Imperfecti*. The imperfect fungi. Form subdivision or form class of the Eumycota or true fungi for which no sexual reproduction is known. Typically mycelial with simple septa. The affinities of many appear to be with the Ascomycotina. Include many saprophytes, e.g. *Aspergillus*, causing food spoilage, *Penicillium*, a source of antibiotics and also plant parasites, e.g. *Fusarium*, *Verticillium* causing wilt diseases etc.

deuterostoma In development, a mouth which arises secondarily, as opposed to a mouth which arises by modification of the blastopore.

deuterotoky Parthenogenesis leading to the production of both males and females.

deutocerebron, deutocerebrum In higher Arthropoda, as Insects and Crustacea, the fused ganglia of the second somite of the head, forming part of the 'brain'.

deviance A statistic measuring the degree of fit of a statistical model by means of comparison with the degree of fit of a more complete model.

deviation The difference between an observation and a fixed value. Statistical measures of dispersion are often based on the deviation of each value in a set of observations from their common mean.

deviation IQ See **intelligence quotient**.

dew claw In Dogs, the useless claw on the inner side of the limb (esp. the hind limb) which represents the rudimentary first digit.

dexiotropic Twisting in a spiral from left to right; spiral cleavage.

dextral See **dextrorse**.

dextran A polyglucose formed by microorganisms in which the units are joined mainly by α–1$\rightarrow$6 links with variable amounts of crosslinking, via α–1$\rightarrow$4, α–1$\rightarrow$3 or α–1$\rightarrow$2. It is hydrolysed by dextranases.

dextrin *Starch gum*. A term for a group of intermediate products obtained in the transformation of starch into maltose and *d*-glucose. Dextrins are obtained by boiling starch alone or with dilute acids. They do not reduce **Fehling's solution**. Crystalline dextrins have been obtained by the action of *Bacillus macerans*.

dextrorse Helical, twisted or coiled in the sense of conventional (right handed) screw thread or of a Z-helix. Cf. **sinistrorse**.

diacoele In Craniata, the third ventricle of the brain.

diadelphous Stamens of a flower fused by their filaments into 2 groups.

diagnosis A formal description of a plant, having special reference to the characters which distinguish it from related species.

diagnosis The determination of the nature of a disordered state of the body or of the mind; the identification of a diseased state.

diagnostic characters Characteristics by which one genus, species, family or group can be differentiated from another.

dialypetalous See **polypetalous**.

diamino-pimelic acid A cell wall constituent of some bacteria and blue-green algae, not known to occur in any other group.

diapause In Insects, a state, which may arise at any stage of the life cycle, in which development is suspended and cannot be resumed, even in the presence of apparently favourable conditions, unless the diapause is first 'broken' by an appropriate environmental change.

diapedesis In Porifera, the passage to the exterior of cells primarily occupying the interior of certain types of larva; in Vertebrates, the passage of blood leucocytes through the walls of blood vessels into the surrounding tissues.

diaphoresis Perspiration.

diaphragm (1) A plate of cells crossing a space. Esp. a plate one cell thick with many intercellular pores through which air but not water may pass between intercellular air spaces in the submerged stems etc. of many aquatic plants. (2) More generally, a transverse partition subdividing a cavity. In Mammals, the transverse partition of muscle and connective tissue which separates the thoracic cavity from the abdominal cavity; in Anura, a fan-shaped muscle passing from the ilia to the oesophagus and the base of the lungs; in some Arachnida, a transverse septum separating the cavity of the prosoma from that of the abdomen; in certain Polychaeta, a strongly developed transverse partition dividing the body cavity into two regions.

diaphysis The shaft of a long limb bone.

diapophyses A pair of dorsal transverse processes of a vertebra, arising from the neural arch.

diapsid Said of skulls in which the supra- and infra-temporal fossae are distinct.

diarthrosis A true (as opposed to a fixed) joint between two bones, in which there is great mobility; a cavity, filled with a fluid, generally exists between the two elements.

diastase An enzyme capable of converting starch into sugar. Produced during the germination of barley in the process of malting.

diastema (1) An equatorial modification of protoplasm preceding cell division. (2) A gap in a jaw where there are no teeth, as in Mammals lacking the canines.

diaster In cell division, a stage in which the daughter chromosomes are situated in two groups near the poles of the spindle, ready to form the daughter nuclei.

diastole Rhythmical expansion, as of the heart, or of a contractile vacuole.

diatoms See **Bacillariophyceae**.

diatropism A tropism in which a plant part becomes aligned at right angles to the source of the orientating stimulus; e.g. rhizomes are typical *diagravitropic*. Cf. **plagiotropic, orthotropic**.

dibranchiate Having 2 gills or ctenidia.

DIC Abbrev. for *Differential Interference*

Contrast. See **differential interference contrast microscope.**

dicentric Having two centromeres.

dichasium, dichasial cyme An inflorescence in which the main stem ends in a flower and bears, from near the flower, two lateral branches also ending in a flower. These branches may in turn bear further similar branches and so on. See **cymose inflorescence, monochasium.**

dichlamydeous Having distinct calyx and corolla. Cf. **perianth.**

dichlorophenoxyacetic acid *2,4-D*. Synthetic **auxin** used as a selective herbicide and in media for tissue culture.

dichocephalous Said of ribs which have 2 heads, a tuberculum, and a capitulum. Cf. **holocephalous.**

dichogamy The maturation of the anthers at a different time from the stigmas in the same flower, *protandry* or *protogyny*. Cf. **homogamy.**

dichoptic Having the eyes of the two sides distinctly separated.

dichotomy Bifurcation of an organ, by the division of an apical cell or meristem into 2 equal parts each growing into a branch. Common in algae and *Selaginella*. Cf. **false dichotomy.**

dichromatism Colour blindness in which power of accurate differentiation is retained for only 2 bands of colour in the spectrum.

Dick test A test for immunity against the toxin of *Streptococcus pyogenes* which causes scarlet fever. A small amount of toxin injected into the skin causes an area of redness after 6 or more hours in persons who do not have antibodies against the toxin.

dicliny Having unisexual flowers, either male and female on different individual plants (dioecy) or both on one plant (monoecy). Cf. **hermaphrodite.** adj. *diclinous*.

Dicotyledones *Magnoliopsida*. The dicotyledons, or dicots, the larger of the two classes of **angiosperms** or flowering plants. Trees, shrubs and herbs of which characteristically the embryo has two cotyledons, the parts of the flowers are in twos or fives or multiples of these numbers, and the leaves commonly are netveined. Contains ca 165 000 spp. in 250 families usually divided among 6 subclasses or superorders; Magnoliidae, Hamamelidae, Caryophyllidae, Dilleniidae, Rosidae and Asteridae. Cf. **monocotyledons.** See **angiosperms.**

dicty-, dictyo- Prefixes from Gk. *diktyon*, net.

dictyosome *Golgi body*. An element of the **golgi apparatus.**

dictyostele A type of **solenostele**, typical of many ferns, e.g. *Dryopteris*, in which overlapping leaf gaps dissect the vascular cylinder into anastomosing strands (meristeles) each with xylem surrounded by phloem amd endodermis.

dicyclic Having the perianth in two whorls.

didactyl Having all the toes of the hind feet separate, as in many Marsupialia. Cf. **syndactyl.**

didymous Formed of two similar parts, partially attached; twinned.

didynamous Having 2 long and 2 short stamens.

dieback Necrosis of a shoot, starting at the apex and progressing proximally.

diel *Daily*. Synonymous with *diurnal* in the sense that it pertains to a period of 24 hours, not in the sense of *by day* rather than *by night*.

diencephalon In Vertebrates, the posterior part of the fore-brain connecting the cerebral hemispheres with the midbrain.

difference threshold The amount by which a given stimulus must be increased or decreased in order for a subject to perceive a *just noticeable difference*, JND.

differential absorption ratio The ratio of concentration of radioisotope in different tissues or organs at a given time after the active material has been ingested or injected.

differential interference contrast microscope Abbrev. *DIC microscope*. A version of the **interference microscope** in which, in effect, the reference beam is separated laterally from the 'specimen' beam by less than the resolving power of the microscope. The brightness of any area of the image therefore depends on the *rate of change* of optical path (refractive index times thickness for transmitted light) in one particular direction across the corresponding part of the specimen. The image appears to show the surface relief of the specimen (as if it were lit at a low angle from the side) but, with transmitted light, the 'heights' do not necessarily represent high points on the cell surface but rather points at which the refractive index or thickness is greater. DIC is useful for the observation of unstained cells and compared to **phase contrast microscopy** there is less disturbance from structures over- or underlying the plane of focus, a relative freedom from haloes, making resolution of detail, esp. in the z-direction, easier and greater sensitivity to small changes in optical path. The *Nomarski microscope* is an example.

differential stain A stain which picks out details of structure by giving to them different colours or different shades of the same colour.

differentiation (1) Those qualitative changes in morphology and physiology occurring in a cell, tissue or organ as it develops from a meristematic, primordial or unspecialized state into the mature or specialized state. (2) Removing excess stain from some of the structures in a microscope preparation so that the whole can be seen more clearly.

diffuse growth Growth where cells divide throughout the tissue. Cf. **apical growth, intercalary** growth, **trichothallic growth.**

diffuse placentation Having the villi developed in all parts of the placenta, except the poles, as in Lemurs, most Ungulates and Cetacea.

diffuse porous Said of wood having the vessels (pores) distributed evenly thoughout a growth ring or changing in diameter gradually across it, e.g. birch, evergreen oaks. Cf. **ring porous.**

diffuse tissue A tissue consisting of cells which occur in the plant body singly or in small groups intermingled with tissues of distinct type.

digametic Having gametes of two different kinds.

digastric Of muscles, having fleshy terminal portions joined by a tendinous portion, as the muscles which open the jaws in Mammals.

digenesis (1) **Alternation of generations.** (2) The condition of having two hosts: said of parasites. adj. *digenetic.*

digenetic reproduction   See **sexual reproduction.**

di George's syndrome See **thymic hypoplasia.**

digestion The process by which food material ingested by an organism is rendered soluble and assimilable by the action of enzymes. adj. *digestive.*

digestive gland Gland(s) present in many Invertebrates and Protochordata, in which intracellular ingestion and absorption take place, as opposed to the alimentary canal proper.

digit A finger or toe, one of the free distal segments of a pentadactyl limb.

digitate Palmate.

digitigrade Walking on the ventral surfaces of the phalanges only, as terrestrial carnivores.

digitule Any small fingerlike process.

digoneutic Producing offspring twice a year.

dihybrid The product of a cross between parents differing in two characters determined by single genes; an individual heterozygous at two gene loci.

dikaryon Fungal hypha or mycelium in which 2 nuclei of different genetic constitution (and different mating type) are present in each cell (or hyphal segment). adj. *dikaryotic.* See **dikaryophase.**

dikaryophase That part of the life cycle of an ascomycete or basidiomycete in which the cells are dikaryotic, i.e. between **plasmo-**gamy and **karyogamy.**

dilambdodont Said of teeth in which the paracone and metacone are V-shaped, well separated and placed near the middle of the tooth, as in some *Insectivora.*

dilator A muscle which, by its contraction, opens or widens an orifice. Cf. **sphincter.**

Dilleniidae Subclass or superorder of dicotyledons. Trees, shrubs and herbs, polypetalous or sympetalous, stamens (if numerous) developing centrifugally, mostly syncarpous, often with parietal placentation. Contains ca 24 000 spp. in 69 families including Malvaceae, Cruciferae and Ericaceae.

dimerous Having two members of a given sort.

dimorphic Existing in 2 distinct forms. Also *dimorphous.*

dimorphism The condition of having 2 different forms, as animals which show marked differences between male and female (sexual dimorphism), animals which have 2 different kinds of offspring, and colonial animals in which the members of the colony are of 2 different kinds.

dinitrogen fixation Same as **nitrogen fixation.**

Dinoflagellata See **Dinophyceae.**

Dinophyceae The *dinoflagellates.* Mesokaryotic algae. Motile cells have 2 laterally inserted flagella, one lying in a transverse groove around the cell and helically coiled, the other lying in a longitudinal groove and posteriorly directed. Chloroplasts with chlorophyll a and c, and **peridinin**, thylakoids in threes, envelope of 3 membranes; store starch in the cytoplasm. Phototrophs include flagellate, colonial, coccoid, palmelloid and a few filamentous sorts. Marine (see **red tide**) and fresh water. Phagotrophic, parasitic and various symbiotic sorts occur.

diocoel The lumen of the diencephalon.

dioecious Having the sexes in separate individuals, n. *dioecism.*

diocotrus In female Mammals, the growth period following metoestrus.

dioptric mechanism A mechanism, consisting of the cornea, aqueous humour, lens and vitreous humour, by which the images of external objects may be focused on the retina of the eye, in Chordata. An analogous mechanism in the ommatidia of Arthropods.

diphasic Of certain parasites, having a life cycle which includes a free active stage. Cf. **monophasic.**

diphtheria See panel.

diphtheria toxin See panel.

diphtheria toxoid See panel.

diphycercal Said of a type of tail fin (found in Lung fish, adult Lampreys, the young of all Fish, and many aquatic Urodela) in which the vertebral column runs horizontally, the fin being equally developed above and below it.

---

diphtheria, toxins and toxoid

An infection, usually air-borne, with the bacillus *Corynebacterium diphtheriae*. Bacilli are confined to the throat, producing local necrosis ('pseudomembrane'), but a powerful **exotoxin** causes damage esp. to heart and nerves.

The *toxin* has one part of the molecule which attaches to a surface component on susceptible cells such as heart muscle and another which interferes with protein synthesis within them. Toxin is effectively neutralized by antitoxin, which is used to treat severe infections in unimmunized persons.

The *toxoid* is the toxin treated with formaldehyde so as to destroy toxicity but not alter its capacity to act as antigen. Used for active immunization against diphtheria. It is usually used after adsorption onto aluminium hydroxide, which acts as an **adjuvant**, and in combination with tetanus toxoid and often with *Bordatella pertussis* vaccine.

---

diphygenic Having 2 different modes of development.

diphyletic Of dual origin; descended from 2 distant ancestral groups.

diphyodont Having 2 sets of teeth; a deciduous or milk dentition and a permanent dentition, as in Mammals.

diplobiont A plant which includes in its life cycle at least 2 kinds of different individuals; if the species is dioecious, there are 3 kinds of individuals. adj. *diplobiontic*.

diploblastic Having 2 primary germinal layers, namely, ectoderm and endoderm.

diplococcus A coccus which divides by fission in one plane, the two individuals formed remaining paired.

diplogangliate, diploganglionate Having paired ganglia.

diplohaplont Organism in which there is an *alternation* of haploid and diploid *generations*. Cf. **haplont**, **diplont**.

diploid Possessing two sets of chromosomes, one set coming from each parent. Most organisms are diploid. Cf. **haploid**.

diploidization The fusion within the vegetative mycelium of two haploid nuclei to give a diploid nucleus in some fungi.

diplonema A stage in the meiotic division (**diplotene** stage) at which the chromosomes are clearly double.

diplont Organisms in which only the zygote is diploid, meiosis occurring at its germination and the vegetative cells being haploid. Cf. **haplont**, **diplohaplont**.

diplophase The period in the life cycle of any organism when the nuclei are diploid. Cf. **haplophase**.

Diplopoda A class of Arthropoda having the trunk composed of numerous double somites, each with 2 pairs of legs; the head bears a pair of uniflagellate antennae, a pair of mandibles, and a gnathochilarium representing a pair of partially fused maxillae; the genital opening is in the third segment behind the head; vegetarian animals of retiring habits. Millipedes.

diplospondyly The condition of having two vertebral centra, or a centrum and an inter centrum, corresponding to a single myotome. adjs. *diplospondylic, diplospondylous*.

diplostemonous Having twice as many stamens as there are petals, with the stamens in 2 whorls, the members of the outer whorl alternating with the petals.

diplotene The fourth stage of meiotic prophase, intervening between pachytene and diakinesis, in which homologous chromosomes come together and there is condensation into tetrads.

diplozoic Bilaterally symmetrical.

Dipnoi, Dipneusti An order of Sarcopterygii, in which the air bladder is adapted to function as a lung, and the dentition consists of large crushing plates. Lung fish.

diprotodont Having the first pair of upper and lower incisor teeth large and adapted for cutting, the remaining incisor teeth being reduced or absent; pertaining to the Diprotodontia. Cf. **polyprotodont**.

Diptera An order of Insects, having one pair of wings, the hinder pair being represented by a pair of club-shaped balancing organs or halteres; the mouth parts are suctorial; the larva is legless and sluggish. Flies, Gnats and Midges.

directing stimulus The stimulus which, though not releasing a component of species-specific behaviour, is important in determining the direction of a response.

direct metamorphosis The incomplete metamorphosis undergone by exopterygote insects, in which a pupal stage is wanting.

disc See **disk**.

discal Pertaining to or resembling a disk or disklike structure; a wing cell of various Insects.

disclimax A stable community which is not

the climatic or edaphic community for a particular place, but is maintained by man or his domestic animals, e.g. a desert produced by overgrazing, where the natural climax would be grassland. The name derives from *disturbance climax*.

Discolichenes A group of lichens in which the fungus is a Discomycete.

Discomycetes A class of fungi in the Ascomycotina in which the fruiting body (ascocarp) is usually an apothecium. Includes the Lecanorales (*lichen*-forming fungi), and many saprophytic and mycorrhizal sorts, e.g. the morels and the truffles.

discontinuous distribution I s o l a t e d distribution of a species, as the Tapir, which is found in the Malay Peninsula and Sumatra, and again in Central and South America.

discontinuous variation A sudden alteration in otherwise smoothly varying characters in a group of organisms over e.g. a geographical range.

discriminant analysis A method of assigning observations to groups on the basis of values of observations previously obtained from each group.

discrimination In animal behaviour, the ability to respond to different patterns of stimulation, often tested for by using a conditioning procedure. See **discrimination training**. In the social psychology of humans, a term denoting behaviour towards people or groups of people based on their inclusion in a particular group, e.g. gender or race.

discrimination training Learning to respond to certain stimuli that are reinforced, and not to others that are not reinforced.

discus proligerus See **zona granulosa**.

disharmony See **hypertely**.

disinfection The destruction of pathogenic bacteria, usually with an antiseptic chemical or **disinfectant**.

disinfestation The destruction of insects, esp. lice.

disjunct Interrupted, disconnected, not continuous, e.g. having deep constrictions between the different tagmata of the body.

disjunction The separation during meiotic **anaphase** of the two members of each pair of homologous chromosomes.

disjunctor A portion of wall material forming a link between the successive conidia in a chain, and serving as a weak plane where separation may occur.

disk *Disc*. (1) A fleshy outgrowth from the receptacle of a flower, surrounding or surmounting the ovary and often secreting nectar. (2) The central part of a capitulum. (3) Any flattened circular structure.

disk floret (1) Usually, one of the florets occupying the central part of the capitulum, whatever its morphology. (2) Sometimes, a *tubular floret*. Cf. **ray floret**.

disomic Relating to two homologous chromosomes or genes.

dispermy Penetration of an ovum by two spermatozoa.

dispersal The active or passive movement of individual plants or animals or their disseminules (seeds, spores, larvae etc.) into or out of a population or population area. It includes emigration, immigration and migration. Should not be confused with *dispersion*.

dispersion The distribution pattern in an animal or plant population, this being random, uniform (more regular than random), or clumped (see **aggregation**). Should not be confused with *dispersal* or with *distribution*, which refers to the species as a whole, although the dispersion of a population can be described as following a random, or **Poisson distribution**. In *statistics*, the extent to which observations are dissimilar in value, often measured by standard deviation, range etc.

displacement In psychoanalytic theory, a *defence mechanism* involving the transfer of emotional energy from an unacceptable object to a safer one, so that gratification of a need comes from a source that is personally or socially more acceptable. In dreams, e.g. one image may be over exaggerated and another, more central image, minimized in its affective quality.

displacement activity The performance of a behaviour pattern which is apparently irrelevant to the situation in which it occurs; common in conflict situations.

display behaviour Species specific patterns of either sound or movement, often stereotyped in form, and which serve a great variety of communicative functions, e.g. in courtship or agonistic behaviour.

dissemination The spread or migration of species, usually by means of seeds, spores and larvae.

disseminule A propagule.

dissociation, dissociative disorder A n unconscious **defence mechanism** in which a group of psychological functions are separated from the remainder of the person's activities. In extreme cases this may result in a *dissociative disorder*, e.g. amnesia, fugue, **multiple personality**.

dissymmetrical See **asymmetrical**.

distal Far apart, widely spaced; pertaining to or situated at the outer end; farthest from the point of attachment. Cf. **proximal**.

distichous Arranged in 2 diametrically opposite rows.

distinct Plant members not joined to one another.

distraction display Behaviour, esp. of some female birds, which is sometimes similar to that of an injured individual (hence called *injury feigning*), generally a response to a predator threatening the eggs or young, and usually effective in diverting its attention.

distribution The occurrence of a species, considered from a geographical point of view, or with reference to altitude or other factors. Sometimes used as equivalent to *dispersal*. Should not be confused with *dispersion* which refers to individuals. In *statistics* the partition of observations into intervals by value; the set of frequencies of observations in a set of intervals; a generic term for mathematical formulae giving probabilities related to values of random variables. See **cumulative distribution function**.

distribution factor A *modifying factor* used in calculating biological radiation doses, which allows for the nonuniform distribution of an internally-absorbed radioisotope.

distribution-free methods Methods of statistical analysis which under certain conditions do not depend on the probability distribution generating the observations.

disuse atrophy Wasting of a part as a result of diminution or cessation of functional activity.

dithiothreitol (CHOH.CH$_2$SH)$_2$. A mild reducing agent often used to reduce protein disulphide bonds.

ditrematous Of hermaphrodite animals, having the male and female openings separate; of unisexual forms, having the genital opening separate from the anus.

diuresis The excretion of urine, esp. in excess.

divaricate Spreading widely apart, forked, divergent.

divergence, divergent evolution Where the same basic structure has evolved to give organs of different form and function. See **homology**. Cf. **convergent evolution**.

divergent With the apices wider apart than their bases.

divergent thinking Thinking which is productive and original, involving the creation of a variety of ideas or solutions which tend to go beyond conventional categories (De Bono). See **convergent thinking**.

diversity An index of the number of species in a defined area, often represented mathematically. *Alpha diversity* is on a local scale, *beta diversity* on a regional scale. See **richness**.

diverticulum (1) Saccular dilatation of a cavity or channel of the body. (2) Lateral outgrowth of the lumen of an organ. (3) Pouchlike protrusion of the mucous membrane of the colon through the weakened muscular wall. (4) A pouchlike side branch on the

mycelium of some fungi. pl. *diverticula*.

division Highest taxonomic rank used in the classification of plants (equivalent to the zoologist's **phylum**), ranking above *class*; the names end in -phyta or, for fungi, -mycota.

dizygotic twins Twins produced from two fertilized eggs. They may be the same or different sexes and are genetically equivalent to full sibs. syn. *fraternal twins*. Cf. **monozygotic twins**.

DNA binding proteins In prokaryotes, promoters, repressers etc. In eukaryotes, similar proteins, excluding the histones.

DNA, deoxyribose nucleic acid See panel.

DNA library Mixture of cloned DNA sequences derived from a single source, like a mouse or a chromosome, and containing ideally all, but in reality most, of the sequences from that source.

DNP Abbrev. for *DiNitroPhenyl*, a commonly used **hapten**.

doctrine of specific nerve energies The assertion that qualitative differences in sensory experience depend on which nerve is stimulated and not on the physical attributes of the stimulus.

dolichol phosphate A long chain unsaturated lipid with a terminal pyrophosphate found in the membranes of the **endoplasmic reticulum**. The core oligosaccharide for N-glycosylation of proteins is constructed on a dolichol phosphate molecule prior to its donation to the nascent polypeptide chain.

doliiform, dolioform Barrel shaped.

domain Applied to immunoglobulins, it is the three-dimensional structure formed by a single homology region of an immunoglobulin heavy or light chain, i.e. V$_L$, C$_L$, V$_H$, CH1, CH2, CH3 or CH4.

domatium A cavity in a plant inhabited by commensal mites or insects.

dome See **apical dome**.

dominance hierarchy An aspect of the social organization of various species, usually referring to aggressive interactions, in which certain individuals predictably dominate, or are dominated by, other members of the group.

dominant Describes a gene (allele) which shows its effect in those individuals who received it from only one parent, i.e. in heterozygotes. Also describes a character due to a dominant gene. Cf. **recessive**.

dominant The species which because of its number or size determines the character of a plant community or vegetation layer. Several species can be *co-dominant*.

Domin scale A ten-point scale used in estimating canopy cover. The scale is not linear.

dormancy A state of temporarily reduced but detectable metabolic activity, as in

DNA, deoxyribose nucleic acid

In its double-stranded form the genetic material of most organisms and organelles, although phage and viral genomes may use single-stranded DNA, single-stranded RNA or double-stranded RNA. The two strands of DNA form a double-helix, the strands running in opposite directions, as determined by the sugar-phosphate 'backbone' of the molecule.

The four bases project towards each other like the rungs of a ladder, with a purine always pairing with a pyrimidine, according to the *base-pairing rules*, in which thymidine pairs with adenine and cytosine with guanine. In its B molecular form the helix is 2.0 nm in diameter with a pitch of 3.4 nm (10 base pairs).

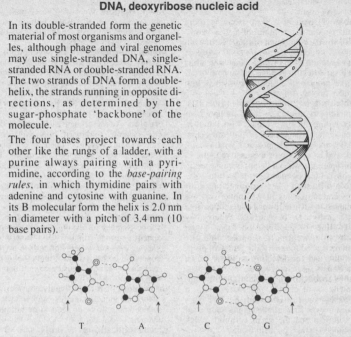

T   A   C   G

The DNA double helix (top right) and the two base pairs (above).

The filled circles are C atoms, the open are N (large) and H (small) and the double circles are O. The arrows indicate the helix attachment sites which are about 1.1 nm apart. T,A,C and G are the base abbreviations. The dashed lines show the hydrogen bonds.

seeds.

dormin See **abscisic acid**.

dorsal (1) Of the surface of flattened thalloid plants which faces away from the substrate. (2) **Abaxial**. The dorsal surface of a leaf is, therefore, the lower surface, but the term is often, by analogy with zoological usage, applied to the upper. (3) Of that aspect of a bilaterally symmetrical animal which is normally turned away from the ground.

dorsalis An artery supplying the dorsal surface of an organ.

dorsal suture The midrib of the carpel in cases where dehiscence occurs along it.

dorsal trace The median vascular supply to a carpel.

dorsiferous Said of animals which bear their young on their back.

dorsifixed Said of an anther which is attached rigidly by its back to the filament. See **basifixed, versatile**.

dorsigrade Walking with the backs of the digits on the ground, as Sloths.

dorsiventral Having structural differences between its dorsal and ventral sides.

dorsum In Anthozoa, the sulcular surface; in Arthropoda, the tergum or notum; in Vertebrates, the dorsal surface of the body or back.

dose General term for quantity of radiation. See **absorbed dose, dose equivalent, effective dose equivalent, collective dose equivalent, genetically significant dose**.

dose equivalent The quantity obtained by multiplying the absorbed dose by a factor to allow for the different effectiveness of the various ionizing radiations in causing harm to tissues. Unit Sievert (*Sv*).

dosemeter Instrument for measuring **dose** and used in radiation surveys, hospitals, laboratories and civil defence. It gives a measure of the radiation field and dosage experienced. Also *dosimeter*.

dose rate The **absorbed dose**, or other dose, received per unit time.

dose reduction factor A factor giving the

reduction in radiation sensitivity for a cell or organism which results from some chemical protective agent.

double (1) Said of a flower having more than the normal number of petals, commonly by the conversion of stamens (or even carpels) to petals, i.e. *petalody*. (2) Also of the capitulum of the Compositae which has (in a superficially similar way) some or all of the tubular disc florets converted into ligulate florets. Ant. *single*.

doublebind communication Mutually contradictory messages which set up a conflict within the individual, who simultaneously receives messages indicating that what is meant is the opposite of what is said. Believed by some to be a causative factor in schizophrenia, when a parent is consistently giving doublebind messages to a child.

double diffusion Test for detecting antigens and antibodies in which the two are arranged to diffuse toward one another in a gel (usually agar or agarose). Where they meet, lines of precipitation form. Since each antigen forms a separate line the method is used for analysing purity of preparations, and for detecting antigens which share common determinants.

double embedding A technique for embedding small objects, otherwise liable to distortion or disorientation, e.g. the specimen may be first orientated and embedded in celloidin and the small celloidin block embedded in hard paraffin wax.

double fertilization The process, characteristic of angiosperms, in which 2 male nuclei enter the embryo sac. One fuses with the egg cell nucleus to form the zygote which develops into the embryo; the other typically fuses with the 2 polar nuclei to form a triploid nucleus from which the endosperm derives.

double-image micrometer A microscope attachment for the rapid and precise measurement of small objects, like cells or particles. The principle is that 2 images of the object are formed by means which allow the separation of the images to be varied and its magnitude read on a scale. By setting the 2 images edge-to-edge (a very precise setting) and then interchanging them, the difference of the scale readings is a measure of the diameter of the object. The double image may be formed by birefringent crystals, prisms, an interferometer-like system or a vibrating mirror.

down feathers See **plumulae**.

Down's syndrome A form of mental retardation caused by a chromosomal abnormality, *trisomy 21*; the main features are moderate to severe retardation, a small round head, slanting eyes and minor abnormalities of hands and feet. These children also often have congenital heart lesions. Formerly *Mongolism*.

downy mildew One of several plant diseases of e.g. vines, onions, lettuce, caused by biotrophic fungi of the family Peronosporaceae. See **Oomycetes**.

dream interpretation In psychoanalytic theory, a technique for understanding the individual's unconscious life by focusing on dream content and attempting to unravel its hidden meaning, which reflects unconscious wishes and conflicts. See **Freud's theory of dreams**.

D region A short sequence of amino acids in the variable region of immunoglobulin heavy chains which is coded by one or another of several separate DNA **exons**. This contributes to antibody diversity (hence the use of *D*).

drip tip A marked elongation of the tip of the leaf, said to facilitate the shedding of rain from the surface of the leaf.

drive A motivational concept used to describe changes in responsiveness to a consistent external stimulus as a function of varying *internal states*. Formerly a popular concept in both psychology and ethology, its usefulness is now considered problematic (e.g. *hunger* as a drive).

drive-reduction hypothesis In learning theory, the idea that reinforcing stimuli must reduce some drive or need in an animal if learning is to occur.

dromaeognathous In Birds, said of the type of palate with large basipterygoid processes springing from the basisphenoid, and the vomers large and flat and connected posteriorly with the palatines, which do not articulate with the rostrum.

drone In social Bees (Apidae), a male.

drop Premature abscission of fruit esp. (e.g. June drop) when half grown or (pre-harvest drop) when almost mature.

Drosophila melanogaster The common fruitfly, a dipteran used extensively for genetic experiments because its giant **salivary gland chromosomes** and other biological characteristics are very suitable for studies of chromosome organization and **gene mapping**.

drug Any substance, natural or synthetic, which has a physiological action on a living body, either when used for the treatment of disease or the alleviation of pain or for purposes of self indulgence, leading in some cases to progressive addiction.

drupe A fleshy fruit with one or more seeds each surrounded by a stony layer (the endocarp), e.g plum.

drupel A small drupe, usually one of a group forming an aggregate fruit, e.g. raspberry.

druse A globose mass of crystals of calcium oxalate (ethandioate) which forms around a central foundation of organic material; in some plant cells.

dry deposition Deposition of gaseous materials on natural surfaces involving turbulent transport. Applied to *pollutant gases*.

dry fruit A fruit in which the pericarp does not become fleshy at maturity.

dry rot (1) One of a number of plant diseases, e.g. of stored potatoes, in which a lesion dries out as it forms. (2) The rotting of timber by the fungus *Serpula (Merulius) lacrymans*, so that it becomes dry, light and friable, with a cracked appearance. Cf. **wet rot**.

duct A tube formed of cells; a tubular aperture in a nonliving substance, through which gases and liquids or other substances (such as spermatozoa, ova, spores) may pass. Also *ductus*.

ductless glands Masses of glandular tissue which lack ducts and discharge their products directly into the blood; as the lymph glands and the endocrine glands.

ductule A duct with a very narrow lumen; a small duct; the fine terminal portion of a duct.

ductus arteriosus A blood vessel important in foetal development linking the *pulmonary artery* to the *aorta*. It closes at birth. In some cases there is abnormal persistence known as *patent ductus arteriosus* where blood flows from the aorta to the pulmonary artery creating an abnormal shunt.

ductus caroticus In some Vertebrates, a persistent connection between the systemic and carotid arches.

ductus Cuvieri See **Cuvierian ducts**.

ductus ejaculatorius In many Invertebrates, as the Platyhelminthes, a narrow muscular tube forming the lower part of the vas deferens and leading into the copulatory organ.

ductus endolymphaticus In lower Vertebrates and the embryos of higher Vertebrates, the tube by which the internal ear communicates with the surrounding medium.

ductus pneumaticus In physostomous Fish, a duct which connects the gullet with the air bladder.

dulosis Among Ants (*Formicoidea*), an extreme form of social parasitism in which the work of the colony of one species is done by captured 'slaves' of another species called *amazons*. Also *slavery, helotism*.

duodenum In Vertebrates, the region of the small intestine immediately following the pylorus, distinguished usually by the structure of its walls. adj. *duodenal*.

duplex A double stranded part of a nucleic acid molecule.

duplication Doubling of a gene or a larger

segment of a chromosome, by a variety of mechanisms including unequal crossing over and fusion between a chromosomal fragment and a whole chromosome of the same sort.

duplicident Having two pairs of incisor teeth in the upper jaw; as Hares and Rabbits.

dura mater A tough membrane lining the cerebro-spinal cavity in Vertebrates.

duramen See **heart wood**.

dwarf male A male animal greatly reduced in size, and usually in complexity of internal structure also, in comparison with the female of the same species; such males may be free living but are more usually carried by the female, to which they may be attached by a vascular connection in extreme cases, as some kinds of deep-sea Angler Fish.

dwarf shoot See **short shoot**.

dyad Half of a tetrad group of chromosomes passing to 1 pole at meiosis.

dyenin The protein which forms columns of side arms along the peripheral **microtubules** of cilia and mediates the movement of the microtubules relative to each other, causing the cilia to bend. It has ATPase activity.

dynamic psychology Refers to a school of thought that assumes the primary importance of inner and subjective mental and emotional events in the explanation of behaviour. See **psychodynamics**.

dys- Prefix from Gk. *dys-*, in English mis-, un-.

dysadaptation Marked reduction in rapidity of adaptation of the eye to suddenly reduced illumination, as in vitamin A deficiency.

dysarthria Difficult articulation of speech, due to a lesion in the brain.

dyscrasia Any disordered condition of the body, esp. of the body fluids.

dysgenic Causing, or tending towards, racial degeneration.

dysgraphia Inability to write, as a result of brain damage or other cause.

dyskinesia A term applied to any one of a number of conditions characterized by involuntary movements which follow a definite pattern, e.g. tics.

dyslexia *Word blindness*; great difficulty in learning to read, write or spell, which is unrelated to intellectual competence and of unknown cause.

dysmelia Misshapen limbs.

dysplasia Abnormality of development.

dyspnea, dyspnoea Laboured or difficult respiration.

dystrophic Said of a lake in which the water is rich in organic matter, such as humic acid, but this consists mainly of undecomposed plant fragments, and nutrient salts are sparse.

E

E, EA, EAC Abbrevs. for *Erythrocyte, Erythrocyte with Antibody on its surface,* and *Erythrocyte with Antibody and Complement*; the latter describing components which have become attached following activation (e.g. EAC 1423). Such red cells are used to detect Fc receptors or complement receptors on other cells.

ear Strictly, the sense organ which receives auditory impressions; in Insects, various tympanic structures on the thorax or forelegs; in some Birds and Mammals, a prominent tuft of feathers or hair close to the opening of the external auditory meatus; in Mammals, the pinna; more generally, any ear-like structure.

early replicating regions Those parts of chromosomes which are replicated early in **S-phase**.

early wood *Spring wood*. The wood formed in the first part of a growth layer during the spring, typically with larger cells with thinner cell walls than the **late wood**.

EBNA *Epstein-Barr virus Nuclear Antigen.* Antigen detected in the nuclei of B cells and tumour cells in conditions associated with infection by the Epstein-Barr virus, such as *infectious mononucleosis, Burkitt's lymphoma* and *nasopharyngeal carcinoma*.

ecad A species with distinctive forms which depend simply on the environment rather than on genotypic differences. Cf. **ecotype**.

eccrine Said of a gland whose product is excreted from its cells.

ecdemic Foreign; not *indigenous* or *endemic*.

ecdysis The act of casting off the outer layers of the integument, as in *Arthropoda*.

ECFA *Eosinophil Chemotactic Factor of Anaphylaxis.* A peptide released from mast cells which causes eosinophils to move into the site from the bloodstream. Local accumulation of eosinophil granulocytes takes place where type-I allergic reactions occur.

echin-, echino- Prefixes from Gk. *echinos,* hedgehog, meaning spiny.

echinococcus A bladderworm possessing a well-developed bladder containing daughter bladders, each with numerous scolices.

Echinodermata A phylum of radially symmetrical marine animals, having the body-wall strengthened by calcareous plates; there is a complex coelom; locomotion is usually carried out by the tube feet, which are distensible finger-like protrusions of a part of the coelom known as the water-vascular system; the larva is bilaterally symmetrical and shows traces of metamerism. Starfish, Sea Urchins, Brittle Stars, Sea Cucumbers and Sea Lilies.

Echinoidea A class of Echinodermata having a globular, ovoid or heart-shaped body which is rarely flattened; there are no arms; the tube feet possess ampullae and occur on all surfaces, but not in grooves; the anus is aboral or lateral, the madreporite aboral; there is a well-developed skeleton; free-living forms. Sea Urchins.

Echiuroidea A phylum of sedentary marine worm-like animals, in which nearly all trace of metamerism has been lost in the adult; the body is sac shaped, and feeding is effected by an anterior non-retractile proboscis, bearing a ciliated groove leading to the mouth.

echoic memory Refers to the brief retention of auditory information, in an unprocessed or *echo* form; fades within 2–6 seconds. Cf. **iconic memory**.

echolalia Aimless repetition of words heard without regard for their meaning, occurring in disease of the brain or in insanity; often seen in catatonic schizophrenia and autistic children.

echolocation Means of locating objects in conditions of poor visibility; involves the production of high frequency sounds and the detection of their echoes.

ECHO viruses *Entero Coxsackie Human Orphan viruses.* A group of Coxsackie viruses, often responsible for enteritis.

eclosion The act of emergence from an egg or pupa case.

ecocline A *cline* occurring across successive zones of an organism's habit.

ecological efficiency Ratios between the amount of energy flow at different points along a food chain, e.g., the *primary* or *photosynthetic efficiency* is the percentage of the total energy falling on the earth which is fixed by plants, this being approximately 1%.

ecological factor Anything in the environment which affects the growth, development and distribution of plants, and therefore aids in determining the characters of a plant community.

ecological indicators Organisms whose presence in a particular area indicates the occurrence of a particular set of water and soil conditions, temperature zones etc. Large species with relatively specific requirements are most useful in this way, and numerical relationships between species, populations, or whole communities are more reliable than a single species.

ecological niche The position or status of an organism within its community or ecosystem. This results from the organism's structural adaptations, physiological responses, and innate or learned behaviour. An organism's niche depends not only on where it lives but also on what it does.

ecological pyramids Diagrams in which the producer level forms the base and successive trophic levels the remaining tiers. They include the *pyramids of numbers, biomass* and *energy.*

ecological succession See **succession.**

ecology The scientific study of the interrelations between living organisms and their environment, including both the physical and biotic factors, and emphasizing both interspecific and intraspecific relations; the scientific study of the distribution and abundance of living organisms (i.e. exactly where they occur and precisely how many there are), and any regular or irregular variations in distribution and abundance, followed by explanation of these phenomena in terms of the physical and biotic factors of the environment.

econometrics The application of statistical methods to economic phenomena.

ecophysiology Branch of physiology concerned with how organisms are adapted to their natural environment.

Eco R1 A type of **restriction enzyme** derived from a strain of *E. coli,* strain R, which cleaves double stranded DNA at a specific sequence.

ecospecies An **ecotype** sufficiently distinct to be given a subspecific name.

ecosystem Conceptual view of a plant and animal community, emphasizing the interactions between living and non-living parts, and the flow of materials and energy between these parts. Ecosystems are usually represented as flow diagrams, showing the path of these flows between producers, consumers and decomposers.

ecotone A transitional zone between two habitats.

ecotype A form or variety of any species possessing special inherited characteristics enabling it to succeed in a particular habitat. Cf. **ecad.**

ectethmoid One of a pair of cartilage bones of the Vertebrate skull, formed by ossification of the ethmoid plate.

ecthoraeum The thread of a nematocyst.

ectoblast See **epiblast.**

ectoderm The outer layer of cells forming the wall of a gastrula; the tissues directly derived from this layer.

ectogenesis Development outside the body.

ectogenous Independent; self supporting.

ectolecithal Said of ova in which the yolk is deposited peripherally.

ectomorph One of Sheldon's somatotyping classifications; ectomorphs are delicately built, not very muscular, and are withdrawn

and intellectual. See **somatotype theory.**

-ectomy Suffix from Gk. *ektomē,* cutting out, used esp. in surgical terms.

ectomycorrhiza See **ectotrophic mycorrhiza.**

ectoparasite A parasite feeding on the internal tissues of the host, but having the greater part of its body and its reproductive structures on the surface. Also *epiparasite.*

ectophloedal Living on the outside of bark.

ectophloic Having phloem on the side of the xylem nearer the periphery of the organ, but not on the side nearer the centre.

ectopia, ectopy Displacement from normal position. Adj. *ectopic.*

ectoplasm A layer of clear non-granular cytoplasm at the periphery of the cell adjacent to the **plasma membrane.**

Ectoprocta A phylum of Metazoa with the anus outside the lophophore; with a coelomic body cavity and the lophophore retractable into a tentacle sheath and without excretory organs. Also *Bryozoa.*

ectotrophic mycorrhiza A mycorrhiza in which there is on the outside of the root a well-developed layer of fungal mycelium the hyphae of which interconnect with hyphae both within the root cortex and also ramifying through the soil. Most trees form an ectotrophic mycorrhizal association often with a basidiomycete.

ectozoon See **ectoparasite.**

ectromelia An infectious disease of mice, due to a virus.

eczema Itching, inflammatory skin condition in which papules, vesicles and pustules may be present together with oedema, scaling or exudation. Although the immediate cause may not be known underlying hypersensitivity to food (e.g. milk proteins) or an environmental allergen is often detectable in atopic persons. Allergens include chemical agents, plant poisons and materials used in trades.

edaphic climax A climax community of which the existence is determined by some property of the soil.

edaphic factor Any property of the soil, physical or chemical, which influences plants growing on that soil.

Edentata An order of primitive terrestrial Mammals characterized by the incomplete character of the dentition; the testes are abdominal; phytophagous or insectivorous forms. Sloths, Ant eaters, Armadillos.

edentulous, edentate Without teeth.

edriophthalmic Having sessile eyes, as some Crustacea.

EDTA *Ethylene Diamine Tetra-Acetic (ethanoic) acid, diamino ethane tetracetic acid.* $CH_2N(CH_2COOH_2)$. A **chelating agent** frequently used to protect enzymes from inhibition by traces of metal ions and as an inhibitor of metal dependent proteases because of its ability to combine with metals. It is also used in special soaps to remove metallic contamination. Also *complexone*.

eel Any fish of the Anguillidae, Muraenidae or other family of the Anguilliformes. The name is extended to other fish of similar form, e.g. Sand Eel.

eel grass Species of *Zostera*, grass-like monocotyledons which grow in the sea mostly around or below the low-water mark, used for sound insulation and the correction of acoustic defects.

effective dose equivalent The quantity obtained by multiplying the dose equivalents to various tissues and organs by the risk weighting factor appropriate to each organ and summing the products. Unit Sievert (Sv).

effective energy *Effective wavelength.* The quantum energy (or wavelength) of a monochromatic beam of X-rays or γ-rays with the same penetrating power as a given heterogeneous beam. Its value depends upon the nature of the absorbing medium.

effector A tissue complex capable of effective response to the stimulus of a nervous impulse, e.g. a muscle or gland.

effector neurone A motor neurone.

efferent Carrying outwards or away from; as the *efferent branchial vessels* in a Fish, which carry blood away from the gills, and *efferent nerves*, which carry impulses away from the central nervous system. Cf. **afferent**.

effort syndrome *Soldier's heart.* A condition in which the subject complains of palpitations, breathlessness and chest pain, often after exercise, in the absence of heart disease. Thought to be due to psychological factors.

effusion An abnormal outpouring of fluid into the tissues or cavities of the body, as a result of infection, or of obstruction to bloodvessels or lymphatics.

egest To throw out, to expel; to defaecate, to excrete. n.pl. *egesta*.

egg See **ovum**.

egg apparatus The egg and the two synergidae in the embryo sac of an angiosperm.

egg, egg cell A nonflagellated, female gamete usually larger than the male gamete with which it fuses.

egg nucleus The female pronucleus.

egg tooth A sharp projection at the tip of the upper beak of young birds and some mammals, by means of which they break open the egg-shell.

ego Originally, a term in philosophy, denoting the existence of a sense of self; in psychoanalytic theory, the rational, reality oriented level of personality which develops in childhood, as the child gathers awareness of and comes to terms with, the nature of the social and physical environment; represents reason and common sense, in contrast to the **id**.

egocentrism In Piaget's theory, a form of thinking most typically found in young children, in which the individual perceives and comprehends the world from a totally subjective point of view, being unable to differentiate between the objective and subjective components of experience.

ego psychology Those Freudian theorists who emphasize ego processes, such as reality perception, learning and conscious control of behaviour, and who argue that the ego has its own energy and autonomous functions, not derived from the **id**; in contrast to *instinct theory*, which states that all mental energy is ultimately derived from the id.

eidetic imagery The ability to reproduce on a dark screen, or when the eyes are closed, a vividly clear and detailed picture or visual memory image of previously seen objects. Commonly present in children up to 14 years, and occasionally persisting into adult life.

ejaculatory duct See **ductus ejaculatorius**.

elaeodochon See **oil gland**.

elaiosome An outgrowth from the surface of a seed, containing fatty or oily material (often attractive to ants) and serving in seed dispersal.

Elasmobranchii A subclass of Gnathostomata, highly developed usually predacious fishes with a cartilaginous skeleton and plate-like gills. Includes Sharks, Skates, Rays.

elastase The proteolytic enzyme secreted by the pancreas which digests *elastin*.

elastic fibres See **yellow fibres**.

elastic fibrocartilage See **yellow fibrocartilage**.

elastic tissue A form of connective tissue in which elastic fibres predominate.

elastin A protein which is the major component of the elastic fibres of the **extracellular matrix**. The molecules provide an extensible network by being heavily cross linked in a random coiled configuration. See **scleroproteins**.

Electra Complex See **Oedipus** and **Electra Complexes**.

electric organ A mass of muscular or nervous tissue, modified for the production, storage, and discharge of electric energy; occurring in Fish.

electrocardiogram *ECG.* A record of the electrical activity of the heart.

electrochemical potential μ. A measure of the (Gibbs) free energy per mole associated with a given ionic species under given conditions and hence of its relative ability to

---

electron microscope

Any of a number of devices, consisting essentially of an evacuated tube in which a beam of electrons is caused to interact with a specimen. The electron beam is made by accelerating electrons through a potential difference of from 1–1500 kV in an *electron gun* and focused by electrical or magnetic fields produced by *electron lenses*.

An image is formed either directly by focusing the electrons that pass through the specimen onto a fluorescent screen, or indirectly, by using the information carried by e.g. the x-rays or secondary electrons which are emitted during the interaction of the primary electrons with the specimen. Abbrev. *EM*.

See **scanning electron microscope** and **transmission electron microscope** for more detail and illustrations.

---

do work. An ion moving passively will tend, e.g. to diffuse across a membrane until, at equilibrium, its electrochemical potential will be the same on both sides.

For the *j* th ion:

$$\bar{\mu}_j = m_j + z_j F\ \psi$$

where μ = **chemical potential (1)**, z = valency, F = the **Faraday** constant, ψ = electrical potential.

electroconvulsive therapy A form of therapy in which an electric current is passed through the brain, resulting in convulsive seizures, used primarily in the treatment of depression. Abbrev. *ECT*.

electrocyte A cell, usually muscle but sometimes nerve, which is specialized to generate an electric discharge.

electroencephalogram *EEG*. A record of the electrical activity of the brain.

electroencephalograph Instrument for study of voltage waves associated with the brain; effectively comprises a sensitive detector (voltage or current), a d.c. amplifier of very good stability and an electronic recording system. Abbrev. *EEG*.

electrogenic pump An ion-translocating pump which causes a net transfer of charge across a membrane and therefore an electrical potential difference across it.

electron micrograph A photomicrograph of the image of an object, taken by substituting a photographic plate for the fluorescent viewing screen of an **electron microscope**.

electron microscope See panel.

electron transfer chain See panel.

electron volts The unit of energy for the photon. Used in radiology and expressed in thousands (kev) or millions (mev) of electron volts.

electrophysiology The study of electrical phenomena associated with living organisms, particularly nervous conduction.

electroplaque Large, flat disc shaped elec-

trocyte. Usually stacked in series to produce a substantial voltage pulse.

electroporation Method of introducing foreign DNA or chromosomes into cells by subjecting them to a brief voltage pulse, which transiently increases membrane permeability, allowing uptake of DNA or chromosomes from the surrounding buffer.

electroreceptor Sense organ specialized for the detection of electric discharges. Found in a variety of Fish particularly Mormyrids, Gymnotids and some Elasmobranchs.

electrotaxis See **galvanotaxis**.

electrotropism See **galvanotaxis**.

elementary bodies Particles present in cells of the body in virus infections.

eleutherodactyl Having the hind toe free.

elevator A muscle which by its contraction raises a part of the body. Cf. **depressor**.

elfin forest See **Krummholz**.

ELISA See **enzyme-linked immunosorbent assay**.

elytra In *Coleoptera*, the hardened, chitinized fore wings which form horny sheaths to protect the hind wings when the latter are not in use; in certain *Polychaeta*, plate like-modifications of the dorsal cirri, possibly for respiration. adjs. *elytroid, elytriform*.

EM See **electron microscope**.

emarginate Notched; esp. of a petal, with a small indentation at the apex.

emasculation In plant breeding, the surgical removal of the stamens from a flower, usually before pollen is shed, to prevent self pollination when cross pollination is planned.

embedding The technique of embedding biological specimens in a supporting medium, such as paraffin wax or plastics like epoxy resin, in preparation for sectioning with a microtome.

embolic gastrulation Gastrulation by invagination.

embolism The blocking of vessels. In

electron transfer chain

The energy derived from the oxidation of glucose by the **tricarboxylic acid cycle** is harnessed for energy requiring processes in the cell in the form of ATP. As the amount of energy liberated from an oxidation step is proportional to the difference in reduction potentials of the oxidized and reduced components of the reaction, the energy from glucose can be released in suitably sized packets for ATP synthesis by the transfer of electrons from glucose to their ultimate acceptor, oxygen, along a series of components with appropriate spans of reduction potentials.

This series of components is known as the *electron transfer* (or *transport*) *chain*. The transfer of electrons from NADH$^+$ to oxygen involves three large multiprotein complexes, composed of *flavoproteins, cytochromes* and other metaloproteins, togther with the smaller components of *ubiquinone* (also known as *coenzyme Q*) and cytochrome c which shuttle between the larger complexes.

Thus the first complex (NADH–ubiquinone reductase) accepts the electrons from NADH$^+$ and passes them on to ubiquinone; the next complex (ubiquinone–cytochrome c reductase) accepts these electrons from the oxidized ubiquinone and passes them to cytochrome c which is then re-oxidized by the third large complex (cytochrome oxidase) on the last stage of their journey to the terminal oxygen. The stepwise increase of the standard reduction potentials of the three complexes ensures that the energy is released as three packets each of sufficient magnitude to convert ADP to ATP and thus account for the synthesis of the 3 molecules of ATP which is associated with the oxidation of each NADH$^+$. Succinate–ubiquinone reductase acts as a second point of entry by also being able to reduce ubiquinone although, in this case, as the electrons enter the chain after the NADH$^+$–ubiquinone reductase complex, only 2 molecules of ATP are synthesized.

The electron transfer chain is located in the inner membrane of the **mitochondrion** where it is so orientated that the passage of electrons is coupled to a flow of protons across the membrane, a process which leads to the actual synthesis of the ATP. See **oxidative phosphorylation** for diagram.

The flow of electrons and hence the oxidation of the food requires a supply of ADP for conversion to ATP. This 'respiratory control' ensures that the energy source is consumed only when energy is required.

plants, a xylem **conduit** blocked by a bubble of air as a result of damage or following **cavitation**. See **tylosis**. In animals, a blood vessel blocked usually by a blood clot or thrombus from a remote part of the circulation.

embolomerous A type of vertebra consisting of a neural arch resting on two notochordal centra, an anterior intercentrum and a posterior pleurocentrum. Found in Labyrinthodontia.

embolus A clot or mass formed in one part of the circulation and impacted in another, to which it is carried by the blood stream.

emboly Invagination: the condition of pushing in or growing in. adj. *embolic*.

embryo (1) A plant at an early stage of development, e.g. within a seed. (2) An immature animal in the early stages of its development, before it emerges from the egg or from the uterus of the mother. In man the term is restricted to the stages between two and eight weeks after conception. adj. *embryonic*.

embryo culture The aseptic culture on a suitable medium of an embryo excised at an early stage. The technique is useful in plant breeding in cases where, as in some hybrids, the embryos abort if left in the ovule. See **plant cell culture**.

embryogenesis (1) The processes leading to the formation of the embryo. (2) The production of **embryoids** in **plant cell culture**.

embryogeny See **embryogenesis**.

embryoid Embryo-like structure, which may grow into a plantlet. See **plant cell culture, anther culture, totipotency**.

embryology The study of the formation and development of embryos.

embryonic fission See **polyembryony**.

embryonic tissue Meristematic tissue.

embryophyte A member of those plant groups in which an embryo, dependent at least at first on the parent plant, is formed, i.e. the bryophytes and vascular plants.

embryo sac The female gametophyte in angiosperms, formed within the ovule by the enlargement of the functional megaspore and containing usually 8 nuclei.

emergence (1) An outgrowth from a plant, derived from epidermal and cortical tissues which does not contain vascular tissue or develop into a stem or leaf. (2) The appearance above ground of germinating seedlings, *pre-emergence, post-emergence*. (3) In animals, an epidermal or subepidermal outgrowth. (4) In Insects, the appearance of the imago from the cocoon, pupa case or pupal integument.

emersed Raised above or rising out of a surface, esp. growing up out of water.

Emerson enhancement effect The more than additive effect on the rate of photosynthesis (in plants and algae) of illuminating simultaneously with far red light ($\lambda > 680$ nm) and light of shorter wavelength ($\lambda < 680$ nm), indicating the existence of the two **photosystems**, I and II.

emesis The act of vomiting.

emigration A category of population dispersal covering one way movement out of the population area. Cf. **immigration, migration**.

emissary Passing out, as certain veins in Vertebrates which pass out through the cranial wall.

empyema Accumulation of pus in any cavity of the body, esp. the pleural.

emunctory Conveying waste matter from the body; any organ or canal which does this.

enamel The external calcified layer of a tooth, of epidermal origin and consisting of elongate hexagonal prisms, set vertically on the surface of the underlying dentine; enamel also occurs in certain scales.

enamel cell See **ameloblast**.

enarthrosis A ball-and-socket joint.

enation (1) Generally an outgrowth as those on leaves caused by some viruses. (2) A nonvascularised, spine-like outgrowth from the axis of some primitive vascular plants, e.g. *Zosterophyllum*. See **enation theory**.

enation theory A theory that regards the *microphylls* of the *Lycopodiales* etc. as simple enations that have become vascularized and therefore different from megaphylls which are regarded as having evolved from branch systems.

encephalitogen Substances present in extracts of brain which when administered with a potent adjuvant such as *complete Freund's adjuvant* cause experimental allergic encephalitis. The active material is myelin basic protein.

encephalography Radiography of the brain after its cavities and spaces have been filled with air, or dye, previously injected into the space round the spinal cord.

encephalon The brain.

encephalospinal See **cerebrospinal**.

enceph-, encephalo- Prefixes from Gk. *enkephalos*, brain.

encoding In memory research, the process of transforming material into a form easily stored and retrieved. See **chunking**.

encounter group A general term for a range of group therapies with the general aim of increasing personal awareness and encouraging creative and open relations with others. Procedures vary widely, though all tend to emphasize free and candid expression of feeling and thought within the group.

encysted Enclosed in a cyst or a sac.

encystment The formation of a walled nonmotile body from a swimming spore.

encystment The formation by an organism of a protective capsule surrounding itself. Also *encystation*. v. *encyst*.

endarch Having the protoxylem on the side of the xylem nearer to the centre of the axis, the usual condition in e.g. angiosperm stems. Cf. **exarch, mesarch**.

endemic (1) Said of a species or family confined to a particular region, e.g. *Primula scotica*, native only in the north of Scotland. (2) Said of a disease permanently established in moderate or severe form in a defined area. Also *indigenous*.

endergonic Said of a biochemical reaction which requires energy and is, therefore, unable to proceed spontaneously. Cf. **exergonic**.

end labelling The enzymatic attachment of a radioactive atom to the end of a DNA or RNA molecule.

endo- Prefix from Gk. *endon*, within.

endobiotic (1) Growing inside another plant. (2) Formed inside the host cell.

endoblast See **hypoblast**.

endocardiac Within the heart.

endocardium In Vertebrates, the layer of endothelium lining the cavities of the heart.

endocarp A differentiated innermost layer of a pericarp, usually woody in texture like the hard outside of a peach stone.

endochondral Situated within or taking place within cartilage; as *endochondral ossification*, which begins within the cartilage and works outwards.

endocoelar See **splanchnopleural**.

endocranium In Insects, internal processes of the skeleton of the head, serving for muscle attachment.

endocrine Internally secreting; said of certain glands, principally in Vertebrates, which pour their secretion into the blood, and so can affect distant organs or parts of the body. See **hormone**.

endocrinology The study of the internal secretory glands.

endocuticle The laminated inner layer of the insect cuticle.

endocytosis The entry of material into the cell by the invagination of the **plasma membrane**. Material can enter in the fluid phase of the resulting vesicle or attached to its membrane.

endoderm The inner layer of cells forming the wall of a gastrula and lining the archenteron; the tissues directly derived from this layer. Also *entoderm*.

endodermis (1) The innermost layer of the cortex in roots and stems, sheathing the stele, one cell thick. In roots typically and sometimes in stems, there are no radial intercellular spaces and all anticlinal walls have a **casparian strip**. In some roots the cell walls become thickened later. See **starch sheath**. (2) A similar layer, with casparian strip or the like, elsewhere, e.g. surrounding the vascular tissues in the pine leaf.

endogamy The production of a zygote by the fusion of gametes from 2 closely related parents or from the same individual. Cf. exogamy. See **autogamy**, **inbreeding**.

endogenous (1) A general term, used to suggest that the causes of some physical or psychological conditions are due to internal and probably physical factors. (2) In plants and animals, processes or organs which originate or develop within them. (3) In higher animals, said of metabolism which leads to the building of tissue or to the replacement of loss by wear and tear. Cf. **exogenous**.

endogenous rhythm A rhythm in movement or in a physiological process which depends on internal rather than external stimuli. It will often persist under constant environmental conditions and may show *entrainment*. See **circadian rhythm**.

endoglycosidase Enzyme which hydrolyses the internal glycosidic bonds of polysaccharides or oligosaccharides, converting them to disaccharides and terminal dextrins.

endolithic Growing within rock like some algae do in limestone.

endolymph In Vertebrates, the fluid which fills the cavity of the auditory vesicle and its outgrowths (semicircular canals etc.).

endolymphangial Situated within a lymphatic vessel.

endolymphatic Pertaining to the membranous labyrinth of the ear in Vertebrates, or to the fluid contained therein.

endomitosis Process, occurring naturally in some differentiating cells or when induced by e.g. colchicine, in which the chromosomes divide without cell division giving

double the original number of chromosomes in the cell. See **restitution nucleus**.

endomorph One of Sheldon's somatotyping classifications; endomorphs are soft and round; they are described as loving comfort, affection and are sociable. *Endomorphy* is thus characterized by general heaviness or stoutness of build and is said to be associated with more or less extraversive personality traits; sociability, hedonism etc. See **somatotype theory**.

endomysium The intramuscular connective tissue which unites the fibres into bundles.

endoneurium Delicate connective tissue between the nerve fibres of a funiculus.

endonuclease An enzyme which cuts a polynucleotide chain internally.

endoparasite A parasite living inside the body of its host.

endopeptidase An enzyme which cuts a polypeptide chain internally, not just removing terminal peptides.

endophyte A plant living inside another plant, but not necessarily parasitic on it.

endophytic Occurring within a plant, e.g. the hyphae of a symbiotic fungus.

endophytic mycorrhiza Same as **endotrophic mycorrhiza**.

endopite The inner ramus of a biramous arthropod appendage.

endoplasmic reticulum See panel.

endopolyploid The product of **endomitosis**.

Endoprocta See **Entoprocta**.

Endopterygota Subclass of the Insecta with complete metamorphosis and a larval form in which the wings develop internally. Also *Holometabola*. Cf. **Exopterygota**.

endorhachis In Vertebrates, a layer of connective tissue which lines the canal of the vertebral column and the cavity of the skull.

endorphins Peptides synthesized in the pituitary gland which have analgesic properties associated with their affinity to the opiate receptors in the brain.

endoscopic embryology In plants, the condition when the apex of the developing embryo points towards the base of the archegonium.

endoscopy Any technique for visual inspection of internal organs. Modern instruments are usually flexible fibre-optic devices with additional facilities for biopsy and cauterization.

endoskeleton In Craniata, the internal skeleton, formed of cartilage or cartilage bone. In Arthropoda, the endophragmal skeleton, i.e. hardened invaginations of the integument forming rigid processes for the attachment of muscles and the support of certain other organs.

endosome A cytoplasmic vesicle derived

---

endoplasmic reticulum

A series of flattened membranous tubules and *cisternae* in the cytoplasm of eukaryotic cells which can either show a smooth profile, *smooth endoplasmic reticulum*, or be decorated with ribosomes, *rough endoplasmic reticulum* as in the drawing. It is the site of synthesis of lipids and some proteins. The membrane is continuous with that of the nuclear envelope. Abbrev. *ER*.

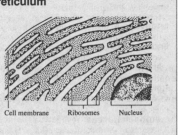

Cell membrane   Ribosomes   Nucleus

---

from a **coated vesicle** by removal of the protein coat. See **receptor mediated endocytosis**.

endosperm Tissue formed within the embryo sac, usually triploid (cf. **double fertilization**), serving in the nutrition of the embryo and often increasing to form a storage tissue in the mature seed, e.g. cereals.

endospermic, endospermous Said of a mature seed having conspicuous endosperm. syn. *albuminous*.

endospore (1) The innermost layer of the wall of a spore; *intine*. (2) A very resistant thick-walled spore formed within a bacterial cell. (3) A spore produced within or by the division of the contents of the parent cell.

endosporic Developing within the spore.

endostyle In some Protochordata and in the larvae of Cyclostomata, a longitudinal ventral groove of the pharangeal wall, lined by ciliated and glandular epithelium. adj. *endostylar*.

endosymbiosis, endocytobiosis Symbiosis in which one organism (prokaryote or eukaryote) lives inside the cells of another eukaryote, the association behaving typically as an organism.

endosymbiotic hypothesis Hypothesis postulating that the chloroplasts, mitochondria and other plastids of most eukaryotes evolved from endosymbiotic prokaryotes, which were able to photosynthesize or respire aerobically. Some subsequent transfer of genes from endosymbiont to host must also have occurred.

endothecium (1) The *fibrous layer* in the wall of an anther. (2) The inner tissues in the young sporophyte of bryophytes, giving rise to the sporogenous tissue and/or the columella. Cf. **amphithecium**.

endotheliochorial placenta See **vasochorial placenta**.

endothelium Pavement epithelium occurring on internal surfaces, such as those of the serous membranes, blood vessels and lymphatics.

endotoxin Toxin retained within, if not part of, the bacterial protoplasm, and only released on lysis or death of the bacterial cells. Examples of diseases caused by endotoxins are typhoid fever and cholera. Endotoxin is therefore a syn. for the lipopolysaccharides derived from Gram-negative bacteria which are powerful activators of macrophages and cause release of **interleukin I** and **tumour necrosis factor**. Very small amounts are sufficient to cause fever and malaise in humans, and this accounts for the unpleasant effects of vaccines containing whole Gram negative bacteria such as TAB vaccine.

endotoxin shock Syndrome following administration of endotoxin or systemic infection with endotoxin producing bacteria. Characterized by prostration, hypotension, fever and leucopaenia.

endotrophic mycorrhiza A mycorrhiza in which the fungal hyphae grow between and within the cells of the root cortex and connect with hyphae ramifying though the soil but which do not form, like in *ectotrophic* mycorrhiza a thick mantle on the surface of the root. Vesicular-arbuscular **mycorrhizas** and the mycorrhizas of orchids and of the Ericaceae are endotrophic.

endozoic (1) Living inside an animal. (2) Said of the method of seed dispersal in which seeds are swallowed by some animal and voided after having been carried for some distance.

end plate A form of motor nerve ending in muscle.

endysis The formation of new layers of the integument following ecdysis.

energy balance A quantitative account of the exchanges of energy between organisms and their surroundings.

enhancement effect Same as **Emerson enhancement effect**.

enhancer DNA sequence which can stimulate transcription of a gene while being at a distance from it.

enkephalins Penta peptides isolated from

the brain which have the same N-terminal amino acid sequences as **endorphins** and share their analgesic properties.

enrichment A method of increasing the proportion of cells with a mutation which cannot be selected directly. Mutants unable to grow in a given medium, are tolerant of agents like penicillin which only kill growing cells.

ensiform Shaped like the blade of a sword.

ensiform process See **xiphisternum**.

enteral Parasympathetic.

enter-, entero- Prefixes from Gk. *enteron*, intestine.

Enterobacteriaceae A family of bacteria belonging to the order Eubacteriales. Gram-negative rods; aerobes; carbohydrate fermenters; many saprophytic. Includes some gut parasites of animals and some blights and soft rots of plants, e.g. *Salmonella typhi* (typhoid), *Escherichia coli* (some strains cause enteritis in calves and infants).

enterocoel A coelome formed by the fusion of coelomic sacs which have separated off from the archenteron. In Chaetognatha, Echinodermata and Cephalochorda.

enteron The single body cavity of Coelenterata; it corresponds to the archenteron of a gastrula; in higher forms, the alimentary canal. See **archenteron**. adj. *enteric*.

Enteropneusta See **Hemichordata**.

enterosympathetic Said of that part of the autonomic nervous system which supplies the alimentary tract.

entire Said of the margin of a flattened organ when it is continuous, being neither toothed nor lobed.

ento- Prefix from Gk. *entos*, within.

entoderm See **endoderm**.

entogastric Within the stomach or enteron.

entomology The branch of zoology which deals with the study of insects.

entomophagous Feeding on insects.

entomophily Pollination by insects. adj. *entomophilous*.

Entoprocta A phylum of Metazoa in which the anus is inside the circlet of tentacles; mainly marine forms. Also *Endoprocta*.

entovarial Within the ovary.

entozoic Living inside an animal.

entozoon An animal parasite living within the body of the host. adj. *entozoic*.

entrainment The process whereby an endogenous clock-driven rhythm is synchronized to the rhythm of environmental events. See **Zeitgeber**.

entry portal Area through which a beam of radiation enters the body.

enucleate (1) Lacking a nucleus. (2) To remove, e.g. by microdissection, the nucleus from a cell.

enucleation The removal of the nucleus from a cell, e.g. by microdissection. adj. and v. *enucleate*.

enuresis A lack of bladder control past the age when such control is normally achieved.

environment (1) The physical and chemical surroundings of an object, e.g. the temperature and humidity, the physical structures, the gases. (2) When applied to human societies, the cultural, aesthetic and any other factors which contribute to the quality of life, are included in the definition.

environmental variance Variation of quantitative character due to non-genetic causes.

enzyme See panel.

enzyme-linked immunosorbent assay ELISA. An assay method in which antigen or antibody are detected by means of an enzyme chemically coupled either to antibody specific for the antigen or to anti-Ig which in turn will bind to the specific antibody. Either the antigen or the antibody to be detected is attached to the surface of a small container or to plastic beads, and the specific antibody allowed to bind in turn. The amount bound is subsequently measured by addition of a substrate for an enzyme which develops a colour when hydrolysed. Commonly used enzymes are *horse radish peroxidase* and *alkaline phosphatase*.

eosinophil Any cell whose protoplasmic granules readily stain red with the dye eosin, particularly a granulocyte in the blood and the oxyphil cells in the *pars anterior* of the pituitary gland.

eosinophilia Increase in numbers of eosinophil leucocytes in the blood above the normal levels (up to 400/mm^3 in humans). Usually associated with repeated immediate type hypersensitivity reactions. Unexplained very high levels of eosinophil leucocytes sometimes occur in subjects with eosinophilic cardiac myopathy and their granule contents cause necrosis of heart muscle cells.

eosinophil leucocyte Polymorphonuclear leucocyte with large eosinophil granules in the cytoplasm containing cationic proteins. Other granules contain peroxidase. The cells have Fc receptors for IgE and for complement. The granules are secreted when the cells are activated (e.g. via these receptors). In parasite infections such as *schistosomiasis* in which IgE antibodies are made, eosinophils attach themselves to the parasites and the granule contents are secreted onto the parasites and kill them.

epapophysis A median process of a vertebral neural arch.

epaxial *Epaxonic*. Above the axis, esp. above the vertebral column, therefore, dorsal; as the upper of two blocks into which the myotomes of fish embryos become divided. Cf. **hypaxial**.

epencephalon See **cerebellum**.

---

enzyme

Most chemical reactions in living organisms only proceed sufficiently fast if mediated by catalytic proteins known as enzymes. Metabolism therefore depends upon the balanced and coordinated effects of these proteins controlling the rates of both synthetic and degradative reactions.

The catalytic effect depends on a restricted region of the protein molecule, known as the *active centre* of the enzyme, which consists of a specific and unique configuration of a few amino acid side chains brought into confluence by the *secondary* and *tertiary* folding of the protein into its *native conformation*. The *substrate* of the enzyme is bound at this site when the catalytic events take place. As the integrity of the active centre depends upon the stability of the native conformation, enzymes will only retain activity under conditions when this conformation is stable. Hence extremes of temperature and pH, organic solvents and heavy metals will all inhibit activity by denaturing the protein. Enzymes may also be inhibited by molecules which react specifically with the active centre. Such poisons may bind to the active centre because they are structurally related to the proper substrate and can be displaced by high concentrations of the substrate. Thus the dicarboxylic acid, malonic acid, will inhibit succinate oxidase which normally converts another dicarboxylic acid, succinic acid to fumaric acid and the inhibition will be reversed by an excess of succinate. This reversible inhibition is termed *competitive inhibition*. In other instances, e.g. most nerve gases, the poison is unrelated to the substrate and the inhibition is termed *non-competitive*.

An enzyme's activity is restricted to a specific set of metabolic reactions depending on the nature of the chemical bond to be modified and the ability of the enzyme to bind to the substrate. For example, enzymes which catalyse the hydrolysis of peptide bonds, *peptidases*, usually have their activity restricted to peptide bonds adjacent to certain amino acids in the polypeptide sequence of the substrate protein.

---

ependyma In Vertebrates, the layer of columnar ciliated epithelium, backed by neuroglia, which lines the central canal of the spinal cord and the ventricles of the brain. adj. *ependymal*.

Ephedra Sea grape, a genus of jointed, all but leafless, desert plants of the Gnetaceae. Source of *ephedrine*.

ephemeral A plant which completes its life cycle in a short period, weeks rather than months.

Ephemeroptera An order of Insects, in which the adults have large membranous forewings and reduced hindwings, and the abdomen bears two or three long caudal filaments; the adult life is very short and the mouthparts are reduced and functionless; the immature stages are active aquatic forms. May flies.

epibasal half The anterior portion of an embryo.

epibiosis Relationship in which one organism lives on the surface of another without causing it harm. Plant epibionts are *epiphytes*, animal epibionts, *epizoites*.

epiblast The outer germinal layer in the embryo of a metazoan animal, which gives rise to the ectoderm. Cf. **hypoblast**.

epiblem rhizodermis The outermost cell layer (epidermis) of a root.

epiboly Overgrowth; growth of one part or layer so as to cover another. adj. *epibolic*.

epicalyx An extra, calyx-like structure below and close to the real calyx in some flowers, e.g. many Rosaceae, including strawberry and many Malvaceae.

epicardium In Vertebrates, the serous membrane covering the heart; in *Urochordata*, diverticula of the pharynx, which grow out and surround the digestive viscera like a perivisceral cavity. adj. *epicardial*.

epicarp The superficial layer of the pericarp, esp. when it can be stripped off as a skin. Also *exocarp*.

epicoele In Craniata, the cerebellar ventricle or cavity of the cerebellum.

epicondyle The proximal part of the condyle of the humerus or femur.

epicormic shoot A shoot growing out, adventitiously or from a dormant bud, from a tree trunk or substantial woody branch.

epicotyl Either (1) the part of the shoot of an embryo or seedling above the cotyledon(s), i.e. the whole of the plumule, or (2) the stem between the cotyledon(s) and the first leaf or leaves, i.e. the first internode of the plumule.

epicritic Pertaining to sensitivity to slight tactile stimuli.

epicuticle (1) In plants, the layer of waxes including long chain (C_{20}) alkanes, alcohols, acids and esters on the surface of the cuticle. Cf. **bloom**. (2) The thin outermost layer of the insect cuticle consisting of lipid and protein. See **cuticulin**.

epidemic An outbreak of an infectious disease spreading widely among people at the same time in any region. Also adj.

epidemiology The study of disease in the population, defining its incidence and prevalence, examining the role of external influences such as infection, diet or toxic substances, and examining appropriate preventative or curative measures.

epidermal, epidermatic Relating to the epidermis.

epidermis Those layers of the integument which are of ectodermal origin; the epithelium covering the body. More specifically in plants, a layer, usually one cell thick, forming a skin over the young shoots and roots of plants, continuous except where perforated by stomata (on shoots) or over lateral or adventitious roots; often carries hairs. Eventually replaced by the periderm in woody plants.

epididymis In the male of Elasmobranchii, some Amphibians and Amniota, the greatly coiled anterior end of the Wolffian duct, which serves as an outlet for the spermatozoa.

epigamic Attractive to the opposite sex; as *epigamic colours*.

epigastric Above or in front of the stomach; said of a vein in Birds, which represents the anterior part of the anterior abdominal vein of lower Vertebrates.

epigeal, epigaeous (1) Germinating with the cotyledons appearing above the surface of the ground. (2) Living on the soil surface.

epigenesis The theory, now universally accepted, that the development of an embryo consists of the gradual production and organization of parts, as opposed to the theory of *preformation*, which supposed that the future animal or plant was already present complete, although in miniature, in the fertilized egg.

epiglottis In Bryozoa, the epistome; in Insects, the epipharynx; in Mammals, a cartiliaginous flap which protects the glottis.

epignathous Having the upper jaw longer than the lower jaw, as in Sperm Whales.

epigynous Having the calyx, corolla and stamens inserted on the top of the inferior ovary.

epilimnion The warm upper layer of water in a lake. Cf. **hypolimnion**.

epilithic Growing on a rock surface.

epimysium The investing connective-tissue coat of a muscle.

epinasty **Nastic movement** in which the upper side of the base of an organ grows more than the lower, resulting in a downward bending of the organ as in the petals of an opening flower or the leaves of a tomato plant with water-logged roots. Cf. **hyponasty**.

epinephros See **suprarenal body**.

epineural In Echinodermata, lying above the radial nerve; in Vertebrates, lying above or arising from the neural arch of a vertebra.

epineurium The connective tissue which invests a nerve trunk, uniting the different funiculi and joining the nerve to the surrounding and related structures.

epiparasite See **ectoparasite**.

epipelic Growing on mud or sand.

epipetalous Attached to or inserted upon the petals, as stamens are in many flowers.

epipharynx In Insects, the membranous roof of the mouth which in some forms is produced into a chitinized median fold, and in Diptera is associated with the labrum, to form a piercing organ; in Acarina, a forward projection of the anterior face of the pharynx. adj. *epipharyngeal*.

epiphloeodal, epiphloeodic Growing on the surface of bark.

epiphragm In Gastropoda, a plate, mostly composed of calcium phosphate, with which the aperture of the shell is sealed during periods of dormancy.

epiphyllous Growing upon, or attached to, the upper surface of a leaf; sometimes, growing on any part of a leaf.

epiphysis A separate terminal ossification of some bones, which only becomes united with the main bone at the attainment of maturity; the pineal body; in Echinoidea, one of the ossicles of Aristotle's lantern. adj. *epiphysial*.

epiphysis cerebri See **pineal organ**.

epiphyte An organism which is attached to another with benefit to the former but not to the latter. See **nest epiphyte, tank epiphyte**.

epiphytotic A widespread outbreak of disease among plants, by analogy with epidemic.

epipleura In Coleoptera, the reflexed sides of the elytra; in Birds, the unicate process; in bony Fish, upper ribs formed from membrane bone; in Cephalochordata, horizontal shelves of membrane arising from the inner sides of the metapleural folds, and forming the floor of the atrial cavity.

epiploic foramen See **Winslow's foramen**.

epiploon In Mammals, a double fold of serous membrane connecting the colon and the stomach (the great omentum). adj. *epiploic*.

epipubic In front of or above the pubis; pertaining to the epipubis.

episematic Serving for recognition; as *episematic colours*.

episepalous (1) Borne on the sepals. (2) Placed opposite to the sepals.

episodic memory Refers to personal memories, tied to a particular time and place. Cf. semantic memory.

episome A self-replicating element able to grow independently of the host's chromosome, but also to integrate into it. Often termed a *plasmid* in prokaryotes.

epispore The outermost layer of a spore wall, often consisting of a deposit forming ridges, spines or other irregularities of the surface.

epistatic Describes a gene or character, whose effect overrides that of another gene with which it is not allelic; analogous to *dominant* applied to genes at different loci. More generally, *epistasis* exists when the effect of two or more non-allelic genes in combination is not the sum of their separate effects.

epistemics The scientific study of the perceptual, intellectual and linguistic processes by which knowledge and understanding are acquired and communicated.

epistomatal, epistomatic A leaf having stomata on the upper surface only. Cf. amphistomatal, hypostomatal.

epistropheus The axis vertebra.

epithalamus In the Vertebrate brain, a dorsal zone of the thalamencephalon.

epithelium A compact layer of cells lining a cavity or covering a surface. Characterized by the arrangement of the cells covering a free surface, by the presence of a basement membrane underlying the lowermost layer of cells, and by the small amount of intercellular matrix; often secretory, the tissues lining the alimentary canal and blood vessels etc. adjs. *epithelial, epitheliomorph*

epitokous Said of the heteronereid stage in Polychaeta.

epitope Term used to describe an antigenic determinant in a molecule which is specifically recognized by an antibody combining site or by the antigen receptor of a T cell.

epitrichium A superficial layer of the epidermis in Mammals, which consists of greatly swollen cells and is found on parts of the body devoid of hair. adj. *epitrichial*.

epixylous Growing on wood.

epizoic (1) Growing on a living animal, e.g. *epibiosis*. (2) Having the seeds or fruits dispersed by animals.

epizoon An animal which lives on the skin of some other animal; it may be an **ectoparasite** or a **commensal**. adj. *epizoan*.

Epstein-Barr virus First virus to be implicated in human cancer because of its association with Burkitt's lymphoma.

equatorial Situated or taking place in the equatorial plane; as the *equatorial furrow* which precedes division of an ovum into upper and lower blastomeres, and the *equatorial plate*, which, during mitosis, is the assembly of chromosomes on the spindle in the *equatorial plane*.

equi- Prefix from L. *aequus*, equal.

equilibration In Piaget's theory, the motivational mechanism underlying intellectual development; contradictory explanations of perceived events produce a state of disequilibrium, and this acts as a motivation to reorganize thinking on a higher cognitive level.

equipotent See totipotent.

Equisetales The horsetails. Order of ca 20 species of the genus *Equisetum*, cosmopolitan except for Australia and New Zealand. Also fossils from the Upper Carboniferous onwards. The sporophyte has roots and rhizomes, and an aerial stem which bears whorls of very small fused leaves and of branches. The stems are photosynthetic. The gametophytes are thalloid and photosynthetic.

equivalve Said of bivalves which have the two halves of the shell of equal size.

ER Abbrev. for *Endosplasmic Reticulum*.

erect Set at right angles to the part from which it grows.

erection The *turgid* condition of certain animal tissues when distended with blood; an upright or raised condition of an organ or part. adj. *erect*.

ergastic substances Non-protoplasmic substances; storage and waste products; starch, oil, crystals, tannins.

ergate A sterile female ant or worker.

ergatogyne An apterous queen ant.

ergatoid Resembling a worker; said of sexually perfect but wingless adults of certain social Insects.

ergonomics The application of various human studies to the area of work and leisure; includes anatomy, physiology and psychology.

ergotism A condition characterized by extreme vasoconstriction leading to gangrene and convulsions; due to eating the grains of cereals which are infected by the ergot fungus *Claviceps purpurea*. Formerly known as *St. Anthony's Fire*.

Ericaceae Family of ca 3000 spp. of dicotyledonous flowering plants (superorder Dilliniidae). Mostly shrubs, mostly calcifuge, cosmopolitan. Includes the heaths (*Erica* spp.) and heather (*Calluna*) and a number of ornamental plants, e.g. Rhododendron.

ericaceous Heatherlike.

ericeticolous Growing on a heath.

ericoid Having very small tough leaves like those of heather.

eriophorous Having a thick cottony covering of hairs.

erogenous zones Sensitive areas of the body, the stimulation of which arouses sexual feelings and responses. They function as substitutes for the genital organs and are associated with stages of development in childhood.

eros In psychoanalytic theory, the constructive life instinct, the urge for survival and procreation.

erumpent Developing at first beneath the surface of the substratum, then bursting out through the substratum and spreading somewhat.

erythema A superficial redness of the skin, due to dilation of the capillaries.

erythroblast A nucleated mesodermal embryonic cell, the cytoplasm of which contains haemoglobin, and which will later give rise to an erthrocyte. See **megaloblast**.

erythroblastosis foetalis *Haemolytic disease of the newborn*. Disease of the human foetus due to immunization of the mother. Escape of foetal red cells into the maternal circulation during pregnancy (or during a previous pregnancy) elicits antibodies in the mother. If these are IgG, and can cross the placental membranes into the foetal circulation, they cause haemolysis of the foetal red cells. This may be sufficient to cause stillbirth or anaemia and severe jaundice with brain damage. *Rhesus antigens* are the commonest cause. See **rhesus blood group system**.

erythrocyte One of the red blood corpuscles of Vertebrates; flattened oval or circular disklike, cells (lacking a nucleus in Mammals), whose purpose is to carry oxygen in combination with the pigment haemoglobin in them, and to remove carbon dioxide.

erythrophore A chromatophore containing a reddish pigment.

erythropoiesis The formation of red blood cells.

erythropoietin A substance produced by the juxtaglomerular cells of the kidneys which stimulates the production of red blood cells by the bone marrow.

erythropterin A red heterocyclic compound deposited as a pigment in the epidermal cells or the cavities of the scales and setae of many Insects.

escape A plant growing outside a garden and derived from cultivated specimens, surviving but not well naturalized.

escape behaviour Refers to defensive behaviour against a predator, often involves specialized evasive manoeuvres or structures, e.g. a special exit burrow.

escape conditioning A procedure in which escape behaviour regularly terminates a negative reinforcement or aversive stimulus.

esophagus See **oesophagus**.

essential element An element necessary for the growth and reproduction of a plant which cannot be replaced by another element. See **macronutrient, micronutrient, deficiency disease**.

essential oils Volatile **secondary metabolites** formed mainly in oil glands, rarely in ducts; mostly terpenoids also aliphatic and aromatic esters, phenolics and substituted benzene hydrocarbons, responsible for the odours of many aromatic plants (and steam distilled as perfumes from some). Some appear to deter insects or herbivores, others to be **allelopathic**.

essential organs The stamens and carpels of a flower.

esterase An enzyme which catalyses the hydrolysis of ester bonds.

ethephon *2-chloroethylphosphoric acid*, which breaks down rapidly in water to yield *ethylene* and is used to promote controlled ripening of fruit.

ethidium bromide 3,8-diamino-5-ethyl-6-phenylphenanthridinium bromide. A fluorescent reagent which binds to double stranded DNA and RNA and is used for their detection after fractionation in a gel matrix. A mutagen.

Ethiopian region One of the primary faunal regions into which the surface of the globe is divided; it includes all of Africa and Arabia south of the tropic of Cancer.

ethmo- Prefix from Gk. *ethmos*, sieve.

ethmohyostylic In some Vertebrates, having the lower jaw suspended from the ethmoid region and the hyoid bar.

ethmoidalia A set of cartilage bones (*ethmoids*) forming the anterior part of the brain case in the Vertebrate skull.

ethmoturbinal In Mammals, a paired bone or cartilage of the nose, on which are supported the folds of the olfactory mucous membrane.

ethogram Refers to attempts to draw up a complete behavioural inventory for a particular species; the term has its origin in early instinct theories of animal behaviour and is rare in modern usage, because of the methodological complexities involved in categorizing units of behaviour.

ethology Describes an approach to the study of animal behaviour in which attempts to explain behaviour combine questions about its immediate causation, development, function and evolution.

ethylene The gas *ethene*; a plant **growth substance** produced esp. in wounded, diseased, ripening and senescent tissues, interacting with auxin and promoting e.g. fruit ripening, leaf abscission and epinasty.

Ethylene and substances that release it (see *ethephon*) are used commercially to regulate the ripening of fruit, esp. stored fruit.

etioplast The form of **plastid** that develops in plants grown in the dark; colourless and containing a *prolamellar body*, rapidly converted to a chloroplast on illumination.

eu- Prefix from Gk. *eu*, well, good.

Eubacteriales One of the two main orders of the true bacteria, distinguished from the *Pseudomonadales* by the peritrichous flagella of the motile members. Spherical or rod-shaped cells, having no photosynthetic pigments; not acid fast; readily stained by aniline dyes.

euchromatin Dispersed **chromatin** visible by microscopy of eukaryotic cell nuclei.

eugamic Pertaining to the period of maturity.

eugenics Study of the means whereby the characteristics of human populations might be improved by the application of genetics.

euglenoid movement A type of movement undergone by many Protozoa, which possess a definite body form, by means of contractions of the protoplasm stretching the pellicle. Also called *metaboly*.

Euglenophyceae Small group of eukaryotic algae. Chloroplasts with chlorophyll a and b, thylakoids in threes, chloroplast envelope of 3 membranes. Storage polysaccharide paramylon (β, $1 \rightarrow 3$ glucan). Flagellated, usually 2 flagella, sometimes palmelloid. Phototrophs (often auxotrophic) and heterotrophs (both osmotrophic and phagotrophic). Fresh water and marine. See **pellicle**.

eukaryote A higher organism; literally those which have 'good nuclei' in their cells, i.e. animals, plants and fungi in contrast to the prokaryotic cells of bacteria and blue-green algae. Eukaryotes possess a nucleus bounded by a membranous nuclear envelope and have many cytoplasmic organelles. Their nuclear DNA is complexed with histones to form chromosomes. Cf. **prokaryote**. adj. *eukaryotic*.

eumerism An aggregation of similar parts.

Eumycota Division containing the true fungi. Eukaryotic heterotrophic, walled organisms, typically with hyphae, or single cells (e.g. yeasts) or chains of cells. The walls usually contain chitin (chitosan in Zygomycotina) as a major constituent. Comprises the subdivisions Mastigomycotina, Zygomycotina, Ascomycotina and Basidiomycotina.

Euphausiacea An order of shrimp-like Malacostraca which are filter feeding and pelagic. They are a major source of food for some Whales. Krill.

Euphorbiaceae Family of ca 7000 spp. of dicotyledonous flowering plants (superorder Rosidae). Trees, shrubs and herbs, mostly tropical, with unisexual flowers, and usually with latex. Includes *Hevea* the source of most natural rubber, manioc (*Manihot utilissima*, cassava or tapioca) and the very large genus *Euphorbia*, the spurges (1600 spp.), the members of which include C_3 plants, C_4 plants and leafless, stem-succulent CAM plants. See **cyathium**.

euphotic zone See **photic zone**.

euploid, euploidy Having a chromosome complement consisting of one or more whole *chromosome sets* and, therefore, haploid, diploid or polyploid. Cf. **aneuploid**.

eupyrene Spermatozoa which are normal, typical. Cf. **apyrene**, **oligopyrene**.

eusporangium A sporangium of a vascular plant in which the wall develops from superficial cells, and the sporogenous tissue from internal cells of the sporophyll or sporangiophore. The wall is usually more than one cell thick at maturity. Cf. **leptosporangiate**.

Eustachian tube In land Vertebrates, a slender duct connecting the tympanic cavity with the pharynx.

Eustachian valve In Mammals, a rudimentary valve separating the openings of the superior venae cavae from that of the inferior vena cava.

eustele A stele in which the primary vascular tissue is organized into discrete vascular bundles surrounding a pith; typical of dicotyledons and gymnosperms.

eustomatous With a well-defined mouth or opening.

Eutheria An infraclass of viviparous Mammals in which the young are born in an advanced stage of development; no marsupial pouch; an allantoic placenta occurs; the scrotal sac is behind the penis, the angle of the lower jaw is not inflexed and the palate is imperforate. The higher Mammals.

eutrophic Said of lakes, mires or soils which are relatively rich in available plant mineral nutrients.

evaginate Not having a sheath.

evagination Withdrawal from a sheath; the development of an outgrowth; eversion of a hollow ingrowth; an outgrowth; an everted hollow ingrowth. Cf. **invagination**.

evapotranspiration The total water loss from a particular area, being the sum of evaporation from the soil and transpiration from vegetation.

evocation *Flora evocation*. The initial event at the root apex in response to the arrival of the floral stimulus which commits the apex to the subsequent formation of flower primordia.

evolute Having the margins rolled outwards.

evolution Changes in the genetic composition of a population during successive generations. The gradual development of more complex organisms from simpler ones.

evolutionarily stable strategy A strategy such that, if most members of a population adopt it, there is no *mutant* strategy that would give higher reproductive fitness.

exalbuminous Said of a mature seed, lacking endosperm; non-endospermic.

exarch Of a xylem strand, having the protoxylem of the side towards the outside of the axis (the normal condition in e.g. angiosperm roots.)

excision repair Enzymatic DNA repair process in which a mismatching DNA sequence is removed and the gap filled by synthesis of a new sequence complementary to the remaining strand.

excitable tissue That which responds to stimulation, e.g. muscle or nervous tissue.

excitation The contraction of muscle resulting from nervous stimulation. Cf. **inhibition**. adj. *excitatory*.

excoriation Superficial loss of skin.

excreta Poisonous or waste substances eliminated from a cell, tissue or organism. adjs. *excrete*, *excreted*. n. *excretion*.

excurrent (1) Running out, as when the midrib of a leaf is prolonged into a point. (2) With a single main axis or trunk and subordinate laterals, as in trees, e.g. pine, esp. when young. (3) Carrying an outgoing current; said of ducts, and, in certain *Porifera*, of canals leading from the apopyles of the flagellated chambers to the exterior or to the paragaster.

ex-, e- Prefixes from L. *ex*, *e*, out of.

exergonic Said of a biochemical reaction which is accompanied by the release of energy, and is therefore capable of proceeding spontaneously. Cf. **endergonic**.

exfoliation The process of falling away in flakes, layers or scales, as some bark in plants.

exhalant Emitting or carrying outwards a gas or fluid; as the *exhalant siphon* in some Mollusca.

exhibitionism Describes behaviour in which sexual gratification is obtained through displaying the genitals to members of the opposite sex; by extension all behaviour motivated by the pleasure of self display.

exine The outer part of the wall of a pollen grain or embryophyte spore, from the patterns of the surface of which it is often possible to identify the genus or even species of plant from which the pollen has come. Cf. **intine**. See **sporopollenin**, **pollen analysis**.

exinguinal In land Vertebrates, outside the groin.

exit portal The area through which a beam of radiation leaves the body. The centre of the exit portal is sometimes called the *emergent ray point*.

exobiology The study of putative living systems which probably must exist elsewhere in the universe.

exocardiac Outside the heart.

exocarp See **epicarp**.

exoccipital A paired lateral cartilage bone of the Vertebrate skull, forming the side wall of the brain case posteriorly.

exocoelar See **somatopleural**.

exocoelom The extraembryonic coelom of a developing Bird, Reptile or Elasmobranch.

exocrine Said of glands, the secretion of which is poured into some cavity of the body, or on to the external surface of the body by ducts. Cf. **endocrine**.

exocuticle The layer of the insect cuticle, between the epicuticle and endocuticle, which becomes hardened and darkened in most Insects.

exocytosis The exit of material from the cell by fusion of internal vesicles with the **plasma membrane** which either voids their contents to the exterior or introduces new surface material into the plasma membrane.

exodermis The outermost layer or layers of the cortex of some roots with more or less thickened and/or suberized cell walls; a specialized hypodermis.

exogamete A gamete which unites with one from another parent.

exogamy The production of a zygote by the fusion of gametes from 2 unrelated parents. Cf. **endogamy**, **outbreeding**, **allogamy**.

exogenous (1) Resulting from causes external to an organism. (2) Produced on the outside of another organ or developed from tissues at or near the surface (as leaf primordia are). (3) Growing by the addition of new layers at or near the surface. (4) In higher animals, said of metabolism which leads to the production of energy for activity. Cf. **endogenous**.

exon See panel.

exonuclease An enzyme which cleaves nucleic acids from a free end. It can thus digest a linear molecule by steps. Cf. **endonuclease**.

exopodite The outer ramus of a crustacean appendage.

Exopterygota A subclass of Insects in which wings occur, although sometimes secondarily lost; the change from young form to adult is gradual, the wings developing externally; the young form is usually a nymph.

exoscopic embryology In plants, the condition when the apex of the embryo is turned towards the neck of the archegonium.

exoskeleton Hard supporting or protective structures that are external to and secreted by the ectoderm, e.g. in Vertebrates, scales, scutes, nails and feathers, in Invertebrates, the carapace, sclerites etc.

exospore (1) The outermost layer of the

exon

That part of the transcribed nuclear RNA of eukaryotes which forms the mRNA after the excision of the **introns**. In eukaryotes many genes, both those for polypeptide products and for ribosomal and transfer RNA, are split into an alternating series of regions all of which are transcribed into a high molecular weight RNA. During processing, some of these regions, called intervening sequences or *introns* and which may form the largest proportion of the molecule, are excised and the remaining regions (the *exons*) join together to make the messenger, ribosomal or transfer RNA.

The presence of introns helps to explain the apparent excess of DNA over that required for the putative number of genes in eukaryotes and there is some evidence that the different exons specify different domains in the protein structure. The presence of introns might therefore facilitate re-assortment of domains during evolution.

Diagram relating the primary gene product to the mature messenger RNA in eukaryotic cells.

A = high molecular weight nuclear RNA, B = the RNA just before the excision of the introns, C = the mature message.

wall of a spore; **exine**. (2) A spore formed by the extrusion of material from the parent cell.

exotic Not native. *ecdemic*.

exotoxin A toxin released by a bacterium into the medium in which it grows. Frequently very toxic, e.g. neurotoxins which destroy cells of the nervous system, haemolytic toxins which lyse red blood cells.

expectancy, principle of Refers to a cognitive theory of animal and human learning which posits that the anticipation of events, and esp. of rewards, is an important aspect of many learning events.

expectation The average value resulting from an infinite series of repetitions of an experiment or observations of a random variable.

experimental allergic encephalomyelitis An auto immune disease produced by injections into mice of proteins present in brain and spinal cord together with complete Freund's adjuvant. After a few days acute encephalomyelitis develops, accompanied by demyelination and progressive paralysis. The main factor is T cell sensitization against myelin basic protein but antibodies may also be involved. A similar condition has been described in humans following immunization against rabies using a crude brain-derived vaccine which is now obsolete.

experimental embryology The experimental study of the physiology and mechanics of development.

expiration The expulsion of air or water from the respiratory organs.

explant A piece of tissue or an organ removed for experimental purposes, from a plant esp. to start a **plant cell culture**.

explantation In experimental zoology, the culture, in an artificial medium, of a part or organ removed from a living individual, tissue culture. n. *explant*.

exploratory behaviour A form of **appetitive behaviour**, by which an animal gains information about its environment; some forms of exploratory behaviour are goal linked; others seem motivated by a general curiosity mingled with mild fear and is not terminated by any apparent end goal.

exponential growth A stage of growth occurring in populations of unicellular microorganisms when the logarithm of the cell number increases linearly with time.

export The process of transferring proteins across a membrane either into the medium or into another cellular compartment.

exposure dose See **dose**.

exposure learning Changes in behaviour that result from an individual being exposed to an object or situation, under circumstances in which no consistent response apart from investigatory or exploratory behaviour is elicited by that situation, and with no obvious reward. See **observational learning**.

expression vector A **vector** in genetic manipulation work, which is specially constructed so that a large amount of the protein product, coded by an inserted sequence, can be made.

exserted Protruding.

extensin The major structural protein of the primary cell walls of higher plants, rich in hydroxyproline and highly insoluble owing to dimerization of tyrosine residues.

extensor A muscle which by its contraction straightens a limb or a part of the body. Cf. **flexor**.

external digestion A method of feeding, adopted by some Coelenterata, Turbellaria, Oligochaeta, Insects and Araneida, in which digestive juices are poured on to food outside the body and imbibed when they have dissolved some or all of the food.

external respiration See **respiration, external**.

external secretion A secretion which is discharged to the exterior, or to some cavity of the body communicating with the exterior. Cf. **internal secretion**.

exteroceptor A sensory nerve ending, receiving impressions from outside the body. Cf. *interoceptor*.

extinction In **classical conditioning**, the weakening of the tendency of a conditioned stimulus to elicit a conditioned response by unreinforced presentations of the conditioned stimulus. In **operant conditioning**, a decline in the tendency to perform the operant response, as a result of the unreinforced occurrences of the response.

extra- Prefix from L. *extra*, beyond, outside of, outwith.

extracellular Located or taking place outside a cell. Cf. **intracellular**, **intercellular**.

extracellular enzyme One secreted out of the cell into the intercellular space or e.g. the lumen of the gut.

extracellular matrix A non-cellular matrix of proteins and glycoproteins surrounding cells in some tissues. It can be extensive as in cartilage and connective tissue and calcified as in bone.

extra-chromosomal DNA (1) Non-integrated viral DNA and other **episomes**. (2) DNA of cytoplasmic organelles, i.e. mitochondrial and chloroplastal DNA.

extra-chromosomal inheritance See cytoplasmic inheritance.

extra-embryonic In embryos developed from eggs containing a great deal of yolk, as those of Birds, pertaining to that part of the germinal area beyond the limits of the embryo.

extra-floral nectary A nectary occurring on or in some part of a plant other than a flower. Also *extra-nuptial nectary*.

extra-nuptial nectary See **extra-floral nectary**.

extrasensory perception Perception of phenomena without the use of the ordinary senses.

extravasation The abnormal escape of fluids, as blood or lymph, from the vessels which contain them. v. *extravasate*.

extraversion/introversion In Jung's theory, a characterization of personality according to a tendency to focus on the external and objective aspects of experience or on internal and subjective ideas. Eysenck uses extraversion-introversion as a trait dimension along which individuals can be placed and uses the popular notion of extravert as *outgoing* and introvert as *withdrawn*.

extrinsic Said of appendicular muscles of Vertebrates which run from the trunk to the girdle or the base of the limb. Cf. **intrinsic**.

extrorse Directed or bent outwards; facing away from the axis; esp. of stamens opening towards the outside of the flower. Cf. **introrse**.

extrovert An extrusible proboscis, found in certain aquatic animals. See **lophophore**.

exudation pressure See **root pressure**.

exumbrella The upper convex surface of a medusa. adj. *exumbrellar*.

exuviae The layers of the integument cast off in ecdysis.

exuvial Pertaining to, or facilitating, ecdysis.

eye The sense organ which receives visual impressions.

eyepiece graticule Grid incorporated in the eyepiece for measuring objects under the microscope. Special type used in particle-size analysis consists of a rectangular grid for selecting the particles and a series of graded circles for use in sizing the particles. Also *micrometer eyepiece*, *ocular micrometer* (US).

eye spot *Stigma*. Usually orange or red spot composed of droplets containing carotenoids. Found in motile cells of many algae, phytoflagellates and lower animals, and presumed to act as a screen on one side of a photoreceptive area in the detection of the direction of illumination in phototaxis.

eye stalk A paired stalk arising close to the median line on the dorsal surface of the head of many Crustacea, bearing an eye.

F

F Symbol for *filial generation* in work on inheritance; usually distinguished by a subscript either F_1 or F1, first filial generation; F_2 or F2, second filial generation. Cf. **P**.

F1 hybrid Crop or strain variety, characterized by unusual vigour and uniformity, produced by crossing two selected inbred lines. Cf. **heterosis, cytoplasmic male sterility, nick**.

Fabaceae See **Leguminosae**.

Fab fragment Fragment of immunoglobulin molecule obtained by hydrolysis with papain. It consists of one light chain linked to the N-terminal part of the adjacent heavy chain. Two Fab fragments are obtained from each molecule. Each contains one antigen-binding site, but none of the heavy chain Fc.

F(ab′)₂ fragment Fragment of immunoglobulin obtained by pepsin digestion. It contains both Fab fragments plus a short section of the hinge region of the heavy chain Fc. It behaves as a bivalent antibody but lacks properties associated with the Fc fragment (e.g. complement activation, placental transmission).

facet One of the corneal elements of a compound eye; a small articulatory surface.

facial Pertaining to or situated on the face; the seventh cranial nerve of Vertebrates, supplying the facial muscles and tongue of higher forms, the neuromast organs of the head and snout in lower forms, and the palate in both.

facilitated diffusion The rapid permeation of solutes into cells by interaction of the solutes with specific carriers which facilitate their entry. Facilitated diffusion is distinguished from **active transport** because it does not allow the entry of solutes against their concentration gradients.

facilitation The augmented response of a nerve due to prestimulation; the activation of physiological and behavioural response resulting from nonspecific stimulation from a conspecific. See **social facilitation**.

FACS See **fluorescence activated cell sorter**.

factor Obs. meaning *gene*.

factor analysis A statistical method for studying the interrelations between various test elements or between different behaviours, in order to discover whether certain correlations allow an explanation in terms of one or more common factors.

factor B, factor D Components involved in the alternative pathway of complement activation.

factor H, factor I A glycoprotein and an enzyme respectively which together inactivate C3b (activated complement C3) and act as a damper on the activated complement.

facultative Able but not obliged to function in the way specified; a facultative *anaerobe* can grow in and perhaps use free oxygen, but will also survive and perhaps grow in its absence. Cf. **obligate**.

facultative heterochromatin Chromatin condensed in some cell types but not in others and which is not expressed when condensed despite containing **coding sequences**. Cf. **constitutive heterochromatin**.

facultative parasite A parasite able to live saprophytically and be cultured on laboratory media. Cf. **obligate parasite**. See **necrotroph**.

faeces The indigestible residues remaining in the alimentary canal after digestion and absorption of food.

Fagaceae Family of ca 100 spp. of dicotyledonous flowering plants (superorder Hamamelidae). Mostly trees, often dominant in broad-leaved forests of esp. the N. hemisphere. Flowers typically in catkins. Includes oak, beech, chestnut and southern beech.

fairy ring A ring, usually in grass, in which the plants near the periphery are green and healthy and those near the centre less so. It will persist and expand for many years and is associated with the mycelium of the fungus which forms fruiting bodies at the ring's periphery.

falcate, falciform Sickle-shaped.

falciform ligament In higher Vertebrates, a peritoneal fold attaching the liver to the diaphragm.

falcula A sharp curved claw. adj. *falculate*.

Fallopian tube In Mammals, the anterior portion of the Müllerian duct; the oviduct.

fallout Particulate matter in the atmosphere, which is transported by natural turbulence, but which will eventually reach the ground by sedimentation or dry or wet deposition. Applied esp. to material from nuclear explosions, when the fall out is radioactive.

false amnion See **chorion**.

false annual ring A second ring of xylem formed in one season, following the defoliation of the tree by the attacks of insects or other accident; oaks are liable to this, as they may be completely stripped of leaves by the oak tortrix.

false fruit A fruit formed from other parts of the flower in addition to the carpels, e.g. the strawberry where the receptacle becomes fleshy.

false pregnancy See **pseudocyesis**.

false ribs *Floating ribs*. In higher Vertebrates, ribs which do not reach the sternum.

false septum A replum.

false tissue See **pseudoparenchyma**.

---

fatty acids

A term for the whole group of saturated and unsaturated monobasic aliphatic carboxylic acids. The lower members of the series are liquids of pungent odour and corrosive action, soluble in water: the intermediate members are oily liquids of unpleasant smell, slightly soluble in water. The higher members from C_{10} upwards, are mainly solids, insoluble in water, but soluble in ethanol and in ethoxyethane.

Medically, saturated fatty acids with no double bonds are linked with the development of **atheroma**. In contrast the polyunsaturated fatty acids, linoleic, linolenic and arachidonic are termed essential fatty acids as they must be included in the diet and may have a preventative role against atheroma and are required for the synthesis of prostaglandins.

---

falx Any sickle-shaped structure. adjs. *falciform, falcate*.

falx cerebri A strong fold of the *dura mater*, lying in the longitudinal fissure between the two cerebral hemispheres.

family A group of similar genera of taxonomic rank below **order** and above **genus**; with plants, the names usually end in -aceae.

family therapy Psychotherapy which regards the family as a unit and as the object of therapy, rather than its individual members, so that roles and attitudes within the family can be explored and changed.

Fanconi's anaemia Human clinical syndrome in which spontaneous rearrangements preferentially involving non-homologous chromosomes occur at an increased rate.

fang The grooved or perforate poison tooth of a venomous serpent; one of the cuspidate teeth of carnivorous animals, esp. the canine or carnassial.

fantasy, phantasy *General*, sequences of private mental images, sometimes in anticipation of possible events, but sometimes irrational and referring to extremely unlikely possibilities. In *psychoanalytic theory*, an imaginary episode, operating on either a conscious or unconscious level, in which the subject is a central figure and which fulfils a conscious or unconscious wish.

farinose Covered with a mealy powder.

-farious Suffix meaning arranged in so many rows.

farmer's lung Respiratory disease due to hypersensitivity to spores of a thermophilic bacterium, *Micropolyspora faeni*, present in the dust of mouldy hay. Mainly occurs among farmers in damp areas. Characterized by attacks of breathlessness coming on some hours after inhaling the dust. A combination of Type III and Type IV hypersensitivity reactions are probably involved. After several seasons can result in pulmonary fibrosis.

far-red light In the context of plant responses mediated by *phytochrome*, light of wavelength around 730 nm.

fascia Any bandlike structure; esp. the connective-tissue bands which separate muscles, nerves etc. adj. *fascial*.

fasciation An abnormal condition, resulting from damage, infection or mutation, in which a shoot (or other organ) grows broad and flattened, resembling several shoots fused laterally.

fascicle A vascular *bundle*.

fascicular cambium The flat strand of cambium between xylem and phloem in a vascular bundle.

fasciculus A small bundle, as of muscle or nerve fibres.

fasciola A narrow band of colour; a delicate lamina in the Vertebrate brain.

Fasciola hepatica (Trematoda, Malacocotylea). The *liver fluke*. The adult lives in the liver of sheep, and the secondary host, entered by the *miracidium* larva, and in which the *redia* and *cercaria* larvae develop, is a water snail, usually of the genus *Limnaea*. Gastropoda, Pulmonata.

fastigiate Having the branches more or less erect and parallel, e.g. Lombardy poplar.

fat See **adipose tissue**.

fat body In Insects, a mesodermal tissue of fatty appearance, the cells of which contain reserves of fat and other materials and play an important part in the metabolism of the animal; in Amphibians, highly vascular masses of fatty tissue associated with the gonads.

fatigue The condition of an excitable cell or tissue which, as a result of activity, is less ready to respond to further stimulation until it has had time to recover.

fatty acids See panel.

fauna A collective term denoting the animals occurring in a particular region or period. pl. *faunas* or *faunae*. adj. *faunal*.

faunal region An area of the earth's surface characterized by the presence of certain species of animals.

faveolate, favose Resembling a honeycomb in appearance. *Favous*, said of a surface pitted like a honeycomb. *Favus*, a hexagonal pit or plate.

Fc fragment Fragment of immunoglobulin obtained by papain hydrolysis representing the C-terminal halves of the two heavy chains linked by disulphide bonds. It has no antibody activity but contains the sites involved in complement activation and placental transmission, and some of the Gm allotype markers.

Fc receptor Receptor present on the plasma membranes of cells which bind the Fc fragment of immunoglobulin. Neutrophils, mononuclear phagocytes, eosinophils, B lymphocytes have receptors for Fc of IgG. These may differ for different IgG subclasses. Mast cells, basophil leucocytes and eosinophils have receptors for Fc of IgE.

fear In animal behaviour, a motivational state aroused by specific stimuli and which normally gives rise to avoidance, defensive or escape behaviour.

feathers Epidermal outgrowths forming the body covering of Birds; distinguished from scales and hair, to which they are closely allied, by their complex structure, and by the possession of a vascular core which at first projects from the surface.

febrifuge Against fever; a remedy which reduces fever.

febrile Pertaining to, produced by, or affected with fever.

fecundity The number of young produced by a species or individual.

feedback inhibition The ability of a later product of a chain of biochemical reactions to lower the synthesis of an earlier metabolite, e.g. by binding with its **promoter**.

female An individual whose gonads produce ova and thus said of (1) the larger and less motile gamete, the egg; (2) plants in which gametophytes and their reproductive structures produce eggs but not male gametes; (3) sporophytes and their reproductive structures which produce the megaspores and hence seeds; (4) individual seed plants or flowers which have functional carpels but not functional stamens. Cf. **male, hermaphrodite**.

female pronucleus The nucleus remaining in the ovum after maturation.

femur The proximal region of the hind limb in land Vertebrates; the bone supporting that region; the third joint of the leg in Insects, Myriapoda and some Arachnida. adj. *femoral*.

fen Vegetation developed naturally on waterlogged land, forming a peat that is neutral or alkaline from the dead parts of tall grasses, sedges and herbs. See **carr**. Cf. **bog**.

fenestra An aperture in a bone or cartilage, or an opening between two or more bones.

fenestra metotica In a typical chondrocra-

nium, an opening behind the auditory capsule, through which pass the ninth and tenth cranial nerves and the internal jugular vein.

fenestra ovalis In Vertebrates in which the middle ear is developed, the upper of two openings in the cochlea. Cf. **fenestra rotunda**.

fenestra pro-otica In a typical chondrocranium, an opening in front of the auditory capsule, through which pass the fifth, sixth, and seventh cranial nerves.

fenestra rotunda In Vertebrates in which the middle ear is developed, the lower of two openings in the cochlea. Cf. **fenestra ovalis**.

fenestrate, fenestrated With window-like perforations; perforated or having translucent spots.

fenestra tympani See **fenestra rotunda**.

fenestra vestibuli See **fenestra ovalis**.

feral Of a domesticated animal which has reverted to the wild.

fermentation See panel.

ferns Pteridophytes of the class *Filicopsida*.

ferredoxin Non-haem iron-sulphur protein, component of the electron transport system in photosynthesis.

ferritin Protein which functions as an iron store in the liver. As the central iron core is visible in the electron microscope ferritin can be used as a tag for the localization of proteins in electron microscopy.

fertile Able to produce asexual spores and/or sexual gametes.

fertile flower (1) A flower with functional carpels and/or stamens. (2) Sometimes a female flower.

fertilisin A substance which is present in the cortex of an ovum which increases sperm motility.

fertility The number or percentage of eggs produced by a species or individual which develop into living young. Cf. **fecundity**.

fertilization The union of 2 sexually differentiated gametes to form a zygote.

fertilization cone A conical projection of protoplasm arising from the surface of an ovum containing many microtubules which are thought to facilitate the entry of the sperm.

fertilization tube See **conjugation tube**.

fetishism Sexual gratification in which an object, or some part of the body, is the main source of sexual arousal, to the exclusion of the person as a whole.

fever The complex reaction of the body to infection, associated with a rise in temperature. Less accurately, a body temperature above the normal.

fibre An elongated sclerenchyma cell, typically tapering at both ends, with a thick secondary wall containing steeply helical or longitudinal cellulose microfibrils, lignified or not and with or without a living protoplast

---

fermentation

During **respiration** of 6-carbon glucose the molecule is cleaved to 3-carbon pyruvic acid by glycolysis with the net production of 2 molecules of ATP by **substrate level phosphorylation** and the reduction of one molecule of **NAD**. Subsequent metabolism can follow several routes but in all cases the NAD must be regenerated by the reoxidation of $NADH^+$ before glycolysis can continue. In aerobic respiration the pyruvate is fully degraded to carbon dioxide and water by the **tricarboxylic acid cycle** and the $NADH^+$ oxidized aerobically. See **electron transfer chain**. In the absence of oxygen a less efficient means must be employed to regenerate NAD at the expense of the reduction of some other metabolite. Such a balanced reduction and oxidation of the **coenzyme** by pairs of metabolites is known to biochemists as fermentation. Several substrates may be used in different circumstances to reoxidize $NADH^+$, e.g. in *alcoholic fermentation* oxidation of the coenzyme is achieved by the reduction of acetaldehyde to produce ethanol, in *lactate fermentation* some of the pyruvate itself is reduced to lactate. Many other end products, including fatty acids and longer chain aliphatic alcohols can be produced by bacterial fermentations.

Fermentation is also used as a term for an industrial process in which large numbers of microorganisms or isolated cells from higher organisms are grown under controlled conditions, both aerobic and anaerobic, in a large fermentor to obtain biochemicals synthesized by the cells.

---

at maturity. Bundles of such fibres constitute some economically important textile fibres, e.g. flax, jute, hemp, sisal.

fibre tracheid A cell of the secondary xylem intermediate between a **libriform fibre** and a tracheid.

fibril, fibrilla Any minute threadlike structure, such as the longitudinal contractile elements of a muscle fibre. adjs. *fibrillar, fibrillate*.

fibrin Protein which forms the fibrin blood clot. The monomer spontaneously associates into long insoluble fibres which are subsequently stabilized by covalent bonds between the polypeptide side chains.

fibrinogen The soluble precursor of **fibrin**. It is converted to fibrin under the influence of thrombin by proteolytic cléavage of terminal peptides.

fibroblasts Flattened connective-tissue cells of irregular form, believed to be responsible for the secretion of the white fibres; lamellar cells, fibrocytes.

fibrocartilage A form of cartilage which has white or yellow fibres embedded in the matrix.

fibrosis The formation of fibrous tissue as a result of injury or inflammation of a part, or of interference with its blood supply.

fibrous layer In the wall of the anther in angiosperms, a layer of cells below the epidermis with uneven cell walls probably responsible, because of the way they shrink on drying, for the opening of the mature anther. Also *endothecium*.

fibrous root system Root system composed of many roots of roughly equal thickness and length, as in grasses. Cf. **taproot system**.

fibrous tissue A form of connective tissue consisting mainly of bundles of white fibres; any tissue containing a large number of fibres.

fibrovascular bundle A vascular bundle accompanied, usually on its outer side, by a strand of sclerenchyma.

fibula In Tetrapoda, the posterior of the two bones in the middle division of the hind limb.

fibulare A bone of the proximal row of the tarsus, in line with the fibula, in Tetrapoda.

ficoll hypaque A proprietary mixture of a large polysaccharide and a dense synthetic organic molecule used in radiography, but the density can also be adjusted to allow separation of different cells by centrifugation. Frequently used for separating mononuclear cells and granulocytes from blood.

fidelity The degree of restriction of a species to a particular situation, a species with high fidelity having a strong preference for a particular type of community.

field capacity Amount of water held in a given soil by capillarity against drainage by gravity to a water potential or suction of about -0.05 bars.

field resistance Resistance shown by a plant to natural infection in the field which is more dependent on the environment and the nature of pathogen and vector than is inoculation in laboratory or greenhouse.

filament Any fine threadlike structure. (1) Generally a chain (unbranched or branched) of cells joined end on end. (2) A fungal hypha. (3) The stalk of a stamen in angiosperms. (4) The axis of a down feather.

filar micrometer Modified eyepiece graticule with movable scale or movable cross hair.

Filicales *Polypodiales*. The leptosporangiate ferns. Order of ca 9000 extant species of Filicopsida. (Also fossils from the Carboniferous onwards.) Sporophyte may have long, creeping, horizontal rhizomes with fronds at intervals, short more or less upright stems with a rosette of fronds, or grow into *tree ferns*. Leaves are circinate in bud and usually pinnately compound. Many are poisonous, some carcinogenic; the expanding fronds (croziers) of a few spp. are eaten; some are cultivated as ornamentals.

Filicopsida *Pteropsida, Polypodiopsida*. The ferns, a class of Pteridophytes from the Devonian onwards. The sporophyte usually has roots, a rhizome or stem and spirally arranged often pinnately-compound leaves (fronds, *megaphylls*) with circinate venation. (The paleozoic sorts show little distinction of stem and leaf.) Sporangia typically borne marginally or abaxially on leaves, mostly homosporous. Spermatozoids multiflagellate. Includes Marattiales (eusporangiate), Filicales and Salviniales (floating, heterosporous aquatica).

filiform Threadlike, e.g. *filiform antenna*.

film badge Small photographic film used as radiation monitor and dosimeter. Normally worn on lapel, wrist or finger and sometimes partly covered by cadmium and tin screens so that exposure to neutrons, and to beta and gamma rays, can be estimated separately.

filoplumes Delicate and hairlike contour feathers with a long axis and few barbs, devoid of locking apparatus at the distal end.

filopodium A long filamentous spike containing actin filaments found at the surface of some cells.

filter feeders In benthic and planktonic communitites, detritus feeders which remove particles from the water. In benthic communities they usually predominate on sandy bottoms. Cf. **deposit feeders**.

fimbria Any fringing or fringe-like structure; the delicate processes fringing the internal opening of the oviduct in Mammals; the ridge of fibres running along the anterior edge of the hippocampus in Mammals; the processes fringing the openings of the siphons in Molluscs.

fimbriate, fimbriated Having a fringed margin.

fimicolous Growing on or in dung.

fin In Fish, some Cephalopoda, and other

aquatic forms, a muscular fold of integument used for locomotion or balancing; supported in the case of Fish by internal skeletal elements.

fine grained Esp. of a small-scale pattern of environmental or resource heterogeneity. Often used in relation to the foraging activities of animals.

fin rays In Fish, the distal skeletal elements which support the fins; in Cephalochorda, the rods of connective tissue supporting the dorsal and ventral fins.

firefly Beetles, family Lampyridae, with light-producing organs whose main function is in attracting a mate.

first ventricle In Vertebrates, the cavity of the left lobe of the cerebrum.

fission Asexual reproduction of some unicellular organisms in which the cell divides into two more or less equal parts as in the fission yeast *Schizosaccharomyces*. Also *binary fission*. Cf. **budding**.

fissiped Having free digits. Cf. **pinnatiped**.

FITC Abbrev. for *Fluorescein IsoThioCyanate* used for making antibodies fluorescent for use in fluorescent antibody techniques.

fitness (1) In natural history, the degree to which an organism is adapted to its environment and can therefore survive the struggle for existence. (2) In evolutionary ecology, the extent to which an individual passes on its genes to the next generation. *Darwinian fitness* is the number of offspring of an individual, or the number relative to the mean (*relative fitness*), that live to reproduce in the next generation; effectively, the number of genes passed on. *Inclusive fitness* is the same as *Darwinian fitness*, but including those genes that the individual shares with its relatives and are passed on by them.

fixation See panel.

fixed action pattern Abbrev. *FAP*. Species specific movements, recognized by their relatively high degree of stereotype, although they may show variation in their orientation aspects. Originally all FAPs were assumed to be innate, but it is now recognized that such patterns are often influenced by environmental factors during development.

fixed interval schedule See **interval schedule of reinforcement**.

fixed ratio schedule See **ratio schedule of reinforcement**.

flabellate, flabelliform Shaped like a fan.

flaccid A cell or cells in a tissue which are not *turgid* but limp. Turgor pressure is zero as a result of water loss. See **plasmolysis**.

flagellar root A group of microtubular structures lying under the plasmalemma close to the basal body in flagellated cells.

Flagellata See **Mastigophora**.

flagellate (1) Having flagella. (2) An organ-

fixation

In *psychoanalytic theory,* a defence mechanism caused by acute anxiety and frustration during development; the individual is temporarily or permanently halted at a particular psychosexual stage of growth and this is reflected in an unevenness of personality development (e.g. fixation at the oral stage may result in a compulsive eating or talking disorder).

In *biology* generally, see **carbon fixation, nitrogen fixation, phosphate fixation.**

In *microscopy*, preparative treatment, esp. before embedding and sectioning, aimed to stabilize (or fix) structures against subsequent treatments; e.g. for electron microscopy, with glutaraldehyde which crosslinks proteins and osmium tetroxide which stabilises lipids, or, for light microsopy, with mixtures containing ethanol, acetic acid, chromium trioxide, mercuric ions etc. which may precipitate proteins.

In *animals*, the action of certain muscles which prevent disturbance of the equilibrium or position of the body or limbs; the process of attachment of a free-swimming animal to a substratum, on the commencement of a temporary or permanent sessile existence.

---

ism in which the body form is a unicell or colony of cells with flagella. (3) Bearing a long thread-like appendage. (4) A member of the Mastigophora.

flagellin The protein monomer which is assembled into a helical array to form the filament of a bacterial flagellum.

flagellum Long hair-like extension from the cell surface whose movement is used for locomotion. In sperm, protozoa and eukaryote algae its internal structure is very similar to that of cilia, but bacterial flagella have a unique structure and a rotatory action. More specifically (1) in Arthropoda, a filiform extension to an appendage, e.g. a crustacean limb, an insect antenna. (2) In Gastropoda, one of the male genitalia.

flag leaf The top-most leaf on a culm, just below the ear, in such cereal grasses as wheat and barley.

flame cell See **solenocyte**.

flashbulb memory Clear recollections that people sometimes have of the events surrounding a dramatic event, e.g. the assassination of a head of state.

flash colours See **startle colours**.

flavescent Becoming yellow; yellowish.

flavones Yellow pigments occurring widely in plants, derivatives of *flavone*.

flavonoids A large group of **secondary metabolites** of bryophytes and vascular plants, based on 2-phenylbenzopyran, often as glycosides. Some sorts are pigments and others may be *phytoalexins*, insecticides or of no known function. Many are significant in chemotaxonomy.

flavoproteins Proteins which serve as electron carriers in the electron transfer chain by virtue of their prosthetic group, *flavine adenine dinucleotide*.

flexor A muscle which, in contracting, bends a limb or a part of the body. Cf. **extensor**.

flexuose, flexuous Zig-zag, wavy.

floating ribs See **false ribs**.

flocculation Formation of floccules in a precipitin test or of agglutinated bacteria in an agglutination test for flagellar antigens of *Salmonella* species.

floccus In Birds, the downy covering of the young forms of certain species; in Mammals, the tuft of hair at the end of the tail; more generally, a tuft.

flock A group of birds that remains together because of social attraction between individuals, rather than, e.g. because of a shared interest in some environmental feature (an *aggregation*). See **schooling** and **herding**.

flooding A conditioning technique in which extinction of avoidance and fear responses is achieved by confronting the anxiety-producing stimulus, often using only imaginal exposure and verbal description by the therapist.

flora (1) The plants of a particular region habitat or epoch. (2) A catalogue or description of such plants.

floral diagram A conventional plan of the arrangement of the parts of a flower as seen in cross section.

floral envelope The calyx and corolla, or perianth.

floral formula A representation of the structure of a flower by means of the letters P (perianth) or K (calyx) and C (corolla), A (androecium) and G (gynoecium), of figures (numbers of parts), of parentheses (conna-

fluorescence activated cell sorter

Instrument in which cells or chromosomes in a suitable medium have their fluorescence or absorption measured as they pass down a fine tube. They are then ejected from an aperture which causes the stream to break up into fine droplets which pass between electrostatic deflector plates. Depending on the amount or nature of the fluorescence, the droplet containing the cell or chromosome is given an electrical charge which causes it to be diverted into an appropriate reservoir. The sorter can also be used to measure the amount of fluorescence and hence e.g. DNA content of the cells.

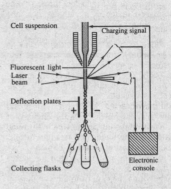

In the figure the laser beam energy is measured after passing through each cell, either directly and as emitted fluorescence . The levels of these signals can then be made to give an appropriate charge to the incoming cell suspension.

tion), of horizontal brackets (adnation) and of a line above or below the G (superior or inferior ovary), thus P 3+3 A 3+3 G*(3)* is the tulip.

floral leaf (1) A bract or bracteole. (2) A sepal or petal.

floral mechanism The arrangement of the flower parts to ensure either cross- or self-pollination.

flore pleno With *double flowers*. Abbrev. *fl pl.*

floret An individual flower in a crowded inflorescence. In grasses, typically consists of an ovary (with 2 styles) and 3 stamens enclosed by 2 bracts (lemma and palea).

florigen A hypothetical plant hormone inferred to be induced in leaves, to move to the shoot apices and there to cause the initiation of flowers.

flow cytometry Sorting cells or chromosomes, to which a fluorochrome has been stoichiometrically bound, on the basis of size. They are passed through a beam of light and directed into one of two flasks depending on how much light they emit or scatter.

flower The reproductive structure in angiosperms consisting of a shoot axis bearing, as lateral members traditionally interpreted as modified leaves, one or more of sepals and petals, or tepals, stamens and carpels.

flowering plants The angiosperms, comprising both the monocotyledons and the dicotyledons.

flow-sorted chromosomes Those sorted on the basis of size by **flow cytometry**.

fl. pl. Abbrev. for *flore pleno*.

fluid mosaic Model of cell membrane structure postulating a bimolecular leaflet of lipid, interspersed with various specific proteins which lie at the surfaces of, or in, the membrane, the lipid being fluid and the proteins able to move laterally.

fluke (1) A semipopular name for worms belonging to the group Trematoda. (2) The tail of a Cetacean.

fluorescein isothiocyanate Fluorochrome commonly conjugated with antibodies for use in **indirect immunofluorescence**.

fluorescence activated cell sorter See panel.

fluorescence microscopy Light microscopy in which the specimen is irradiated at wavelengths which will excite the natural or artificially introduced fluorochromes. An optical filter absorbs the exciting wavelengths but transmits the fluorescent image which can be studied normally.

fluorimeter One used for measuring the intensity of fluorescent radiation.

fluorography The photography of fluoroscopic images.

fluoroscope Measurement system for examining fluorescence optically. Fluorescent screen assembly used in fluoroscopy.

flush Area of land, fed by ground water which may be oligotrophic or eutropic and, in the latter, may enrich the soil locally.

fluviatile Of streams and rivers, *fluvial*.

fluviomarine Able to live in rivers and in the sea, as the Salmon.

fluvioterrestrial Found in rivers and on their banks, as the Otter.

flying-spot microscope A type of light microscope in which the object is scanned in two dimensions by a light spot formed by the diminished image of a cathode ray tube placed at the eyepiece plane of a compound microscope. All the transmitted energy can be collected by a photomultiplier and an image, suitable for electronic analysis, reconstructed using the timing circuits driving the CRT. The **scanning electron microscope** is an analogous instrument. See **confocal microscope.**

foetal membranes In Reptiles, Birds and Mammals, outgrowths from the embryo, or the extraembryonic tissue, which surround and protect the foetus and facilitate respiration. See **allantois, amnion, chorion**.

foetus A young mammal within the uterus of the mother, or in oviparous animals, the young within the egg, from the commencement of organogeny until birth. adj. *foetal*.

foldback DNA Sequence complementarity which allows a single-stranded molecule to form secondary structure. **Hairpin DNA** is one form with a minimal loop at its end.

foliaceous, foliose (1) Flat and leaflike. (2) Bearing leaves.

foliar feeding Supplying mineral elements in solution by watering them onto the foliage; useful where soil conditions prevent uptake through the roots as in *lime-induced chorosis*.

foliar gap, trace See **leaf gap, leaf trace**.

foliose (1) Leafy, having the body differentiated into stem and leaves. Cf. **thallose**. (2) Said of lichens with a flattened thallus growing over the substratum, but rather loosely attached. Cf. **crustose, fruticose.**

follicle (1) Spherical accumulations of lymphocytes (largely B cells) present in lymphoid tissues which become enlarged after antigenic stimulation to become secondary follicles or **germinal centres**. (2) Any small saclike structure, such as the pit surrounding a hair root. adjs. *follicular, folliculose*. See **Graafian follicle**. (3) A dry, dehiscent, many-seeded fruit formed from one carpel, dehiscing along one side only, as in *Delphinium*, where there is a group of follicles.

following response Refers to the fact that newly hatched presocial birds will follow a moving object during a fairly brief period soon after hatching; under natural conditions this would be the parents or siblings. See **imprinting**.

fontanelle In Craniata, a gap or space in the roof of the cranium.

food allergy The presence of **allergens** in food that provoke hypersensitivity reactions. Infants may be allergic to bovine milk; many other allergies have been de-scribed. *Food idiosyncrasy* describes adverse effects occurring after certain foods without an immunological basis.

food body A soft mass of cells, containing oil and other nutrient substances, attached to the outside of the seed coat; it is eaten by ants, which drag the seed along, leave it when they have eaten the food body, and so assist dispersal.

food chain Sequence of organisms dependent on each other for food. The number of links in the chain is usually only four or five. See **food webs.**

food groove A groove along which food is passed to the mouth; median and ventral in Branchiopoda; along the edge of each ctenidium in Bivalvia.

food pollen Pollen, formed by some flowers, which attracts insects; it may be incapable of bringing about fertilization and may be formed in special anthers. Insects seeking food pollen help in conveying good pollen to other flowers.

food vacuole In the cytoplasm of some *Protozoa*, a space surrounding a food particle and filled with fluid.

food webs The interlocking patterns formed by a series of interconnected food chains.

foot (1) A locomotor appendage; in Crustacea, any appendage used for swimming or walking; in Arachnida, Myriapoda and Insects, the tarsus; in Echinodermata, the podia (see **podium**); in Mollusca, a median ventral muscular mass, used for fixation or locomotion; in land Vertebrates, the podium of the hind limb, or of all limbs in Tetrapoda. (2) In plants, the basal part of an embryo, a developing sporophyte or a spore-producing body, embedded in the parental tissues and serving to absorb nutrients.

foot rot One of a number of plant diseases, caused by a variety of fungi, in which the primary symptom is the death of the roots and stems around or below soil level.

foraging Behaviour that involves searching for, capturing and consuming food.

foramen An opening or perforation, esp. in a chitinous, cartilaginous or bony skeletal structure.

foramen lacerum An opening of the Vertebrate skull in the side of the brain case, which is situated between the alisphenoid and the orbitosphenoid, and through which pass the third, fourth, fifth ophthalmic, and sixth cranial nerves.

foramen magnum The main opening at the back of the Vertebrate skull, by which the spinal cord issues from the brain case.

foramen triosseum A large aperture between the three bones of the pectoral girdle in Birds.

Foraminifera An order of Sarcodina, the

members of which have numerous fine anastomosing pseudopodia and a shell which is usually calcereous; the ectoplasm is sometimes vacuolated.

forb Any herb other than a grass.

forbidden clone Clones of lymphocytes reactive with 'self' antigens. According to the clonal selection theory they should have been eliminated during the early maturation of lymphocytes, and hence forbidden. The term is of historical interest, and is sometimes used, but the actual mechanisms by which 'self reactivity' is avoided are more complicated.

forceps In Dermaptera, the pincer-shaped cerci; in Arachnida and Crustacea, the opposable distal joints of the chelae; in Echinodermata, the distal opposable jaws of pedicellariae. adj. *forcipulate*.

fore-brain In Vertebrates, that part of the brain which is derived from the first or anterior brain vesicle of the embryo, comprising the olfactory lobes, the cerebral hemispheres and the thalamencephalon; the first or anterior brain vesicle itself.

fore-gut That part of the alimentary canal of an animal which is derived from the anterior ectodermal invagination or stomodaeum of the embryo.

forest (1) Natural vegetation in which the dominant species are trees, with crowns that touch each other to form a continuous canopy. (2) Almost any vegetation with trees, including plantations. (3) In Britain, areas which were formerly with trees and used for hunting but which may now be treeless.

forfex A pair of pincers, as of an earwig.

formal operations In Piagetian theory, the fourth and final major stage of cognitive development, occurring during adolescence; it reflects a transition from logic bound to the real and concrete to a logic capable of dealing with abstract events. See **concrete operations**.

form genus See **form taxon**.

formol toxoid Any toxoid prepared by formaldehyde treatment of a toxin.

form taxon An artificial taxon, e.g. a *form genus*, intended to provide a name for morphologically similar but possibly unrelated organisms or parts of organisms when it is not possible to determine their correct taxonomic position. Thus an *organ genus* in palaeobotany provides names for fossil leaves, spores, pieces of cuticle etc. where it is not (yet) possible to reconstruct the whole plant. Again in Deuteromycetes, it provides names for fungi with no, or no known, sexual reproduction.

fornix In the brains of higher Vertebrates, a tract of fibres connecting the posterior part of the cerebrum with the hypothalamus.

Forssman antigen, antibody A glyco-lipid antigen present on tissue cells of many species, and on the red cells of some species such as sheep. Forssman antibody may be elicited by immunization with the antigen, but often present as a natural antibody in humans. Sheep red cells coated with Forssman antibody are used in complement fixation tests.

fossa A ditchlike or pitlike depression, as the *glenoid fossa*.

fossa rhomboidalis See **fourth ventricle**.

fossette In general, a small pit or depression; in some Arthropoda, the socket which receives the base of the antennule.

fossil The relic or trace of some plant or animal which has been preserved by natural processes in rocks of the past.

fossorial Adapted for digging, as the mole.

fourth ventricle In Vertebrates, the cavity of the hind-brain.

fovea A small pit or depression. adjs. *foveate, foveolar, foveolate*.

fovea centralis In the Vertebrate eye, the areas of greatest visual resolution, seen as a small depression at the centre of the **macula lutea**.

foveola Same as *fovea*.

fraction 1 protein See **ribulose 1,6-bis-phosphate carboxylase oxygenase**.

fractionation A system of treatment commonly used in radiotherapy in which doses are given daily or at longer intervals over a period of 3 to 6 weeks.

fragile-X syndrome Hereditary human condition involving mental subnormality in which a portion of the X-chromosome tends to show breaks in appropriately prepared lymphocytes.

fragmentation (1) See **amitosis**. (2) The breaking off of a portion from the main body of a chromosome. (3) In plants a form of vegetative or asexual reproduction in which pieces of the parent become detached and grow into new individuals, esp. in many filamentous algae.

frass Faeces; excrement.

free Not joined to another member.

free-air dose A dose of radiation, measured in air, from which secondary radiation (apart from that arising from air, or associated with the source) is excluded.

free association A psychoanalytic technique for probing unconscious ideas and feelings; the individual verbalizes whatever thought comes to mind, without structuring or censoring their content.

free cell formation Free nuclear division followed by delimitation of separated cells without all parental cytoplasm being used, as in ascospore formation.

free central placentation Placentation in which the placentas develop on a central column or projection which arises from the base of a unilocular ovary and is not connected by septa to the ovary wall, as in many Caryophyllaceae, Primulaceae.

free floating anxiety See **generalized anxiety disorder**.

freemartin In cattle, a sterile female intersex occurring as co-twin with a normal bull calf.

free recall A test of memory in which the subject is required to produce as many learned items of the learned task as possible, without regard for the order in which it was first learned.

free space The *apoplast*, esp. in estimates of the fraction it forms (the apparent free space) of a tissue.

freeze-drying A method of fixing tissues by freezing sufficiently rapidly as to inhibit the formation of ice crystals, e.g. by immersion in isopentane cooled to $-190°C$, followed by dehydration in vacuo.

freeze etch, freeze fracture A technique for specimen preparation for electron microscopy in which tissue is frozen (strictly vitrified) by very rapid cooling to below *minus* $100°C$, fractured and a surface replica made (for microscopical examination) either immediately (freeze fracture) or after allowing some water to sublime (freeze etch). See **carbon replica technique**.

freeze substitution Specimen preparation for electron microscopy in which tissue is rapidly 'frozen' (see **freeze etch**) and then the ice dissolved out at low temperatures with suitable solvents (often containing a fixative) before embedding in the normal way.

frequency The number of any given species in an area, or the percentage of sample squares which contain the species.

frequency distribution See **distribution**.

frequency table A table classifying a set of observations by the number of occurrences of particular values or types.

Freudian slip An error in speech, which Freud believed revealed unconscious ideas or wishes.

Freud's theory of dreams Holds that the dream is a disguised transformation of unconscious wishes. The *manifest content* of the dream refers to the images we remember upon waking, and is a transformation of repressed wishes and ideas, the *latent dream content*; dream interpretation consists of working out the nature and meaning of the transformation.

Freund's adjuvant A water in oil emulsion into which is incorporated an antigen for the purpose of immunization against it. The *complete form* has heat-killed mycobacteria added to it and this form of adjuvant is very effective for eliciting both T and B cell immunity. It is not used in humans because it is liable to cause suppurating granulomas.

frond (1) A large compound leaf, esp. of a fern, cycad or palm. (2) A more or less leaf-like thallus of lichen, liverwort or alga. (The term is imprecise in either usage.)

frons In Insects, an unpaired sclerite of the front of the head; in higher Vertebrates, the front of the head above the eyes. adj. *frontal*.

frontal (1) A paired dorsal membrane bone of the Vertebrate skull, lying between the orbits. (2) Pertaining to the frons.

frontal lobes The front part of the cortex. It is thought to be involved in immediate memory.

frontal plane The median horizontal longitudinal plane of an animal.

frontal sinuses Air cavities, connected with the nasal chambers, extending into the frontal bones, in Mammals.

fructification Any seed- or spore-bearing structure like the large spore-bearing structures of many fungi. Also *fruiting body*.

frugivorous Fruit eating.

fruit (1) The structure that develops from the ovary of an angiosperm as the seeds mature, with (false fruit) or without (true fruit) associated structures. (2) Sometimes, any of the various structures associated with the mature seed of a gymnosperm, esp. e.g. the fleshy cone scales of the juniper 'berry'. (3) A fructification.

fruiting body, fruit body See **fructification**.

frustration (1) In animal behaviour, a motivational state assumed to occur in situations in which an animal's actions do not lead to an expected outcome (e.g. a food reward). Often the initial response to a frustrating situation is to increase the persistence and intensity of behaviour; continued frustration often leads to **redirected behaviour**. (2) In human behaviour, it refers to either the prevention of activity that is directed towards a goal, or to the psychological state that results from being prevented from reaching a goal.

frustule The silicified cell wall of a diatom (Bacillariophyceae) consisting of two halves which fit one (the hypotheca) inside the other (the epitheca) like the two halves of a petri dish (together, in some usages, with the cell itself).

frutescent, fruticose (1) Shrubby. (2) Said of lichens which are attached to the base, branching and more or less bushy, either erect or pendant. Cf. **crustose, foliose**.

F-scale A scale constructed to measure the **authoritarian personality**; contains questions designed to measure proposed aspects of authoritarianism, e.g. submission to authority, admiration for power etc.

fucivorous Seaweed-eating.

fucoxanthin $C_{40}H_{56}O_6$ or $C_{40}H_{60}O_6$; the main carotenoid found in brown algae.

Brown crystalline solid; mp 168°C. It is a *xanthophyll*; a major **accessory pigment** in most members of the Heterokontophyta and responsible for their brownish colours (e.g. the brown algae).

fugue A condition, related to amnesia, in which the individual takes a sudden and unexpected trip from home, assuming a new identity and forgetting the past, in an attempt to escape from overwhelming stress.

fuliginous Soot-coloured.

fumigation Disinfection by means of gas or vapour.

function The normal activity of a biological structure, as *digestive function* or *ribosomal function*.

functional Carrying out normal activities; active (as opposed to *passive*).

functional psychosis A severe mental disorder which cannot be attributed to any certain physical pathology.

fungi Heterotrophic, eukaryotic organisms reproducing by spores, and their allies. See **Eumycota, Myxomycota**.

fungicide A substance that kills fungal spores and/or mycelium.

Fungi imperfecti Older name for Deuteromycotina.

fungistatic Preventing the growth of a fungus without killing it.

funicle The stalk by which the ovule (and seed) is attached to the placenta in angiosperms.

funiculus In some Invertebrates (as Bryozoa), thickened strands of mesoderm attaching the digestive organs to the body wall; more generally, any small cord, as a tract of nerve fibres in the central nervous system. adj. *funicular*.

funnel A modified part of the foot in Cephalopoda, protruding from the mantle cavity and acting as an exhalent channel. In Annelida, a nephrostome.

fur In Mammals, the thick undercoat of short, soft, silky hairs.

furca Any forked structure; in Vertebrates, a divergence of nerve fibres; in some Arthropods, a pair of divergent processes at the end of the abdomen.

furcula In Collembola, the leaping apparatus, consisting of a pair of partially fused appendages arising from the fourth abdominal somite; in Birds, the partially fused clavicles, the wishbone; more generally, any forked structure.

6-furfurylaminopurine See **kinetin**.

furrowing Cell division by intucking the plasmalemma to pinch the cell into two. Cf. **cell plate**.

fusiform Elongated, broadest in the middle and tapering towards both ends; shaped like a spindle.

fusiform initials More or less elongated initial cells, in the cambium, giving rise to all the cells of the secondary xylem and phloem except for the ray cells. Cf. **ray initials**.

G

G Symbol for giga, i.e. 10^9.

G1 Period of growth in the **cell cycle** between the end of cell division and the beginning of DNA synthesis.

G2 As for G1 but between the end of DNA synthesis and the beginning of the next division.

Gadiformes An order of mainly marine and often deep-water Osteichthyes with elongated body and long dorsal and anal fins. Some are economically important. Cod, Haddock and Grenadiers.

Gaia Theory proposed by J. E. Lovelock in 1979 concerning the role of biota in maintaining a climatic homeostasis.

Gal Abbrev. for *Galactose*.

galactobolic Refers to the action of neurohypophysial peptides which contract mammary myoepithelium and so eject milk.

galactophorous See **lactiferous**.

galactopoiesis Increase in milk secretion.

galactose A hexose of the formula CH_2.OH.(CHOH)$_4$.CHO; thin needles; mp 166°C; dextrorotatory. It is formed together with D-glucose by the hydrolysis of milk sugar with dilute acids. Stereoisomeric with glucose, which it strongly resembles in properties. Present in certain gums and seaweeds as a polysaccharide *galactan* and as a normal constituent of milk.

galactosis See **lactation**.

galeate, galeiform Shaped like a helmet or a hood.

gall An abnormal localized swelling or outgrowth, usually of characteristic shape which follows an attack by a parasite or pest.

gall bladder In Vertebrates, a lateral diverticulum of the bile duct in which the bile is stored.

Galliformes An order of ground-living Birds with feet well adapted for running. Game birds (which seek their food, berries, seeds, buds, and insects, on the ground). Brush turkeys, Currassows, Turkeys, Pheasants, Partridge, Grouse and Quail.

galvanic skin response A change in the electrical resistance of the skin, recorded by a polygraph, widely used as an index of autonomic reaction. Abbrev. *GSR*.

galvanotaxis Tendency of organisms to grow or move into a particular orientation relative to an electric current passing through the surrounding medium. Also *galvanotropism, electrotropism, electrotaxis*.

games, theory of A mathematical formalization of decision and strategic processes involving the probabilities and values of various outcomes of action choices for the decision makers.

gametangium Any cell or organ within which gametes are formed, e.g. antheridium, oögonium, archogonium.

gametes Reproductive cells which will unite in pairs to produce zygotes; germ cells. adj. *gametal*.

gametogenesis The formation of gametes from gametocytes; *gametogeny*.

gametophore The branch of a moss which bears the sex organs.

gametophyte The characteristically haploid generation in the life cycle of a plant showing **alternation of generations**, producing the gametes, forming the zygote which germinates to give the **sporophyte**.

gam-, gamo- Prefixes from Gk. *gamos*, marriage, union.

gamma camera See **scintillation camera**.

gamma detector A radiation detector specially designed to record or monitor gamma radiation.

gamma globulin Describes the serum proteins which on electrophoresis have the lowest anodic mobility at neutral pH. These are mainly immunoglobulins. The term was used to describe immunoglobulins until more specific means of distinguishing them were developed.

gamma-ray source A quantity of matter emitting γ-radiation in a form convenient for radiology.

gamocyte In Protozoa, a phase developing from a trophozoite and giving rise to gametes.

gamone Any chemical substance released by a gamete or hypha that is attractive to another appropriate gamete or hypha in sexual reproduction, e.g. malic acid in ferns, called *sirenin*.

gamopetalous Having a corolla consisting of a number of petals united by their edges. Cf. **polypetalous**.

gamophyllous Having the perianth members united by their edges. Cf. **polyphyllous**.

ganglion A plexiform collection of nerve fibre terminations and nerve cells. pl. *ganglia*.

ganglion impar The unpaired, most posterior of the abdominal sympathetic ganglia.

ganglioside Glycolipids derived from cerebroside by the addition of complex oligosaccharide chains.

ganoid Formed of, or containing, ganoin. Said of fish scales of rhomboidal form, composed of an outer layer of ganoin and an inner layer of isopedin; (fish) having these scales.

ganoin A calcareous substance secreted by the dermis and forming the superficial layer of certain fish scales; it was formerly supposed to be homologous with enamel.

gape The width of the mouth when the jaws are open.

gap junction Junction between cells which allows direct communication between cells by molecules which can diffuse through pores in the junction. The flow is controlled by the opening or closing of these pores.

gaseous exchange See **respiration**.

gas exchange The uptake and output of gases, esp. of carbon dioxide and oxygen in photosynthesis and respiration.

gas gland A structure in the wall of the air bladder in certain Fish which is capable of secreting gas into the bladder. See **rete mirabile**.

Gasserian ganglion In Vertebrates, a large ganglion of the fifth cranial nerve, near its origin.

gaster The abdomen proper in Hymenoptera, being the region posterior to the first abdominal segment which, in many members, is constricted.

gastero-, gastr-, gastro- Prefixes from Gk. *gaster*, gen. *gastros*, stomach.

Gasteromycetes The puffballs, earth stars and stinkhorn fungi, a class of Basidiomycotina in which the hymenium is enclosed until after the spores have matured. Most are saprophytic in soil.

gas transport The transport by the blood of oxygen from the site of *external respiration* to cells where it is needed for aerobic respiration (see **respiration, aerobic**), this frequently involving a *respiratory pigment*, and the transport away from the respiring cells of any carbon dioxide produced.

gastric Pertaining to, or in the region of, the stomach.

gastric juice Human gastric juice consists principally of water (99.44%), free HCl (0.02%), and small quantities of NaCl, KCl, $CaCl_2$, $Ca_3(PO_4)_2$, $FePO_4$, $Mg_3(PO_4)_2$ and organic matter including enzymes.

gastrin Polypeptide hormone secreted by specialized cells in the stomach mucosa which stimulates the secretion of acid and pepsin by other cells in the mucosa.

gastrocentrous Said of vertebrae in which the centrum is composed largely of the pleurocentrum and the intercentrum is reduced or absent; in all *Amniota*.

gastrocnemius In land Vertebrates, a muscle of the shank.

gastrocoel See **archenteron**.

Gastropoda A class of Mollusca with a distinct head bearing tentacles and eyes, a flattened foot, a visceral hump which undergoes torsion to various degrees and is often coiled, always bilaterally symmetrical to some extent, with the shell usually in one piece. Snails, Slugs, Whelks etc. Also *Gasteropoda*.

Gastrotricha A class of phylum Aschelmin-

thes. Small, free-living worm-like aquatic animals with ciliated tracts on the cuticle which has scales and bristles.

gastrovascular Combining digestive and circulatory functions, as the canal system of Ctenophora and the coelenteron of Cnidaria.

gastrula In development, the double walled stage of the embryo which succeeds the blastula.

gastrulation The process of formation of a gastrula from a blastula during development.

gas vacuole Structure in the cells of some planktonic blue-green algae which provide buoyancy and are composed of many small, more or less cylindrical, gas-filled vesicles.

Gause's principle The idea that closely related organisms do not co-exist in the same niche, except briefly. See **competitive exclusion principle**.

Gaussian distribution See **normal distribution**.

G-banding See **banding techniques**.

geitonogamy Fertilization involving pollen and ovules from different flowers on the same individual plant (*ramet*) or from the same clone (*genet*); a form of *allogamy* (1). See **self pollination, self fertilization, inbreeding**.

gel The inert matrix in which polynucleotides and polypeptides can be separated by electrophoresis.

gel diffusion tests Precipitin tests in which antigen and antibody are placed separately in a gel medium (commonly agar) and allowed to diffuse towards one another and to form lines of precipitate where they meet in suitable proportions.

gemma (1) A bud that will give rise to a new individual, e.g. the multicellular structure in algae, pteridophytes and esp. bryophytes. (2) Same as **chlamydospore**. pl. *gemmae*.

gemma cup Cup-like structure in which gemmae are borne in some mosses and liverworts.

gommation Budding; gemma formation.

gemmiferous, gemmiparous Producing gemmae.

gemmule In fresh water Porifera, an aggregation of embryonic cells within a resistant case, which is formed at the onset of hard conditions when the rest of the colony dies down, and which gives rise to a new colony when conditions have once more become favourable.

gender identity The individual's sense of being a man or a woman.

gender role The set of attitudes and behaviour a given culture considers appropriate for each sex.

gene The hereditary determinant of a specified difference between individuals. Can refer either to a particular *allele* or to all alleles at a particular *locus*. Molecular analysis has shown that a specific sequence or parts of a sequence of DNA can be identified with the classical gene. See **cistron**.

genealogy The study of the development of plants and animals from earlier forms.

gene bank A collection of plants or, more often of seeds, cell cultures, frozen pollen etc. of species of known or potential use to man, and esp. of landraces that may contain genes of use in crop breeding.

genecology The branch of ecology which seeks genetic explanations of the patterns of distribution of plants and animals in time and space.

gene dosage Number of copies of a gene in a given cell or individual organism.

gene-for-gene concept The concept that there are corresponding genes for virulence and resistance in pathogen and host respectively.

gene frequency The proportion of all representatives of a particular gene in a population that contain the specified allele.

gene mapping See **mapping**.

gene number The total number of different **coding sequences** which are transcribed into RNA in an individual or species.

generalist Organism with many food sources, or able to live in many habitats. See **specialist**.

generalization In learning, when the learned behaviour occurs in situations similar to, but not identical with, that in which learning first occurred.

generalized anxiety disorder A chronic state of diffuse, unfocused anxiety, in the absence of specific symptoms such as are found in phobic reactions, and without any associated specific stimuli or objects.

general paresis A psychosis of organic origin (produced by syphilitic infection), in which there is a progressive deterioration of cognitive or motor functions, culminating in death.

general sexual dysfunction In women, the absence or weakness of the physiological changes normally accompanying the excitement phase of sexual response.

generation The individuals of a species which are separated from a common ancestor by the same number of broods in the direct line of descent.

generative cell A cell of the male gametophyte in pollen grain or pollen tube. In gymnosperms it divides to give the sterile (stalk) cell and the spermatogenous (body) cell. In angiosperms it divides to give the two male gametes. See **double fertilization**.

generator potential An electrical potential arising in a sensory neuron as a result of a sensory stimulus.

genesis The origin, formation or development of some biological entity. adj. *genetic*.

genet A genetic individual, the product of one zygote; either an individual plant grown from a sexually-produced seed or all the individuals produced by vegetative reproduction from one such plant. Cf. **ramet**.

gene therapy Colloquial term for the substitution of a functional for a defective gene as a treatment for a *genetic disease*.

genetically significant dose The dose that, if given to every member of a population prior to conception of children, would produce the same genetic or hereditary harm as the actual doses received by the various individuals.

genetic code See panel.

genetic correlation Measure of the extent to which the variation of two different quantitative characters is caused by the same genes.

genetic drift The process by which *gene frequencies* are changed by the chances of random sampling in small populations.

genetic engineering Colloq. for **genetic manipulation**.

genetic equilibrium The situation reached when the frequencies of genes and of genotypes in a population remain constant from generation to generation.

genetic manipulation See panel.

genetics The study of heredity; of how differences between individuals are transmitted from one generation to the next; and of how the information in the genes is used in the development and functioning of the adult organism.

genetic spiral In phyllotaxis, a hypothetical line through the centres of successive leaf primordia at the shoot apex.

genetic variance Measure of the variation between individuals of a population due to differences between their *genotypes*.

-gen, -gene Suffixes meaning generating, producing.

gen-, geno- Prefixes from Gk. *genos*, race, descent.

genial Pertaining to the chin.

genicular Pertaining to, or situated in, the region of the knee.

geniculate Bent rather suddenly, like the leg at the knee, as *geniculate antennae*.

geniculate ganglion In Craniata, the ganglion from which the seventh cranial nerve arises.

geniohyoglossus The muscle which moves the tongue in Vertebrates.

genital atrium In Platyhelminthes and some Mollusca, a cavity into which open the male and female genital ducts.

genitalia, genitals The gonads and their ducts and all associated accessory organs.

genital stage In psychoanalytic theory, the final phase of psychosexual development in which sexual pleasure involves not only

genetic code

The rules which relate the four bases of the DNA or RNA with the 20 amino acids found in proteins. There are 64 possible different 3-base sequences (triplets) using all permutations of the four bases. One triplet uniquely specifies one amino acid (except for AUG when acting as an initiating codon in bacteria), but each amino acid can be coded by up to 6 different triplet sequences. The code is therefore *degenerate*. **Initiating codons** specify the start of a polypeptide chain and the triplets known as *Ochre, Amber* and *Opal* are **stop codons** which terminate the chain. Initiating codons are AUG and GUG in bacteria with the former specifying the amino acid *n*-formylmethionine at the beginning of the chain and methionine within it. In eukaryotes, AUG is the only initiator and always translates as methionine.

The evidence suggests that the code is universal, applying from the simplest to the most evolutionarily advanced organism, although minor variations have been found particularly in the mitochondrial DNAs.

In the following table the amino acids and the bases are specified by their three and one letter symbols respectively.

Ala	GCU, GCC, GCA, GCG		Lys	AAA, AAG
Arg	CGU, CGC, CGA, CGG		Met	AUG
	AGA, AGG		Phe	UUU, UUC
Asn	AAU, AAC		Pro	CCU, CCC, CCA, CCG
Asp	GAU, GAC		Ser	UCU, UCC, UCA, UCG
Cys	UGU, UGC			AGU, AGC
Glu	GAA, GAG		Thr	ACU, ACC, ACA, ACG
Gln	CAA, CAG		Trp	UGG
Gly	GGU, GGC, GGA, GGG		Tyr	UAU, UAC
His	GAU, GAC		Val	GUU, GUC, GUA, GUG
Ile	AUU, AUC, AUA		*OCHRE*	UAA
Leu	UUA, UUG, CUU, CUC		*AMBER*	UAG
	CUA, CUG		*OPAL*	UGA

gratification of one's own impulses but attention to and pleasure in, the social and physical pleasure of one's mate. Freud was referring to heterosexual patterns, and to sexual pleasure derived mainly through the genitalia.

genome The totality of the DNA sequences of an organism or organelle.

genomic DNA Nuclear chromosomal DNA.

genomic library DNA library derived from a whole, single **genome**.

genotype The particular alleles at specified loci present in an individual; the genetic constitution. adj. *genotypic*. Cf. **phenotype**.

genu A kneelike structure, i.e. a bend in a nerve tract; more particularly, part of the corpus callosum in Mammals.

genus A taxonomic rank of closely related forms, which is further subdivided into species and therefore below **family** and above **species**. pl. *genera*. adj. *generic*.

genys In Vertebrates, the lower jaw.

geobiotic Terrestrial; living on dry land.

geocarpy The ripening of fruits underground, the young fruits being pushed into the soil by a post-fertilization curvature of the stalk.

geocline A **cline** occurring across topographic or spatial features of an organism's range.

geographical race A collection of individuals within a species, which differ constantly in some slight respects from the normal characters of the species, but not sufficiently to cause them to be classified as a separate species, and which are peculiar to a particular area.

geophagous Earth eating.

geophilous Living on, or in, the soil.

geophyte Herbs with perennating buds below the soil surface. It includes plants with bulbs, corms, rhizomes etc. See **Raunkiaer system**.

geotaxis The locomotory response of a motile organism or cell to the stimulus of gravity. adj. *geotactic*. Cf. **geotropism**.

geotropism The reaction of a plant member

genetic manipulation

The term for the procedures with which it is now possible to combine DNA sequences from widely different organisms *in vitro*, often with great precision. Two major advances have made these procedures possible: the discovery of **restriction enzymes** and the ability to construct suitable **vectors** (such as plasmids) into which a DNA sequence can be inserted to form a hybrid molecule. Such hybrid molecules are able to multiply in a rapidly growing host (bacterium or yeast). A necessary further part of the procedure is the selection of those host cells which contain hybrid molecules. One way to do this is to arrange for the inserted DNA to disrupt a vector sequence which normally prevents growth (*positive selection*).

Restriction enzymes which cut e.g. eukaryote DNA at specific sites enable a sequence such as part of a gene to be identified and purified. They are also the main way in which vectors can be constructed so that they are stripped of unnecessary functions while still being able to replicate after the donor DNA has been incorporated.

Genetic manipulation, or its loose synonym, *cloning*, should not be confused with the procedures long practised in horticulture and microbiology in which exact copies of a plant or bacterium are routinely propagated. See **plant genetic manipulation**. Cf. **clone**.

'Gene manipulation' has the following legal definition in the UK. 'The formation of new combinations of heritable material by the insertion of nucleic acid molecules, produced by whatever means outside the cell, into any virus, bacterial plasmid or other vector system so as to allow their incorporation into a host organism in which they do not naturally occur but in which they are capable of continued propagation.' See **polymerase chain reaction** for a method of amplifying a DNA sequence without using living organisms.

or sessile animal to the stimulus of gravity, shown by a growth curvature; cells become more elongated on one side than the other, tending to bring the axis of the affected part into a particular relation to the force of gravity. adj. *geotropic*. Cf. **geotaxis**. Also *gravitropism*.

germ The primitive rudiment which will develop into a complete individual, as a fertilized egg or a newly formed bud; a unicellular micro-organism.

germ band In Insects, a ventral plate of cells, produced in the egg by cleavage, which later gives rise to the embryo.

germ cells *Germinal cells*. In Metazoa, the reproductive cells. Gametes; spermatozoa and ova, or the cells which give rise to them.

germinal aperture, germinal pore See **germ pore**.

germinal centre An aggregation of lymphocytes, mainly B-cells with numerous blast forms undergoing cell division, which develops from a primary follicle in response to antigenic stimulation. A prominent feature is the presence of follicular dendritic cells bearing antigen-antibody complexes closely entangled with the B-cells, and of tingible body macrophages (so called because they contain remnants of the nuclei of dead lymphocytes). Germinal centres are thought to be sites at which B-memory cells are produced with receptors which recognize antigens in the complexes. The cell death which occurs there may represent B cells which have undergone non-viable mutations or which have receptors which bind *idiotypic antibodies*.

germinal disk The flattened circular region at the top of a megalecithal ovum, in which cleavage takes place.

germinal epithelium A layer of columnar epithelium which covers the stroma of the ovary in Vertebrates.

germinal layers See **germ layers**.

germination The beginning of growth in a spore, seed, zygote etc. esp. following a dormant period.

germ layers The three primary cell layers in the development of Metazoa, i.e. ectoderm, mesoderm and endoderm.

germ line The cells whose descendants give rise to the *gametes*.

germ nucleus See **pronucleus**.

germ pore A thin area in the wall of a spore or pollen grain though which the germ or pollen tube emerges at germination.

germ tube The hypha or other tubular outgrowth that emerges from a spore at germination (e.g. pollen tube).

gerontic Pertaining to the senescent period in the life history of an individual.

gerontology The scientific study of the processes of aging.

Gestalt German word meaning *whole*. Gestalt psychology is an approach to perception and other cognitive skills, which stresses the need to understand the underlying organization of these functions, and believes that to dissect experience into its 'constituent bits' is to lose its essential meaning.

gestalt therapy A therapeutic approach, developed by Fritz Perls, which focuses on the present manifestations of past conflict, often using role playing and other *acting out* techniques in order to help individuals gain insight into themselves and their behaviour. Not to be confused with the work of the Gestalt school of psychology, which focuses primarily on cognitive functions.

gestation In Mammals, the act of retaining and nourishing the young in the uterus; pregnancy.

giant cells Cells of unusual size, as the myeloplaxes of bone marrow; certain cells of the excitable region of the cerebrum; certain cells sometimes found in lymph glands; large multinucleate cells of the thymus and of the spleen. **Osteoclasts.**

giant fibres In many Invertebrates (e.g. Annelida, Cephalopoda and some Arthropoda) enlarged motor axons in the ventral nerve cord which transmit impulses very rapidly and initiate escape behaviour.

gibberellic acid A **gibberellin** obtained from cultures of the fungus *Gibberella fuijikuroi* used commercially e.g. to promote rapid, even malting of barley.

gibberellin Any of a large group of terpenoid plant *growth substances*, synthesized mainly in young leaves and promoting stem elongation and, sometimes, flowering, germination and the utilization of reserves as in germinating barley grains, e.g. *gibberellic acid*.

gibbous (1) Swollen, esp. if swollen more on one side than another. (2) With a pouch.

giga- Prefix used to denote 10^9 times, e.g. a gigawatt is 10^9 watts.

gigantism Abnormal increase in size, often associated with polyploidy.

gill (1) A membranous respiratory outgrowth of aquatic animals, usually in the form of thin lamellae or branched filamentous structures. (2) One of the vertical plates, bearing the hymenium, on the undersurface of the cap of the fruiting body of the mushroom and other agarics.

gill arch In Fish, the incomplete jointed skeletal ring supporting a single pair of gill slits; one segment of the branchial basket.

gill bars See **gill rods.**

gill basket In Fish and Cyclostomes, the skeletal ring supporting a single pair of gill slits; one segment of the branchial basket.

gill book The booklike respiratory lamellae of Xiphosura borne by the opisthosoma, of which they represent the appendages. Also *book gill.*

gill clefts See **gill slits.**

gill cover See **operculum.**

gill pouch One of the pouchlike gill slits of Cyclostomes and Fish.

gill rakers In some Fish, small processes of the branchial arches, which strain the water passing out via the gill slits and prevent the escape of food particles.

gill rods In Cephalochorda, skeletal bars which support the pharynx. Also *gill bars.*

gill slits In Chordata, the openings leading from the pharynx to the exterior, on the walls of which the gills are situated. Also *brachial clefts, gill clefts.*

ginger-beer plant A symbiotic association of a yeast and a bacterium, which ferments a sugary liquid containing oil of ginger, giving ginger beer. Often known popularly as *Californian bees* or by similar names.

gingival In Mammals, pertaining to the gums.

ginglymus An articulation which allows motion to take place in one plane only; a hinge joint. adj. *ginglymoid.*

Ginkgoales Order of gymnosperms, abundant world wide in the Mesozoic. There is one extant species, *Ginkgo biloba*, a Chinese tree with small, fan-shaped leaves. Zooidogamous.

girder Usually longitudinal strand of mechanical tissue, often T or I shaped in cross section, giving strength and esp. stiffness to a plant part as in a grassleaf.

girdle In Vertebrates, the internal skeleton to which the paired appendages are attached, consisting typically of a U-shaped structure of cartilage or bone with the free ends facing dorsally. In *Amphineura*, the part of the mantle which surrounds the shells

gizzard See **proventriculus.**

glabrescent, glabrate (1) Almost hairless. (2) Becoming hairless.

glabrous Having a smooth hairless surface.

gland A structure at or near the surface of a plant or a single or an aggregate of epithelial cells in animals, specialized for the elaboration of a secretion useful to the organism or of an excretory product. adj. *glandular.*

gland cell A unicellular gland, consisting of a single goblet-shaped epithelial cell producing a secretion, usually mucus.

glandular epithelium, tissue Epithelial tissue specialized for the production of secretions.

glans (1) A nut. (2) A glandular structure.

glans penis A dilatation of the extremity of

the mammalian penis.

glaucescent (1) Somewhat glaucous. (2) Becoming glaucous.

glaucous Greyish- or bluish-green; covered with a bloom as a plum is.

gleba The spore-bearing tissue enclosed within the peridium of the fructification of *Gasteromycetes*, and in truffles.

glenoid Socket shaped; any socket-shaped structure; as the cavity of the pectoral girdle which receives the basal element of the skeleton of the fore-limb.

glenoid fossa In Mammals, a hollow beneath the zygomatic process.

gley, glei soil A soil that is permanently or periodically waterlogged and, therefore anaerobic, characterized by its blue-grey colours (due to ferrous iron) often mottled with orange-red (ferric iron).

glia Same as *neuroglia*. From Gk. *gliā*, meaning glue.

gliding Form of movement in algae in which cells or filaments, without flagella, move against a solid surface.

gliding growth Same as **sliding growth**.

Gln Symbol for **glutamine**.

globoid A rounded inclusion of phytin in a protein body.

globus Any globe-shaped structure; as the *globus pallidus* of the Mammalian brain. adj. *globate*.

glochidiate Bearing barbed hairs or bristles.

glomerate Collected into heads.

glomerulonephritis Kidney diseases in which the major lesion is in the glomeruli. The capillary walls contain deposits of IgG and often of activated complement components. These are due to deposition of immune complexes from the circulation, or to antibodies against some component present in the glomerular capillary wall. The glomeruli are infiltrated with inflammatory cells. This condition is reversible, but, if it becomes chronic, gradual obliteration of the glomeruli occurs resulting in renal insufficiency.

glomerulus A capillary blood plexus, as in the Vertebrate kidney; a nestlike mass of interlacing nerve fibrils in the olfactory lobe of the brain. adj. *glomerular*.

glossa In Vertebrates, the tongue; any tongue-like structure. adj. *glossate, glossal*.

gloss-, glosso- Prefixes from Gk. *glossa*, tongue.

glossopharyngeal Pertaining to the tongue and the pharynx; the ninth cranial nerve of Vertebrates, running to the first gill cleft in lower forms, to the tongue and the gullet in higher forms.

glottis In higher Vertebrates, the opening from the pharynx into the trachea.

glucagon A hormone released by the alpha

cells of the **islets of Langerhans** which raises the blood sugar by stimulating the breakdown of *glycogen* in the liver to *glucose*.

gluco- See terms prefixed **glyc-**.

glucose The commonest **aldohexose**, with the formula $C_6H_{12}O_6$ and the major source of energy in animals. Starch and cellulose are condensation polymers of glucose. Also *dextrose, grape sugar*.

glucose-6-phosphate dehydrogenase An enzyme that catalyses the dehydrogenation of glucose-6-phosphate to 6-phosphogluconolactone, an important pathway in carbohydrate metabolism. Deficiency may cause the development of *haemolytic anaemia*.

glume (1) One of the two bracts at the base of each **spikelet** in the inflorescence of a grass. (Old name: *sterile glume*.) (2) The bract subtending the flower in the Cyperaceae (sedges and allies). (3) *Flowering glume* is an old name for **lemma**.

glutamic acid *2-amino-pentane-1,6-dioic acid*. The *L*-isomer is an amino acid and constituent of proteins, and is classed as acidic as it has two carboxylic acid groups. Symbol Glu, short form E.

Side chain:　　$\overset{\text{HO}}{\underset{\text{O}}{\diagdown}}\text{C–CH}_2\text{–CH}_2\text{–}$

See **amino acid**.

glutamine The 6-amide of glutamic acid. Symbol Gln, short form Q.

Side chain:　　$\overset{\text{H}_2\text{N}}{\underset{\text{O}}{\diagdown}}\text{C–CH}_2\text{–CH}_2\text{–}$

See **amino acid**.

gluteal Pertaining to the buttocks.

gluten The reserve protein of wheat grain, a mixture of gliadin and glutenin. Gluten intolerance is important in the development of **coeliac disease**.

gluteus In land Vertebrates, a retractor and elevator muscle of the hind limb.

Gly Symbol for **glycine**.

glycine *Aminoethanoic acid*, the simplest of the naturally occurring amino acids. Symbol Gly, short form G.

Side chain: H—. See **amino acid**.

glycocalyx The carbohydrate layer on the outer surface of animal cells which is composed of the oligosaccharide termini of the membrane glycoproteins and glycolipids.

glycogen Storage polysaccharide, α, $1{\rightarrow}4$ linked with frequent α, $1{\rightarrow}6$ branches, existing as small granules in blue-green algae and bacteria and in the cytoplasm of eumycete fungi and animals but not in plants or eukaryotic algae.

glycolipid A glycosylated lipid. See **cerebroside, ganglioside**.

glycolysis The sequence of reactions which

converts glucose to pyruvate with the concomitant net synthesis of two molecules of ATP. Under aerobic conditions it is the prelude to the complete oxidation of glucose *via* the **tricarboxylic cycle**.

glycophyte A plant which will grow only in soils containing little sodium chloride or other sodium salt. Most plants are glycophytes. Cf. **halophyte**.

glycoprotein A protein with covalently linked sugar residues. The sugars may be bound to OH side chains of the polypeptide *(O-linked)* or to the amide nitrogen of asparagine side chains *(N-linked)*.

glycosyltransferase An enzyme which catalyses the transfer of sugars onto a protein or another sugar side chain. These enzymes bring about the glycosylation of glycoproteins.

glyoxylate cycle Series of reactions resulting in the formation of succinate from acetyl CoA, enabling carbohydrates to be made from fatty acids as in germinating oil-storing seeds. See **glyoxysome**.

glyoxysome, glyoxisome Cytoplasmic organelles of plant cells similar to **peroxisomes** but also containing enzymes of the *glyoxylate cycle*, a cyclic metabolic pathway that generates succinate from acetate. They are abundantly present in e.g. the endosperm or cotyledons of oil rich seeds.

Gm allotype An allotypic determinant (recognizable by specific antibodies) due to amino acid substitutions at various positions in the constant region of human IgG heavy chains. At least 25 different allotypes are known, useful in genetic studies.

gnathic Pertaining to the jaws.

gnathites Mouth parts, esp. those of Insects.

gnathobase In Arthropoda, a masticatory process on the innerside of the first joint of an appendage.

gnathopod In Arthropoda, any appendage modified to assist in mastication.

Gnathostomata A superclass of Chordata, including all Vertebrate animals with upper and lower jaws; comprises a wide range of animals, from Fishes to Tetrapods. Cf. **Agnatha**.

gnathostomatous Having the mouth provided with jaws.

gnathotheca The horny part of the lower beak of Birds.

Gnetopsida Class of the gymnosperms of ca 80 spp. Mostly trees, shrubs, lianas and switch plants. Leaves reticulately veined or scales. Xylem with vessels. Reproductive structures organized into compound strobili. The micropyle projects as a long tube. Siphonogamous. Three genera: *Gnetum*, *Ephedra* and *Welwitschia* (the last a curious turnip-like plant with 2 strap-shaped leaves, of the Namibian desert.)

gnotobiotic Of a known or defined environment for living organisms, as a sterile culture inoculated with one or a few strains of bacteria. It also describes an environment in which animals can be reared and in which all the living microbes are known. This avoids antigenic stimulation by casual infections. If all living microbes are absent the environment is said to be *germ free*.

goal-directed behaviour Implies that the individual has some model of the goal situation, and that discrepancies between the current and goal situation are used to guide behaviour; conscious awareness of the goal or complex cognitive abilities are not necessarily associated with goal-directed behaviour, which probably occurs in many species.

goblet cell A goblet- or flask-shaped epithelial gland cell, occurring usually in columnar epithelium.

goitre Morbid enlargement of the thyroid gland as in *Basedow's disease*.

Golgi apparatus A cytoplasmic organelle consisting of a stack of plate-like **cisternae** often close to the nucleus. It is the site of protein glycosylation.

golgi body See **golgi apparatus**.

Golgi's organs In Vertebrates, a type of sensory nerve ending occurring in tendons near the point of attachment of muscle fibres, stretch receptors.

gomphosis A type of articulation in which a conical process fits into a cavity; as the roots of teeth into their sockets.

gonad A mass of tissue arising from the primordial germ cells and within which the spermatozoa or ova are formed; a sex gland, ovary or testis. adj. *gonadal*.

gonal Forming or giving rise to a gonad, as the *gonal ridge*.

gon-, gono- Prefixes from Gk. *gonos*, seed, offspring.

goni- Prefixes from Gk. *gonia*, angle; *gony*, knee. Not to be confused with prefixes **gon-**, **gono-**.

gonidium (1) A cell which gives rise to an asexual daughter colony in *Volvox*. (2) Any algal cell in the thallus of a lichen.

gonoblast A reproductive cell.

gonochorism Sex determination.

gonochoristic Having separate sexes.

gonocoel That portion of the coelom, the walls of which give rise to the gonads; hence, the cavity of the gonads.

gonoduct A duct conveying genital products to the exterior; a duct leading from a gonad to the exterior.

gonopods The external organs of reproduction in Insects, associated with the 8th and 9th abdominal segments in females and with the 9th only in males. In Chilopoda, a pair of modified appendages borne on the 17th

graft-versus-host disease, reaction

GVH. A **reaction** of transfused or transplanted lymphocytes against *host* antigens and therefore a major complication of bone marrow and other forms of grafting. It occurs when a tissue graft (notably bone marrow) contains T lymphocytes which can respond to antigens present in the recipient which are not identical with those of the donor. If the recipient is unable to suppress the donor lymphocytes, because of immunological immaturity or to immunosuppression due to X-irradiation or to drugs, they cause severe damage to the recipient which begins in the skin and gut and may progress to death.

(genital) body segment.

gonopore The aperture by which the reproductive elements leave the body.

gonosome In colonial animals, all the individuals concerned with reproduction.

Graafian follicle A vesicle, containing an ovum surrounded by a layer of epithelial tissue, which occurs in the ovary of Mammals.

gracilis A thigh muscle of land Vertebrates.

gradient analysis Method of ordination in which vegetation samples or stands are plotted on axes representing environmental variables.

gradient of reinforcement The curve which describes the declining effectiveness of reinforcement as the interval between the response and reinforcement increases.

graft chimera A more or less stable **periclinal chimera** in which the skin and core (typically) derive from different species. Graft chimeras arise, rarely, from the junction of stock and scion, in interspecific grafts, can be propagated only vegetatively and exhibit characteristics between the two species. Graft chimeras are indicated by a + sign before the specific or generic name (intra- and inter-generic chimeras respectively). The most famous of these is probably + *Laburnocytisus adami* which consists of a skin of *Cytisus purpureus* over a core of *Laburnum anagyroides*.

graft hybrid In some usages, *graft chimera*, in others, *burdo*.

grafting The placing for propagation or experiment of a piece of one plant (scion) on to a piece, usually with roots, of another (stock) so that the tissues may unite and growth follow.

graft-versus-host disease See panel.

graft-versus-host reaction See panel.

grain (1) See **caryopsis**. (2) The pattern on the surface of worked wood due to variations in the size, shape, arrangement and composition of the cells forming the wood.

gramicidin A heterogeneous group of ionophores. Thus gramicidin A is an open chain polypeptide while gramicidin S is a cyclic peptide.

graminacious, gramineous Relating to grasses.

Graminae *Poaceae*. The grass family, ca 9000 spp. of monocotyledonous flowering plants (superorder Commelinidae). Annual or perennial herbs, sometimes woody (the bamboos), often tufted or rhizomatous; cosmopolitan and represented in most habitats of earth. The aerial stems are usually hollow and bear leaves, sheathing the stem at the base and with flat, long and narrow blades, in two ranks. The flowers are relatively inconspicuous and wind pollinated (see **floret** and **spikelet**). Many tropical grasses are C_4 plants, including maize. The grasses are extremely important economically for food (all the cereals — rice, wheat, oats, maize etc. and sugar cane), for fodder, and also for some constructional and furniture-making materials (bamboos) and in lawns, sportsfields etc.

graminicolous Living on grasses, esp. of parasitic fungi.

graminivorous Grass eating.

Gram-negative bacteria Those bacteria which fail to stain with Gram's reaction. The reaction depends on the complexity of the cell wall and has for long determined a major division between bacterial species. Cf. **Gram-positive bacteria**.

Gram-positive bacteria The comparative simplicity of the cell wall of some bacterial species allows them to be stained by Gram's procedure. See **Gram-negative bacteria**.

grand period of growth The period in the life of a plant, or of any of its parts, during which growth begins slowly, gradually rises to a maximum, gradually falls off, and comes to an end, even if external conditions remain constant.

granulocyte A general term describing polymorphonuclear leucocytes.

granuloma A localized accumulation of macrophages around the site of some continuing stimulus, such as a persisting antigen which causes delayed type hypersensitivity and the release of lymphokines chemotactic to monocytes (which turn into macrophages). The macrophages may become tightly compressed and their cell membranes often fuse together so as to form multinucleated giant cells.

granum A stack, rather like a pile of coins, of ca 5–50 **thylakoids** (or disks) forming one of say 40–60 such bodies in the **stroma** of a chloroplast in vascular plants, bryophytes and some green algae.

grass See **Gramineae**.

grassland Natural, semi-natural or farm vegetation in which the dominant species are grasses. Major grasslands of the world include pampas, prairies and savannas.

graveolent Having a strong rank odour.

gravid Pregnant; carrying eggs or young.

graviperception The perception of gravity by plants.

gravitropism See **geotropism**.

gray SI unit of absorbed dose. Symbol Gy. See **absorbed dose**.

green algae The *Chlorophyta*.

green glands The antennal excretory glands of decapod Crustacea.

green manure A method of increasing soil organic matter by planting a crop on temporarily free land and ploughing it in while still green.

gregaria phase In locusts (Orthoptera), a phase characterized by high activity and gregarious tendencies, differing morphologically from the **solitaria phase** with which, under natural conditions, it alternates.

grey matter The centrally-situated area of the central nervous system, mainly composed of cell bodies. Cf. **white matter**.

grit cell A **stone cell** occurring in a leaf or in the flesh of a fruit e.g. pear.

grooming Refers to maintenance of and attention to all aspects of the body surface; it can be performed by individuals to their own bodies, or between members of a dyad or larger social group.

ground meristem Partly differentiated meristematic tissue (**primary meristem**) derived from the apical meristem and giving rise to ground tissue.

ground tissue Tissue other than vascular tissue, epidermis and periderm, mostly parenchyma and collenchyma of e.g. pith and cortex.

group selection A form of natural selection proposed to explain the evolution of behaviour which appears to be for the long-term good of a group or species, rather than for the immediate advantage of the individual.

group therapy Any form of psychotherapy which involves several persons at the same time, such as a small group of patients with similar psychological or physical problems who discuss their difficulties under the chairmanship of e.g. a doctor. They thus learn from the experiences of others and teach by their own.

growing point Apical meristem.

growth An irreversible change in an organism accompanied by the utilization of material, and resulting in increased volume, dry weight or protein content; increase in population or colony size of a culture of micro-organisms.

growth curvature A curvature in an elongated plant organ, caused by one side growing more than the other.

growth form See **life form**.

growth hormones Substances having growth-promoting properties, e.g. pituitary growth hormone; in plants, *auxins*.

growth inhibitor A substance that inhibits plant growth at low concentrations, esp. an endogenous substance, of which the best characterized is **abscisic acid**.

growth in soft agar The ability of cells in culture to grow in a low-concentration gel of agar. It is one of the properties which distinguish a **transformed cell** from normal cells which can only grow in culture in contact with a substrate.

growth movement Movement of a plant part brought about by differential growth. Cf. **turgor movement**.

growth regulator A **growth substance**, esp. one of the synthetic types.

growth retardant A synthetic substance used to retard the growth of a plant, e.g. to stop the sprouting of stored onions or to restrict the height of grain crops, e.g. maleic hydrazide.

growth ring A recognizable increment of wood (secondary xylem) in a cross section of a stem; most commonly an **annual ring**, but under some conditions more than one (or no) growth ring is formed within one year. See **early wood, late wood**.

growth room A room in which plants are grown under controlled artificial lighting, photoperiod, temperature etc.

growth substance One of a number of substances (sometimes called *hormones*) formed in plants, or a synthetic analogue thereof, which have specific effects on plant growth or development at low concentrations (say 10 μM). *Abscisic acid, auxins, cytokinins, ethylene, gibberellins*.

GSR Abbrev. for *Galvanic Skin Response*.

guanine Purine base occurring in DNA and RNA, pairing with cytosine.

Structure: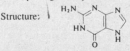

See **DNA, genetic code**.

guanophore See **xanthophore**.

guard cell One of a pair of specialized epidermal cells which surround and by increase and decrease in their turgor, open and close a stoma, thus regulating **gas exchange**.

gubernaculum In Mammals, the cord supporting the testes, in the scrotal sac; in Hydrozoa, an ectodermal strand supporting the gonophore in the gonotheca; in Mastigophora, a posterior flagellum used in steering. adj. *gubernacular*.

guest An animal living and/or breeding in the nest of another animal, as a Myrmecophile in an Ant's nest.

guild A division or category of a plant species from one area, made on the basis of similar *phenology* and *morphology*. See **synusia**. Also, of an animal community, a group of species within the same trophic level which exploits a common resource, in a similar manner.

gula In Vertebrates, the upper part of the throat.

gular In some Fish, a bone developed between the rami of the lower jaw; in Chelonia, an anterior unpaired element of the plastron.

gullet The oesophagus; in Protozoa the cytopharynx.

gum Viscid plant secretion exuding naturally or on wounding, soluble or swelling in water. Mostly complex, branching polysaccharides. Some; e.g. gum arabic, alginic acid, agar, are economically important.

gum lac An inferior type of lac, containing much wax, produced by some lac Insects (Hemiptera) in Madagascar.

gummosis A pathological condition of plants characterized by the conspicuous secretion of gums.

gums In higher Vertebrates, the thick tissue masses surrounding the bases of the teeth.

gustatory calyculus See **taste bud**.

gut The alimentary canal.

gutta A patch of colour or other marking, resembling a small drop, on the surface of an animal. pl. *guttae*. adj. *guttulate*.

guttation The exudation of drops of water (containing some solutes) from an uninjured part of a plant, commonly under conditions of high humidity, typically from **hyda-**

thodes at tips or margins of leaves.

guttural Pertaining to the throat.

GVH See **graft-versus-host reaction**.

gymno- Prefix from Gk. *gymnos*, naked.

gymnocyte A cell without a cell wall.

Gymnomycota See **Myxomycota**.

gymnosperms Group (classified as a division Gymnospermophyta or a class Gymnospermae, or regarded as polyphyletic) containing those **seed plants** in which the ovules are not enclosed in carpels, the pollen typically germinating on the surface of the ovule. There is no double fertilization and the xylem is vessel-less (except in the Gnetales). Contains ca 700 extant spp., often classified as 3 classes: Cycadopsida, Coniferopsida and Gnetopsida.

gynaecium See **gynoecium**.

gynandrism See **hermaphrodite**.

gynandromorph An animal exhibiting male and female characters.

gynandromorphism The occurrence of secondary sexual character of both sexes in the same individual.

gynandrous Having the stamens and style united.

gyn-, gyno-, gynaeco- Prefixes from Gk. *gyne*, gen. *gynaikos*, woman.

gynobasic Said of a style which appears, because of the infolding of the ovary wall, to be inserted at the base of the ovary, e.g. Labiatae.

gynodioecious Said of a species having some individual plants with hermaphrodite flowers only and others with female flowers only. Cf. **dioecious**.

gynoecium, gynaecium The female part of a flower, consisting of one or more carpels. Cf. **androecium**.

gynomonoecius Said of a species having all the plants bearing both female and hermaphrodite flowers. Cf. **monoecious**.

gynophore An elongation of the receptacle between androecium and gynoecium.

gynospore Same as **megaspore** in heterosporous plants, Cf. **androspore**.

gyrus A ridge between two grooves; a convolution of the surface of the cerebrum.

H

habenula A strap-like structure; in particular, a nerve centre of the diencephalon.

habit The chacteristic growth and form of a plant or the established normal behaviour of an animal species. As a general term in learning theory, it refers to learned patterns of behaviour which are very consistent and predictable in particular situations. Its specific usage varies in different branches of psychology, e.g. referring to compulsive behaviour in clinical psychology, to a particular cognitive style in cognitive psychology.

habitat The normal locality inhabited by a plant or animal, particularly in relation to the effect of its environmental factors.

habituated culture A plant cell culture that has developed an ability to synthesize auxin since its isolation and can now grow in its absence.

habituation An aspect of learning in which there is a decrease in responsiveness as a consequence of repeated stimulation; the habituated response reappears if the stimulus is withheld for a long time.

hadrom, hadrome (1) The conducting elements and associated parenchyma in xylem tissue. Cf. **leptom**. (2) The *hydroids* of mosses.

Haeckel's law See **recapitulation theory**.

haemad Situated on the same side of the vertebral column as the heart.

haemagglutinin A substance which agglutinates red blood cells. This may be a specific antibody, or a **lectin**, or a component of certain viruses (e.g. influenza or measles) by which they bind to cell surfaces.

haemal arch A skeletal structure arising ventrally from a vertebral centrum, which encloses the caudal blood vessels.

haemal canal The space enclosed by the centrum and the haemal arch of a vertebra, through which pass the caudal blood vessels.

haemal, haematal, haemic Pertaining to the blood or to blood vessels.

haemal ridges See **haemapophyses**.

haemal spine The median ventral vertebral spine formed by the fusion of the haemapophyses, below the haemal canal.

haemal system The system of vessels and channels in which the blood circulates.

haemapoiesis, haemopoiesis *Haematopoiesis, haematosis, haematogenesis*. The process of forming new blood.

haemapophyses A pair of plates arising ventrally from the vertebral centrum, and meeting below the haemal canal to form the haemal arch and spine. Also *ridges*.

haematobium An organism which lives in blood. adj. *haematobic*.

haematoblast A primitive blood cell, which may develop into an erythrocyte or a leucocyte; a blood platelet.

haematocele An effusion of blood localized in the form of a cyst in a cavity of the body.

haematochrome A red or orange pigment accumulated in the cells of some green algae usually when nitrogen starved, as *Chlamydomonas nivalis*, responsible for 'red snow'.

haematocrit A graduated capillary tube of uniform bore in which whole blood is centrifuged, to determine the ratio, by volume, of blood cells to plasma.

haematogenesis See **haemopoiesis**.

haematogenous Having origin in the blood.

haematophagous Feeding on blood.

haematozoon An animal living parasitically in the blood.

haemic See **haemal**.

haemocoel A secondarily formed body cavity derived from the blood vessels which replaces the coelom in Arthropods and Molluscs.

haemocyanin A blue respiratory pigment, containing copper, in the blood of *Crustacea* and *Mollusca*. It has respiratory functions similar to haemoglobin.

haemocytes The corpuscles found floating in haemolymph.

haemocytoblast A stem cell in bone marrow or in other haemopoietic tissues.

haemocytolysis See **haemolysis**.

haemocytometer An apparatus consisting of a special glass slide with a grid of lines engraved on the bottom of a shallow rectangular trough so that if a coverslip is placed over the trough the grid demarcates known volumes. Cells from a well-mixed suspension are introduced into the space and the number in the grid squares counted under the microscope. Used for blood counts, mitotic counts etc.

haemodialysis The restoration of diffusable chemical constituents of the blood towards normal, by passing blood across a cellulose membrane which has on the other side a fluid containing **electrolytes** at the desired concentration. Principally used to restore body chemistry in patients with kidney failure, *artificial kidney*.

haemoglobin, hemoglobin The red pigment of blood whose major function is the transport of oxygen from the lungs to the tissues. It is a protein of 4 polypeptide chains each bearing a haem prosthetic group which serves as an oxygen binding site.

haemoglobinometer An instrument for measuring the percentage of haemoglobin in the blood.

haemolymph The watery fluid containing leucocytes, believed to represent blood, found in the haemocoelic body cavity of certain Invertebrates.

haemolysin (1) Antibody capable of lysing red cells in the presence of complement. (2) A bacterial toxin which lyses red cells.

haemolysis *Haemocytolysis*. The lysis of red blood cells.

haemolytic anaemia Anaemia due to an abnormal increase in the rate of destruction of circulating erythrocytes. This can result from the presence of antibodies against the erythrocytes (e.g. against the Rhesus antigen as in **erythroblastosis foetalis**, or to auto-antibodies); or from over activity of mononuclear phagocytes in association with grossly enlarged spleen (hypersplenism); or from metabolic abnormalities of the erythrocytes such as glucose-6-phosphatase deficiency aggravated by some drugs.

haemolytic plaque assay A method used to detect and enumerate individual cells secreting antibody in vitro. Sheep red cells (treated when necessary so as to bind the antibody) are mixed with a cell suspension to be assayed in a thin layer of agarose, and incubated so that they can secrete antibody which diffuses onto the red cells. Complement is then added and cells secreting antibody are revealed by the presence of an area of haemolysis around them.

haemophilia A hereditary bleeding tendency, with deficiency of a normal blood protein (factor VIII, anti-haemophilic globulin) preventing normal clotting. The classic **sex-linked** disorder. Clinically indistinguishable from *Christmas disease* where there is a factor IX deficiency and is sometimes called *h.B.*

haemorrhage Bleeding; escape of blood from a ruptured blood vessel.

haemotropic Affecting blood.

hair A slender, elongate structure, mostly composed of **keratins**, arising by proliferation of cells from the Malpighian layer of the epidermis in Mammals; in plants, a **trichome**; more generally, any threadlike growth of the epidermis.

hair follicle See **follicle**.

hairpin loop A short double-stranded region made possible in a single-stranded nucleic acid molecule because of the complementarity of neighbouring sequences. Common in tRNA.

hair plates Groups of articulated sensory hairs occurring near the joints of the appendages in insects and acting as proprioceptors.

hali-, halo- Prefixes from Gk. *hals*, salt.

haliplankton The plankton of the seas.

hallucination A perceptual experience that occurs in the absence of any appropriate external stimulus.

hallux In land Vertebrates, the first digit of the hind-limb.

halobiotic Living in salt water, esp. in the sea.

halo effect In the perception of other people, the tendency to generalize an impression of one characteristic of a personality to other, unrelated aspects of the personality.

halolimnic Originally marine but secondarily adapted to fresh water.

halophile A freshwater species capable of surviving in salt water.

halophilic bacteria Salt-tolerant bacteria occurring in the surface layers of the sea, where they are important in the nitrogen, carbon, sulphur and phosphorus cycles. Many are pigmented or phosphorescent and if present in quantity may colour the surface water. Some halophilic bacteria, e.g. halobacterium, are able to live in salted meats.

halophyte A plant able to grow where the soil is rich in sodium chloride or other sodium salts. Cf. **glycophyte**.

halophytic vegetation A population of halophytes, e.g. on mangrove swamp, salt marsh or alkaline soil.

halosere A *sere* that starts on land emerging from the sea, e.g. a salt marsh.

halteres A pair of capitate threads which take the place of the hind wings in *Diptera*, and assist the insect to maintain its equilibrium while flying; balancers.

Hamamelidae Subclass or superorder of dicotyledons. Mostly woody plants, the perianth poorly developed, or none, flowers often unisexual, often borne in catkins, often wind pollinated. Contains ca 3400 spp. in 23 families including Betulaceae and Fagaceae. (Includes the Amentiferae.)

hand monitor Radiation monitor designed to measure radioactive contamination on the hands of an operator, or to be held in hand.

hanging drop preparation A preparation for the microscope in which the specimen, in a drop of medium on the undersurface of a coverslip, is suspended over a hollow-ground slide, to which it is sealed to prevent evaporation.

hapanthous See **hapaxanthic**.

hapaxanthic *Hapanthous*. Flowering and fruiting once and then dying. See **monocarpic**.

haplo- Prefix from Gk. *haploos*, single, simple.

haplobiont A plant which has only one kind of individual or form in its life history. adj. *haplobiontic*.

haplodiploidy A means of sex determination where females develop from fertilized eggs and are therefore diploid and the males from unfertilized eggs and are haploid, e.g. Honeybees.

haplodont Having molars with simple crowns.

haploid Of the reduced number of chromosomes characteristic of the germ cells of a species, equal to half the number in the somatic cells. Cf. **diploid**.

haploidization In the **parasexual cycle** of fungi, the progressive loss of chromosomes from the diploid set by occasional nondisjunction until stable haploid nuclei are formed.

haplont Organism in which only the gametes are haploid, meiosis occurring at their formation and the vegetative cells being diploid. Cf. **diplont**.

haplophase The period in the life cycle of any organism when the nuclei are haploid. Cf. **diplophase**.

haplostele A protostele in which the solid central core of xylem is circular in cross section.

haplostemonous Having a single whorl of stamens. Cf. **diplostemonous**.

haplotype A particular set of *alleles* at several very closely linked loci.

haploxylic Said of a leaf containing one vascular strand.

hapten A substance which can combine with antibody but cannot itself initiate an immune response unless it is attached to a carrier molecule. Most haptens are small molecules (e.g. dinitrophenyl). Haptens are often conjugated chemically to carrier proteins for experimental purposes, since they provide easily recognized antigenic determinants.

hapteron A holdfast, i.e. a unicellular or multicellular organ attaching a plant to the substrate.

haptonema Appendage arising between the flagella of the motile cells of the Haptophyceae sometimes serving for temporary attachment to a surface.

Haptophyceae *Prymnesiophyceae*. Class of eukaryotic algae, flagellated and palmelloid sorts (often interconvertible). Motile cells usually with 2 equal, smooth, flagella and a **haptonema**. Some sorts or stages bear **coccoliths**. Mostly phototrophic, some heterotrophic (both osmotrophic and phagotrophic). Mostly marine.

haptotropism *Thigmotropism*. A **tropism** like that of a tendril coiling round its support in which differential growth is determined by touch.

hard bast Sclerenchyma present in phloem.

hardening Increasing resistance to cold as temperatures are gradually lowered either naturally or as the result of horticultural

practice. Analogous hardening to drought, heat, wind etc. occurs.

Harder's glands In most of the higher Vertebrates, an accumulation of small glands near the inner angle of the eye, closely resembling the lacrimal gland.

hard palate In Mammals, the anterior part of the roof of the buccal cavity, consisting of the horizontal palatine plates of the maxillary and palatine bones covered with mucous membrane.

hard radiation Qualitatively, the more penetrating types of X-, beta- and gamma-rays.

Hardy-Weinberg law The gene frequencies in a large population remain constant from generation to generation if mating is at random and there is no selection, migration or mutation. If two alleles A and a are segregating at a locus, and each has a frequency of p and q respectively, then the frequencies of the genotypes AA, Aa, and aa are p^2, $2pq$ and q^2 respectively.

harvest spider Common name for Arachnids of the order *Opiliones*. Also *harvestmen*.

Hashimoto thyroiditis A disease of the thyroid gland, characterized by chronic inflammatory changes due to infiltration by lymphocytes, plasma cells and macrophages. The gland becomes enlarged and hard. Autoantibodies against thyroglobulin and other thyroid antigens are usually present in the blood.

hashish See **cannabis**.

hastate Shaped like an arrow head with narrow basal lobes pointing outwards (i.e. halberd-shaped).

Hatch-Slack pathway Pathway of carbon in some sorts of C_4 plants. See C_4

haustellate Mouthparts modified for sucking, as in many Insects.

haustellum In *Diptera*, the distal expanded portion of the proboscis. adj. *haustellate*.

haustorium An outgrowth from a parasite which penetrates a tissue or cell of its host and acts as an organ for absorbing nutrients.

Haversian canals Small channels pervading compact bone and containing blood vessels.

Haversian lamellae In compact bone, the concentrically arranged lamellae which surround a Haversian canal.

Haversian spaces In the development of bone, irregular spaces formed by the internal resorption of the original cartilage bone.

Haversian system In compact bone, a Haversian canal with surrounding lamellae.

Hawthorne effect The observation that experimental subjects who are aware that they are part of an experiment often perform better than totally naive subjects. See **demand characteristics**.

hay fever *Allergic rhinitis.* Acute nasal catarrh and conjunctivitis in atopic subjects, caused by inhalation of allergens (usually pollens). Due to an immediate hypersensitivity (Type 1) reaction resulting from combination of cell–IgE antibody with the causative allergen. There is a tendency for this type of sensitization to be inherited.

head A dense inflorescence of sessile flowers, usually a **capitulum**.

health physics Branch of radiology concerned with health hazards associated with ionizing radiations, and protection measures required to minimize these. Personnel employed for this work are *health physicists* or *radiological safety officers.*

heart A hollow organ, with muscular walls, which by its rhythmic contractions pumps the blood through the vessels and cavities of the circulatory system.

heart attack The common term for a *myocardial infarction.*

heart wood *Duramen.* The inner, older layers of wood in the trunk or branch of a tree or shrub. Compared to the surrounding sap wood, it is denser and darker, the cell walls are impregnated with resins, phenolics etc. and it no longer functions in storage or conduction. It is more prone to rot in the tree but more durable as timber.

heat The period of sexual excitation.

heath Vegetation type consisting of evergreen woody shrubs growing on acid soil. In northern Europe the species are largely members of the *Ericaceae*, but the term heath is often used more widely to cover dwarf shrub communities in other parts of the world.

heat shock protein A class of protein produced in excess when the organism or culture is subject to an elevated temperature. Mechanisms such as those controlling protein synthesis may be studied with such procedures.

heat spot An area of the skin sensitive to heat owing to the presence of certain nerve endings beneath the skin.

heavy (H) chain Polypeptide chain in immunoglobulins which together with the light chain makes up the complete molecule. Each heavy chain consists of a variable region and a constant region composed of three or four domains, depending on the class. See **immunoglobulin.**

heavy metal (1) In electron microscopy, metal of high atomic number used to introduce electron density into a biological specimen by staining, negative staining or shadowing. (2) In plant nutrition, metals of moderate to high atomic number, e.g. Cu, Zn, Ni, Pb, present in soils due to an outcrop or mine spoil, preventing growth except for a few tolerant species and ecotypes.

hebephrenia *Hebephrenic schizophrenia, disorganized schizophrenia.* A type of schizophrenia characterized by incoherence of thought, odd and childlike behaviour, and inappropriate emotional expression, occurring about puberty.

hecto- Metric prefix meaning ×100.

hectocotylized arm In some male *Cephalopoda*, one of the tentacles modified for the purpose of transferring sperm to the female.

helical coil model Model of **metaphase** chromosome organization which envisages that the primary DNA helix is packed by secondary and higher orders of coiling.

helical thickening Secondary wall deposited in the form of a helix in the tracheids and vessel elements of xylem.

helicoid Coiled like a flat spring.

helicotrema An opening at the apex of the cochlea, by which the scala vestibuli communicates with the scala tympani.

helio- Prefix from Gk. *helios*, sun.

heliophyte See **sunplant.**

heliotaxis Response or reaction of an organism to the stimulus of the sun's rays. adj. *heliotactic, heliotropic.*

heliotropism See **heliotaxis.** More particularly, response by growth curvature.

helix A line, thread, wire or other structure curved into a shape such as it would assume if wound in a single layer round a cylinder; a form like a screw thread. Very common in biological macromolecules, e.g. DNA.

Helminthes A name formerly used in classification to denote a large group of worm-like Invertebrates now split up into *Platyhelminthes*, *Nematoda* and smaller groups.

helophyte Marsh plant with perennating buds below the surface of the marsh. See **Raunkiaer system.**

helper T lymphocyte Often termed *helper T cell.* A thymus derived lymphocyte which cooperates with B lymphocytes to enable them to produce antibody when stimulated by antigen or by some polyclonal mitogens. Helper T cells release lymphokines causing differentiation and growth of B cells. They also influence the generation of cytotoxic and suppressor T cells.

hem-, hema-, hemo- US for *haem-, haema-, haemo-.*

hemi- Prefix from Gk. *hemi*, half.

Hemiascomycetes A class of fungi in the Ascomycotina in which no ascocarps are formed. Mostly unicellular or with poorly developed mycelium. Includes the yeasts, *Saccharomyces* and *Schizosaccharomyces*, and some plant parasites causing, e.g. peach leaf curl.

hemibranch The single row of gill lamellae or filaments, borne by each face of a gill arch in Fish; a gill arch with respiratory lamellae or filaments on one face only.

hemicelluloses A group of polysaccharides in the matrix of plant cell walls; homo- and hetero-polymers, linear and branched, of xylose, glucose and other sugars.

Hemichorda A subphylum of Chordata, lacking any bony or cartilaginous skeletal structures; no tail or atrium; having a reduced notochord in the pre-oral region, and a superficial central nervous system; the three primary coelomic cavitites persist in the adult. Also *Hemichordata, Protochordata.*

hemicryptophyte Plant with buds at soil level. See **Raunkiaer system.**

hemignathous Having jaws of unequal length.

Hemimetabola See **Exopterygota.**

hemimetabolic Having an incomplete or partial metamorphosis.

hemiparasite See **partial parasite.**

hemipenes In Squamata, the paired eversible copulatory organs.

Hemiptera An order of Insecta, having two pairs of wings of variable character; the mouth parts are symmetrical and adapted for piercing and sucking; in some forms, the females are wingless; many are ectoparasitic, others feed on plant juices. Bugs, Cicadas, Aphids, Plant Lice, Scale Insects, Leaf Hoppers, White Flies, Black Flies, Green Flies, Cochineal Insects.

hemisphere One of the cerebral hemispheres (see **cerebrum**).

hemizygous Having only one representative of a gene or a chromosome, as are male mammals which have only one *X-chromosome.*

hemoglobin See **haemoglobin.**

Henle's loop The loop formed by a uriniferous tubule when it enters the medulla of the Mammalian kidney and turns round to pass upwards again to the cortex.

hepat-, hepato- Prefixes from Gk. *hepar,* gen. *hepatos,* the liver.

hepatic *n.* A liverwort. See **Hepaticopsida.** *adj.* Pertaining to the liver.

hepatic artery In Craniata, a branch of the coeliac artery which conveys arterial blood to the liver.

hepatic duct In Craniata, a duct conveying the bile from the liver and discharging into the intestine as the common bile duct.

Hepaticopsida, Hepaticae Class of the Bryophyta containing ca 10 000 spp. The gametophyte is thalloid or leafy with unicellular rhizoids and the capsule (sporophyte) without a columella.

hepatic portal system In Craniata, the part of the vascular system which conveys blood to the liver; it consists of the **hepatic portal vein** and the **hepatic artery.**

hepatic portal vein In Craniata, the vein which conveys the blood from the alimentary canal to the liver.

hepatopancreas In many Invertebrates (as Mollusca, Arthropoda, Brachiopoda), a glandular diverticulum of the mesenteron, frequently paired, consisting of a mass of branching tubules, and believed to carry out the functions proper to the liver and pancreas of higher Vertebrates.

hepatoportal system See **hepatic portal system.**

herb (1) A plant that does not develop persistent woody tissues above ground. (2) A plant that is used for medicinal purposes or for flavour.

herbaceous A soft and green plant or one without persistent woody tissues above ground.

herbaceous perennial A perennial plant with a perennating structure at or below ground level and producing aerial shoots that die at the end of the growing season.

herbarium A collection of dried plants; by extension, the place where such a collection is kept.

hercogamy Physical arrangement of anthers and stigma so that pollen is not transferred from one to the other in the absence of an insect visit.

herding The control of the movements of animals by individuals of the same or different species.

hereditary Inherited; capable of being inherited; passed on or capable of being passed on from one generation to another.

hereditary angioneurotic oedema Disease characterized by recurrent episodes of transient oedema of skin and mucous membranes, and due to absence or functional inactivity of *Cl⁻-esterase inhibitor.* It is inherited as a Mendelian dominant. See **C1⁻-inhibitor.**

heredity The relation between successive generations, by which characters persist.

heritability Measure of the degree to which the variation of a character is inherited. Ranges from 0 to 100%. It is the proportion of *additive genetic variance* in the total *phenotypic variance.* Usually symbolized by H^2.

hermaphrodite In man, one whose reproductive organs are anatomically ambiguous, so that they are not exclusively male or female. Said of flowers with male and female organs functional in the one flower. Also *monoclinus.* Cf. **unisexual, dioecious.** Generally having both male and female reproductive organs in one individual, thus producing both male and female gametes. n. *hermaphroditism.*

heter-, hetero- Prefixes from Gk. *heteros,* other, different.

hetero-auxin Old name for indole acetic acid; auxin.

heteroblastic (1) Having a marked mor-

heteroduplex DNA

A duplex DNA made by **renaturing** (or *annealing*) single DNA strands from different sources. Depending on salt concentration and temperature heteroduplexes can form under high *stringency* between long perfectly matched complementary strands with relatively short mismatched single-stranded regions or at low stringency with the heteroduplex held by many short double stranded regions only a few base pairs long. It is therefore possible to measure the degree of similarity between DNA sequences and hence possibly evolutionary relatedness.

In addition, enzymes which cut only single-stranded DNA and electron microscopy can be used to analyze the size of the remaining duplex regions and to 'map' DNA from two sources.

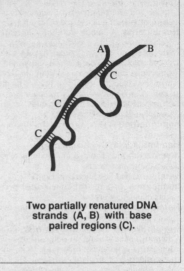

Two partially renatured DNA strands (A, B) with base paired regions (C).

phological difference between the first-formed structures, like leaves on a seedling, and those formed later. See **juvenile phase**. Cf. **homoblastic**. (2) A plant whose seeds vary in the conditions they require for germination.

heteroblastic Showing indirect development.

heterocercal Said of a type of tail fin, found in adult Sharks, Rays, Sturgeons and many other primitive Fish, in which the vertebral column bends abruptly upward and enters the epichordal lobe, which is larger than the hypochordal lobe.

heterochlamydeous Having a distinct calyx and corolla.

heterochromatin Relatively dense chromatin visible by microscopy in eukaryotic cell nuclei. Generally contains DNA sequences inactive in transcription. See **facultative-, constitutive-**.

heterocoelous Said of vertebral centra in which the anterior end is convex in vertical section, concave in horizontal section, while the posterior end has these outlines reversed.

heterocotylized arm See **hectocotylized arm**.

heterodactylous Of Birds, having the first and second toes directed backwards, the third and fourth forwards, as the Trogons.

heterodont Of teeth, having different forms adapted to different functions.

heterodromous Having two kinds of asymmetric flowers, one the mirror image of

the other and sometimes associated in pairs.

heteroduplex DNA See panel.

heteroecious Said of a parasite which requires two, usually unrelated, host species to complete different stages of its life cycle. Cf. **autoecious**.

heterogamete See **anisogamete**.

heterogametic sex The sex that is heterozygous for the *sex-determining chromosome*, the male in mammals, the female in birds; its gametes determine the sex of the progeny. Cf. **homogametic**.

heterogamous Producing unlike gametes or flowers.

heterogenesis See **metagenesis, abiogenesis**. adj. *heterogenetic*.

heterogenous summation Refers to the fact that where a response is influenced by stimulus characters acting through more than one sensory modality, their effects may supplement each other, e.g. parts of the stimulus presented successively may produce the same response as when they are presented simultaneously.

heterogeny Cyclic reproduction in which several broods of parthenogenetic individuals alternate with one or more broods of sexual forms.

heterogony Reproduction by both parthenogenesis and **amphigony**.

heterokaryon Somatic cell hybrid containing separate nuclei from different species. Also *heterokaryote*.

heterokaryosis Coexistence of genetically

different nuclei in a common cytoplasm, esp. in a fungal hypha.

heterokont, heterokontan Having flagella of unequal length or different type.

Heterokontophyta A division of eukaryotic algae, characterized by: chlorophyll a and usually c; mostly with fucoxanthin; triplet thylakoids; chloroplast ER; mitochondrial microvilli; **heterokont** flagellation; storing β, 1→3 glucans in the cytoplasm. Contains the classes Xanthophyceae, Chrysophyceae, Bacillariophyceae, Raphidophyceae and Paeophyceae.

heterolecithal With unequally distributed yolk.

heteromastigote Having one or more anterior flagella directed forwards and a posterior flagellum directed backwards.

heteromerous Said of a lichen thallus with the algal cells confined to a distinct layer, with pure fungus above and below. Cf. **homoiomerous**.

heterometabolic Having incomplete metamorphosis.

heteromorphic, heteromorphous (1) Existing in or having more than one form. (2) See **alternation of generations**.

heteronomous Subject to different laws, esp. of growth and specialization. Cf. **autonomous**.

heterophil antigen Antigens occurring on cells of many different animal species, plants and bacteria which are sufficiently similar to elicit antibodies which cross react extensively, e.g. *Forssman antigen*.

heterophylly An individual plant with two or more different forms of leaf as in the submerged, floating and aerial leaves of many water plants.

heteroplasma In tissue culture, a medium prepared with plasma from an animal of a different species from that from which the tissue was taken. Cf. **autoplasma, homoplasma**.

heteroplastic In experimental zoology, said of a graft which is transplanted to a site different from its point of origin, e.g. epithelial cells of cornea to a skin site. In medicine, *heteroplasty* is the operation of grafting on one person body tissue removed from another.

heteropycnosis Differential stainability of bands of chromosomes.

heteroscedastic In statistics, having unequal variances.

heterosexual Sexual interest directed at members of the opposite sex.

heterosis The difference between the mean of a quantitative character in a crossbred generation and the mean of the two parental strains. Also *heterozygous advantage*. See **hybrid vigour**.

heterospory The production of more than one type of spore, typically megaspores and microspores. adj. *heterosporous*. Cf. **homospory**.

heterostyly The condition in which individuals of a species have style and thus stigma lengths falling into 2 or more distinct classes. This causes the anthers to be placed low in flowers with high stigmas and vice versa. Heterostyly appears to promote cross-pollination. Cf. **homostyly, illegitimate pollination, pin, thrum**.

heterothallism The condition in which there are 2 (or more) mating types with sexual reproduction only successful between individuals of a different type. Also *self-incompatibility*. adj. *heterothallic*. Cf. **homothallism**.

heterotrichous Having cilia or flagella of two or more different kinds.

heterotrophic Said of organisms which require carbon in organic form, as do all animals (except phytoflagellates), fungi, some algae, parasitic plants and most bacteria. Cf. **autotrophic**.

heterotypic Differing from the normal condition. Cf. **homotypic**.

heterotypic division The first (reductional) of two nuclear divisions in meiosis in which the number of chromosomes is halved. The second (equational) is a *homotypic division*.

heterozygosis The condition of being *heterozygous*.

heterozygosity The proportion of individuals in a population that are *heterozygous* at a specified locus, or at a number of loci averaged.

heterozygote An individual with two different alleles at a particular locus, the individual having been formed from the union of gametes carrying different alleles. adj. *heterozygous*. Cf. **homozygote**.

hex- Prefix from Gk. *hex*, six.

Hexactinellida A class of Porifera, which is usually distinguished by the possession of a siliceous skeleton composed of triaxial spicules and by large thimble-shaped flagellated chambers.

hexamerous Having parts in sixes.

hexapod Having 6 legs.

Hexapoda See **Insecta**.

hexarch Having 6 strands of protoxylem.

H-2 histocompatibility system The major histocompatibility system in the mouse. H-2 genes determine the major histocompatibility antigens on the surface of somatic cells and also the immune response (Ir) genes. The antigens in a given strain of mice are controlled by alleles within the H-2 locus. The H-2 system corresponds closely to the **HLA system** of humans, and much of our understanding of the latter was derived from studies in mice.

hibernation The condition of partial or

complete torpor into which some animals relapse during the winter season. v. *hibernate*.

hidrosis Formation and excretion of sweat.

high endothelial venule Specialized venules in the thymus dependent area of lymph nodes characterized by prominent high endothelial lining cells. Recirculation of lymphocytes from blood to lymph takes place through the walls of these venules.

high energy phosphate compounds Phosphate compounds with a high negative free energy of hydrolysis. Endergonic metabolic reactions are driven by coupling them with the exergonic hydrolysis of these phosphate esters, the most common example being the hydrolysis of ATP.

higher-order conditioning A form of conditioning in which a conditioned stimulus from earlier training serves as an unconditioned stimulus.

Hill reaction The light-driven transport of electrons from water to some acceptor other than CO_2 (e.g. ferricyanide) with the production of oxygen, by isolated chloroplasts or chloroplast-containing cells.

hilum In plants, a scar or mark, esp. (1) on the testa where the stalk was attached to the seed; (2) the central part of a starch grain around which the starch is deposited. In animals, a small depression in the surface of an organ, which usually marks the point of entry or exit of blood vessels, lymphatics or an efferent duct. See *hilus*.

hind brain In Vertebrates, that part of the brain which is derived from the third or posterior brain vesicle of the embryo, comprising the cerebellum and the medulla oblongata; the posterior brain vesicle itself.

hind-gut That part of the alimentary canal of an animal which is derived from the posterior ectodermal invagination or proctodaeum of the embryo.

hinge The flexible joint between the 2 valves of the shell in a bivalve Invertebrate, such as a pelecypod Mollusc or a Brachiopod; any similar structure; a joint permitting of movement in one plane only.

hinge ligament The tough uncalcified elastic membrane which connects the 2 valves of a bivalve shell.

hinge region A flexible region of immunoglobulin heavy chains near the junction of the Fab and Fc portions. This flexibility allows the angle between the arms bearing the antigen combining sites to vary widely and so accommodate different dispositions of the antigen.

hippocampus In the Vertebrate brain, a tract of nervous matter running back from the olfactory lobe to the posterior end of the cerebrum. adj. *hippocampal*.

hirsute Hairy; having a covering of stiffish hair or hairlike feathers.

hirudin An anticoagulin, present in the salivary secretion of the leech, which prevents blood clotting by inhibiting the action of thrombin on fibrinogen.

Hirudinea A class of Annelida, the members of which are ectoparasitic on a great variety of aquatic and terrestrial animals; they possess anterior and posterior suckers, and most of them lack setae; hermaphrodite animals with median genital openings; the development is direct. Leeches.

His Symbol for **histidine**.

hispid Coarsely and stiffly hairy as in many Boraginaceae; having a covering of stiffish hair or hair-like feathers.

histidine *2-Amino-3-imidazolepropanoic acid.* The *L*-isomer is a 'neutral' amino acid and constituent of proteins. It is a precursor of **histamine**. Symbol His, short form H.

Side chain: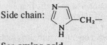

See **amino acid**.

histiocyte A macrophage found within the tissues in contrast to those found in the blood (monocytes).

histochemistry The chemistry of living tissues.

histocompatibility antigen Genetically determined antigens present on the surface of nucleated cells, including blood leucocytes. Coded for by *MHC* and other genes. They are responsible for the differences between genetically non-identical individuals which cause rejection of **allografts**. See **major histocompatibility complex**.

histocompatibility testing Tests whereby donor and recipient are matched as closely as possible prior to tissue grafting in humans.

histocompatible Compatible for the purpose of homografting. Used of pairs or sets of individuals who share identical **histocompatibility antigens**.

histogen One of three meristems (**dermatogen, periblem, plerome**) at the shoot or root tip that give rise exclusively to particular tissues in that organ (epidermis, cortex, stele + pith, respectively). The concept of discrete histogens has been replaced more recently by the **tunica-corpus concept**.

histogenesis Formation of new tissues.

histogram A graphical representation of class frequencies as rectangles against class interval, the value of frequency being proportional to the area of the corresponding rectangle.

histology The study of the minute structure of tissues and organs.

histolysis Dissolution and destruction of tissues.

histones Basic proteins involved in the

packaging of DNA in the eukaryotic nucleus to form **chromatin**, which is folded into **nucleosomes**, the first level of chromosome organization above the DNA helix. There are five types of histone molecule, four of which have been highly conserved in sequence throughout eukaryotic evolution.

histozoic Living in the tissues of the body, amongst the cells.

HIV Abbrev. for *Human Immunodeficiency Virus.*

HLA-A, HLA-B, HLA-C Histocompatibility antigens, each coded by different loci in the MHC genes, and for which there are numerous allelic products at each locus. These antigens belong to Class I and have similar structures, consisting of a membrane bound polypeptide to which is attached noncovalently β2 microglobulin. Class I antigens are present on almost all nucleated cells. Cytotoxic T cells recognize antigens in association with Class I molecules.

HLA-D Histocompatibility antigens coded for by separate loci in the MHC gene complex. There are three distinct loci — DP, DR and DQ — each of which has several alleles. These antigens belong to Class II, and are dimers composed of two different membrane bound polypeptide chains. HLA-D antigens are normally absent from most nucleated cells but present on B lymphocytes, dendritic (interdigitating) cells, and macrophages stimulated by **interferon**. T-helper cells recognize antigens in association with Class II molecules.

Hn RNA, heterogeneous nuclear RNA Population of RNA molecules in the nucleus including the precursors of mature messenger RNA, which are eventually found in the cytoplasm.

hoarding Refers to the storing of food or other items in the animal's home or territory; occurs in small mammals and in some bird species.

hock The tarsal joint of a Mammal.

Hodgkin's disease One of a group of diseases named lymphomas which involve lymphoid tissues. It is characterized by destruction of the normal architecture of lymph nodes and replacement with reticular cells, lymphocytes, neutrophil and eosinophil leucocytes, and an unusual kind of giant cell with two nuclei. This is accompanied by deficient cell-mediated immunity although antibody formation is normal.

Hoechst 33258 TN for a DNA-specific stain, used in chromosome **banding**.

hol- See **holo-**.

Holarctic region One of the primary faunal regions into which the surface of the globe is divided. It includes North America to the edge of the Mexican plateau, Europe, Asia (except Iran, Afghanistan, India south of the Himalayas, the Malay peninsula), Africa north of the Sahara, and the Arctic islands.

holdfast Any single-celled or multicellular organ other than a root, which attaches a plant (esp. an alga) to the substrate.

holobenthic Passing the whole of the life cycle in the depths of the sea.

holoblastic Said of ova which exhibit total cleavage.

holobranch In Fish, a branchial arch carrying two rows of respiratory lamellae or filaments, one on the posterior and one on the anterior face.

holocarpic Having the whole thallus transformed at maturity into a sporangium or a sorus of sporangia.

holocentric chromosome One lacking a localized **centromere**, and along whose full length spindle **microtubules** attach. Found in some plants, protozoa and certain classes of insect.

holocephalous Said of single-headed ribs.

hologamy The condition of having gametes which resemble in size and form the ordinary cells of the species; union of such cells.

holo-, hol- Prefixes from Gk. *holos*, whole.

holomastigote Having numerous flagella scattered evenly over the body.

Holometabola See **Endopterygota**.

holometabolic Insects showing a complete metamorphosis, as members of the *Endopterygote*. Cf. **hemimetabolic**. n. *holometabolism.*

holophytic Living by photosynthesis, like most plants which are not parasitic, saprophytic or phagocytic etc. Cf. **holozoic**.

holostyly A type of jaw suspension in which the upper jaw fuses with the cranium, the hyoid arch playing no part in the suspension; characteristic of *Holocephali*.

Holothuroidea A class of *Echinodermata* having a sausage-shaped body without arms; the tube feet possess ampullae and may occur on all surfaces; the anus is aboral, the madreporite internal; the skeleton is reduced to small ossicles embedded in the soft integument; free-living mud feeders. Sea Cucumbers.

holotrichous Bearing cilia of uniform length over the whole surface of the body.

holotype The one specimen designated as the nomenclatural type in a published description of the species.

holozoic Living by ingestion or phagocytosis and digestion, like most animals; not *osmotrophic*. n. *holozoon*. Cf. **holophytic**.

homeostasis The tendency for the internal environment of an organism to be maintained constant. The analogous tendency of plant and animal populations to remain constant as a result of density-dependent mechanisms operating on birth rate, survival or death rate.

homeotic mutants Mutants in *Drosophila* which effect large-scale changes in development, e.g. the substitution of a leg for a wing.

homeotypic division See **mitosis**.

home range A definite area to which individuals, pairs, or family groups of many types of animals restrict their activities.

homing behaviour The navigational behaviour in a number of species, ranging from returning home after a daily foraging or other excursion, to the more complex navigational task involved in large migrations.

homo- Prefix from Gk. *homos;* same.

homoblastic A species in which the first formed leaves in a seedling or shoot are very like those formed later. Cf. **heteroblastic**. Animals showing direct embryonic development, i.e. originating from similar cells.

homocercal Said of a type of tail fin, found in all the adults of the higher Fish, in which the vertebral column bends abruptly upwards and enters the epichordal lobe, which is equal in size to the hypochordal lobe.

homochlamydeous Having a perianth consisting of members all of the same kind, not distinguishable into sepals and petals.

homodont Said of teeth which all have the same characteristics.

homoeomerism In metameric animals, the condition of having all the somites alike. Cf. **heteromerism**. adj. *homoeomeric*.

homoeosis In metameric animals, the assumption by a merome of the characters of the corresponding merome of another somite. Cf. **heterosis**.

homogametic Having all the gametes alike.

homogametic sex The sex that is *homozygous* for the *sex-determining chromosomes*. Cf. **heterogametic sex**.

homogamy (1) The simultaneous maturation of the anthers and stigmas in a flower. Cf. **dichogamy**, **protandry**, **protogyny**. (2) Inbreeding, usually due to isolation.

homogenesis The type of reproduction in which the offspring resemble the parents.

homogeny Individuals or parts thereof which are **homologous**. adj. *homogenous*.

homograft Graft of tissue from one individual to another of the same species.

homoiohydric Plants able to regulate water loss and to remain hydrated for some time (hours, days or years) when the external water supply is restricted, e.g. most terrestrial vascular plants, few of which can survive desiccation. Cf. **poikilohydric**.

homoiomerous Said of a lichen thallus which has an even distribution of algal cells through its thickness. Cf. **heteromerous**.

homoioplastic, homoplastic In experimental zoology, said of a graft which is transplanted to a site identical with its point

of origin, e.g. a skin graft to a skin site.

homoiothermal See **warm-blooded**.

homokaryon **Somatic cell hybrid** containing separate nuclei from the same species.

homologous Of the same essential nature, and of common descent.

homologous chromosomes Chromosomes that pair with each other during synapsis at *meiosis*, so that one member of each pair is carried by every gamete.

homologous organs Organs which are equivalent morphologically and of common evolutionary origin but which may be similar or dissimilar in appearance or function.

homologous theory of alternation The hypothesis that the sporophyte is of a similar nature to the gametophyte and thus that vascular plants evolved from algae with an **isomorphic alternation of generation**. Also *homologous alternation of generation*. Cf. **antithetic theory of alternation**.

homologous variation The occurrence of similar variations in related species.

homology (1) Morphological equivalence, common evolutionary origin. n. *homologue*. Cf. **analogy**. (2) Refers to the degree of similarity of DNA or peptide sequences.

homomorphic Said of chromosome pairs which have the same form and size.

homomorphous Alike in form.

homoplasma In tissue culture, a medium prepared with plasma from another animal of the same species as that from which the tissue was taken. Cf. **autoplasma**, **heteroplasma**.

homoplastic Of the same structure and manner of development but not descended from a common source. n. *homoplasty*.

homopolymer DNA or RNA strand whose nucleotides are all of the same kind. Usually made enzymatically from a single nucleotide precursor.

homoscedastic Having the same variance (applied to sets of observations).

homosexuality Sexual interest directed at members of one's own sex.

homospory A species which produces only one type of a spore. adj. *homosporous*. Cf. **heterospory**.

homostyly The condition in which all the styles are the same length. Cf. **heterostyly**.

homothallism The condition in which successful fertilization can take place between any two gametes from the same organism. Also *self compatibility*. adj. *homothallic*. Cf. **heterothallic**. It is analogous to self compatibility in flowering plants.

homothermous See **warm-blooded**.

homotypic Conforming to the normal condition. Cf. **heterotypic**.

homozygosis The condition of being *homozygous*.

homozygote An individual whose two

genes at a particular locus are the same allele, the individual having been formed by the union of gametes carrying the same allele. adj. *homozygous*. Cf. **heterozygote**.

homunculus A dwarf of normal proportions; a mannikin or little man created by the imagination; a miniature human form once believed by animalculists to exist in the spermatozoon.

honeycomb bag See **reticulum**.

honey dew A sweet substance secreted by certain *Hemiptera*; emitted through the anus.

honey guide, nectar guide Lines or dots (sometimes visible only in UV photographs), on the perianth, that direct a pollinating insect to the nectar in a flower.

hoof In Ungulates, a horny proliferation of the epidermis, enclosing the ends of the digits.

hookworms Parasitic strongyloid nematodes with hooklike organs on the mouth for attachment to the host. Humans are attacked chiefly by the genus *Ankylostoma*, which penetrates the bare feet and induces a form of anaemia. See **ankylostomiasis**.

horizon (1) The more or less coloured visual impression experienced subjectively by blind persons. (2) A layer in a soil distinguishable from others by colour, hardness, inclusions or other visible or tangible properties.

hormone In animals, a substance released by an *endocrine gland* into the bloodstream which carries it to remote sites in the body. There are many kinds with a wide range of activating and repressing functions; their synthetic derivatives are often used therapeutically. See **steroid hormone**. In plants, a substance which at low concentrations (say < 10 μM) regulates some process in cells other than the ones in which it is made, e.g. abscisic acid, auxins, cytokinins, ethylene, gibberellins and, in some usages, their synthetic analogues. See **growth substance**.

horn Keratin, one of the pointed or branched hard projections borne on the head in many Mammals; any conical or cylindrical projection of the head resembling a horn; in some Birds, a tuft of feathers on the head; in some *Gastropoda*, a tentacle; in some Fish, a spine. adjs. *horned, horny*.

hornwort See **Anthocerotopsida**.

horseradish peroxidase Enzyme commonly conjugated with antibodies for use in *immunoassays*. Catalyses deposit of dye at the site of binding. Abbrev. *HRP*.

host An organism which supports another organism (parasite) at its own expense; in molecular biology that in which a plasmid or virus can replicate.

host range The range of species or strains of bacteria which will support the replication of a plasmid or virus.

hot spot Region of a chromosome peculiarly susceptible to mutation or recombination.

housekeeping gene Term for genes specifying those functions common to all cells and their metabolism, in contrast to those needed only in differentiated cells.

HRP See **horseradish peroxidase**.

human immunodeficiency virus The retrovirus responsible for **acquired immunodeficiency syndrome**, several types of which have been recognized. Abbrev. *HIV*, with a number added to denote the type.

humeral In Vertebrates, pertaining to the region of the shoulder; in Insects, pertaining to the anterior basal angle of the wing; in *Chelonia*, one of the horny plates of the plastron.

humerus The bone supporting the proximal region of the fore-limb in land Vertebrates. adj. *humeral*.

humic acids Complex organic acids occurring in the soil and in bituminous substances formed by the decomposition of dead vegetable matter.

humicole, humicolous Growing on soil or on humus.

humification The formation of humus during the decomposition of organic matter in soils.

humoral immunity Specific immunity attributable to antibodies as opposed to cell-mediated immunity.

humour, humor A fluid; as the *aqueous humour* of the eye.

humus Amorphous, black or dark-brown material, a mixture of macromolecules based on benzene carboxylic and phenolic acids, often complexed with clays, which results from the decomposition of organic matter in soils. See **mull, mor, moder**.

humus plant A flowering plant, often with little chlorophyll, depending for much of its nutrition on a mycorrhizal association with a fungus growing in a rich humus.

hyal , hyalo Prefixes from Gk. *hyalos*, clearstone, glass.

hyaline Translucent and colourless; without fibres or granules, e.g. *hyaline cartilage*.

hyaloid Clear, transparent; as the *hyaloid membrane* of the eye which envelops the vitreous humour.

hyaloplasm (1) The ground substance or matrix of the cytoplasm, between the organelles. (2) Cytoplasm.

hybrid Offspring of a *cross* between two different strains, varieties, races or species. (n. or adj.)

hybrid antibodies Antibody molecules in which the two antigen combining sites are of different specificities. This can be achieved by recombining half molecules of two specific antibodies in vitro, or may also occur when cells from two different antibody pro-

---

hybridoma

B cell hybridoma. A cell line obtained by the fusion of a myeloma cell line, which is able to grow indefinitely in culture, with a normal, antibody secreting B cell. The resulting cell line has the properties of both partners, and continues to secrete the antibody product of the normal B cell. By choosing a myeloma which has ceased to make its own immunoglobulin product but has retained the machinery for doing this, the hybridoma secretes only the normal B cell antibody. Since the cell line is cloned the antibody is monoclonal.

T cell hybridoma. A cell line obtained by fusion of a T lymphoma cell line with a normal T lymphocyte. The resulting hybridoma may retain the properties of the lymphoma but acquires the antigen recognition specificity of the normal T cell. Such hybridomas provide a source of homogeneous T cell antigen receptors, and are also useful for recognizing specific antigens which stimulate T cells such as histocompatibility antigens.

---

ducing cell lines are fused to make a **hybridoma**. A distinct kind of hybrid antibody is made from the variable region of a monoclonal antibody from one species joined to the constant region of immunoglobulin of another species. This can be achieved by insertion of hybrid genes into a myeloma cell line, with the aim of producing antibodies with the general attributes of human Ig but with a desired antigen binding capacity borrowed from a mouse hybridoma.

hybridization (1) The formation of a new organism by normal sexual processes or more recently by protoplast fusion. (2) When DNA or RNA molecules from two different sources are annealed or renatured together, they will form hybrid molecules, whose stability is some measure of their sequence relatedness.

hybridoma See panel.

hybrid sterility The lack of fertility of some interspecific hybrids which results from lack of homology between chromosome sets and thus abnormal segregation, which may be overcome by chromosome doubling. See **amphidiploid**.

hybrid vigour The increase in desirable qualities such as growth rate, yield, fertility, often exhibited by *hybrids*, i.e. favourable *heterosis*.

hydathode Structure through which water is exuded from uninjured plants.

hydatid cyst A large sac or vesicle containing a clear watery fluid and encysted immature larval *Cestoda*.

hydranth In *Hydrozoa*, a nutritive polyp of a hydroid colony.

hydraulic capacity The amount of water held by a soil between **field capacity** and the **permanent wilting point**, a measure of available water.

hydr-, hydro- Prefixes from Gk. *hydor*, gen. *hydatos*, water.

hydrocoel In *Echinodermata*, the **water-vascular system**.

hydrofuge Water-repelling; said of certain hairs possessed by some aquatic insects and used for retaining a film of air.

hydrogen bacteria Chemosynthetic bacteria which obtain the energy required for carbon dioxide assimilation from the oxidation of hydrogen to water. Organic compounds may also be oxidized by these species under suitable conditions.

hydrogen-bonding Hydrogen atoms in the groups –O-H and –N-H can form non-covalent bonds with a nitrogen or oxygen atom. The bonding is ionic and weak, but hydrogen bonding between purines and pyrimidines contributes to the stability of the DNA double helix. The planar stacking of the bases and cooperation between adjacent hydrogen bonds making for a very stable and rigid structure.

hydroid (1) Water-conducting cell in the stems of some moss gametophytes, e.g. *Polytrichum*. (2) An individual of the asexual stage in Coelenterata which show alternation of generations.

Hydromedusae See **Hydrozoa**.

hydrophily (1) Living in water. (2) Pollination by water.

hydrophobia Literally fear of water, a symptom of rabies. Used synonymously with *rabies*.

hydrophyte (1) A plant with leaves partly or wholly submerged in water. (2) Water plant with perennating buds at the bottom of the water (a sort of **cryptophyte**). See **Raunkaier system**.

hydroponics Technique of growing plants without soil for experimental purposes and sometimes crops. The roots can be in either a nutrient solution or in an inert medium percolated by such a solution, *water culture*, *sand culture*. See **nutrient film technique**.

hydropote A gland-like structure found on the submerged surfaces of the leaves of many water plants.

hydropyle A modified area of the serosal cuticle of the developing egg of some *Orthoptera* for the uptake or loss of water.

hydrosere A sere beginning with submerged soil, as at the margin of a lake.

hydrostatic skeleton A body form maintained by muscles acting upon a fluid-filled cavity, usually the coelom as in many soft-bodied invertebrates.

hydrotaxis Response or reaction of an organism to the stimulus of moisture. adj. *hydrotactic.*

hydrotropism A tropism in which the orientating stimulus is water.

hydroxyproline *4-hydroxy-pyrrole-2-carboxylic acid.* An imino acid formed by the post-translational hydroxylation of proline residues within protein molecules, common in cell-wall proteins.

Hydrozoa A class of *Cnidaria*, in which alternation of generations typically occurs; the hydroid phase is usually colonial, and gives rise to the medusoid phase by budding.

hygro- Prefix from Gk. *hygros*, wet, moist.

hygrochastic Said of a plant movement caused by the absorption of water.

hygrophyte A plant of wet or waterlogged soils.

hygroscopic movement *Imbibitional movement, hygrometric movement.* That caused by changes in moisture content of unevenly thickened cell walls e.g. of *elators* or the awns of the grains of some grasses.

hylophagous Wood-eating.

hymen In Mammals, a fold of mucous membrane which partly occludes the opening of the vagina in young forms.

hymenium The fertile layer containing the asci or basidia in Ascomycetes and Basidiomycetes.

Hymenomycetes The mushrooms, toadstools, agarics and bracket fungi, a class of the Basidiomycotina in which the hymenium is exposed at the time of spore formation. Many are saprophytic in soil, dung, leaf litter etc. some are parasitic, e.g. honey fungus *Armillaria mellea*, many form ectotrophic mycorrhizas with trees.

hymenophore Any fungal structure which bears a hymenium.

Hymenoptera An order of Endopterygota having usually two almost equal pairs of transparent wings, which are frequently connected during flight by a series of hooks on the hindwing; mandibles always occur but the mouthparts are often suctorial; the adults are usually of diurnal habit; the larvae show great variation of form and habit. Saw flies, Gall flies, Ichneumons, Ants, Bees, Wasps.

hyo- Prefix from Gk. *hyoeides*, U-shaped.

hyoid In higher Vertebrates, a skeletal apparatus lying at the base of the tongue,

derived from the hyoid arch of the embryo.

hyoid arch The second pair of visceral arches in lower Vertebrates, lying between the mandibular arch and the first branchial arch.

hyoideus In Vertebrates, the post-trematic branch of the facial (seventh cranial) nerve, which runs to the mucosa of the mouth and to the muscles of the hyoid region, and, in aquatic forms, to the neuromast organs of the region below and behind the orbit.

hyomandibular In Craniata, the dorsal element of a hyoid arch.

hyomandibular nerve In Craniata, the post-trematic branch of the seventh cranial nerve; it divides into anterior and posterior branches.

hyostyly A type of jaw suspension, found in most Fish, in which the upper jaw is attached to the cranium anteriorly by a ligament, posteriorly by the hyomandibular. adj. *hyostylic.*

hypanthium The flat or concave receptacle of a perigynous flower.

hypapophyses Paired ventral processes of the vertebrae of many higher *Craniata*.

hypaxial Below the axis, esp. below the vertebral column, therefore ventral; as the lower of two blocks into which the myotomes of fish embryos become divided.

hyper- Prefix from Gk. *hype*, above.

hyperdactyly The condition of having more than the normal number (five) of digits, as in Cetacea.

hyperdiploidy The condition where the full chromosome complement is present, as well as a portion of one chromosome which has been translocated.

hypergammaglobulinaemia Term used to describe clinical conditions in which the concentration of immunoglobulins in the blood exceeds normal limits. May result from continuous antigenic stimulation in chronic infections, from autoimmune diseases, or from abnormal proliferation of the B cells which occurs in *Waldenstrom macroglobulinaemia* or in *myelomatosis*. The term came into use before immunoglobulins were identified and accepted.

hyperkinetic state, hyperactive state In children, characterized by motor restlessness, poor attention and excitability; often associated with learning difficulties.

hypermetamorphic Of Insects, passing through two or more sharply distinct larval instars. n. *hypermetamorphosis.*

hyper-osmotic A solution of greater osmolarity than that within a cell (or any other solution). Water will move out of a cell suspended in a hyperosmotic solution unless the osmotic balance is reversed by a flow of solute into the cell.

hyperparasitism The condition of being parasitic on a parasite. n. *hyperparasite.*

hyperphalangy The condition of having

more than the normal number of phalanges, as in Whales.

hyperpharyngeal Above the pharynx, as the *hyperpharyngeal band* in *Tunicata*.

hyperplasia Excessive multiplication of cells of the body; an overgrowth of tissue due to increase in the number of tissue elements; generally, overgrowth and usually pathological. adj. *hyperplastic*. Cf. **hypertrophy**.

hyperploid Having a chromosome number slightly exceeding an exact multiple of the haploid number.

hypersensitivity Condition in which a pathogen kills its host cells so quickly that the spread of infection is prevented.

hyperstomatal Having stomata only on the upper surface. Cf. **amphistomatal**, **hypostomatal**.

hypertely The progressive attainment of disproportionate size, either by a part or by an individual.

hypertension *High blood pressure*. A chronic elevation of blood pressure due to constricted arteries; often a consequence of intense and persistent stress.

hypertonic solution A solution with a water potential lower than that of a cell suspended in it; water passes into the cell until the water potentials are equalized. See **isotonic**.

hypertrophy Abnormal, usually pathological, enlargement of an organ or cell. Cf. **hyperplasia**.

hypha (1) The mycelium of a fungus which is a branching, filamentous structure with apical growth; the tubular cytoplasm contains the nuclei and may be divided by septa. (2) Elongated tubular cell in the thallus of some algae.

hyphopodium A more-or-less lobed outgrowth from a hypha, often serving to attach an epiphytic fungus to a leaf.

hypnagogic imagery Hallucinatory type of phenomena or fantasy occurring in the drowsy state just before falling asleep.

hypnagogic state The transitional state of consciousness while falling asleep.

hypnosis A temporary, trance-like state, induced by certain verbal or non-verbal procedures (hypnotism) and characterized by heightened suggestibility both during and after hypnosis. See **post-hypnotic suggestion**.

hypnospore A thick-walled, non-motile, resting spore. See **aplanospore**.

hypnotic Of the nature of, or pertaining to, hypnosis; a drug which induces sleep.

hypo- Prefix from Gk. *hypo*, under.

hypoblast The innermost germinal layer in the embryo of a metazoan animal, giving rise to the endoderm and sometimes also to the mesoderm. Cf. **epiblast**.

hypobranchial The lowermost element of a branchial arch.

hypobranchial space The space below the gills in Decapoda.

hypocercal Said of a type of caudal fin, found in Anaspida and Pteraspida, in which the vertebral column bends downwards and enters the hypochordal lobe which is larger than the epichordal lobe.

hypochondriasis A disorder characterized by excessive preoccupation with bodily functions and sensations with the false belief that the latter indicate bodily disease; associated with a number of psychiatric syndromes.

hypocone A fourth cusp arising on the cingulum on the postero-internal side of an upper molar tooth, producing a quadritubercular pattern.

hypocotyl The part of the axis of the embryo between the radicle and the cotyledon(s), and the region of the seedling which derives from it.

hypodermis, hypoderm (1) A layer, one or more cells thick immediately below the epidermis and differing morphologically from the underlying ground tissue. (2) In Arthropoda and other Invertebrates with a distinct cuticle, the epithelial cell layer underlying the cuticle, by which the cuticle is secreted. adj. *hypodermal*.

hypogammaglobulinaemia Condition in which the concentration of immunoglobulins in the blood is much lower than normal. In the infantile sex-linked form there is a maturation defect in B cells. A late acquired form has various causes, one of which is excessive activity of suppressor T cells. A clinically similar condition may result from displacement of lymphoid tissues by lymphoma or leukaemic cells. There is greatly increased susceptibility to bacterial, but not to viral infections. Treatment is by regular administration of immunoglobulin concentrates prepared from the pooled blood of normal persons which contains antibodies against most environmental pathogenic microbes.

hypogeal, hypogaeous (1) Living beneath the surface of the ground. (2) Germinating with the cotyledons remaining in the soil.

hypoglossal Underneath the tongue; the 12th cranial nerve of Vertebrates, running to the muscles of the tongue.

hypoglottis In Vertebrates, the under part of the tongue; in *Coleoptera*, part of the labium.

hypognathous Having the under jaw protruding beyond the upper jaw; having the mouth parts directed downwards.

hypogynous Having the calyx, corolla and stamens inserted on the convex receptacles beside or beneath the base of the ovary, e.g.

buttercup (*Ranunculus*). See **superior**. Cf. **perigynous, epigynous**.

hypohyal The lowermost element of a hyoid arch.

hypolimnion The cold lower layer of water in a lake. Cf. **epilimnion**.

hyponasty The greater growth of the lower side which causes the upward bending of an organ. See **nastic movement**. Cf. **epinasty**.

hyponome In *Cephalopoda*, the funnel by which water escapes from the mantle cavity.

hypo-osmotic A solution of lower osmolarity than that within a cell (or any other solution). Water will move into a cell suspended in a hypo-osmotic solution until the water potentials across the membrane are equalized or the cell bursts. The osmotic imbalance may also be reduced by loss of solute from the cell.

hypopharyngeal Below the pharynx, as the *hypopharyngeal* groove of *Cephalochorda*.

hypophloedal Growing just within the surface of bark.

hypophysis Generally, a downwardly growing structure. In plants, the suspensor cell closest to the embryo. In Cephalochorda, the olfactory pit; in Vertebrates, the pituitary body. adj. *hypophysial*.

hypoplasia Abnormal, usually pathological, under development of a tissue because the cells are smaller or fewer. adj. *hypoplastic*. Cf. **hyperplasia, hypertrophy**.

hypoploid Having a chromosome number a little less than some exact multiple of the haploid number.

hyposensitization Administration of a graded series of doses of an allergen to atopic subjects suffering from immediate type hypersensitivity to it. This must be done with great care in order to avoid anaphylactic reactions. The aim is to increase the level of specific IgG antibodies and/or to diminish the level of IgE antibodies in order that subsequent contact with the allergen under natural conditions causes less severe reactions.

hypostatic Opposite of *epistatic*; analogous to *recessive* applied to genes at different loci.

hypostoma, hypostome In some Cnidaria, the raised oral cone; in Insects, the labrum; in Crustacea, the lower lip or fold forming the posterior margin of the mouth; in some Acarina, the lower lip formed by the fusion of the pedipalpal coxae.

hypostomatal Leaf etc. with stomata only on the lower surface. Cf. **amphistomatal, epistomatal**.

hypostomatous Having the mouth placed on the lower side of the head, as Sharks.

hypotarsus In Birds, the fibulare.

hypothalamus In the Vertebrate brain, the ventral zone of the thalamencephalon. In Man the part of the brain which makes up the floor and part of the lateral walls of the third ventricle. The mammillary bodies, tuber cinerium, infundibulum, neurohypophysis and the optic chiasma are also part of the hypothalamus.

hypotonic solution A solution with a water potential greater than that of a cell suspended in it; water will flow out of the cell until the water potentials are equalized. See **isotonic**.

hypotrichous Having cilia, principally on the lower surface of the body.

hypso- Prefix from Gk. *hypsos*, height.

hypsodont Of a Mammalian tooth with a high crown and deep socket. Cf. **brachyodont**.

hypsophyll A non-foliage leaf inserted high on a shoot, e.g. a floral bract. Cf. **cataphyll**.

Hyracoidea An order of small Eutherian Mammals having 4 digits on the fore limb and 3 on the hind limb, pointed incisor teeth with persistent pulps, lophodont grinding teeth, no scrotal sac and 6 mammae; terrestrial African forms. Dassies, Hyraxes.

hysteranthous Said of leaves which develop after the plant has flowered.

hyster-, hystero- Prefixes from Gk *hysteria*, womb or *hysteros*, later.

I

IAA Abbrev. for *Indole-3-Acetic Acid*.

Ia antigens Class II histocompatibility antigens with functions similar to human **HLA-D antigens** on mouse B cells, macrophages and accessory cells.

IBA Abbrev. for *Indole-3-Butyric Acid*.

ichthyic Pertaining to, or resembling, Fish.

ichthy-, ichthyo- Prefixes from Gk. *ichthys*, fish.

ichthyopterygium A paddle-like fin, or limb, used for swimming, e.g. pectoral or pelvic fin of Fish.

ichthyosis *Xerodermia*. A disease characterized by dryness and roughness of the skin, resembling fish scales, due to lack of secretion of the sweat and the sebaceous glands.

iconic memory A transient visual trace that fades rapidly after removal of the stimulus. Cf. **echoic memory**.

id A term, originally introduced by Groddeck and later used by Freud, to denote the sum total of the primitive instinctual forces in an individual. It subserves the pleasure-pain principle, in which the activities of the organism are concerned with the immediate increase of pleasurable and reduction of painful stimuli. It is dominated by blind impulsive wishing and forms a major part of the unconscious mind.

ideas of reference A characteristic of some mental disorders, notably schizophrenia, in which the individual perceives irrelevant and independent environmental and social events as relating to himself or herself ('people are looking at me').

identification In psychoanalytic theory, the way in which an individual incorporates (introjects) the values, standards, sexual orientation, and mannerisms of the same sex parent, as part of the development of the **super ego**. It can also be used to describe the influence of any relevant and powerful figure for the internalization of external norms.

idio- Prefix from Gk. *idios*, peculiar, distinct.

idioblast A cell of clearly different properties to the others in the tissue, as a stone cell in pear fruit.

idiogram *Karyogram*. A diagram (or photomontage) of the chromosome complement of a cell, conventionally arranged to show the general morphology including relative sizes, positions of centromeres etc.

idiopathy Any morbid condition arising spontaneously, having no known origin. adj. *idiopathic*.

idiothermous See **warm-blooded**.

idiotope Antigenic determinant on immunoglobulin molecules characteristic of the product of a single clone or a small minority of clones, and associated with or part of the antigen binding site.

idiot savant A child who, despite generally diminished skills, shows astonishing proficiency in one isolated skill ('foolish wise one').

idiotype Set of one or more idiotopes by which a clone of immunoglobulin forming cells can be distinguished from other clones. Some idiotypes appear to be unique to an individual animal; others are found in many members of the same animal species.

Ig See panel for the various classes. Ig is the general abbreviation for **immunoglobulin**, under which term there is information on structure and genetic control.

IL-1, IL-2 etc. See **interleukin-1, interleukin-2** etc.

Ile Symbol for **isoleucine**.

ileum In Vertebrates, the posterior part of the small intestine.

iliac region The dorsal region of the pelvic girdle in Vertebrates.

iliac veins In Fish, the paired veins from the pelvic fins, draining into the lateral veins.

ilio- Prefix which refers to that part of the pelvic girdle of a Vertebrate known as the *ilium*; used in the construction of compound terms, e.g. *iliofemoral*, pertaining to the ilium and the femur.

ilium A dorsal cartilage bone of the pelvic girdle in Vertebrates. adj. *iliac*.

illegitimate pollination The transfer of pollen in a way the floral structure appears to discourage, e.g. from the anther of one **pin** flower to the stigma of another.

illegitimate recombination Recombination between species whose DNA shows little or no homology, facilitated by duplicate sequences casually present.

illusion A surprising perceptual experience, due to the fact that some aspect of the relation between the physical stimulus and the individual's perception of it, violates normal expectation.

imaginal bud, disk One of a number of masses of formative cells which are the principal agents in the development of the external organs of the imago, during the metamorphosis of the *Endopterygota*.

imago The form assumed by an Insect after its last ecdysis, when it has become fully mature; final instar. adj. *imaginal*. pl. *imagines*.

imbibition (1) Uptake of water where the driving force is a difference of **matric potential** rather than osmotic. (2) Uptake of water and swelling by seeds, the first step in germination.

imbricate Plant organs, arranged so that the edges overlap like the tiles on a roof. See **aestivation**, **vernation**. Cf. **valvate**.

imitation See **observational learning**.

immediate hypersensitivity Antibody mediated hypersensitivity characteristically due to release of histamine and other vasoactive substances. Also *type 1 reaction*.

immigration A category of population dispersal covering one way movement into the population area. Cf. **emigration, migration**.

immobilized culture The use of plastic foam, beads or sheets to let cells grow on the surfaces and interstices in close contact with the circulating medium. Combines the advantages of surface and suspension culture. See **plant cell culture**.

immune Protected against any particular infection; someone who is in this state, i.e. has **immunity**. v. *immunize*, to make immune against infection. n. *immunization*.

immune adherence Adherence of antigen-antibody complexes or antibody coated microbes which contain bound C3b or C4b to complement receptors on erythrocytes or platelets of some species and on macrophages and polymorphs.

immune bodies See **antibodies**.

immune complex Macromolecular complex formed by antigen and antibody linked by their combining sites. The size of the complex depends upon the ratio of antigen to antibody. At **optimal proportions** the complexes come out of solution as precipitates, but in antigen excess they may remain in solution. Such complexes if present in the blood stream may be deposited in the walls of small blood vessels or in renal glomeruli, and there activate complement. This elicits an inflammatory reaction (Type 3) and is the cause of *immune complex diseases*.

immunity A state of not being susceptible to the invasive or pathogenic effects of potentially infective microbes or to the effects of potentially toxic antigenic substances. Due to natural or non-specific mechanisms or to specific acquired immunity, or to both.

immunization Administration either of antigen to produce active immunity or of antibody to confer passive immunity, and thereby to protect against the harmful effects of antigenic substances or microbes.

immunoblot Technique used to detect the presence of a polypeptide which is antigenic to the specific antibody used in the assay. A protein mixture is denatured and separated on the basis of size by **SDS gel electrophoresis**, then blotted on to a sheet of nitrocellulose, preserving the pattern due to size. The proteins in the pattern are reacted with the specific antibody, then with an **AP-** or **HRP**-conjugated second antibody against the first which after development, produces visible staining at the site of binding. Also *western blot*.

immunofluorescence Technique in which antigen or antibody is conjugated to a fluorescent dye and then allowed to react with the corresponding antibody or antigen in a tissue section or a cell suspension. This enables the location of antibodies or antigens in or on cells to be determined by fluorescence microscopy.

immunogen A substance that stimulates humoral and/or cell mediated specific immunity when introduced into the body.

immunoglobulin See panel.

immunoglobulin genes See panel.

immunological memory Describes the fact that antibody and cell mediated responses occur more rapidly and are quantitatively greater after a second exposure to antigen (provided that the interval is more than a few days and less than several years). Due to the stimulation on first exposure to the antigen of increased numbers of B and T cells capable of recognizing the antigen, and which have reverted to a resting long lived form ('memory cells'). These are available to respond on second exposure to the antigen.

immunological tolerance A state in which an animal fails to respond to an antigen capable of inducing humoral or cell mediated immunity, which has been induced by prior administration of the same antigen by a route or in a form such as either to eliminate potentially responsive lymphocytes or to induce specific suppressor T cells. Can result from administration of antigens in very early life (before the immune system is fully developed).

immunosorbent Use of an insoluble preparation of an antigen to bind specific antibodies from a mixture, so that the antibodies can later be eluted in pure form.

immunosuppression Artificial suppression of immune response by the use of drugs which interfere with lymphocyte growth (antimetabolites), or by irradiation or by antibodies against lymphocytes. A state of immunosuppression can also exist as a result of infections (such as by **human immunodeficiency virus**, HIV or cytomegalovirus) which damage lymphocytes.

immunotoxin Molecule toxic to cells coupled to antibody specific for some antigen present at the surface of a particular type of cell, whereby the toxin can be selectively directed to cells of that type.

imperfect flower A flower in which either the stamens or the carpels are lacking or, if present, nonfunctional.

imperfect fungi The Deuteromycotina.

imperfect stage Stage in the life cycle of e.g. a fungus in which the organism can only reproduce asexually if at all.

immunoglobulin, Ig classes

IgA. The major immunoglobulin present in mucosal secretions, where it appears as a dimer linked by a peptide made separately by epithelial cells, which assists to transport it across the mucosa. In the blood IgA is present in monomeric and polymerised forms. It is mainly synthesized by B cells in the lymphoid tissues of the gut and respiratory tract, and IgA antibodies constitute the main humoral defence mechanism against microbes on mucosal surfaces.

IgD. Immunoglobin present only in very low concentrations in blood, but present at the surface of B lymphocytes where it probably functions as an antigen receptor.

IgE. Immunoglobulin the Fc region of which binds very strongly to a receptor on the surface of mast cells and basophils. Normally present in blood only in very low concentrations, but the concentration is increased in atopic subjects, and in infection by several helminth parasites.

IgG. The major immunoglobulin in humans and most species from amphibians upwards (but not in fish). IgG fixes complement and crosses the placenta (i.e. can be passed from mother to foetus). Although mainly present in body fluids some is also present at mucosal surfaces. In humans there are four varieties (subclasses) differing from each other in respect of parts of their heavy chains and the number of disulphide bonds which link the heavy chains together. This confers different biological properties, the functions of which are not fully understood. The subclasses are termed IgG1, IgG2, IgG3 and IgG4. Similar subclasses are present in mice and rats but not in rabbits.

IgM High molecular weight immunoglobulin, consisting in humans of five basic units (of two light and two heavy chains) arranged as a pentamer and joined together by disulphide bonds and a small link peptide (J chain). A monomeric form is present on the cell surface of B lymphocytes, where it acts as the earliest form of antigen receptor. In some species the form present in the blood may be a monomer, in others a tetramer and in others a hexamer. IgM activates complement very effectively but does not cross the human placenta. IgM antibodies are the first to be synthesized and released after a primary antigenic stimulation, and are the predominant form made in response to many bacterial capsular polysaccharides. Since their pentameric structure enables them to attach firmly to antigenic sites situated close together on a surface, they provide effective protection against many bacteria.

imperforate Lacking apertures, esp. of shells; said of Gastropod shells which have a solid columella.

implant A graft of an organ or tissue to an abnormal position.

implantation In Mammals, the process by which the blastocyst becomes attached to the wall of the uterus.

implosive therapy See **flooding**.

impregnation The passage of spermatozoa from the body of the male into the body of the female.

impression formation A traditional area of research in social psychology, referring to the issue of how information about other individuals is integrated into a unified impression, often on the basis of very little information.

imprinting An aspect of learning in some species, through which attachment to the im-

portant parental figure develops; it involves a narrowing of the stimuli effective for several filial responses to those first encountered during a short period after birth. See **response, sensitive period**.

in- Prefix from L. into, not.

inbred The condition of the offspring produced by *inbreeding*. See **inbreeding coefficient**.

inbred line A strain which has been *inbred* over many generations and whose **inbreeding coefficient** is nearly 100%. All members are genetically identical and *homozygous* at all loci, or very nearly so. Cf. **isogenic**.

inbreeding The mating together of individuals that are related by descent. The offspring are *inbred* to an extent depending on the degree of relationship. See **inbreeding coefficient**. Inbreeding produces *homozygosis*.

inbreeding coefficient A measure of the

immunoglobulin, immunoglobulin genes

A family of proteins all of which have a similar basic structure, made up of two **light** and two **heavy chains** linked together by disulphide bonds so as to form a Y shaped molecule with two flexible arms (see figure). The detailed shape of the ends of the arms, the *variable region*, varies from one molecule to another within the family, so providing a wide range of antigen binding sites. The structure of the rest of the molecule is relatively constant from one molecule to another, and generally similar, but members are divided into immunoglobulin classes and subclasses determined by the amino acid sequences of their heavy chains. See **IgA**, **IgD** etc. All antibodies are immunoglobulins, and the same antigen binding sites can be present on the arms of immunoglobulins belonging to any of the classes.

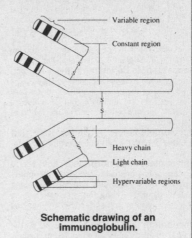

Labels: Variable region; Constant region; Heavy chain; Light chain; Hypervariable regions

Schematic drawing of an immunoglobulin.

Each polypeptide chain is coded for by several genes, each having a number of alleles, which are spliced together before transcription. Heavy chains are coded by V (variable region), D (diversity) and J (joining) genes which are linked to genes for the C (constant) region of one or other class. Light chains are coded by a different set of V and J genes joined to genes for one or other of the light chain constant regions. Splicing together of the V, D and J genes prior to joining the C region genes appears to take place more or less at random during the early stages of B cell development. This results in an enormous variety of possible immunoglobulins, of which any individual B cell can make one.

degree to which an individual is *inbred*. Ranges from 0 when the parents are unrelated to 100% when the parents for many generations back have been related. Usually symbolized by F.

inbreeding depression The reduction of desirable characters such as growth rate, yield, fertility, consequent on the *homozygosis* produced by *inbreeding*, esp. in those which normally outbreed.

incentive learning A motivational concept that refers to the expectation of rewards or punishments from the environment. The high or low incentive value of a goal is reflected in the amount of energy the organism will expend to obtain it.

incept The rudiment of an organ.

incertae sedis Of uncertain taxonomic position.

incest taboo A strong negative social sanction which forbids sexual relations between members of the same immediate family, found in all human societies.

incidence The frequency with which new cases of a given disease presents in a particular period for a given population. Different from **prevalence**.

incidental learning Learning without trying to learn. Cf **intentional learning**.

incipient plasmolysis The state of a plant cell in which the **turgor** pressure is zero but the protoplast is in contact with the cell wall all round. Loss of water will result in **plasmolysis**; uptake of water will generate turgor.

incisiform Shaped like an incisor tooth.

incisors The front teeth of Mammals; they have a single root, are adapted for cutting, and are the only teeth borne by the premaxillae in the upper jaw.

inclusion bodies Particulate bodies found in the cells of tissue infected with a virus.

inclusive fitness The number of copies of an individual's genes passed on to the next generation plus an additional number passed on by relatives as a result of the individual's behaviour towards those relatives. Allied to the concept of **kin selection**.

incompatibility (1) The consistent failure of fertilization or hyphal fusion between particular combinations of individual plants, algae, fungi etc. See **self incompatibility**. (2) Interaction between stock and scion resulting in the failure of the graft either (a) immediately, leading to the rapid death of at least the scion, or (b) after some years of apparently successful growth, resulting in fracture across the graft union. (3) The relationship between a plant and a pathogen to which the plant is not susceptible.

incompatible behaviours Behaviour patterns which cannot occur simultaneously, because of reciprocal inhibition in the case of reflexes for example, but also due to psychological factors, such as the limits of attention.

incomplete flower A flower in which either or both the calyx and corolla are lacking.

incomplete metamorphosis In Insects, a more or less gradual change from the immature to the mature state, a pupal stage being absent and the young forms resembling the parents, except in the absence of wings and mature sexual organs.

incubation Behaviour which maintains the eggs of birds and other species in a fairly stable thermal and gaseous environment; most bird species accomplish it by sitting on the eggs, but other methods are also used by various species of birds and insects, e.g. mound building, sunning.

incubous Said of the leaf of a liverwort when its upper border (the border towards the apex of the stem) overlaps the lower border of the next leaf above it and on the same side of the stem.

incurrent Carrying an ingoing current; said of ducts and, in certain *Porifera*, of canals leading from the exterior to the spongocoel.

incus In Mammals, an ear ossicle, derived from the quadrate; more generally, any anvil-shaped structure. pl. *incudes*.

indeciduate Said of Mammals in which the maternal part of the placenta does not come away at birth.

indefinite (1) Numerous but not fixed in number. (2) Monopodiol growth. (3) A racemose inflorescence.

indehiscent Not opening spontaneously to release seeds, spores etc. Cf. **dehiscence**.

index case The first or original case of a disease. A term used in the **epidemiology** of infectious disease. In genetics synonymous with the *proband* or *propositus*.

Indian hemp See **cannabis**.

indicator (1) The presence of a species which gives an indication of features of the habitat, or method of land management by growing well or badly, e.g. the stinging nettle which indicates a high level of available phosphorus in the soil. (2) Plants which react to a particular pathogen or environmental factor with obvious symptoms and may, therefore, be used to identify that pathogen or factor.

indicator species analysis Multivariate statistical technique to enable classification of vegetation on the basis of the presence or absence of key species.

indigenous Native; not imported.

indigestion A condition, marked by pain and discomfort, in which the normal digestive functions are impeded. Also *dyspepsia*.

indirect immunofluorescence A technique in which a specific antibody is first bound to its antigen. A fluorochrome visible under **fluorescence microscopy** and conjugated to a second antibody specific to the first is then used to detect the presence of the first antibody and therefore the original antigen.

indirect metamorphosis The complex change characterizing the life cycles of Endopterygota; the young are larvae and the imago is preceded by a pupal instar.

individual A single member of a species; a single zooid of a colony of Coelenterata or Polyzoa; a single unit or specimen.

individual distance A spatial relationship between members of a flock (or other social group, e.g. a school or herd) which is maintained through the two conflicting tendencies of aggression/avoidance and of social attraction.

indol-3-butyric acid A synthetic plant growth regulator with auxin-like activity used esp. in **rooting compounds**.

indole-3-acetic acid *Auxin, heteroauxin, IAA.* The commonest naturally-occurring plant growth substance of the **auxin** type.

inducer An agent which increases the transcription of specific genes.

inducible enzyme An enzyme which is formed only in response to an inducing agent, often its substrate. Cf. **constitutive enzyme**.

induction The production of a definite condition by the action of an external factor.

indumentum (1) The hairy covering of a plant. (2) A covering of hair and feathers.

indusium In some Insects, a third embryonic envelope lying between the chorion and the amnion in the early stages of development of the egg; a cerebral convolution of the brain in higher Vertebrates; an insect larva case. adjs. *indusiate, indusiform.*

industrial melanism **Melanism** which has developed as a response to blackening of trees etc. by industrial pollution. This favours melanic forms, esp. among moths which rest on trees during the day.

inequipotent Possessing different potentialities for development and differentiation.

inequivalve Having the 2 valves of the shell unequal.

infection The invasion of body tissue by living micro-organisms, with the consequent production in it of morbid change; a diseased condition caused by such invasion; the infecting micro-organism itself.

inferior Lower; under; situated beneath, e.g. the *inferior* rectus muscle of the eyeball or an ovary having perianth and stamens inserted round the top, i.e. *epigynous*. The ovary appears to be sunk into and fused with the receptacle. Cf. **superior**.

inferiority complex A concept, first proposed by Adler, referring to the repressed and powerful conviction of inferiority, whose basis lies in the universal experience of infantile helplessness and dependancy; these feelings become repressed during development, and are a powerful dynamic force in determining adult personality and often, character disorders.

inferior vena cava See **postcaval vein**.

infertility Inability to produce offspring.

infestation The condition of being occupied or invaded by parasites, usually parasites other than bacteria.

inflorescence (1) Flowering branch (or portion of the shoot above the last stem leaves) including its branches, bracts and flowers. See **racemose, cymose, mixed, capitulum, umbel, panicle.** (2) In bryophytes, a group of antheridia or archegonia and associated structures.

infra- Prefix from L. *infra*, below.

inframarginal Below the margin; a marginal structure; in *Chelonia*, one of certain plates of the carapace lying below the marginals; in *Asteroidea*, one series of ossicles situated on the lower margin of each ray.

infraorbital foramen In Mammals, a foramen on the outer surface of each maxilla for the passage of the second division of the fifth cranial nerve.

infraorbital glands In Mammals, one of the four pairs of salivary glands.

infundibulum A funnel-shaped structure; in Vertebrates, a ventral outgrowth of the brain; a pulmonary vesicle; in Cephalopoda, the siphon; in Ctenophora, the flattened gastric cavity. adj. *infundibular*.

ingestion The act of swallowing or engulfing food material (*ingesta*) so that it passes into the body. v. *ingest*.

ingluvies An oesophageal dilatation of Birds; the crop.

ingroup, outgroup A set of related concepts used in studies of intergroup relations; a person's ingroup is the group the individual perceives as the group they belong to; and the *outgroup* is the group to which one does

not belong, and feels no allegiance to.

inguinal Pertaining to, or in the region of, the groin, e.g. the *inguinal canal* through which the testes descend in male mammals.

inhalant Pertaining to, or adapted for, the action of drawing in a gas or liquid, e.g. the *inhalant* siphon in some *Mollusca*.

inhibition The stopping or deceleration of a metabolic process. Cf. **excitation**. adj. *inhibitory*.

inhibitor A substance which reduces or prevents some metabolic or physiological process. Also *growth inhibitor*.

inhibitory Said of a nerve, whose action upon other nerves tends to render them less liable to stimulation.

initial (1) A dividing cell in a meristem; one of the daughters or its progeny adding to the tissues of the plant, the other remaining in the meristem and repeating the process. (2) Cell in the earliest stages of specialization.

initiator codon The RNA sequence AUG or rarely GUG which always specify the first amino acid of a protein.

injected Having the intercellular spaces filled with water or other fluid.

ink See **ink sac**.

inkblot test See **Rorschach test**.

ink sac In some *Cephalopoda*, a large gland, opening into the alimentary canal near the anus, which secretes a dark-brown pigment (sepia).

innate Refers to behaviour which normally occurs in all members of a species despite natural variation in environmental influences. The term is sometimes used to imply genetic determinism, but it is now accepted that the development of innate behaviour is complex, and often involves both endogenous and environmental factors. In plants (1) sunken into or originating within the thallus or (2) an anther joined by its base to the filament.

innate capacity for increase See **r**. Abbrev. r_{max}.

innate releasing mechanism *IRM*. In classical ethological models of motivation, a term referring to the notion that there is a specific relation between a particular stimulus and a particular response, such that a specific external stimulus (see **sign stimulus**) releases the motivational energy underlying the discharge of a specific response. The term is currently used in a general descriptive fashion, without the early implications about underlying mechanisms, which are more complex than was initially supposed.

inner glume Same as **palea**.

innervation The distribution of nerves to an organ.

innocent (1) Not malignant; not cancerous. See **tumour**. (2) A heart murmur of no pathological significance.

innominate Without a name, e.g. the *in-*

nominate artery of some Mammals, which leads from the aortic arch to give rise to the carotid artery and the subclavian artery; the *innominate vein* of Cetacea, Edentata, Carnivora and Primates, which leads across from the jugular-subclavian trunk of one side to that of the other; the *innominate bone*, which is the lateral half of the pelvic girdle.

inoculation The placing of cells, spores etc. on or in a potential host, soil or culture medium.

inquiline A guest animal living in the nest of another animal, or making use of the food provided for itself or its offspring by another animal.

insanity A medical legal term used in the defence of individuals who have committed crimes while their capacity for rational thought and behaviour was seriously impaired.

Insecta A class of mainly terrestrial mandibulate Arthropoda which breathe by tracheae; they possess uniramous appendages; the head is distinct from the thorax and bears one pair of antennae; there are 3 pairs of similar legs attached to the thorax, which may also bear wings; the body is sharply divided into head, thorax and abdomen. Also *Hexapoda*.

insecticides Natural (e.g. derris, pyrethrins) or synthetic substances for destroying insects. The widely used synthetic compounds are broadly classified according to chemical composition, viz. chlorinated (e.g. **DDT**), organophosphates (e.g. **malathion**), carbamates, dinitrophenols. Some of these have properties now considered undesirable, like persistence in *food chains*. Newer compounds like the modified pyrethrins have greater stability than the natural substance and still retain their effectiveness. See **contact-**, **fumigants**, **stomach-**.

Insectivora An order of small, mainly terrestrial Mammals having numerous sharp teeth, tuberculate molars, well-developed collar bones, and plantigrade unguiculate pentadactyl feet; insectivorous. Shrews, Moles and Hedgehogs. Sometimes divided into two orders: Lipotyphla and Menotyphla.

insemination The approach and entry of the spermatozoon to the ovum, followed by the fusion of the male and female pronuclei.

insertion (1) A stretch of chromosome that has been inserted into another. (2) The place where one plant member grows out of another, or is attached to another. (3) The manner of such an attachment. (4) The point or area of attachment of a muscle, mesentery or other organ.

insertion element DNA sequence a few hundred bases long which can be naturally inserted into genomic DNA. Does not code

for mRNA but can inactivate a coding sequence by its presence.

insessorial Adapted for perching.

insight learning A form of intelligent activity that involves the apprehension of relations between the elements of a problem; it often appears to occur suddenly, as a function of cognitive rather than behavioural effort, and for this reason is often contrasted with a more passive trial and error mode of learning or problem solving.

in-situ hybridization The identification by **autoradiography** in a fixed cell or section of the place where a radioactive or similar **probe** binds. Can identify a complementary sequence or an antigen.

inspiration The drawing in of air or water to the respiratory organs.

instar The form assumed by an Insect during a particular stadium.

instinct Refers to a variety of related notions about the causes of behaviour, and its usage has undergone many historical changes, referring sometimes to the hypothetical motivational forces that might impel it (e.g. Descartes), to reflexive behaviour (Darwin), to the irrational aspects of human behaviour (Freud), and most recently, in classical ethology, to the explanation of innate behaviour patterns. Among modern thinkers it is generally agreed that the concept has little descriptive or explanatory value, and tends to be avoided, although the term still crops up in the popular literature. See **innate**.

institution A type of relationship or pattern of relationships characteristic of a given society; institutionalized practices of customs within a society may be social, economic, legal etc. See **institutionalization**.

institutionalization A syndrome of apathy and withdrawal resulting from long periods in any understimulating institution, e.g. a mental hospital.

insulin A protein hormone which is produced by the β islet cells of Langerhans of the pancreas. Widely used in the treatment of **diabetes mellitus**, insulin injection results promptly in a decline in blood glucose concentration and an increase in formation of products derived from glucose. Insulin was the first protein to have its polypeptide sequence completely established and it consists of two peptide chains joined by two disulphide bridges. Insulin was formerly entirely produced from pancreatic extracts from pig and oxen, but can now be made from a **genetic engineered** strain of the bacterium, *Escherischia coli*.

integral dose See **dose**.

integrated virus A viral DNA sequence which has *integrated* into the host's chromosomes at one or more sites.

integration The insertion of one DNA

sequence into another by recombination.

integument A covering layer of tissue, esp. the skin and its derivatives or, in plants, a layer or tissue in an ovule surrounding and often more or less fused with the **nucellus** which develops into the testa. See **micropyle, testa**. adj. *integumented*.

intelligence quotient Abbrev. *IQ*. The score on an intelligence test. In early versions it was obtained by multiplying by 100 the ratio of *mental age* to *chronological age* to find out whether a child's mental age was ahead or behind its chronological age; in newer versions it is read off a table constructed from norms for individuals of the same age. The *deviation IQ*.

intensifying screen (1) Layer or screen of fluorescent material adjacent to a photographic surface, so that registration by incident X-rays is augmented by local fluorescence. (2) Thin layer of lead which performs a similar function for high-energy X-rays or gamma-rays, as a result of ionization produced by secondary electrons.

intentional learning Learning when informed that there will be a later test of learning.

intention movement The incomplete phases of behaviour patterns that often occur in conflict situations; many are assumed to provide the phylogenetic basis for the evolution of threat and courtship displays.

interambulacrum In Echinodermata, esp. Echinoidea, the region intervening between two ambulacral areas.

interbranchial septa The stout fibrous partitions separating the branchial chambers in Fish.

intercalare A cartilage or ossification lying between the basiventrals, or between the basidorsals of the vertebral column.

intercalary Placed between other bodies or the ends of a stem, filament, hypha etc. See **intercalate**.

intercalary meristem A meristem located somewhere along the length of a plant member and which divides to give *intercalary growth*, as at the base of a grass leaf.

intercalate To add, to insert, as in *intercalated somite*. adj. *intercalary*.

intercalating dyes Chemical compounds with a high affinity for DNA whose molecules *intercalate* between the planar base pairs of the DNA helix. They are used for visualizing DNA, changing its density and inducing breakage. Often mutagenic and carcinogenic.

intercellular Between cells. Cf. **intracellular, extracellular**.

intercellular spaces The interconnecting spaces between cells in a tissue, air filled in vascular plants, and providing for gas exchange.

intercentra See **hypapophyses**.

interchange The mutual transfer of parts between two chromosomes.

interchondral Said of certain ligaments and articulations between the costal cartilages.

interclavicle In Vertebrates, a bone lying between the clavicles, forming part of the pectoral girdle.

intercostal Between the ribs.

interdorsal An intercalary element lying between adjacent basidorsals of the vertebral column.

interfasicular cambium Vascular cambium between the vascular bundles of the stem and joined to the cambium in the bundles to make a complete cylinder.

interfasicular region Tissue between the vascular bundles of a stem. Also *medullary ray, pith ray*.

interference The usually negative effect that the presence of one **chiasma** has on the probability of a second occurring in its vicinity.

interference microscope Microscope in which the phase changes caused by differences in optical path (refractive index times thickness for transmitted light) within the specimen can be measured or made visible as differences in brightness or colour in the image. The light is split into two beams, one passing through the specimen, the other, the *reference beam*, ideally through empty medium near the specimen. The two beams, with suitably manipulated phase, are then made to interfere at the image plane. Areas of specimen which have similar optical paths appear in the image similarly bright or coloured. Cf. **differential interference contrast microscope**. Because the refractive index of an aqueous solution is nearly proportional to concentration of solutes, the microscope can be used to estimate the dry mass (to 'weigh') microscopic objects. See **phase contrast microscopy**.

intorforon See panel.

interkinesis See **interphase**.

interleukin See panel.

intermediate filaments Cytoplasmic filaments of *intermediate* diameter from 7 to 11 nm. There are several different subclasses, e.g. *desmin, keratin, lamin, vimentin*, each with a characteristic cellular distribution. They are all constructed of proteins possessing a rod-like structure built from four α-helical domains.

intermediate host In the life history of a parasite, a secondary host which is occupied by the young forms, or by a resting stage between the adult stages in the primary host.

intermedium A small bone of the proximal row of the basipodium, lying between the tibiale and fibulare, or between the radiate and ulnare.

---

interferon

A group of proteins with antiviral activity originally identified as products released by cells in response to virus infection. When taken up by other cells the replication of virus within those cells is inhibited by a mechanism which blocks the translation of viral messenger RNA. Interferons have since been found to be released from cells by other agents also, including immunological stimuli. α-interferons are released from leucocytes and β-interferons from fibroblasts in response to viral infection. γ-interferon is distinct, is released by activated macrophages and by B lymphocytes and is regarded as a **lymphokine**. It can up-regulate the expression of MHC class II antigens and down-regulate the expression of MHC class I antigens, and activates **NK cells** to increase their cytocidal capacity. It also increases the sensitivity of B cells to growth and differentiation factors. Interferons have been used as anti-cancer agents in some clinical trials.

---

intermittent reinforcement Schedules of reinforcement in which only some responses are reinforced; it can be based on *ratio* or *interval reinforcement*.

internal image Since an antibody made in response to an **epitope** of an antigen has a binding site (idiotope) complementary to the epitope, antibodies raised against that idiotope will have binding sites which have a similar shape to the epitope on the original antigen, i.e. have an internal image of that antigen. Anti-idiotypic antibodies have been shown to elicit antibodies against an antigen in an animal which was never exposed to the antigen itself.

internal phloem *Intraxylary phloem*. Primary phloem on the centric side of the xylem as in the Solanaceae.

internal respiration See **respiration, internal**.

internal secretion A secretion which is poured into the blood vessels, or into the canal of the spinal cord; a hormone.

internasal septum See **mesethmoid**.

interneuron A nerve cell within the central nervous system which is neither sensory or motor but communicates between other nerve cells.

internode (1) Between two successive nodes of a stem or hypha. (2) The myelinated part of a nerve between two adjacent nodes of Ranvier.

internuncial Interconnecting, e.g. neurone of central nervous system interposed between afferent and efferent neurones of a reflex arc.

interoceptor A sensory nerve ending, specialized for the reception of impressions from within the body. Cf. **exteroceptor**.

interopercular In Fish, a ventral membrane bone supporting the operculum.

interparietal A median dorsal membrane bone of the Vertebrate skull, situated between the parietals and the supraoccipital.

interphase The period of the **cell cycle** between mitoses.

interpositional growth See **intrusive growth**.

inter-renal body In selachian Fish, a ductless gland which lies between the kidneys and corresponds to the cortex of the suprarenal gland of higher Vertebrates.

interruptedly pinnate A leaf having pairs of large and small leaflets alternating along the rachis. See **pinnate**.

intersegmental membrane The flexible infolded portion of the cuticle between adjacent definitive segments of the Arthropod body which allow freedom of movement.

intersex An individual which exhibits characters intermediate between those of the male and female of the same species. Often due to a chromosomal abnormality.

interspecific Said of an event such as a cross between individuals from separate species. Cf. **intraspecific**.

interstitial Occurring in the interstices between other structures; as the *interstitial* cells of *Cnidaria*, which are small rounded embryonic cells occurring in the interstices between the columnar cells forming the ectoderm and endoderm.

interval schedules of reinforcement Schedules in which reinforcement is given for the first response made after a certain period of time has passed. On fixed interval schedules, the period of time is always the same; on variable interval schedules the period of time fluctuates. Both types are forms of *partial* or *intermittent* reinforcement. Cf. **continuous reinforcement**.

intervening sequence See **intron**.

intervertebral Between the vertebrae; said of the fibro-cartilage disks between the vertebral centra in Crocodiles, Birds and Mammals.

intestine In Vertebrates, that part of the alimentary canal leading from the stomach

---

interleukin

A term used to describe products of macrophages and T lymphocytes which influence the differentiation of themselves or of other cells. Although hard to obtain pure under natural conditions, the genes responsible have been cloned and their products synthesized by recombinant DNA techniques. Abbrev. *IL*.

Interleukin-1 is secreted by activated macrophages. Causes fever by acting as a pyrogen at the thermoregulation centre in the brain, and the synthesis and release of **acute phase proteins** by liver cells. It is also required for activation of T lymphocytes to develop receptors for interleukin-2 and to secrete interleukin-2. It has a direct mitogenic activity for thymocytes. There are two forms of IL-I termed alpha and beta which differ partly in respect of their amino acid sequences, but they have similar biological activities.

Interleukin-2 is a growth factor for T lymphocytes made by T lymphocytes which have been activated by IL-I and a mitogenic agent such as concanavalin A or a specific antigen. It binds to specific receptors on the same or other T cells and is thus able to maintain a population of T cells in continuous growth and differential.

Interleukin-3 is a factor made by activated T lymphocytes which acts to stimulate growth and differentiation of the progenitors of all haemopoietic cells.

Interleukin-4 is another factor made by activated T lymphocytes but it causes resting B lymphocytes to divide.

---

to the anus; in Invertebrates, that part of the alimentary canal which corresponds to the Vertebrate intestine, or was thought by the early investigators so to correspond. adj. *intestinal*.

intine The inner part of the wall of a pollen grain or vascular plant spore. Cf. **exine**.

intoxication The state of being poisoned.

intracellular Within the cell.

intracellular enzyme One secreted and functioning within the cell. Cf. **extracellular enzyme**.

intrafusal A muscle fibre contained within a muscle spindle. Concerned with proprioceptive reflexes.

intrapleural Within the thoracic cavity.

intraspecific Pertaining to interactions within members of a species. Cf. **interspecific**.

intra-vitam staining The artificial staining of living cells. Usually by injection of the stain into a circulatory system.

intraxylary phloem See **internal phloem**.

intrazonal soil Well developed soil in which local conditions have modified the influence of the regional climate. Cf. **zonal soil**, **azonal soil**.

intrinsic Said of appendicular muscles of Vertebrates which lie within the limb itself, and originate either from the girdle or from the limb bones. Cf. **extrinsic**.

introgression *Introgressive hybridization*. Incorporation of genes of one species into the gene pool of another *via* an interspecific hybrid.

introjection In social psychology, the internalization of the norms and values of one's social group, so that the individual comes to be guided by a sense of what is appropriate or inappropriate, rather than by external rewards and punishments. In psychoanalytic theory, the term is synonymous with **identification**.

intromittent Adapted for insertion, as the copulatory organs of some male animals.

intron Genes in eukaryotes are organized in such a way that while the whole sequence is transcribed, only part of it forms the messenger RNA. *Introns* or *intervening sequences*, which are often long, are excised during the maturation of the RNA. Cf. **exons**, the part which is expressed in the protein product.

introrse Directed or bent inwards. Facing towards the axis, esp. stamens opening towards the centre of the flower. Cf. **extrorse**.

introvert (1) In the Jungian theory of behaviour, a person whose libido is inwardly directed resulting in the shunning of interpersonal relations and absorption in egocentric thoughts. Generally, a person more interested in his thoughts and feelings than in the external world and its social activity. n. *introversion*. See **extraversion**. (2) A structure or part of the body which may be involuted or turned inside out, as the proboscis of a Nemertinean worm.

intrusive growth Type of tissue growth in which an elongating cell grows by insertion between other cells. Cf. **symplastic growth**.

intussusception New material inserted into the thickness of an existing cell wall. Cf. **apposition**.

invagination Insertion into a sheath; the development of a hollow ingrowth; the pushing in of one side of the blastula in embolic gastrulation. Cf. **evagination**. adj. *invaginate*.

inv allotypes Allotypic antigenic determinants on the constant region of the kappa chain of human immunoglobulins.

inversion A stretch of chromosome that has been turned round so that the order of the nucleotides in the DNA is reversed.

invertase A plant enzyme which hydrolyses cane-sugar. Also *sucrase*.

Invertebrata A collective name for all animals other than those in the phylum Chordata; i.e. all those animals that do not exhibit the characteristics of vertebrates, namely possession of a notochord or vertebral column, ventral heart, eyes etc. Some animals near the chordate boundary line, e.g. Hemichorda and Urochorda, are sometimes regarded as invertebrates. The term is not used as a scientific classification.

investment The outer layers of an organ or part, or of an animal.

in vitro Literally 'in glass'. Used to describe the experimental reproduction of biological processes in the more easily defined environment of the culture vessel or plate. Cf. **in vivo**.

in vitro **transcription** Use of a laboratory medium without the presence of cells to obtain specific mRNA production from a DNA sequence. Also *cell-free transcription*.

in vitro **translation** The use of pure mRNA, ribosomes, factors, enzymes and precursors to obtain a specific protein product without the presence of cells. The procedure can be coupled to *in vitro* transcription. Also *cell-free translation*.

in vivo Used to describe biological processes occurring within the living organism or cell. Cf. **in vitro**.

involucre (1) Bracts forming a calyx-like structure close to the base of a usually condensed inflorescence, as in the Compositae. (2) A sheath or ring of leaves surrounding a group of archegonia or antheridia in bryophytes.

involuntary muscle   See **unstriated muscle**.

involute Tightly coiled; said of Gastropod shells or the margins of a leaf rolled inwards or upwards, i.e. towards the *adaxial surface*. Cf. **revolute**. See **vernation**.

iodophilic bacteria Bacteria which stain blue with iodine, revealing the presence of starchlike compounds. They occur in large numbers in the rumen where they are associated with cellulose decomposition.

ionophore A compound which enhances the permeation of biological membranes by specific ions, acting either as specific ion carriers or by creating ion channels across the membrane, e.g. *gramicidin, valinomycin*.

IPA Abbrev. for *IsoPentenyl Adenine*.

ipsilateral Pertaining to the same side of the body. Cf. **contralateral**.

I region Region in the murine major histocompatibility complex which contains genes coding for Class II histocompatibility antigens.

Ir gene *Immune response gene*. Found in the I region, and so called because in inbred strains of mice the ability to respond to certain simple peptides depends upon which Class II antigens are expressed on their cells.

irid-, irido- Prefixes from Gk. *iris*, gen. *iridos*, rainbow.

iridocyte A reflecting cell containing guanin, found in the integument of Fish and of certain Cephalopods, to which it gives an iridescent appearance.

iris In the Vertebrate eye, that part of the choroid, lying in front of the lens, which takes the form of a circular curtain with a central opening. adj. *iridial*.

IRM Abbrev. for *Innate Releasing Mechanism*.

iron bacteria Filamentous bacteria, which can convert iron oxide to iron hydroxide, deposited on their sheaths. Important in formation of **bog iron ore**.

irregular (1) Asymmetric, not arranged in an even line or circle. (2) Not divisible into halves by an indefinite number of longitudinal planes.

irritabilty A property of living matter, namely, the ability to receive and respond to external stimuli.

irritant Any external stimulus which produces an active response in a living organism.

ischium A posterior bone of the pelvic girdle in *Tetrapods*. adjs. *ischial, ischiadic*.

iso- Prefix from Gk. *isos*, equal.

iso-agglutination (1) The adhesion of spermatozoa to one another by the action of some substance produced by the ova of the same species. (2) The adhesion of erythrocytes to one another within the same blood group. Cf. **hetero-agglutination**.

iso-antigen Use now mainly confined to blood group antigens. Cf. **allo-antigen**.

isobilateral Divisible into symmetrical halves along two distinct planes, esp. of a leaf which has palisade towards both faces, *unifacial*.

isocercal Said of a type of secondarily symmetrical tail fin in Fish in which the areas of the fin above and below the vertebral column are equal.

isodactylous Having all the digits of a

limb the same size.

isodont Having all the teeth similar in size and form.

iso-electric focusing A technique for separating proteins according to their **iso-electric points**. The mixture of proteins is placed in a pH gradient established by ampholines in an electric field across a liquid or a gel matrix; the proteins move in the electric field until they reach their iso-electric points in the pH gradient where, having lost net charge, they are focused.

iso-enzymes See **isozymes**.

isogamy Sexual fusion of similar gametes. Cf. **anisogamy**, **oögamy**.

isogenetic Having a similar origin.

isogenic Describes a strain in which all individuals are genetically identical but not necessarily homozygous. Cf. **inbred line**.

isokont, isokonton Having two or more flagella of equal length or identical morphology as the whiplash flagella of *Chlamydomonas*.

isolate To establish a pure culture of a micro-organism.

isolating mechanism Anything which prevents the exchange of genetic material between two populations. It can include geography, physiology or behaviour.

isolecithal Said of ova which have yolk distributed evenly through the protoplasm.

isoleucine An essential and non-polarizable amino acid. Symbol Ile, short form I.

Side chain:
$$CH_3{-}CH_2$$
$$\diagdown$$
$$CH{-}$$
$$\diagup$$
$$CH_3$$

See **amino acid**.

Isomastigote Having two or four flagella of equal length.

isomerous Equal numbers as in the parts in two whorls of a flower.

isometric contraction The type of contraction involved when a muscle produces tension but is held so that it cannot change its length.

isomorphic (1) Morphologically similar. (2) See **alternation of generations**.

isomorphic alternation of generations Alternation of generations which are morphologically alike. Also *homologous alternation of generations*. Cf. **heteromorphic alternation of generations**.

isomorphous replacement See **multiple isomorphous replacement**.

isonome Line on a map joining points of equal abundance of a given species of plant.

iso-osmotic A solution with the same osmolarity as that within a cell (or any other solution). When an animal cell is suspended in an iso-osmotic solution there is no net flux of water across the cell membrane unless osmotic balance is destroyed by solute

movement into or out of the cell. Water movement in cells with walls is affected by the hydrostatic pressure exerted by the wall in addition to the osmotic forces. See **isotonic**.

isopedin The thin layer of bone forming the inner layer of the cosmoid scales of *Choanichthyes*.

isopentenyladenine A natural cytokinin.

Isopoda An order of Malacostraca, in which the carapace is absent, the eyes are sessile or borne on immovable stalks; the body is depressed and the legs used for walking; they show great variety of form, size and habit; some are terrestrial, plant feeders or ant guests, others are marine, free-living, and feeding on seaweeds or ectoparasitic on fish. Woodlice etc.

isopodous Having the legs all alike.

Isoptera An order of social Exopterygota living in large communities which occupy nests excavated in the soil or built up from mud and wood; different polymorphic forms or castes occur in each species; the mouth parts are adapted for biting and both pairs of wings, if present, are membranous and can be shed by means of a basal suture; exclusively herbivorous. White Ants; Termites.

isostemonous With stamens in one whorl and equal in number to petals.

isotonic solution A solution with water potential equal to that of a cell suspended in the solution; consequently there is no water flux between cell and solution. In animal cells these conditions are established for non-permeating solutes at iso-osmolarity but in plant cells the contribution of the hydrostatic pressure of the cell wall must be taken into account.

isotonic contraction The type of contraction involved when a muscle shortens while maintaining a constant tension.

isotope therapy Radiotherapy by means of radioisotopes.

isotopic dilution The mixing of a particular nuclide with one or more of its isotopes.

isotype Used to describe a class of immunoglobulin (IgM, IgG etc.) defined by the identity of its **heavy chain** constant region.

isozyme *Isoenzyme*. Electrophoretically distinct forms of an enzyme with identical activities, usually coded by different genes.

isthmus A neck connecting two expanded portions of an organ; as the constriction connecting the mid-brain and the hind-brain of Vertebrates.

iter A canal or duct, as the reduced ventricle of the mid-brain in higher Vertebrates.

iteroparous Reproducing on two or more occasions during a lifetime.

ivory The dentine of teeth, esp. the type of dentine composing the tusks of elephants.

J K

Jacobson's glands In some Vertebrates, nasal glands, the secretion of which moistens the olfactory epithelium.

Jacobson's organ In some Vertebrates, an accessory olfactory organ developed in connection with the roof of the mouth.

James-Lange theory of emotions The theory that emotion is the subjective experience of one's own bodily reactions in the presence of certain arousing stimuli; the stimuli cause certain physiological responses, and the awareness of these responses causes emotion.

jaws In gnathostomatous Vertebrates, the skeletal framework of the mouth enclosed by flesh or horny sheaths, assisting in the opening and closing of the mouth, and usually furnished with teeth of horny plates to facilitate seizure of the prey or mastication; in Invertebrates, any similar structures placed at the anterior end of the alimentary tract.

J chain A polypeptide chain with a high content of the amino acid cysteine, which enables it to form disulphide bonds. It helps to link together IgA molecules into polymeric forms and to hold IgM in pentameric form.

jejunum In Mammals, that part of the small intestine which intervenes between the duodenum and the ileum.

J gene *J exon*. A short sequence of DNA coding for part of the hypervariable region of immunoglobulin light or heavy chains near to the site of joining to the constant region. There are several possible J exons, of which any may be used. The gene for the beta chain of the T cell antigen receptor also includes a different J exon which has similar variability.

JND Abbrev. for *Just Noticeable Difference*.

Johnston's organ In Insects, a sensory structure situated within the second antennal joint, mechanoreceptive in function.

Jordanon species *Microspecies*. One of a number of true-breeding, morphologically slightly different, lines within a complex of largely inbreeding plants. Cf. **Linnaean species**.

jugal A paired membrane bone of the zygoma of the Vertebrate skull, lying between the squamosal and the maxilla.

jugular Pertaining to the throat or neck region, e.g. a *jugular* vein.

jugular nerve The posterior branch of the hyomandibular component of the 7th cranial nerve in Vertebrates; it carries the visceromotor component of the hyomandibular branch.

just noticeable difference See **difference threshold**. Abbrev. *JND*.

juvenile A structure characteristic of the *juvenile phase*.

juvenile hormone See **neotenin**.

juvenile phase Phase before flowering in woody plants; it differs in many attributes including the ease of rooting from cuttings; usually transient but can be maintained as in some cultivated ornamental plants including many conifers.

kappa chain One of the two types of **light chain** of immunoglobulins, the other being the lambda chain. An individual immunoglobulin molecule bears either two kappa or two lambda chains. The proportion of molecules bearing each chain varies in different species. In humans about 60% are of the kappa and 40% of the lambda type.

kary-, karyo- Prefixes from Gk. *karyon*, nucleus. Also *cary-*, *caryo-*.

karyogamy The fusion of the two gametes to form a zygote. It usually follows cytoplasmic fusion but may as in some fungi be followed by a prolonged binucleate stage, *dikaryophase*. See **plasmogamy**.

karyogram See **idiogram**.

karyon The cell-nucleus, only used in compound words. From Gk. for nut or kernel.

karyotype The appearance, number and arrangement of the chromosomes in the cells of an individual.

Kaspar-Hauser experiments Those in which animals are reared in complete isolation from other animals of their own or other species.

kata- Prefix from Gk. *kata*, down. Also *cata-*.

katadromous Of Fish, migrating to water of greater density than that of the normal habitat to spawn, as the Freshwater Eel which migrates from fresh to salt water to spawn. Cf. **anadromous**. Also *catadromous*.

kataplexy The state of imitation of death, adopted by some animals when alarmed.

K-cell A cell generally resembling a lymphocyte which bears Fc receptors by which it can bind to other cells to which are attached antibodies against antigens on the cell surface (thus exposing the Fc portions). K-cells induce lesions in the membrane of the target cell which kill it. This is an example of antibody-dependent cell-mediated cytotoxicity.

keloid A dense new growth of skin occurring in skin that has been injured.

kelp (1) A general name for large seaweeds, esp. *Laminaria* and allies. (2) The ashes from burning such seaweed, formerly a source of soda and iodine.

keratin The class of **intermediate filament** which is characteristic of epithelial cells; α-keratin is a major component of skin; β-keratin exists as a β-sheet and is a major component of *silk*.

keratogenous See **keratin**.

key Set of instructions devised to enable an unknown organism to be identified by a student on the basis of critical characteristics.

keyhole limpet haemocyanin A large copper-containing protein from a particular kind of limpet. Haemocyanins normally function as oxygen carrying molecules, but are widely used as immunogens in immunology since they are likely to be completely foreign to mammals. Abbrev. *KLH*.

kidney A paired organ for the excretion of nitrogenous waste products in Vertebrates.

kidney stones Hard deposits formed in the kidney. The composition varies, and kidney stones have been found to consist of uric acid and urates, calcium oxalate, calcium and magnesium phosphate, silica and alumina, cystine, xanthine, fibrin, cholesterol and fatty acids. Passage of the stones down the ureter may cause severe pain (*renal colic*).

kilo- Prefix denoting 1000; used in the metric system, e.g. 1 kilogram =1000 grams.

kinaesthesis A general term for sensory feedback from muscles, tendons and joints, which inform the individual of the movements of the body or limbs, and the position of the body in space. adj. *kinaesthetic*.

kinase An enzyme which catalyses the phosphorylation of its substrate by ATP. Thus protein kinases phosphorylate proteins and hexose kinases phosphorylate hexose.

kinesin A protein of wide distribution in eukaryotes which is responsible for the movement of organelles to which it is attached along microtubules.

kinesis A simple response to environmental stimuli in which the animal's response is proportional to the intensity of stimulation; it involves a change in speed of movement or rate of turning. Unlike a **taxis**, the animal's body is not oriented to the stimulus although the effected movements often produce a change of position relative to it. Cf. **taxis**.

kinetin 6-*furfurylaminopurine*. A synthetic plant growth regulator of the cytokinin type.

kinetochore Paired structures within the **centromeric** region of metaphase chromosomes, to which spindle **microtubules**

attach. They lie on each side of the **primary constriction** and when viewed with the electron microscope, appear as a trilaminar plate with microtubules entering at regular intervals.

kinetodesma, kinetosome See **kinety**, **basal body**.

kinety A unit of structure in the Protozoa comprising the *kinetosomes* (the basal granules of the cilia and flagella) and the *kinetodesma* (a fine strand running from the kinetosomes). In Flagellata the line of cell division is parallel to the *kinetia* (symmetrical division), and in Ciliata the plane of cleavage cuts across the *kinetia* (percentien division).

kingdom Higher taxonomic rank; composed of a number of **divisions**.

kinin See **cytokinin**.

kinins A class of vasoactive peptides that are associated with local regulation of blood flow, e.g. bradykinin.

kin selection Natural selection for behaviour that lowers an individual's own chance of survival but raises that of a relative.

KLH Abbrev. for *Keyhole Limpet Haemocyanin*.

klinostat, clinostat An apparatus to rotate a plant, e.g. slowly about a horizontal axis in order to cancel out at least some of the effects of gravity in investigations of gravitropism etc.

knee In land Vertebrates, the joint between the femur and the crus.

knot (1) A node in a grass stem. (2) A hard and often resinous inclusion in timber, formed from the base of a branch which became buried in secondary wood as the trunk thickened.

Koch's postulates Four criteria laid down by Koch to show that a disease was caused by a micro-organism: (1) organism must be observed in all cases of the disease; (2) it must be isolated and grown in pure culture; (3) the culture must be capable of reproducing the disease when inoculated into a suitable experimental animal; (4) the organism must be recovered from the experimental disease.

Korsakoff's syndrome, psychosis An irreversible nutritional deficiency, caused by chronic alcoholism and related malnutrition; characterized by severe **anterograde amnesia** and **confabulation**.

Krantz anatomy The leaf anatomy of most C_4 plants, characterized by the bundle sheath cells being large and having conspicuous chloroplasts. The mesophyll cells have less conspicuous choloroplasts.

Krummholz Stunted, wind-trimmed trees between the *timber-line* and the *tree-line* on mountains.

K-strategist Organism which assigns relatively little of its resources to reproduction; characteristic of stable, saturated communities; has low fecundity and high competitive ability. Cf. **R-strategist**.

Kupffer cell The form of macrophages which line the blood sinuses of the liver.

They are phagocytic and remove foreign particles from the blood. They have Fc receptors and can engulf blood cells which have become coated with antibody.

kurtosis The degree to which a statistical distribution is sharply peaked at its centre.

L

labelling theory The view that no actions are inherently deviant or abnormal, and that the label *mental illness* applied to certain behaviours and to individuals, acts as a self-fulfilling prophecy; once applied, the individual's self expectations, and the expectations of others result in behaviour associated with the label, and so perpetuate the condition.

labellum (1) The posterior petal of the flower of an orchid and often the most conspicuous part of the flower. (2) The lower lip of the corolla forming a landing platform for pollinating insects in a number of families, e.g. Labiatae.

labellum A spoon-shaped lobe at the apex of the glassa in Bees; in certain Diptera, a pair of fleshy lobes into which the proboscis (**labium**) is expanded distally. pl. *labella*.

labia Any structures resembling lips; as the lips of the vulva in Primates. See the *sing.* form *labium*. adj. *labial*.

labia majora In Mammals, the two prominent folds which form the outer lips of the vulva.

labia minora In Mammals, the small inner lips of the vulva; nymphae.

labiate (1) With a lip or lips, as the corolla of various Labiatae. (2) Pertaining to a species of the Labiatae.

labium In Insects, the lower lip, formed by the fusion of the second maxillae; in the shells of Gastropoda, the inner or columellar lip of the margin of the aperture. See pl. form **labia**.

labrum In Insects and Crustacea, the platelike upper lip; in the shells of Gastropoda, the outer lip or right side of the margin of the aperture. adj. *labral*.

labyrinth Any convolute tubular structure, esp., the bony tubular cavity of the internal ear in Vertebrates, or the membranous tube lying within it.

labyrinthodont Having the dentine of the teeth folded in a complex manner.

lac A resinous substance, an excretion product of certain Coccid insects (Gascardia, Tachardia etc.) in certain jungle trees; used in the manufacture of shellac. Chief source, India. See **shellac**, for which the name is frequently loosely used.

lachrymal See **lacrimal**.

laciniate Deeply cut into narrow segments; fringed.

lacrimal Pertaining to, or situated near, the tear gland; in some Vertebrates, a paired lateral membrane bone of the orbital region of the skull, in close proximity to the tear gland.

lacrimal duct In most of the higher Vertebrates, a duct leading from the inner angle of the eye into the cavity of the nose; it serves to drain off the secretion of the lacrimal gland from the surface of the eye.

lacrimal gland In most of the higher Vertebrates, a gland situated at the outer angle of the eye; it secretes a watery substance which washes the surface of the eye and keeps it free from dust.

lacrimation Shedding of tears.

lactation The production of milk from the mammary glands of Mammalia.

lacteals In Vertebrates, lymphatics, in the region of the alimentary canal, in which the lymph has a milky appearance, due to minute fat globules in suspension.

lactic Pertaining to milk.

lactiferous Plants or vessels containing latex. Mammals producing milk, vessels carrying milk.

Lactobacillaceae A family of bacteria belonging to the order *Eubacteriales*. Gram-positive cocci or rods; carbohydrate fermenters are anaerobic or grow best at low oxygen tensions. Includes several pathogenic organisms, e.g. *Streptococcus pyogenes* (scarlet fever, tonsillitis, erysipelas, puerperal fever), *Diplococcus pneumoniae* (one cause of pnuemonia).

lactose $C_{12}H_{22}O_{11}+H_2O$, rhombic prisms which become anhydrous at 140°C, mp 201°–202°C with decomposition; when hydrolysed it forms d-galactose and d-glucose. It reduces Fehling's solution and shows mutarotation. It occurs in milk and is not so sweet as glucose.

lacuna A cavity, gap, space or depression such as (1) a space in a tissue caused by the breakdown of the protoxylem, a **leaf gap** or sometimes an intercellular space or (2) one of the cell-containing cavities in bone. adjs. *lacunar, lacunose*.

LAD Abbrev. for *Language Acquisition Device*.

laevo- Prefix from L. *laevus*, left.

lagena In higher Vertebrates, a pocket lined by sensory epithelium and developed from the posterior side of the sacculus, which becomes transformed in Mammals into the scala media or canal of the cochlea.

lagging Slow movement towards the poles of the spindle by one or more chromosomes in a dividing nucleus, with the result that these chromosomes do not become incorporated into a daughter nucleus.

Lagomorpha An order of Eutheria with two pairs of upper, and one of lower, incisor teeth which grow throughout life; there are no canines, and there is a wide diastema between the incisors and the cheek teeth into which the cheeks can be tucked to separate the front part of the mouth from the hind during gnawing. Rabbits and Hares.

lagopodous Having feet covered by hairs or feathers.

LAI Abbrev. for *Leaf Area Index*.

lalling, lallation Babbling speech of infants; lack of precision in the articulatory mechanism of the mouth.

Lamarckism A theory, now discredited, that evolutionary change takes place by the *inheritance* of *acquired characters*, i.e. that characters acquired during the lifetime of an individual (e.g. an athlete's strong muscles) are transmitted to its offspring.

lambda chain One of the two types of light chain of immunoglobulins, the other being the **kappa chain**.

lambda phage A well-studied temperate phage which can either grow in synchrony with its host (*E.coli*) in its **lysogenic** phase or go into a *lytic* phase, when its genome is replicated many times by a *rolling circle* mechanism. An important vector in genetic manipulation procedures.

lamella A structure resembling a thin plate such as a thin plate or layer of cells or a gill of an agaric. adjs. *lamellar, lamellate*.

lamellibranch Having platelike gills, as members of the group, Bivalvia.

lamell-, lamelli- Prefixes from L. *lamella*, thin plate.

lamina (1) The network of lamin proteins which lie on the inner surface of the nuclear membrane. (2) Expanded blade-like part of a leaf, as distinct from petiole and leaf base. (3) Any flattened platelike structure.

lamina propria Layer of connective tissue which supports the epithelium of the digestive tract and with it forms the mucous membrane. The lamina propria is the site at which accumulate lymphocytes and plasma cells, mast cells and accessory cells in immunological reactions involving the gut.

laminarin The storage polysaccharide of the brown algae, a β, $1 \rightarrow 3$ glucan.

lamina terminalis In Craniata, the anterior termination of the spinal cord which lies at the anterior end of the diencephalon.

laminin A large fibrous protein which is a major component of the basal lamina.

lamins A group of three proteins, lamin A, B and C, of the **intermediate filament** type which forms the nuclear lamina.

lampbrush chromosome Chromosome of the first meiotic prophase observed in many **eukaryotes**, which has a characteristic appearance due to the orderly series of lateral loops of chromatin, arranged in pairs on either side of the chromosome axis.

lanate Woolly.

lanceolate Lance-shaped; much longer than broad and tapering at both ends.

lancinating Of pain, acute, shooting, piercing, cutting.

landrace Ancient or primitive cultivar of a crop plant.

Langerhans cell A cell with dendritic shape present in the epidermis and characterized by the presence in the cytoplasm of Birbeck granules, i.e. small racquet-shaped objects detected by electron microscopy, probably composed of some undigested material. Langerhans cells are the form taken in passage through the skin by the type of accessory cell which includes interdigitating or **dendritic cells**. MHC Class II antigens are strongly expressed, and these cells are very effective at antigen presentation to T lymphocytes. Also name used for the spindle-shaped cells in the centre of each acinus of the pancreas.

Langerhans islets Irregular masses of hyaline epithelium cells, unfurnished with ducts, occurring in the Vertebrate pancreas; they are responsible for the elaboration of the hormone insulin.

language acquisition device According to Chomsky, a hypothetical brain structure that enables an individual to learn the rules of grammar on hearing spoken language. Abbrev. *LAD*.

laniary Adapted for tearing, as a canine tooth.

lanuginose Woolly.

lanugo In Mammals, prenatal hair.

lapidicolous Living under stones.

large intestine See **colon**.

larva The young stage of an animal if it differs appreciably in form from the adult; a free-living embryo.

larviparous Giving birth to offspring which have already reached the larva stage.

larvivorous Larva-eating.

laryngotracheal chamber In Amphibians, a small chamber into which the lungs open anteriorly, and which communicates with the buccal cavity by the glottis.

larynx Organ in the throat of man and vertebrates which with the lungs forms the voice.

latency General term for the interval before some reaction; also, the dormancy of a particular behaviour or response.

latency period, stage According to Freud, the fourth psychosexual stage of development, in which sexuality lies essentially dormant, occurring roughly between the ages of six and the onset of puberty.

latent An organism or part in a resting condition or state of arrested development, but capable of becoming active or undergoing further development when conditions become suitable; said also of hidden characteristics which may become evident under the right circumstances.

latent content In psychoanalytic theory,

the unconscious material or hidden meaning of a dream that is being expressed in a disguised fashion through symbols contained in the dream. See **manifest dream content**.

latent learning Learning the characteristics of a situation which is not manifested by any immediate overt behaviour, but which is revealed when the individual is later placed in a situation requiring previously acquired information.

latent period (1) Period of time between stimulation and the first signs of a response. (2) Period of time between infection and the appearance of the symptoms of disease.

lateral Arising from a parent axis or the structure so formed. Cf. **leader**.

lateralization, laterality The organization of brain functions such that each of the cerebral hemispheres control different psychological functions, particularly in verbal and visual spatial skills. In most right handers, the left hemisphere is specialized for language functions, while the right hemisphere is better at various visual and spatial tasks.

lateral line In Fish, a line of neuromast organs running along the side of the body. In Nematoda, a paired lateral concentration of hypodermis containing the excretory canal.

lateral meristem A meristem lying parallel to the sides of the axis, e.g. cambium.

lateral plate In Craniata, a ventral portion of each mesoderm band which surrounds the mesenteron in embryo; in Insects, the paired lateral region of the germ band of the embryo which becomes separated from the opposite member of its pair by the middle plate.

laterigrade Moving sideways, as some Crabs.

laterosphenoid The so called 'alisphenoid' of Fish, Reptiles and Birds (representing an ossification of the wall of the chondrocranium), as distinct from the alisphenoid of Mammals (developed from the splanchnocranium).

late wood The wood formed in the later part of a growth ring, usually with smaller cells with thicker walls than the **early wood**.

latex Fluid, often milky and viscous, exuded from *laticifers* when many plants are cut. It consists of a watery solution containing many different substances including terpenoids which form rubber, alkaloids such as the opium alkaloids, sugar, starch etc. When rubber trees (e.g. *Hevea Brasiliensis*) are tapped a colloidal system is exuded, consisting of caoutchouc dispersed in an aqueous medium, rel.d. 0.99, which forms rubber by coagulation. The coagulation of latex can be prevented by the addition of ammonia or formaldehyde. Latex may be vulcanized directly, the product being known as *vultex*. (2) In synthetic rubber manufacture, latex is the name of the process stream in which the polymerized product is produced.

lati- Prefix from L. *latus*, broad.

laticifer A cell or vessel containing **latex**, present in tissues of many plants.

lattice hypothesis An hypothesis proposed to explain how aggregates are formed when antibodies combine with a soluble antigen, the size and composition of which varies with the ratio of the two components. Divalent antibody molecules combine with antigenic determinants on a polyvalent antigen to form a lattice. When the ratio of the two is such that all the antigen molecules are attached to one another by antibody molecules a macromolecular lattice is formed which comes out of solution. This ratio is known as the optimal proportion. In antigen excess the complexes will be smaller and may remain in solution.

Lauraceae Family of ca 2500 spp. of dicotyledonous flowering plants (superorder Magnoliidae). Mostly woody, tropical and subtropical. Includes avocado, cinnamon, camphor, bay laurel and some timber trees.

law of effect Thorndike's formulation of the importance of reward in learning, which states that the tendency of a stimulus to evoke a response is strengthened if the response is followed by a satisfactory or pleasant consequence, and is weakened if the response is followed by an annoying or unpleasant consequence.

layering (1) See **stratification**. (2) A method of artificial propagation in which stems are pegged down and covered with soil until they root, when they can be detached from the parent plant.

LD 50 The dose of a toxic substance that, administered in a named way, will kill 50% of a large number of individuals of a given species.

leaching The removal of substances from soils by percolating water. Cf. **cheluviation, flush, lessivage, podsol**.

lead equivalent Absorbing power of radiation screen expressed in terms of the thickness of lead which would be equally effective for the same radiation.

leader The younger part of the main stem or a main branch of woody plants. Its branches are called *laterals*. See **spur**.

lead poisoning Chronic poisoning due to inhalation, ingestion, and skin absorption of lead. Recognized as a hazard, both for young children (formerly through sucking lead or lead painted articles or toys), and also industrially. Characterized by anaemia, constipation, severe abdominal pain, and perhaps ultimately renal damage. Lesser degrees now recognized and soft water delivered by lead pipes is a potential hazard.

lead protection Protection provided by

metallic lead against ionizing radiation. Joined with other substances to provide further protection, e.g. lead glass, lead rubber.

leaf (1) In modern vascular plants a lateral organ of limited growth which develops from a primordium at a shoot apex. In angiosperms a leaf typically has a bud in its axil. Most leaves are more or less flat and green and photosynthetic in function; modified leaves include bud scales, bulb scales, many sorts of spine and tendril, bracts and probably sepals and petals, and possibly stamens and carpels. Cf. **sporophyll**. (2) In bryophytes, similar but usually smaller and thinner (mostly 1 cell thick) structures. See **phyllid**.

leaf area index The ratio of the total area of leaves of a plant or crop to the area of soil available to it.

leaf gap Region of parenchyma in the vascular cylinder of a stem above a leaf trace.

leaflet One of the leaf-like units which together make up the lamina of a compound leaf.

leaf mosaic See **mosaic**.

leaf scar The scar, usually covered by a thin protective layer, left on a stem following the abscission of a leaf.

leaf sheath The sheath surrounding the stem at the base of a leaf in grasses.

leaf succulent A plant with succulent photosynthetic leaves. Many are CAM plants, e.g. *Aloe*. Cf. **stem succulent**.

leaf trace A vascular bundle in a stem from its junction with another bundle of the stele to the base of the leaf; a leaf may have one trace or more.

learned helplessness Described by Seligman, a condition characterized by a general sense of powerlessness, which has its origins in a traumatic and inescapable event, but which persists in situations where escape or avoidance is possible. It specifically refers to effects created in the laboratory in animals exposed to uncontrollable events, but is thought by some to be one of the factors underlying depression in humans.

learning set Refers to the increased ability to solve a particular kind of problem as a consequence of previous experience with similar kinds of problems.

learning theory A range of theories which, unlike **depth psychology**, explain behaviour in terms of learning and conditioning. Largely the offspring of *behaviourism*, it takes as its starting point the association of various stimuli and responses. See **behaviour therapy**, **social learning theory**, **S-R theory**.

least distance of distinct vision For a normal eye it is assumed that nothing is gained by bringing an object to be inspected

nearer than 25 cm, owing to the strain imposed on the ciliary muscles if the eye attempts to focus for a shorter distance.

LE cell See **lupus erythematosus cell**.

lecithin See **phosphatidyl choline**.

lecith-, lecitho- Prefixes from Gk. *lekithos*, yolk of egg.

lecithocoel The segmentation cavity of a holoblastic egg.

lectin Proteins derived usually from plants that bind specifically to sugars or to oligosaccharides. Since similar sugar residues are present in glycoproteins or glycolipids on the surface of many cells, lectins cause agglutination of these cells. Some lectins bind specifically to certain cell types or even to cells at a particular stage of differentiation, and are used to identify them or to isolate glycoproteins extracted from them. Several act as polyclonal mitogens for T and/or B lymphocytes (e.g. *Concanavalin A* from Jack beans, *Phytohaemagglutinin* from red kidney beans and *Pokeweed mitogen* from *Phytolacca americana*).

leg In horticulture, a single short trunk from which branches arise, as of a fruit bush that is not managed as a **stool**.

leghaemoglobin A protein, similar to haemoglobin, in the nitrogen-fixing root nodules of leguminous plants, where it is involved in the maintenance of a low concentration of oxygen. See **nitrogenase**.

legume (1) A fruit from one carpel, dehiscent along top and bottom, e.g. pea pod. (2) A member of the Leguminosae. adj. *leguminous*.

Leguminosae *Fabaceae*. The pea family, ca 17 000 spp. of dicotyledonous flowering plants (superorder Rosidae). Trees, shrubs and herbs, cosmopolitan. The flowers have five free petals and a superior ovary of one carpel. The fruit is characteristically a legume or pod. There are three subfamilies (sometimes ranked as families); Mimosoideae, Caesalpinioideae and Papilionoideae, the last with more or less butterfly-shaped or 'pea' flowers. They form root nodules with symbiotic, nitrogen-fixing bacteria, *Rhizobium* spp. Although the seeds of many are poisonous (esp. uncooked), the Papilionoideae include extremely important crops, e.g. the protein-rich seeds and pods of many sorts of peas and beans, lentils and groundnuts, and forage crops, e.g. alfalfa.

lemma The lower of the two bracts enclosing a floret of a grass. Cf. **palea**.

lens In Arthropoda, the cornea of an ocellus or compound eye; in Craniata and Cephalopoda, a structure immediately behind the iris which serves to focus light on to the retina. See **crystalline lens**.

lentic Associated with standing water; inhabiting ponds, swamps etc.

lenticel A small patch of the periderm in which intercellular spaces are present allowing some gas exchange between the internal tissues and the atmosphere.

lenticular Shaped like a double convex lens.

lepido- Prefix from Gk. *lepis*, gen. *lepidos*, a scale.

Lepidoptera An order of Endopterygota, having two pairs of large and nearly equal wings densely clothed in scales; the mouth parts are suctorial, the mandibles being absent and the maxillae forming a tubular proboscis; the larva or caterpillar is active and usually herbivorous, with biting mouth parts. Butterflies and Moths.

lepidote With a covering of scale-like hairs.

lepospondylous Said of vertebral centra in which there is a skeletal ring constricting the notochord in the intervertebral region, with an expansion between each pair of adjacent centra.

lepromin test Test in which killed *Mycobacterium leprae* organisms are injected into the skin of subjects with leprosy. Those with lepromatous leprosy do not react, whereas those with the tuberculoid form of leprosy show a tuberculin type response. Not diagnostic of leprosy but an aid to classification and prognosis.

lepto- Prefix from Gk. *leptos*, slender.

leptocercal, leptocercous Having a long slender tail.

leptodactylous Having slender digits.

leptodermatous Thin skinned.

leptom, leptome The sieve elements and associated parenchyma cells (but not sclerenchyma cells) of the phloem. Cf. **hadrom**.

leptonema See **leptotene**.

leptosporangium A sporangium characteristic of the Filicales to which most modern ferns belong. Cf. **eusporangium**.

leptotene The first stage of meiotic prophase, in which the chromatin thread acquires definite polarity.

Leslie matrix model Specific deterministic model to predict the age structure of a population given the age structure at some past time and the age-specific survival and fecundity rates. Proposed by P. H. Leslie in 1945.

lessivage The process in which clay particles are washed downwards in a soil by water percolating through it. Cf. **leaching**, **cheluviation**.

lethal Causing death. (1) Of an environmental factor, fatal to an organism. (2) Of a genetic factor, causing death often, as in bacteria, only in a particular medium or environment, then a *conditional lethal*.

Leu Symbol for **leucine**.

leucine *2-amino-4-methylpentanoic acid.* A non-polarizable amino acid. Symbol Leu. Short form L.

Side chain: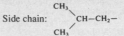

See **amino acid**.

leuco- Prefix from Gk. *leukos*, white. Also *leuko*.

leucoblast A cell which will give rise to a leucocyte.

leucocyte A white blood corpuscle; one of the colourless amoeboid cells occurring in suspension in the blood plasma of many animals.

leucocytosis An increase in the number of leucocytes in the blood.

leucopenia, leucocytopenia Abnormal diminution in the numbers of white cells in the blood.

leucoplast A colourless **plastid**.

leukaemia, leukemia Progressive uncontrolled overproduction of any one of the types of white cell of the blood, with suppression of other blood cells and infiltration of organs such as the spleen and liver. Cancers of the bloodcell forming tissues. Occurs in acute and chronic **myeloid**, **lymphatic**, and **monocytic** varieties.

leuko- See **leuco-**.

leukotrienes Pharmacologically active substances related to **prostaglandins** and generated from arachidonic acid by the action of lipoxygenases. A series of hydroxyicosatetraenoic acids, of which leukotriene B_4 is chemotactic for granulocytes; leukotrienes C_4, D_4 and E_4 (which contain cysteine) have the properties of a 'slow reacting substance of anaphylaxis' (*SRS-A*), i.e. they cause contraction of some types of smooth muscle, esp. bronchial muscle, and increase vascular permeability. Leukotrienes may be released from platelets and various leucocytes when damaged, and SRS-A activity is generated as a result of combination of antigen with IgE antibody on mast cells.

levator See **elevator**.

lewisite $ClCH=CH.AsCl_2$. *2-chloro-ethenyl dichloro arsine*. Dark coloured oily liquid (colourless when pure) with a strong smell of geraniums. Vesicant having lachrymatory and nose irritant action. Used as poison gas.

Leydig's duct See **Wolffian duct**.

L-forms Morphological variants developed from large bodies by prolonged exposure to various treatments. Consist of colonies of filterable bodies with a cytoplasmic matrix. Frequently revert to the normal form. If blood or serum is supplied in their nutrient medium may breed true for some generations.

liana, liane A climbing plant, esp. a woody climber of tropical forests.

libido According to Freud, a motivating force associated with instinctual drives towards survival, pleasure and the avoidance of pain, and which manifests itself in wish fulfilment.

library See **DNA library**.

libriform fibre Fibre in the **xylem**, usually the longest cell type in the tissue with thick walls and slit-like pits.

lichen A symbiont association of a fungus with an alga forming a macroscopic body, a thallus. Lichen-forming fungi are mostly Ascomycotina (see **Discomycetes**), a few are Basidiomycotina. The alga is usually Chlorophyceae, occasionally Cyanophyceae, rarely one of each. Most lichens are very sensitive to air pollution (esp. by SO_2) and the species present can be used as an index of pollution. Estimates suggest 20 000 spp.

Lieberkühn's crypts The simple tubular glands occurring in the mucous membrane of the small intestine in Vertebrates.

lien See **spleen**.

lienal Pertaining to the spleen.

lienogastric Pertaining to, or leading to, the spleen and the stomach, as the *lienogastric artery* and *vein* in Vertebrates.

life cycle The various stages through which an organism passes, from fertilized ovum to the fertilized ovum of the next generation.

life form The overall morphology of a plant categorizing it as an annual herb, a shrub, a succulent etc. See **Raunkiaer system**.

life table Tabulated display of the mortality schedule of a population, first devised by insurance companies and now used by ecologists in the study of wild populations.

ligament A bundle of fibrous tissue joining two or more bones or cartilages.

ligase An enzyme which seals nicks in one strand of a duplex DNA, much used to seal the gaps formed when one DNA sequence is artificially inserted into another.

ligation Joining two linear nucleic acid molecules together by a phosphodiester bond. Such linear molecules often have **sticky ends** which facilitate ligation.

light (L) chain A polypeptide chain present in all immunoglobulin molecules. In most molecules two identical light chains are linked to two heavy chains by disulphide bonds. There are either of two types of light chain, **kappa** and **lambda**, but never both on the same molecule. Light chains have a N-terminal variable region which forms part of the antigen combining site and a C-terminal constant region.

light reactions Those reactions in **photosynthesis** in which light is absorbed, transferred between pigments, and used to generate the ATP and reduced NADP used in the **dark reactions** for CO_2 fixation. See **photosystems I and II**, **photosynthetic pigments**.

light trap A device for catching night-flying moths and/or other insects attracted by light.

lignicole, lignicolous Living on or in wood or on trees, as certain species of Termites.

lignin A complex cross-linked polymer, based on variously substituted p-hydroxyphenyl propane units; a constituent with cellulose etc. of the cell walls of xylem tracheids and vessels and of many sclerenchyma and some parenchyma cells etc. where its major function appears to be to impart rigidity to an otherwise flexible wall.

lignivorous Wood-eating.

ligulate Strap-shaped.

Liliaceae Family of ca 3500 spp. of monocotyledonous flowering plants (superorder Liliidae). Mostly herbs, often with bulbs, corms or rhizomes, cosmopolitan. The flowers are usually showy, with 6 perianth members, 6 stamens and a superior ovary of three carpels. Includes *Allium*, the onions, garlic, leeks, *Colchicum*, the source of **colchicine**, and many ornamental plants.

Liliidae Subclass or superorder of monocotyledons. Mostly herbs, sepals and petals both usually petaloid, mostly syncarpous, ovary superior or more often inferior. Contains ca 28 000 spp. in 17 families including Liliaceae, Amaryllidaceae, Iridaceae and Orchidaceae.

Liliopsida See **Monocotyledones**.

limaciform Sluglike.

limb (1) A jointed appendage, as a leg. (2) The lamina of a leaf. (3) The widened upper part of a petal. (4) The upper, often spreading, part of a sympetalous corolla.

limbous Overlapping.

lime-induced chlorosis That of young leaves induced in (relatively) calcifuge plants by growth on calcareous or overlimed soils (i.e. soils of too high pH), caused apparently by a disturbance in iron metabolism and curable horticulturally by the application of e.g. chelated iron *sequestrene* to leaves or roots.

limicolous Living in mud.

limiting factor The variable, or combination of variables, which limits the rate of growth of an organism or a population, or which limits the rate of a physiological process.

limivorous Mud-eating; as certain aquatic Invertebrates which swallow mud to extract from it the nutritious organic matter that it contains.

limnobiotic Living in fresh water.

limnology The study of lakes.

limnophilous Living in marshes, esp. freshwater marshes.

Lincoln index See **mark and recapture**.

linear (1) A leaf having parallel sides, and at least four to five times as long as broad. (2)

A tetrad of pollen grains in a single row.

lingua Any tonguelike structure; in Insects, the hypopharynx; in *Acarina*, the floor of the mouth.

lingual In Arthropods, pertaining to the lingua; in Molluscs, pertaining to the radula; in Vertebrates, pertaining to the tongue.

lingulate Tongue-shaped.

linkage The tendency of genes, or characters, to be inherited together because the genes are on the same chromosome.

linkage group All the genes known from their linkage to be on the same chromosome.

linkage map A diagram showing the positions of genes on a chromosome or set of chromosomes.

Linnaean (Linnean) system The system of classification and of **binomial nomenclature** established by the Swedish naturalist Linnaeus.

Linnaean species A species defined broadly, often including many varieties. Cf. **Jordanon**.

lipase Enzyme which cleaves the hydrocarbon chains from lipids.

lipid body A cytoplasmic inclusion, esp. in the storage tissue of oil- and fat-rich seeds, typically spherical, 0.1–5 μm diameter, bounded by what appears to be half a unit membrane, and containing lipids.

lipogenous Fat-producing.

lipoplast A lipid body.

lipopolysaccharide *LPS*. Term commonly used to refer to bacterial lipopolysaccharides which constitute the O-antigens of Gram negative bacilli, esp. enterobacteria. Chemically they consist of various antigenically specific polysaccharides linked to lipid A. The latter contains rhamnose linked to galactosamine diphosphate which is esterified with myristic acid, and is responsible for the endotoxin properties common to all LPS. They are also polyclonal mitogens for B lymphocytes.

lipoprotein Complex of protein and lipid in varying proportions, classified according to their increasing density into **chylomicrons**, *very low density lipoproteins (VLDL)*, *low density lipoproteins (LDL)* and *high density lipoproteins (HDL)*. They transport lipids in the plasma between various organs in the body, e.g. the gut, liver and adipose tissues.

liposome Spherical shell formed when mixtures of phospholipids, with or without, cholesterol are dispersed in aqueous solutions. Liposomes are made up of one or several concentric phospholipid bilayers within which other molecules can be incorporated. They simulate many permeability properties of membranes and are used for the administration of certain drugs.

lissencephalous Having smooth cerebral hemispheres.

lithite See **statolith**.

lithocyst See **statocyst**.

lithodomous Living in rocks.

lithogenous Rock-building, as certain Corals.

lithophagous Stone-eating; said of graminivorous Birds which take small stones into the gizzard, to aid mastication; also of certain Molluscs which tunnel in rock.

lithophyte A plant growing on rocks or stones, esp. an alga attached to a rock or stone.

lithotomous Stone-boring, as certain Molluscs.

litter More or less undecomposed fallen leaves and other plant residues at the soil surface.

littoral (1) Of the seashore. (2) The shallower water of lakes where light reaches the bottom and where rooted plants may grow.

littoral zone Faunal zone bounded by the continental shelf, i.e. down to approximately 200 m.

liver In Invertebrates, the digestive gland or **hepatopancreas**; in Vertebrates, a large mass of glandular tissue arising as a diverticulum of the gut, which secretes the bile and plays an important part in the storage and synthesis of metabolites.

liver flukes A group of trematode parasites, esp. *Fasciola hepatica* and *Clonorchis sinensis*, infecting man via various species of water Snail as intermediate hosts, causing damage to the liver and surrounding organs.

liverworts See **Hepaticopsida**.

L+, L0, LR dose of toxin Used in standardizing diptheria toxin and antitoxin. Describes the quantity of toxin which, when mixed with one standard unit of antitoxin, will respectively kill a guinea pig under standardized conditions, or when injected subcutaneously will produce no observable reaction, or when inoculated into the skin will produce a minimal lesion.

Lloyd Morgan's canon The proposal that one should never explain behaviour in terms of a higher mental function if a more simple process will explain it.

loam A soil that is a mixture of sand, silt and clay particles, a desirable texture for horticultural and agricultural uses.

lobe A rounded or flap-like projection. adjs. *lobate, lobed, lobose, lobulate*.

lobed Leaf, petal etc. divided into curved or rounded parts connected to each other by an undivided centre.

lobotomy See **pre-frontal lobotomy**.

lobule, lobulus A small lobe; one of the polyhedral cell masses forming the liver in Vertebrates. adjs. *lobular, lobulate*.

localization An ability of the sense organs of an animal to determine the source of a stimulus in space. Many sense organs can detect direction, but probably only the eyes and ears or lateral line system, can judge distance. See **echolocation**.

locular, loculatous Divided into compartments by septa.

locule, loculus A cavity, esp. one within a sporangium or an ovary, containing the spores or ovules repectively.

loculicidal Said of a dehiscent fruit which opens by splitting down the middle of each compartment of the ovary. Cf. **septicidal, septifragal**.

locus The position on a chromosome occupied by a specified gene and its alleles. pl. *loci*.

locust One of several kinds of winged insects of the family *Acridiidae*, akin to Grasshoppers, highly destructive to vegetation.

lodicules Small scales inserted below the stamens in the florets of most grasses; when the floret is mature the lodicules swell, forcing apart lemma and palea, and allowing stamens and stigmata to grow out.

logistic equation Equation describing the typical increase of a population towards an asymptotic value:

$$\frac{dN}{dt} = rN\left(\frac{K - N}{K}\right)$$

where N is the population size, t is time, r is the innate capacity for increase and K is the maximum or asymptotic value of N.

lomasome A compact mass of membranes in the cytoplasm adjacent to a cell wall, apparently invaginated from the plasmalemma, of obscure function.

lomentum A fruit, usually elongated, which develops constrictions as it matures, finally breaking across these into 1-seeded portions. adj. *lomentose*.

long-day plant A plant that will flower better under conditions of long days and short nights, e.g. spinach, *Spinacia oleracea*. Cf. **short-day plant, day-neutral plant**. See **photoperiodism**.

longi- Prefix from L. *longus*, long.

longicorn Having elongate antennae, as some Beetles.

longipennate Having elongate wings or feathers.

longirostral Having a long beak or rostrum.

longitudinal valve In Amphibians, a large, flaplike valve which traverses the *conus arteriosus* longitudinally and obliquely.

long shoot A shoot of mostly woody plants with relatively long internodes, that extends the canopy. Long shoots typically bear few or no flowers and in some species bear no foliage leaves but only scales (e.g. pines) or spines (e.g. *Berberis*). Cf. **short shoot, growth bud**.

long-term memory According to the 3-store model of memory, a memory system that keeps memories for long periods, has a very large capacity, and stores items in an organized and processed form. See **short term memory, sensory store**.

looming response Refers to the fact that many animals, including humans, respond to a sudden increase in stimulus size by ducking or turning away from it.

loph A crest connecting the cusps of a molar tooth.

lophobranchiate Having tuftlike, crestlike or lobelike gills.

lophodont Of Mammals, having cheek teeth with transverse ridges on the grinding surface.

lophophore A ciliated tentacle used for food gathering; found in some aquatic invertebrates.

lore In Birds, the space between the beak and the eye. adj. *loral*.

Lotka's equations Mathematical relationships between (1) the populations of two species living together and competing for food or space and (2) the situation where one is the predator and the other the prey.

lower quartile The argument of the cumulative distribution function corresponding to a probability of 0.25; (of a sample) the value below which occur a quarter of the observations in the ordered set of observations.

LPS Abbrev. for *LipoPolySaccharide*.

LSD See **lysergic acid diethylamide**.

lucid dreaming A dreaming state in which the dreamer feels awake and in conscious control over dream events.

luciferase An oxidizing enzyme which occurs in the luminous organs of certain animals (e.g. Firefly) and acts on luciferin to produce luminosity. Also *photogenin*.

luciferin A proteinlike substance which occurs in the luminous organs of certain animals and is oxidized by the action of luciferase. Also *photophelein*.

lumbar Pertaining to, or situated near, the lower or posterior part of the back; as the *lumbar* vertebrae.

lumen (1) The space enclosed by the walls of a cell, esp. if the protoplast has disappeared. (2) The central cavity of a duct or tubular organ.

lunar In tetrapods, the middle member of the three proximal carpals in the fore limb, corresponding to part of the astragalus in the hind limb.

lunate, lunulate Crescent-shaped; shaped like the new moon.

lung The respiratory organ in air-breathing Vertebrates. The lungs arise as a diverticulum from the ventral side of the pharynx; they consist of two vascular sacs filled with constantly renewed air.

lung book An organ of respiration in some

Arachnida (Scorpions, Spiders), consisting of an air-filled cavity opening on the ventral surface of the body; it contains a large number of thin vascular lamellae, arranged like the leaves of a book.

lunule, lunula   A crescentic mark, e.g. the white crescent at the root of the nail. adj. *lunular*.

lupus erythematosus cell   A neutrophil leucocyte which contains in its cytoplasm homogeneous masses of phagocytosed nuclear material derived from other dead neutrophils. This occurs when anti-nuclear antibodies are present in the blood which opsonize the released nucleoproteins. LE cells are formed in vitro in blood drawn from subjects with systemic lupus erythematosus, and their presence is of diagnostic significance.

luteal   Pertaining to, or resembling, the *corpus luteum*.

lutein cells   Yellowish coloured cells occurring in the corpus luteum of the ovary; they contain fat-soluble substances, developed from either the follicular cells or the theca interna.

Lycopsida   The Lycopods, a class of Pteridophyta dating from the Early Devonian onwards. The sporophyte usually has roots, stems and leaves (**microphylls**), is protostelic and has lateral sporangia. Includes the Protolepidodendrales, Lycopodiales, Lepidodendrales, Isoetales, Selaginellales.

lymph   A colourless circulating fluid occurring in the lymphatic vessels of Vertebrates and closely resembling blood plasma in composition. adj. *lymphatic*.

lymphatic system   In Vertebrates, a system of vessels pervading the body, in which the lymph circulates and which communicates with the venous system; lymph glands and lymph hearts are found on its course.

lymph gland   An aggregation of reticular connective tissue, crowded with lymphocytes, surrounded with a fibrous capsule, and provided with afferent and efferent lymph vessels.

lymph heart   A contractile portion of a lymph vessel, which assists the circulation of the lymph and forces the lymph back into the veins.

lymph-, lympho-   Prefixes from L. *lympha*, water.

lymphocyte function associated molecules   *LFA-1*, *CD2*, *LAF-3*. Factors present on the surface of leucocytes which are involved in their ability to adhere to other cells and to exert other functions such as killing or phagocytosis, their nature is at present ill defined, but their absence in certain rare subjects is accompanied by defective immunity.

lymphocytes   Spherical cells with a large round nucleus (often slightly indented) and very scanty cytoplasm. Diameter varies between 7 and 12 μm. They are actively mobile. The term is essentially morphological and is used to refer to cells responsible for development of specific immunity, B lymphocytes being associated with humoral and T lymphocytes with cellular immunity. Lymphocytes are the predominant constituents of **lymphoid tissues**. A normal adult human contains about 2×10^{12} lymphocytes.

lymphogenous   Lymph-producing.

lymphoid tissues   Body tissues in which the predominant cells are lymphocytes, i.e. lymph nodes, spleen, thymus, Peyer's patches, pharyngeal tonsils, adenoids, and in birds the caecal tonsils and bursa of Fabricius.

lymphokine   Generic name for proteins (other than antibodies or surface receptors) which are released by lymphocytes, stimulated by antigens or by other means, which act on other cells involved in the immune response. The term includes interferons, interleukins, lymphotoxins, migration inhibition factor.

lymphoma   Neoplastic disease of lymphoid tissues and sometimes involving bone marrow, in which the neoplastic cells originate from lymphocytes or mononuclear phagocytes. Includes Hodgkin's disease, reticulosarcoma, giant follicular lymphoma, lymphatic leukaemia and Burkitt's lymphoma. Clinically divided by histological appearance into two main groups: *Hodgkin's* and *non-Hodgkin's lymphoma*, with the former carrying the more favourable prognosis.

lymphotoxin   *LT*. A lymphokine, produced by T cells, which is cytolytic for non-lymphocytic cells. It has some structural and functional similarity to **tumour necrosis factor**. Syns. *LTα* and *TNFβ*.

lymph sinuses   Part of the **lymphatic system** in some lower Vertebrates, consisting of spaces surrounding the blood vessels and communicating with the coelom by means of small apertures.

lymph vessels   See **lymphatic system**.

lyocytosis   Histolysis by the action of enzymes secreted outside the tissue, as in Insect metamorphosis.

lyra   The psalterium in the brain of Mammals; any lyre-shaped structure, as the lyre pattern on a bone.

lyrate   Shaped like a lyre. A pinnately lobed leaf with a large terminal and small lateral lobes.

lyriform organs   Patches, consisting of well-innervated ridges of chitin, on the legs, palpi, chelicerae, and body of various *Arachnida*; mechanoreceptive in function.

Lys   Symbol for **lysine**.

lyse   To cause to undergo *lysis*.

lysergic acid diethylamide   *Lysergide*.

Compound which, when taken in minute quantities, produces hallucinations and thought processes outside the normal range. The results sometimes resemble schizophrenia. Abbrev. *LSD*.

lysigenic, lysigenous Said of a space formed by the breakdown and dissolution of cells. Also *lysogenous*. Cf. **rexigenous, schizogenous**.

lysin A substance which will cause dissolution of cells or bacteria.

lysine *2,6-diaminohexanoic acid*. A 'basic' amino acid, as it contains two amino groups. The L-isomer is a non-polarizable amino acid and constituent of proteins. Symbol Lys, short form K.
Side chain: $H_2N-CH_2-CH_2-CH_2-CH_2-$
See **amino acid**.

lysis Decomposition or splitting of cells or molecules, e.g. hydrolysis, lysin.

lysogeny That part of the life cycle of a temperate phage in which it replicates in synchrony with its host. Cf. **lytic cycle**.

lysosome A vesicular cytoplasmic organelle containing hydrolytic enzymes that degrade those cellular constituents which become incorporated into the vesicle.

lysozyme Enzyme present in egg white, tears, nasal secretions, on the skin, and in monocytes and granulocytes. It lyses certain bacteria, chiefly Gram-positive cocci, by splitting the muramic acid-β(1-4)-N-acetylglucosamine linkage in their cell walls. The first *enzyme* to have its three-dimensional structure determined.

lyssa See **lytta**.

lytic cycle That part of the life cycle of a temperate phage in which it causes **lytic infection**.

lytic infection Infection of a bacterium by a phage which replicates uncontrollably, destroying its host and eventually releasing many copies into the medium.

lytta In *Carnivora*, a rod of cartilage or fibrous tissue embedded in the mass of the tongue. Also *lyssa*.

M

m Abbrev. for *milli-*.

maceration The process of soaking a specimen in a reagent in order to destroy some parts of it and to isolate other parts.

macr-, macro- Prefixes from Gk. *makros*, large.

macrocyte An abnormally large red cell in the blood.

macrogamete The larger of a pair of conjugating gametes, generally considered to be the female gamete.

macroglia A general term for neuroglial cells, including astroglia, oligodendroglia and ependyma.

macroglobulin Any globulin with a molecular weight above about 400 000 but usually applied to IgM (mol.wt. 900 000) or to a protease inhibitory molecule alpha-2-macroglobulin (mol.wt. 820 000). In macroglobulinaemia there are markedly raised levels of IgM in the blood, due to there being an abnormally proliferating clone of neoplastic plasma cells, such as that found in **Waldenstrom's macroglobulinaemia**.

macromere In a segmenting ovum, one of the large cells which are formed in the lower or vegetable hemisphere.

macronucleus In Ciliophora, the larger of the two nuclei which is composed of vegetative chromatin. Cf. **micronucleus**.

macronutrient Element required in relatively large quantities by living organisms; for plants they are H, C, O, N, K, Ca, Mg, P, S. Cf. **micronutrient**. See **essential element**.

macrophage See panel.

macrophagous Feeding on relatively large particles of food. Cf. **microphagous**.

macrophyll A large leaf. Also *megaphyll*.

macroscopic Visible to the naked eye.

macrosmatic Having a highly developed sense of smell.

macrosome A large protoplasmic granule or globule.

macrosplanchnic Having a large body and short legs; as a Tick.

macrospore See **megaspore**.

macrosporophyll *Megasporophyll*. Leaflike structure bearing mega sporangia.

macrotous Having large ears.

macula acustica Patches of sensory epithelium in the utriculus, sacculus and lagena of the Dogfish and the utricle and saccule of Mammals. In the latter they are organs of static or tonic balance consisting of supporting cells and hair cells, the hairs being embedded in a gelatinous membrane containing crystals of calcium carbonate, i.e. an otolith.

macula lutea See **yellow spot**.

macula, macule A blotch or spot of colour;

---

macrophage

Cell of the **mononuclear phagocyte system**. Derived from blood monocytes which migrate into tissues and differentiate there. They are actively phagocytic, and ingest particulate materials including microbes. They contain lysosomes and an oxidative microbicidal system which becomes functional when the macrophages are activated by ingestion of degradable particles or by lymphokines such as interferon. They contain and secrete large amounts of lysozyme.

Macrophages are motile, although some remain in the same site (e.g. as Kupffer cells in liver sinusoids) for long periods. They have Fc and complement receptors which are involved in attachment and engulfment of antibody-coated materials. The microbicidal activity is much increased by the action of lymphokines secreted by T lymphocytes. They adhere firmly to glass or plastic surfaces, and are often separated in vitro by such means. In tissues at the site of a delayed hypersensitivity reaction they may adhere to one another, and alter their morphology to resemble epithelial cells or fuse together to form multinucleated giant cells. Macrophages are important accessory cells for presentation of antigens to T lymphocytes.

Although most resting macrophages express little or no Class II histocompatibility antigens they are induced to do so by interferon. Antigens ingested by them do not become wholly degraded, and peptide fragments are returned to the cell surface where they are associated with the Class II antigen, and can be recognized by antigen receptors of T cells.

a small tubercle; a small shallow pit.

madreporite In Echinodermata, a calc-areous plate with a grooved surface, perfor-ated by numerous fine canaliculi and situated in an interambulacral position, through which water is passed to the axial sinus. adj. *madreporic*.

Magendie's foramen In Vertebrates, an aperture in the roof of the fourth ventricle of the brain, through which the cerebrospinal fluid communicates with the fluid in the spaces enclosed by the meningeal mem-branes.

maggot An acephalous, apodous, eruciform larva such as that of certain Diptera.

Magnoliidae Subclass or superorder of di-cotyledons. Trees, shrubs and herbs charac-terized by a well-developed perianth, numerous centripetal stamens and an apo-carpous gynoecium. Contains ca 11 000 spp. in 36 families, including the Magnol-iaceae and the Ranunculaceae. More or less synonymous with *ranalian complex*; gener-ally regarded as the most primitive of extant angiosperms.

Magnoliophyta In some classifications a division containing the **angiosperms**. Sub-divided into two classes: the Magnoliopsida (see **Dicotyledones**) and the Liliopsida (see **Monocotyledones**.)

Magnoliopsida See **Dicotyledones**.

major depression A type of **affective dis-order** characterized by major depressive episodes occurring without intervening manic episodes.

major histocompatibility complex The collection of genes coding for the major histocompatibility antigens. Abbrev. *MHC*. See **HLA system**.

malacia Pathological softening of any organ or tissue.

malacology The study of Molluscs.

malacophily Pollination by snails.

Malacostraca Largest subclass within the Crustacea. The body clearly divided into head, thorax and abdomen, often with a carapace and possessing 20 segments. Crabs, Lobsters, Shrimps.

malacostracous Having a soft shell.

malar Pertaining to the **mala**; pertaining to, or situated in, the cheek region of Verte-brates; the **jugal**.

male An individual of which the gonads produce spermatozoa or some correspond-ing form of gamete, i.e. the smaller and motile gamete; the antherozoid or sperma-tozoid in plants. Also said of (1) a gameto-phyte which produces male but not female gametes, or (2) a sporophyte which pro-duces microspores or pollen and hence indi-vidual seed plants or flowers which have functional stamens but not functional car-pels. Cf. **female, hermaphrodite**.

maleic hydrazide See **growth retardant**.

male pronucleus Nucleus of the sperm-atozoon.

male sterility Condition in which viable pollen is not formed. Used by plant breeders to ensure cross pollination esp. in the pro-duction of **F1 hybrid** seed. See **cytoplasmic male sterility**.

malignant Tending to go from bad to worse, esp. cancerous (see **tumour**).

malleolar Pertaining to, or situated near, the malleolus; in Ungulata, the reduced fibula.

malleolus A process of the lower end of the tibia or fibula.

malleus In Mammals, one of the ear ossicles; in Rotifera, one of the masticatory ossicles of the mastax; more generally, any hammer-shaped structure.

Mallophaga An order of the Psocopteroi-dea (Paraneoptera); ectoparasites, usually of Birds, with reduced eyes, flattened form, tarsal claws and biting mouthparts, e.g. bit-ing lice.

Malpighian body, corpuscle In the Ver-tebrate kidney, the expanded end of a uriniferous tubule surrounding a glomeru-lus of convoluted capillaries; in the Verte-brate spleen, one of the globular or cylindrical masses of lymphoid tissue which envelops the smaller arteries.

Malpighian cell A macrosclereid in the epidermis of the testa of a leguminous seed.

Malpighian layer The innermost layer of the epidermis of most Chordates, containing polygonal cells which continually pro-liferate, dividing by mitosis. Also known as *rete Malpighii, stratum germinativum*.

Malpighian tubes In Insects, some *Arach-nida* and Myriapods, tubular glands of ex-cretory function opening into the alimentary canal, near the junction of the mid gut and hind gut.

Malvaceae Family of ca 1000 spp. of di-cotyledonous flowering plants (superorder Dilleniidae). Trees, shrubs and herbs, cos-mopolitan except in very cold regions. The flowers have 5 sepals (sometimes with an epicalyx), 5 free petals, many stamens with the filaments united at the base and 5 fused carpels. Includes cotton, okra and some or-namental plants, e.g. hollyhock.

mamilla A nipple.

mamillary body See **corpus mamillare**.

mamma In female Mammals, the milk gland; the *breast*. adj. *mammary*.

Mammalia A class of Vertebrates. The skin is covered by hair (except in aquatic forms) and contains sweat glands and sebaceous glands; they are homoiothermous; the young are born alive (except in the Prototheria) and are initially nourished by milk; respiration is by lungs and a diaphragm is present; the cir-culation is double, and only the left systemic arch is present; dentition heterodont and diphyodont; there is a double occipital condyle; the lower jaw articulates with the squamosal; the long bones and vertebrae have three centres of ossification; there is an external ear and three auditory ossicles in the middle ear; large cerebral hemispheres.

Man The human race, all living races being included in the genus *Homo*, suborder Anthropoidea of the Primates. Man's distinguishing features include elaboration of the brain and behaviour, including communication by facial gestures and speech; the erect posture; the structure of the limbs (including the opposable thumb) and skull; the dentition with small canines; the long period of postnatal development associated with parental care.

mandible In Vertebrates, the lower jaw; in Arthropoda, a masticatory appendage of the oral somite; in Polychaeta and Cephalopoda, one of a pair of chitinous jaws which lie within the buccal cavity. adj. *mandibular, mandibulate*.

mandibular glands In some Insects, glands opening near the articulation of the mandibles; in some *Lepidopterous larvae*, they function as salivary glands, the true salivary glands secreting silk; also present in the hive Bee and other adult *Hymenoptera*.

mania Mental abnormality where the mood is of extreme elation with speed of thought giving a flight of ideas. Patient has little or no insight and business or family affairs can be seriously damaged.

manic-depressive See **bipolar disorder**.

manifest dream content See **Freud's theory of dreams**.

manna See **honey dew**.

manoxylic wood Secondary xylem in gymnosperms with very wide parenchymatous rays between the small groups of conducting cells, as in cycads. Cf. **pycnoxylic wood**.

mantle In Urochordata, the true body wall lying below the test and enclosing the atrial cavity; in Mollusca, Brachiopoda and Cirripedia, a soft fold of integument which encloses the trunk and which is responsible for the secretion of the shell or carapace.

mantle cavity In Urochorda, the atrial cavity; in Mollusca, Brachiopoda and Cirripedia, the space enclosed between the mantle and the trunk.

Mantoux test A form of **tuberculin test** commonly used in humans to indicate present or past infection with *Mycobacterium tuberculosis* (or previous immunization with BCG). A tuberculin preparation, usually PPD, is injected intracutaneously. A positive test indicates that delayed hypersensitivity is present.

manubrium Any handle-like structure; the

anterior sternebra in Mammals; part of the malleus of the ear in Mammals; the pendant oral portion of a medusa.

manus The podium of the forelimb in land Vertebrates.

map Depending on the level of organization, it can refer to the ordering of genes in an eukaryotic chromosome or to the determination of the arrangement of DNA sequences in a gene or cluster of genes. Thus *gene mapping* or *sequence mapping*.

mapping The determination of the positions and relative distances of genes on chromosomes by means of their **linkage**.

marcescent Withered but remaining attached to the plant.

marginal Said of a species or community which occurs at the boundary between two distinct habitats.

marginal meristem Meristem along margin of a leaf primordium giving rise to the tissues of the blade.

marihuana See **cannabis**.

mark and recapture Method of animal census, in which a marked sample of animals is returned to the population, allowed to mix with unmarked individuals, and then the proportion of the marked individuals in a second sample is used to derive an estimate of the population size. Also called *Lincoln index*.

marrow The vascular connective tissue which occupies the central cavities of the long bones in most Vertebrates, and also the spaces in certain types of cancelled bone.

marsh Vegetation which is seasonally waterlogged and non-peat forming.

Marsupiala The single order included in the Mammalian group Metatheria, and having the characteristics of the subclass, e.g. Opossum, Tasmanian wolf, Marsupial Mole, Kangaroo, Koala bear.

marsupium A pouchlike structure occupied by the immature young of an animal during the later stages of development; as the abdominal pouch of metatherian Mammals. adj. *marsupial*.

mask A prehensile structure of the nymphs of certain Dragonflies (Odonata), formed by the labium.

masochism Sexual gratification through having pain inflicted on oneself.

masseter An elevator muscle of the lower jaw in higher Vertebrates. adj. *masseteric*.

mass-flow hypothesis Hypothesis proposing that translocation in the **phloem** results from a continuous flow of water and solutes, esp. sugars, through the sieve tubes from source to sink, the flow being driven by a hydrostatic pressure difference caused by the osmotic movement of water following the loading of solutes into the sieve tubes at the source end and their removal (unloading) at the sink end.

mast Fruit of beech, oak and other forest trees.

mast cell Cell with basophil cytoplasmic granules similar to but smaller than those of basophil leucocytes in the blood. The granules contain mainly histamine, serotonin and heparin. Mast cells bind the Fc region of IgE antibodies. Reaction with an antigen (or the action of anaphylotoxin) causes extrusion of the granule contents, and the release of various newly formed pharmacologically active substances. These include leukotrienes, **platelet activating factor** and a peptide which is chemotactic for eosinophil leucocytes. There are two populations of mast cells. One is normally present in connective tissues, in the neighbourhood of small blood vessels. The other, *mucosal mast cells*, is induced by IL-3 secreted by T cells. They have sparse granules and are more like lymphocytes, and become prominent in the mucosa of the gut in intestinal nematode infections.

mastication The act of reducing solid food to a fine state of subdivision or to a pulp.

masticatory Pertaining to the trituration of food by the mandibles, teeth, or gnathobases, prior to swallowing.

Mastigomycotina Subdivision or class containing those Eumycota or true fungi in which the spores and/or gametes are motile. Includes the Chytridiomycetes and, in some schemes, the Domycetes.

Mastigophora A class of Protozoa which possess one or more flagella in the principal phase; may be amoeboid but usually have a pellicle or cuticle; often parasitic but rarely intracellular; have no meganucleus; reproduction mostly by longitudinal binary fission; nutrition may be holophytic, saprophytic or holozoic. Also *Flagellata*.

mastoid Resembling a nipple; as a posterior process of the otic capsule in the skull of Mammals.

maternal effect A *non-heritable* influence of a mother on characters in her offspring, e.g. through her milk supply.

maternal immunity Passive immunity acquired by the newborn animal from its mother. In humans and other primates this is mainly by active transport of IgG antibodies across the placenta. In species which have thicker placentas, such as ungulates, antibody is not transferred *in utero* but is acquired from the colostrum, and is absorbed intact from the gut during the first few days of life. The young of birds acquire maternal immunity from antibody in the egg yolk.

mating type A group of individuals, within a species, which cannot breed among themselves but which are able to breed with individuals of other such groups. See **plus strain, minus strain, incompatibility**.

matric potential That component of the **water potential** due to the interaction of the water with colloids and to capillary forces (surface tension). Often important component in soils and cell walls. Symbol ψ.

matrix More or less continuous matter in which something is embedded, e.g. the non-cellulosic substances of the cell wall in which the cellulose microfibrils lie in vascular plant cell walls. adj. *matrix*.

matrix The intercellular ground substance of connective tissues.

matroclinous Exhibiting the characteristics of the female parent more prominently than those of the male parent. Cf. **patroclinous**.

matromorphic Resembling the mother.

matter See **pus**.

maturation Final stages in the development of the germ cells; more generally, the process of becoming adult or fully developed.

maturation divisions The divisions by which the germ cells are produced from the primary spermatocyte oöcyte, during which the number of chromosomes is reduced from the diploid to the haploid number. See **meiosis**.

maturation of behaviour Refers to environmentally stable characteristics, that are irreversible during development despite variation in environmental influences, and which develop with little or no practice.

maxilla In Vertebrates, the upper jaw; a bone of the upper jaw; in *Arthropoda*, an appendage lying close behind the mouth and modified in connection with feeding. pl. *maxillae*. adjs. *maxillary, maxilliferous, maxilliform*.

maxillary Pertaining to a maxilla; pertaining to the upper jaw; a paired membrane bone of the Vertebrate skull which forms the posterior part of the upper jaw.

maxilliped In Arthropoda, esp. Crustacea, an appendage behind the mouth, adapted to assist in the transferance of food to the mouth.

maximum permissible concentration The recommended upper limit for the dose which may be received during a specified period by a person exposed to ionizing radiation. Also called *permissible dose*.

maximum permissible dose rate, flux That dose rate or flux which, if continued throughout the exposure time, would lead to the absorption of the maximum permissible dose.

maximum permissible level A phrase used loosely to indicate *maximum permissible concentration, dose* or *dose rate*.

maze An apparatus consisting of a series of pathways in a more or less complicated configuration, beginning with a starting box, possibly including blind alleys, and ending in a goal box which generally contains a reward, this not being visible from the starting box. The simplest mazes are the T- and Y-mazes.

mean See **expectation**; the arithmetic mean of a sample.

mean lethal dose The single dose of whole body irradiation which will cause death, within a certain period, to 50% of those receiving it. Abbrev. *MLD*.

mean-square error The expectation of the square of the difference between an estimate of a parameter and its true value, taken with respect to the sampling distribution of the estimate.

meatus A duct or channel, as the external auditory *meatus* leading from the external ear to the tympanum.

mechanical tissue Tissues such as collenchyma and sclerenchyma whose primary function is the support of the plant body.

meconium (1) The green coloured faecal material passed by the newborn infant or animal. (2) In certain Insects, liquid expelled from the anus immediately after the emergence of the imago.

mediad Situated near, or tending towards, the median axis.

median The central value in a set of observations ordered by value, dividing the ordered set into two equal parts; the argument of the cumulative distribution function of a random variable corresponding to a probability of one half.

mediastinum In higher Vertebrates, the mesentery-like membrane which separates the pleural cavities of the two sides ventrally; in Mammals, a mass of fibrous tissue representing an internal prolongation of the capsule of the testis.

medical model The conceptualization of mental disorders as a group of diseases analogous to physical diseases.

medi-, medio- Prefixes from L. *medius*, middle.

medium Substance in which an organism or part exists naturally, e.g. *aqueous medium*, or has been placed experimentally. See **culture medium, mountant**.

medulla The central portion of an organ or tissue, as the *medulla* of the Mammalian kidney or a plant thallus; bone marrow and pith. adj. *medullary*. Cf. **cortex.**

medulla oblongata The hind brain in Vertebrates, excluding the cerebellum.

medullary bundle Vascular bundle in the pith.

medullary canal The cavity of the central nervous system in Vertebrates; the central cavity of a shaft bone.

medullary folds In a developing Vertebrate, the lateral folds of a medullary plate,

by the upgrowth and union of which the tubular central nervous system is formed.

medullary plate In a developing Vertebrate, the dorsal platelike area of ectoderm which will later give rise to the central nervous system.

medullary ray *Primary ray*. (1) The interfasicular region. (2) Ray stretching from pith to cortex and deriving from the interfasicular region.

medullary sheath (1) The peripheral layers of cells of the pith. The cells are usually small, sometimes thick-walled, and sometimes more or less lignified. (2) A layer of white fatty substance (myelin) which, in Vertebrates, surrounds the axons of the central nervous system and acts as an insulating coat.

medullated nerve fibres Axons of the central nervous system which are provided with a **medullary sheath**.

medullated protostele Protostele with a central core (medulla) of nonvascular tissue.

medullate, medullated Having a pith.

medusa In metagenetic Coelenterata, a free-swimming sexual individual.

mega- Prefix denoting 1 million, or 10^6, e.g. a frequency of 1 *megahertz* is equal to 10^6 Hz; *megawatt* = 10^6 watts; *megavolt* = 10^6 volts.

megagamete See **macrogamete**.

megalecithal Said of eggs which contain a large quantity of yolk. Cf. **microlecithal**.

megaloblast An embryonic cell which has a large spherical nucleus and of which the cytoplasm contains haemoglobin, which will later give rise to *erythroblasts* by mitotic division within the blood vessels.

megamere See **macromere**.

meganucleus See **macronucleus**.

megaphanerophyte A tree over 30 m high.

megaphyll (1) Leaf, typically associated with a **leaf gap** and often relatively large and with a branching system of veins, assumed to have evolved from a branch system or directly from a **telome** truss, and supposed to be the sort of leaf possessed by seed plants and ferns. (2) Any large leaf. adj. *megaphyllous*. Cf. **microphyll**.

megasporangium A sporangium which contains megaspores. It corresponds to ovule or nucellus of seed plant. See **heterospory**.

megaspore In a heterosporous species, a meiospore potentially developing into a female gametophyte. Also *macrospore*.

megasporophyll A leaf-like structure (or a structure thought to be homologous to a leaf) bearing megasporangia. See **sporophyll**. Cf. **microsporophyll**.

Meibomian glands In Mammals, sebaceous glands on the inner surface of the eyelids, between the tarsi and conjunctiva.

Also *tarsal glands*.

meiomerous Having a small number of parts. n. *meiomery*.

meiosis A process of cell division by which the chromosomes are reduced from the diploid to the haploid number. adj. *meiotic*.

meiospore Spore, commonly one of four, containing a nucleus formed by meiosis and therefore haploid. The spores of bryophytes and vascular plants.

meiotic arrest The waiting at a particular stage of meiosis by the oöcyte until some stimulus is received, usually the entry of the sperm.

Meissner's corpuscles In Vertebrates, a type of sensory nerve found in the skin, in which the nerve breaks up into numerous branches which surround a core of large cells in a connective-tissue capsule. Probably sensitive to touch.

melanism The abnormal situation, caused by over production of melanin, where a proportion of the individuals in an animal population are black or melanic. See **industrial melanism**.

melan-, melano- Prefixes from Gk. *melas*, gen. *melanos*, black

melanoblast A special connective tissue cell containing melanin.

melanophore A chromatophore containing black pigment.

melanosis The abnormal deposit of the pigment melanin in the tissues of the body.

melanosporous Having black spores.

Melastomaceae Family of ca 3000 spp. of dicotyledonous flowering plants (superorder Rosidae). Mostly shrubs and small trees, mostly tropical, important in the South American flora. Of little economic value.

melibiose Naturally occurring di-saccharide based on glucose and α-galactose. (*D-galactosyl-α1→6-D-glucose*).

melliphagus, mellivorous Honey-eating.

melting temperature of DNA See panel.

member Any part of a plant considered from the standpoint of morphology. More generally an organ of the body, esp. an appendage.

membrana A thin layer or film of tissue; a membrane.

membrana tectoria A soft fibrillated membrane overlying Corti's organ.

membrana tympani A thin fibrous membrane forming the tympanum or eardrum.

membrane (1) A thin sheet-like structure, often fibrous, connecting other structures or covering or lining a part. (2) *Cell membrane, unit membrane*; a sheet (say 10 nm thick), composed characteristically of a bimolecular leaflet of lipid and protein, enclosing a cell, an organelle or a vacuole. See **tonoplast, pit membrane**.

membrane filter Thin layer filter made by fusing cellulose ester fibres or by β-bombardment of thin plastic sheets, so that they are perforated by tiny uniform channels. Also *molecular filter*.

membranella In *Ciliophora*, an undulating membrane formed of two or three rows of cilia.

memory span The number of items a person can recall after just one presentation.

memory trace The inferred change in the nervous system that persists after learning or experiencing something.

Mendelian character A character determined by a single gene and inherited according to *Mendel's laws*.

Mendelian genetics See panel.

Mendel's laws See panel.

meninges In Vertebrates, envelopes of connective tissue surrounding the brain and spinal cord. sing. *meninx* (Gk. membrane).

meniscus A small interarticular plate of fibrocartilage which prevents violent concussion between two bones, such as the intervertebral disks of mammals.

menopause The natural cessation of menstruation in women.

mensa The biting surface of a tooth.

menstruation The periodical discharge from the uterus occurring from puberty to menopause. Also *xenomenia*.

mental age A score which was devised by Binet to represent a child's test performance; it indicates the average age (50% of that age group) of those children who pass the same test items as the person whose mental age is being computed, expressed in terms of years and months. See **intelligence quotient**.

mental retardation A condition characterized by a very low intellectual ability and a serious inability to cope with the environment.

mental set The tendency to view new problems in the same fashion as old problems; also a predisposition to perceive or think in one way rather than another.

mentum In higher Vertebrates, the chin; in some Gastropoda, a projection between the head and foot; in Insects, the distal sclerite forming the basal portion of the labium, situated between the submentum and the prementum. adj. *mental*.

mericlinal chimaera A chimaera in which one component does not completely surround the other; an incomplete *periclinal chimaera*.

meridional Extending from pole to pole; as *meridional furrow* in a segmenting egg.

meristele A strand of vascular tissue, enclosed in a sheath of endodermis, forming part of a dictyostele.

meristem A group of actively dividing cells including **initials** and their undifferentiated

melting temperature of DNA

A number of agents will cause double stranded DNA to *melt* or separate into single strands and thus lose its ordered structure. These include heating, raising the pH and adding substances, like sodium perchlorate, which break hydrogen bonds. Melting can be detected by a decrease in viscosity or, more easily, by measuring ultraviolet absorption in a spectrophotometer in which the sample can be heated. This is because the *absorptivity* of single-stranded DNA is some 40% higher than that of its parent duplex. The temperature at which a given DNA is *half* melted is called the *Tm*.

The melting profile also gives information about the base composition and homogeneity because GC base pairs are more stable than AT. The process of melting DNA is more commonly called *denaturation* when it is used as a first step in a hybridization test for sequence homology.

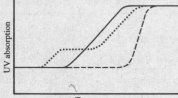

Melting curves of different kinds of DNA.

(——) Heterogeneous DNA like that of a higher organism. (- - - -) Homogeneous DNA of high GC content like that of some bacteria. (........) DNA with a minor AT-rich component.

derivatives. See **apical meristem, intercalary meristem, lateral meristem.**

meristem culture See **plant cell culture.**

meristic Segmented; divided up into parts; pertaining to the number of parts, as in *meristic variation.* See **merome.**

meristic variation Variation in the number of organs or parts; as in the number of body somites of a metameric animal.

mermaid's purse A popular name applied to the horny purselike capsule in which the eggs of certain Selachian Fish (Sharks, Dogfish, Skates, Rays) are enclosed.

mero- Prefix from Gk. *meros*, part.

meroblastic Said of a type of ovum in which cleavage is restricted to a part of the ovum, i.e. is incomplete, usually due to the large amount of yolk. Cf. **holoblastic.**

merogamy Having gametes smaller than the vegetative cells and produced by a special division; union of such gametes.

merogenesis Segmentation, formation of parts.

merogony See **schizogony.**

merome A body somite or segment of a metameric animal.

meroplankton Organisms which spend only part of their life history in the plankton. Cf. **holoplankton.**

merosthenic Having the hind limbs exceptionally well developed, as in Frogs or Kangaroos.

merozoite In *Protozoa*, a young trophozoite which results from the division of the schizont.

merycism *Rumination*. The return, after a meal, of gastric contents to the mouth; they are then chewed and swallowed once more.

mesarch Said of a strand of xylem having the protoxylem in the centre with metaxylem developing centripetally and centrifugally (towards and away from the centre).

mesaxonic foot A foot in which the skeletal axis passes down the third digit, as in *Perissodactyla.* Cf. **paraxonic foot.**

mescaline A hallucinogenic drug derived from the Mexican cactus, mescal.

mesectoderm Parenchymatous tissue formed from ectoderm cells which have migrated inwards.

mesencephalon The midbrain of Vertebrates.

mesenchyma Parenchyma; embryonic mesodermal tissue of spongy appearance. adj. *mesenchymatous.*

mesenchyme Mesodermal tissue, comprising cells which migrate from ectoderm, or endoderm, or mesothelium into the blasto-

Mendel's laws, Mendelian genetics

Laws dealing with the mechanism of inheritance.

The *First law* (law of segregation) states (in modern terminology) that the two alleles received one from each parent are segregated in gamete formation, so that each gamete receives one or the other with equal probability. This results in various characteristic ratios in the progeny depending on the parental genotypes and dominance.

The *Second law* (law of recombination) states that two characters determined by two unlinked genes are recombined at random in gamete formation, so that they segregate independently of each other, each according to the first law. This results in various dihybrid ratios.

Mendelian genetics is that of characters determined by single genes with effects large enough to be easily recognizable.

coele. Cf. *mesothelium*.

mesenteric Pertaining to the mesenteron; pertaining to a mesentery.

mesenteric caeca Digestive diverticula of the mesenteron in many Invertebrates (e.g. in Arachnida, Crustacea, Echinodermata, Insects).

mesenteron See **midgut**.

mesentery In Coelenterata, a vertical fold of the body wall projecting into the enteron. More generally, a fold of tissue supporting part of the viscera. adj. *mesenterial, mesenteric*.

mesethmoid A median cartilage bone of the Vertebrate skull, formed by ossification of the ethmoid plate. Also called *internasal septum*.

mesiad Situated near, or tending towards, the median plane.

mesial, mesian In the median vertical or longitudinal plane.

mesmerism An early name for *hypnosis*.

mes-, meso- Prefixes from Gk. *mesos*, middle.

mesobenthos Fauna and flora of the sea floor, at depths ranging from 100 to 500 fathoms (200 to 1000 m).

mesoblast The mesodermal or third germinal layer of an embryo, lying between the endoderm and ectoderm. adj. *mesoblastic*.

mesoblastic somites In developing metameric animals, segmentally arranged blocks of mesoderm, the forerunners of the somites.

mesocarp The middle, often fleshy, layer of the pericarp.

mesocoele In Vertebrates, the cavity of the midbrain; mid-ventricle; Sylvian aqueduct.

mesoderm See **mesoblast**.

mesogaster In Vertebrates, the portion of the dorsal mesentery which supports the stomach.

mesogloea In Coelenterata, a structureless

layer of gelatinous material intervening between the ectoderm and the endoderm.

mesokaryote *Dinokaryote*. Organism having a nucleus with a nuclear envelope and with chromosomes, but having very little histone associated with the DNA of the chromosomes which remain condensed throughout interphase; the condition in the Dinophyceae. Cf. **prokaryote, eukaryote**.

mesolecithal A type of egg of medium size containing a moderate amount of yolk which is strongly concentrated in one hemisphere, i.e. the lower one in eggs floating in water. Found in Frogs, Urodela, Lungfishes, Lower Actinopterygii and Lampreys. Cf. **oligolecithal, telolecithal**.

mesometrium The mesentery which supports the uterus and related structures.

mesomorph One of Sheldon's somatotyping classifications; mesomorphs are hard, rectangular and well muscled; delight in physical activity, love adventure and power, and are indifferent to people. See **somatotype theory**.

mesonephric duct See **Wolffian duct**.

mesonephros Part of the kidney of Vertebrates, arising later in development than, and posterior to, the *pronephros*, and discharging into the **Wolffian duct**; becomes the functional kidney in adult anamniotes. adj. *mesonephric*.

mesophilic bacteria Bacteria which grow best at temperatures of 20–45°C.

mesophyll The ground tissue of a leaf, located between the upper and lower epidermis and typically differentiated as chlorenchyma. See also **palisade** mesophyll, **spongy mesophyll**.

mesophyte A plant adapted to habitats that are neither very wet nor dry. Cf. **hydrophyte, xerophyte**.

mesorchium In Vertebrates, the mesentery supporting the testis.

mesosternum In Insects, the sternum of the mesothorax; in Vertebrates, the middle part of the sternum, connected with the ribs; the gladiolus.

mesotarsal In Insects, the tarsus of the second walking leg; in land Vertebrates, the ankle joint or joint between the proximal and distal rows of tarsals.

mesothelium Mesodermal tissue comprising cells which form the wall of the cavity known as the coelom. Cf. **mesenchyme**.

mesothorax The second of the 3 somites composing the thorax in Insects. adj. *mesothoracic*.

mesotrochal Having an equatorial band of cilia.

mesovarium In Vertebrates, the mesentery supporting the ovary.

mestome Conducting tissue, with associated parenchyma, but without mechanical tissue.

mestome sheath, mestom sheath An endodermis-like bundle sheath; the inner of two sheaths round the vascular bundle of some grasses, e.g. wheat.

Met Symbol for **methionine**.

meta- Prefix from Gk. *meta*, after.

metabolism The sum total of the chemical and physical changes constantly taking place in living matter. adj. *metabolic*.

metabolite A substance involved in metabolism, being either synthesized during metabolism or taken in from the environment.

metaboly The power possessed by some cells of altering their external form, e.g. as in Euglenida.

metacarpal, metacarpale One of the bones composing the **metacarpus** in Vertebrates.

metacarpus In land Vertebrates, the region of the forelimb between the digits and the carpus.

metachronal rhythm The rhythmic beat of cilia on a cell surface in which the beat of adjacent cilia is slightly out-of-phase. Consequently it is seen as a series of waves passing over the ciliary surface.

metachrosis The ability, shown by some animals (as the Chameleon) to change colour by expansion or contraction of chromatophores.

metacoele In *Craniata*, the cavity of the hindbrain; the fourth ventricle.

metadiscoidal placentation Having the villi at first scattered and then restricted to a disk as in Primates.

metagenesis See **alternation of generations**.

metamere See **merome**.

metamerism *Metameric segmentation*. Repetition of parts along the long axis of an animal. adj. *metameric*.

metamorphosis Pronounced change of form and structure taking place within a comparatively short time, as the changes undergone by an animal in passing from the larval to the adult stage. adj. *metamorphic*.

metanephric duct Ureter of amniote Vertebrates.

metanephridia Nephridia which open into the coelom, the open end (nephrostome) being ciliated. Cf. **protonephridium**.

metanephros In amniote Vertebrates, part of the kidney arising later in development than, and posterior to, the mesonephros; becomes the functional kidney, with a special *metanephric duct*. adj. *metanephric*.

metaphloem The later-formed *primary phloem*, esp. that which matures after the organ has ceased to elongate. Cf. **protophloem**.

metaplasia Tissue transformation, as in the ossification of cartilage.

metaplasis The period of maturity in the life cycle of an individual.

metapodium (1) In Vertebrates, the second podial region; metacarpus or metatarsus; palm or instep. (2) In Insects, that portion of the abdomen posterior to the *petiole*. (3) In Gastropoda, the posterior part of the foot. adj. *matapodial*.

metapophysis In some Mammals, a process of the vertebrae above the prezygapophysis which strengthens the articulation.

metasitism Cannibalism.

metasoma In Arachnida, the posterior part of the abdomen, or hindmost tagma of the body, which is always devoid of appendages. adj. *metasomatic*.

metastasis The transfer, by lymphatic channels or blood vessels, of diseased tissue (esp. cells of malignant tumours) from one part of the body to another; the diseased area arising from such transfer. More rarely in Zoology, transference of a function from one part or organ to another; metabolism.

metatarsal, metatarsale One of the bones composing the **metatarsus** in Vertebrates.

metatarsus In Insects, the first joint of the tarsus when it is markedly enlarged; in land Vertebrates, the region of the hindlimb between the digits and the tarsus.

Metatheria A subclass of viviparous Mammals in which the newly born young are carried in an abdominal pouch which encloses the teats of the mammary glands; an allantoic placenta is usually lacking; the scrotal sac is in front of the penis, the angle of the lower jaw is inflexed, and the palate shows vacuities. Contains only one order, the Marsupiala.

metathorax The third or most posterior of the three somites composing the thorax in Insects. adj. *metathoracic*.

metaxenia Any effect that may be exerted by pollen on the tissues of the female organs.

metaxylem The later-formed *primary xylem*, esp. that which matures after the organ has ceased to elongate; commonly with reticulately thickened or pitted walls. Cf. **protoxylem**.

Metazoa A subkingdom of the animal kingdom, comprising multicellular animals having two or more tissue layers, never possessing choanocytes, usually having a nervous system and enteric cavity, and always showing a high degree of coordination between the different cells composing the body. Cf. **Protozoa, Parazoa**.

metecdysis The period after a moult in *Arthropoda* when the new cuticle is hardening and the animal is returning to normal in its physiological condition.

metencephalon The anterior portion of the hindbrain in Vertebrates, developing into the cerebellum, and, in Mammals, the **pons** (part of the **medulla oblongata**). Cf. **myelencephalon**.

methadone Methadone hydrochloride. An analgesic with similar but less marked properties, both good and bad, than morphine.

methaemoglobin A compound of haemoglobin and oxygen, more stable than oxyhaemoglobin, obtained by the action of oxidizing agents on blood, e.g. nitrites and chlorates.

methionine *2-amino-5-thio-hexanoic acid.* The L-isomer is a non-polarizable amino acid. Symbol Met, short form M.
Structure: CH_3—S—CH_2—CH_2—
See **amino acid**.

methylation Addition of a methyl (–CH_3) group to a nucleic acid base, usually cytosine or adenine. In bacteria, this can protect the site against cleavage by the appropriate restriction enzyme. In eukaryotes, the transcription of certain genes is inhibited by DNA methylation, although the function is not fully understood.

metoecious Heteroecious.

metoestrus In Mammals, the recuperation period after oestrus.

metoxenous Heteroecious.

metric trait Same as **quantitative character**.

MHC Abbrev. for *Major Histocompatibility Complex.*

MHC restriction The recognition by T lymphocytes (cytotoxic or helper cells) of foreign antigen on the surface of another cell, e.g. a virus-infected cell or an accessory cell, only occurs when the antigen is associated with *self antigens* of the major histocompatibility complex. Cytotoxic T lymphocytes usually respond to foreign antigen in association with Class I MHC antigens, whereas helper T lymphocytes respond to foreign antigen in association with Class II MHC antigens. This was deduced from observations in vitro, and is important for understanding how T lymphocytes function. It has no practical relevance to their function in any normal individual, all of whose cells are 'self', but may be relevant when chimerism is produced by bone marrow grafting.

micelle A colloidal sized aggregate of molecules, such as those formed by surface active agents. By extension, a crystalline region, inferred from X-ray diffraction data, within a microfibril of cellulose or similar structure.

micro- Prefix from Gk. *mikros*, small. When it is used of units it indicates the basic unit $\times 10^{-6}$, e.g. 1 microampere (μA) = 10^{-6} ampere. Symbol μ.

microaerophile An organism which grows well at low oxygen concentrations.

microbe An organism which can only be seen under the microscope.

microbody Cytoplasmic organelle of eukaryotes up to ca 1.5 μm diameter, bounded by a single membrane. Types include the **glyoxysome** and **peroxisome**.

microclimate The climate in small places, e.g. very close to organisms or in specific habitats such as nests or on bare ground.

Micrococcaceae A family of bacteria belonging to the order *Eubacteriales*. Grampositive cocci; includes free-living, parasitic, saprophytic and pathogenic species, e.g. *Staphylococcus aureus* (one cause of food poisoning).

microdissection A technique for small-scale dissection, e.g. on living cells, using *micromanipulators* and viewing the object through a microscope. Mechanical, hydraulic or piezo-electric mechanisms are used to translate the movements of a joystick or similar device to that of the needle, knife or pipette which has to be moved in the field of view of the microscope.

micro-environment The environment of small areas in contrast to large areas, with particular reference to the conditions experienced by individual organisms and their parts, e.g. leaves etc.

microfibril Fine fibril, 5–30 nm wide, in a **cell wall**; of cellulose in vascular plants and some algae but of other polysaccharide (e.g. xylans) in other algae. Cf. **matrix**.

microfilament A filament of F-actin, about 6 nm in diameter, found in the cytoplasm usually in long bundles and involved in localized cell contractions and streaming. Cf. **microtubule**. See **stress fibres**.

microfilaria The early larval stage of certain parasitic Nematoda.

microgamete The smaller of a pair of conjugating gametes, generally considered to be the male gamete.

microgametocyte In Protozoa, a stage developing from a trophozoite and giving rise to male gametes.

microglia A small type of neuroglia cell (occurring more frequently in grey matter than in white matter) having an irregular body and freely branching processes; can be phagocytic.

microglobulin Any small globulin. Used in respect of Bence-Jones protein in urine or of β2-*microglobulin*.

micro-incineration A technique for examining the distribution of minerals in slide preparations of tissue sections or cells. The organic material is vaporized by heat and the nature and position of the mineral ash determined by microscopic examination.

microlecithal Said of eggs containing very little yolk. Cf. **megalecithal**.

micromanipulator An instrument used to handle cells seen in a microscope, e.g. to remove a nucleus or inject RNA. The fine movements are controlled indirectly by pneumatic, mechanical or other means.

micromere In a segmenting ovum, one of the small cells which are formed in the upper or animal hemisphere.

micrometer eyepiece See **eyepiece graticule**.

micron Obsolete measure of length equal to one millionth of a metre, symbol μ. Replaced in SI by *micrometre*, symbol μm.

micronucleus In Ciliophora, the smaller of the two nuclei which is involved with sexual reproduction. Cf. **macronucleus**.

micronutrient A *trace element* required in relatively small quantities by living organisms; for plants the micronutrients include Fe, B, Mn, Zn, Cu, Mo, Cl. Cf. **macronutrient**. See **essential element**.

microphage A small phagocytic cell in blood or lymph, chiefly the polymorphonuclear leucocytes (*neutrophils*). adj. *microphagocytic*

microphagous Feeding on small particles of food. Cf. **macrophagous**.

microphanerophyte A phanerophyte, 2 to 8 m high.

microphyll (1) A small leaf, not associated with a leaf gap, assumed to have evolved from an **enation** and supposed to be the sort of leaf possessed by lycopods. (2) Any very small leaf, e.g. of heather or *Tamarix*. adj. *microphyllous*. Cf. **megaphyll**.

Microphyllophyta Division of the plant kingdom, here treated as the class Lycopsida.

Micropodiformes An order of Birds with a short humerus and long distal segments to the wings. Swifts and Hummingbirds.

micropodous Having the foot, or feet, small or vestigial.

micropropagation See **plant cell culture**.

micropterous Having small or reduced fins

or wings.

micropyle (1) A tiny opening in the integument at the apex of an ovule, through which the pollen tube usually enters. (2) The corresponding opening in the testa of the seed. Also called *foramen*. (3) An aperture in the chorion of an Insect egg through which a spermatozoon may gain admittance.

microradiography Exposure of small thin objects to **soft X-rays**, with registration on a fine-grain emulsion and subsequent enlargement up to 100 times. Also used to signify the optical reproduction of an image formed, e.g. by an electron microscope.

microscope See panel.

microsmatic Having a poorly developed sense of smell.

microsomes The membranous pellet obtained by centrifugation of a cell homogenate after removal of the mitochondria and nuclei. It was originally believed to represent a cell organelle but is now known to consist of fragments of endoplasmic reticulum, plasma membrane, Golgi apparatus etc.

microspecies See **Jordanon**.

microsplanchnic Having a small body and long legs, as a Harvestman.

microsporangium The structure within which **microspores** are formed. In seed plants, the pollen sac.

microspore In heterosporous species a **meiospore** able to develop into a male gametophyte; in seed plants, the pollen grain; a small swarm spore or anisogamete of Sarcodina.

microsporocyte A cell which divides to give microspores, i.e. a microspore mother cell.

microsporophyll A leaf-like structure bearing microsporangia. Cf. **megasporophyll**. See **sporophyll**.

microsporophyte A typically diploid cell in a microsporangium, which divides by meiosis to give four microspores.

microtome An instrument for cutting thin sections of specimens, esp. sections 1–10 μm thick for light microscopy. See **ultramicrotome**.

microtubule Tubular structure about 24 nm in diameter formed by the aggregation of **tubulin** dimers and small amounts of associated proteins in a helical array. They function as skeletal components within cells and may be arranged into complex structures such as cilia and mitotic spindles.

microtubule-organizing centre Cytoplasmic site where microtubules are formed and organized; associated with or including centromeres, basal bodies, centrioles etc. Abbrev. *MTOC*.

microvilli (1) Finger-like protrusions of the cell surface. Their length and abundance varies characteristically between cells.

microscope

An instrument used for obtaining magnified images of small objects. The *simple microscope* is a convex lens of short focal length, used to form a virtual image of an object placed just inside its principal focus. The essential elements of the *compound microscope* are two short-focus convex lenses, the objective and the eyepiece mounted at opposite ends of a tube. To obtain the highest resolution, however, a *condenser* is also needed.

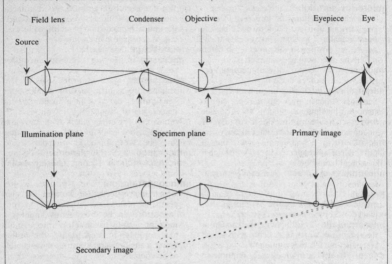

The condenser forms a highly reduced image of the field lens which thus defines the area illuminated at the specimen plane (see lower diagram); the objective forms a magnified image of the specimen or *object* near the eyepiece which in turn allows the eye to see a much enlarged secondary image at a comfortable distance. The two image, the illumination and specimen planes form one set of *conjugate planes* in the microscope. A second set comprises the light source and the planes at which images of it are formed (see A, B and C in the upper diagram). A and B are of particular importance in **phase contrast** and **interference microscopy**. This method of illuminating a microscope is called *Kohler illumination* and is widely used because even illumination can be easily obtained. Another method, called *critical illumination*, has no field lens but requires an even light source. The condenser also provides a high-aperture cone of rays which is essential if the objective is to perform at its highest resolution.

In practice the objective, in particular, comprises many lenses, often of special glass or other materials and additional components are also needed to fold the light path and to provide binocular viewing at a convenient angle.

For most microscopes, the magnifying power is roughly equal to $450/f_o f_e$, where f_o and f_e are the focal lengths of objective and eyepiece in centimetres. See **electron-**, **ultraviolet microscope**.

When present in very high density, as on the apical surface of epithelial cells of the small intestine, they form a *brush border*.(2) Finger-like infoldings of the inner mem-

brane of a mitochondria, as in the brown algae, diatoms etc. See **Heterokontophyta**. Cf. **crista**.

micturition In Mammals, the passing to the

exterior of the contents of the urinary bladder.

midbrain In Vertebrates, that part of the brain which is derived from the second or middle brain vesicle of the embryo; the second or middle brain vesicle itself dorsally comprises the tectum, including optic lobes (*corpora quadrigemina* in Mammals), and laterally the tegmentum and ventrally the *crura cerebri* (cerebral peduncles).

middle ear In Tetrapoda excluding Urodela, Gymnophiona and snakes, the cavity containing the auditory ossicles.

middle lamella Layer of intercellular material, mostly pectic substances, developing from the cell plate and cementing together the primary walls of contiguous cells.

midgut That part of the alimentary canal of an animal which is derived from the archenteron of the embryo.

midrib (1) The main vein or nerve of a leaf. (2) Thickened region down the middle of a thallus.

midriff See **diaphragm**.

MIF Abbrev. for *Migration Inhibition Factor*.

migration Long-distance animal movement, often involving large populations and often seasonal.

migration inhibition factor Lymphokine which acts on macrophages so as to increase their adhesiveness. When released by T lymphocytes *in vivo* (e.g. in the peritoneal cavity) it causes macrophages to clump together. *In vitro* it inhibits their migration (e.g. out of a capillary tube), this can be used as a semi-quantitative means of detecting the existence of delayed type hypersensitivity. Abbrev. *MIF*.

migratory cell See **amoebocyte**.

mildew A plant disease in which fungal mycelium is visible on the surface of the host, e.g. downy mildew, powdery mildew. Also *mould*.

milk glands The mammary glands of a female Mammal; in viviparous Tsetse flies, special uterine glands by which the larva is nourished until it is ready to pupate.

milk teeth In diphyodont Mammals, the first or deciduous dentition.

milli- Prefix from L. *mille*, thousand. When attached to units, it denotes the basic unit $\times 10^{-3}$. Symbol *m*.

millipede Any myriapod of the class Diplopoda, vegetarian cylindrical animals with many joints most of which bear two pairs of legs.

Millipore filter TN for a type of **membrane filter**.

milt The spleen; in Fish, the testis or spermatozoa; to fertilize the eggs.

mimicry The adoption by one species of the colour, habits, sounds or structure of another species. adjs. *mimic, mimetic*.

mineralization The decomposition in soils of organic matter by micro-organisms with the release of the mineral elements (N, P, K, S etc.) as inorganic ions.

mineral nutrient An essential element, other than C, H or O, normally obtained as an inorganic ion taken up e.g. through the roots of a land plant, e.g. N, P, K, Ca. See **macronutrient, micronutrient**.

minimal area Concept used in sampling vegetation; the minimum area which must be searched in order to find (nearly) all of the species. See **species/area curve**.

minus strain One of the two, arbitrarily designated, mating types of a heterothallic species. Cf. **plus strain**. See **heterothallism**.

miosis Contraction of the pupil of the eye.

MIP Abbrev. for *Multiple Isomorphous Replacement*.

miracidium The ciliated first-stage larva of a Trematode.

mire Same as **bog**.

miscarriage Expulsion of the foetus before the 28th week of pregnancy. Loosely, abortion.

mitochondrion See panel.

mitogen Any agent which induces mitosis in cells. In immunology often used to refer to substances such as lectins or lipopolysaccharides which cause a wide range of T or B lymphocytes to undergo cell division.

mitosis The normal process of somatic cell division, in which each of the daughter cells is provided with a chromosome set identical to that of the parent cell. adj. *mitotic*. Cf. **amitosis, meiosis**.

mitospore Spore formed by mitosis and hence having the same number of chromosomes as the parent.

mitotic crossing over The exchange of genetic material between homologous chromosomes during mitosis, resulting in genetic **recombination**. See **crossing over**.

mitotic index The proportion in any tissue of dividing cells, usually expressed as per thousand cells.

mitral Mitre-shaped or *mitriform*; as the *mitral valve*, guarding the left auriculo-ventricular aperture of the heart in higher Vertebrates, or the *mitral layer*; of the olfactory bulb, composed of mitre-shaped cells.

mitral valve See **bicuspid valve**.

mitriform See **mitral**.

mixed Said of nerve trunks containing motor and sensory fibres.

mixed bud A bud containing young foliage leaves and also the rudiments of flowers or of inflorescences.

mixed inflorescence An inflorescence in which some of the branching is racemose and some is cymose.

mitochondrion

A mitochondrion is a mobile cytoplasmic *organelle* of **eukaryotes** visible in the light microscope whose main function is the generation of ATP by **aerobic respiration**. The organelle consists of an outer membrane surrounding an inner membrane enclosing a protein rich matrix. The outer membrane is freely permeable to metabolites and of indeterminate function, the characteristic functions of the mitochondrion being associated with the inner membrane and the matrix proteins. The components necessary for **oxidative phosphorylation**, for the **electron transfer chain** and the proton translocating ATPase, are located on highly convoluted infoldings of the inner membrane called *cristae*. Proton translocating ATPase molecules can be seen in the electron microscope as knob-like structures protruding from the surface of the cristae into the matrix. The enzymes which degrade the energy sources, glucose and fatty acids, by the **tricarboxylic acid cycle** and **beta-oxidation** are distributed between the inner membrane and the matrix.

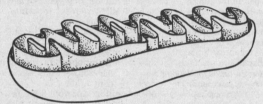

Sectional view of a mitochondrion showing the outer membrane and the infolded cristae of the inner membrane.

The biogenesis of mitochondria is complex. New mitochondria arise by the division of pre-existing mitochondria while the synthesis of the mitochondrial proteins is under the co-ordinated activity of nuclear genes and mitochondrial DNA. The organelles contain all the components required for protein synthesis, ribosomes, tRNA etc. This limited autonomy of the mitochondria together with features like the ribosomal nucleotide sequences and circular DNA, has led to the suggestion of an independent prokaryote existence prior to their incorporation into the cytoplasm of present day eukaryotes. Oxidative phosphorylation occurs across the cytoplasmic plasma membrane in modern prokaryotes. See the **endosymbiotic hypothesis** for an analogous hypothesis concerning chloroplasts.

mixotrophic Combining two or more fundamental methods of nutrition, as certain *Mastigophora* which combine holophytic with saprophytic nutrition, or as in a partial parasite.

MLC mixed lymphocyte culture Lymphocytes from two individuals cultured together for 3 to 5 days, at the end of which the number of dividing cells is measured, usually by incorporation of ^{3}H-thymidine. This provides a measure of the extent to which the histocompatibility antigens of the two differ, and is used to assess the suitability for tissue transplantation.

mnemonics Rules of learning that improve recall (e.g. rhyming).

mobbing A form of harassment directed at predators by potential prey.

mobile element DNA sequence capable of excising itself from one chromosome and then re-integrating itself, or its copies, into different sites in the chromosome set.

modal interval The class interval corresponding to the largest frequency in a tabulation of observations into class intervals.

modality A category of sensation, e.g. touch, smell or sight.

modal value See **mode**.

mode The most frequent value in a set of observations; the value of a random variable at which the corresponding probability density function is a maximum.

modelling In behaviour therapy, the learning of a new behaviour by imitation of a model, and usually overtly or covertly reinforcing the desired behaviour.

moder Form of humus intermediate between **mull** and **mor**.

modifier A gene which influences the effect of another.

modiolus The conical central pillar of the cochlea.

molality The concentration of a solution expressed as the number of moles of dissolved substance per kilogram of solvent.

molarity The concentration of a solution expressed as the number of moles of dissolved substance per dm^3 of solution.

molars The posterior grinding or cheek teeth of Mammals which are not represented in the milk dentition.

mole (1) In chemistry, the amount of substance that contains as many entities (atoms, molecules, ions, electrons, photons etc.) as there are atoms in 12 g of ^{12}C. It replaces in SI the older terms *gram-atom*, *gram-molecule* etc., and for any chemical compound will correspond to a mass equal to the relative molecular mass in grams. Abbrev. *mol*. See **Avogadro number**. (2) **Naevus**. (3) A haemorrhagic mass formed in the Fallopian tube as a result of bleeding into the sac enclosing the embryo.

molecular biology The study of the structure and function of macromolecules in living cells. Notably successful in explaining the structure of proteins, and the role of DNA as the genetic material, but with the ultimate aim of explaining, as completely as possible in molecular terms, the biology of cells and organisms. It is not primarily concerned with metabolic pathways or with the chemistry of natural products.

molecular genetics The study and manipulation of the molecular basis of heredity.

Mollusca Unsegmented coelomate Invertebrates with a head (usually well developed), a ventral muscular foot and a dorsal visceral hump; the skin over the visceral hump (the *mantle*) often secretes a largely calcareous shell and encloses a mantle cavity into which open the anus and kidneys, and in which are the **ctenidia**, originally used for gaseous exchange. There is usually a **radula**; some have haemocyanin as a respiratory pigment; well-developed blood and nervous systems; often with larva of the **trochophore** type. Chitons, Slugs, Snails, Mussels, Whelks, Limpets, Squids, Cuttlefish and Octopods.

moment The expected value of the nth power of a random variable, where n is an integer indexing the set of moments.

monad (1) A flagellated unicellular organism. (2) A single pollen grain, not united with others. Cf. **tetrad**.

monadelphous Having all the stamens in the flower joined together by their filaments, e.g. many Leguminosae.

monandrous (1) Having 1 antheridium. (2) Having 1 stamen.

monarch Having a single strand of protoxylem in the stele.

mongrel The offspring of a cross between varieties or races of a species.

monimostyly In Vertebrates, the condition of having the quadrate immovably united to the squamosal. Cf. **streptostyly**.

mon-, mono- Prefixes from Gk. *monos*, alone, single.

monoamine oxidase An enzyme involved in the inactivation of catecholamine neurotransmitters. Inhibitors of monoamine oxidase are used as antidepressants.

monocardian Having a completely undivided heart.

monocarpellary Having, or consisting of, a single carpel.

monocarpic Dying at the end of its first fruiting season, as do annuals, biennials and some perennials. See **hapaxanthic**. Cf. **polycarpic**.

monocerous Having a single horn.

monochasium, monochasial cyme An inflorescence in which the main stem ends in a flower and bears, below the flower, a lateral branch which itself ends in a flower. This branch may in turn bear further similar branches. Also *cymose inflorescence*. Cf. **dichasium**.

monochlamydeous Having a perianth of one whorl of members.

monochlamydeous chimera A periclinal chimera in which one component is present, at least at the shoot apex, as a single superficial layer of cells.

monoclimax theory The idea, proposed by F. E. Clements in 1916, that all successional sequences in a region lead to a single climax vegetation of a type determined by the climate. Cf. **polyclimax theory**.

monocliny Hermaphrodite flowers. Cf. **dicliny**.

monoclonal antibody Antibody produced by a single clone of cells or a cell line derived from a single cell. Such antibodies are all identical and have unique amino acid sequences. Commonly used to refer to antibody secreted by a hybridoma cell line, but can also refer to the immunoglobulin produced *in vivo* by a B cell clone, such as a plasmacytoma if this has identifiable antibody properties.

monocolpate Said of a pollen grain having a single colpus.

Monocotyledones *Liliopsida*. The monocotyledons, or monocots, the smaller of the two classes of **angiosperm** or flowering plants. Mostly herbs of which the embryo has characteristically one cotyledon, the parts of the flower are in threes or sixes and the leaves often have the main veins parallel;

very few are woody, fewer still have secondary thickening. Contains ca 55 000 spp. in 60 families divided among four subclasses or superorders; Alismatidae, Arecidae, Commelinidae and Liliidae.

monocotyledonous (1) Belonging to the Monocotyledones. (2) Said of an embryo or seed having one cotyledon.

monocule An animal possessing a single eye, as the Water Flea *Daphnia*. adj. *monocular*.

monoculture A culture, crop or plantation with only one species.

monocyclic Stamens or other floral parts in a single whorl.

monocyte A large motile phagocytic cell with an indented nucleus present in normal blood, where it is the blood representative of the **mononuclear phagocyte system**. Monocytes are derived from promonocytes in the bone marrow. They remain in the blood for only a short time and then migrate into the tissues where they become macrophages.

monodactylous Having only a single digit.

monodont Having a single persistent tooth, as the male Narwhal.

monoecious (1) Said of a flowering plant having unisexual flowers but with both sexes on each plant. See **dicliny**. (2) Said of moss gametophytes and algae, producing male and female gametes on the same individual.

monoestrous Exhibiting only one oestrous cycle during the breeding season. Cf. **polyoestrous**.

monogenetic Multiplying by asexual reproduction; showing a direct life history; of parasites, having a single host.

monogerm Said of varieties of sugar beet etc. in which each small dry fruit, sown like a seed, produces a single seedling rather than the group of seedlings typically produced by such fruits.

monogony Asexual reproduction.

monolayer culture A **tissue culture** technique whereby thin sheets of cells are grown, on glass, in a nutrient medium.

monolete Said of a pollen grain or other spore with a single linear scar or aperture.

monomorphic Showing little change of form during its life history; having a single form. Cf. **polymorphic**.

mononuclear phagocyte system A classification of phagocytic cells of which the typical mature form is the macrophage. They are all derived from bone marrow promonocytes, and share in varying degrees the ability to ingest and digest particulate materials, and to express the receptors characteristic of macrophages. However their appearance differs according to the tissue in which they are situated. Macrophages in the alveoli of the lung, the peritoneal cavity, and moving in tissues and in the lymph are free to migrate. Others such as Kupffer cells, tissue histiocytes, osteoclasts and astroglia remain in situ for long periods of time.

monophagous Feeding on one kind of food only, e.g. *Sporozoa*, living always in the same cell, or phytophagous Insects, with only one food plant. Also *monotrophic*.

monophasic Having an abbreviated life cycle, without a free active stage; said of certain Trypanosomes. Cf. **diphasic**.

monophyletic Of a group of species which are descended from a single ancestral species. See **polyphyly**.

monophyletic group, monophyly (1) Generally a group that is descended from a common ancestor. (2) In cladistics a group comprising a common ancestor and all of its descendents. Cf. **polyphyletic group, paraphyletic group**.

monophyodont Having only a single set of teeth, the permanent dentition. Cf. **polyphyodont**.

monoploid True **haploid**.

monopodial growth *Indefinite growth, indeterminate growth*. Pattern of growth in which a shoot continues to grow indefinitely and bears lateral shoots which behave similarly, e.g. in pines and other conifers. See **racemose inflorescence**. Cf. **sympodial growth**.

monosome The unpaired accessory or X-chromosome. See **sex determination**.

monosomy Condition in which a particular chromosome is represented once only in an otherwise diploid complement; a sort of **aneuploidy**.

monospermy Fertilization of an ovum by a single spermatozoon.

monosporous (1) Containing 1 spore. (2) Derived from 1 spore.

monostichous Forming 1 row.

monotocous Producing a single offspring at a birth.

Monotremata The single order included in the Mammalian group Prototheria and having the characteristics of the subclass, e.g. Spiny Ant eater (*Tachyglossus*), Duck-billed Platypus (*Ornithorhynchus*).

monotrophic See **monophagous**.

monotypic A taxonomic group containing only one subordinate member, a family with a single genus or a genus with a single species.

monozygotic twins Twins produced by the splitting in two of a single fertilized egg, or of the early embryo derived from it. syn. *identical twins*. They are always of the same sex and are genetically equivalent to a single individual. Cf. **dizygotic twins**.

Monro's foramen A narrow canal connecting the first or second ventricle with the third ventricle in the brain of Vertebrates.

monster An abnormal form of a species.

Monte-Carlo methods Procedures employed to obtain numerical solutions to mathematical problems by means of random sampling.

moor, moorland Vegetation composed of ericaceous plants and certain grasses, growing on acid soil.

mor Crumbly layer of partially-decomposed plant-like litter occurring under some coniferous trees and woody shrubs, in which the pH is acid and the soil fauna poor. See **mull**, **moder**.

Morgagni's ventricle In the higher Mammals, a paired pocket of the larynx, anterior to the vocal cords, and acting in the Anthropoidea as a resonator.

morph A specific form or shape of an organism, singled out for attention.

morphactins Group of synthetic compounds, based on fluorene-carboxylic acid which have a variety of effects, mostly inhibitory, on plant growth and development.

morphallaxis Change of form during regeneration of parts, as the development of an antenna in certain Crustacea to replace an eye; gradual growth or development.

morphine The principal alkaloid present in opium. Characterized by containing a phenanthrene nucleus in addition to a nitrogen ring. Extensively used as a hypnotic to obtain relief from pain.

morph-, morpho-, -morph Prefixes and suffix from Gk. *morphe*, form.

morpho- See **morph-**.

morphogenesis The origin and development of a part, organ or organism. adj. *morphogenetic*.

morphology (1) The study of the structure and forms of organisms, as opposed to the study of their functions. (2) By extension, the nature of a member. adj. *morphological*.

morphosis The development of structural characteristics; tissue formation. adj. *morphotic*.

mortality The death of individuals in a population. Cf. **natality**.

morula A solid spherical mass of cells resulting from the cleavage of an ovum.

mosaic (1) An individual displaying the effects of different alleles or genes in different parts of the body, but derived from a single embryo, e.g. a tortoiseshell cat. (2) A plant or part of a plant with an irregular distribution of cells of different genetic constitution, resulting from e.g. the mutation of an unstable gene or the sorting out of dissimilar plastids. (3) The disposition in three dimensions of the leaves of a shoot or a plant, resulting from phyllotaxis and leaf morphology, which appears to maximize

interception of light while minimizing mutual shading. (4) A patchy variation of the normal green colour of leaves; usually the result of infection by a virus. Cf. **chimaera.**

mosaic development Development when the eventual fate of cells of the developing embryo is determined, e.g. on formation of the blastomeres in the spirally segmented eggs of some Invertebrates. A *mosaic egg* is thus one whose areas of future functional development are determined in the early stages of cleavage.

mosquito The family Culicidae of the Diptera; have piercing proboscies and suck blood, often transmitting diseases (e.g. malaria, yellow fever, elephantiasis) while doing so; larvae and pupae aquatic, e.g. *Anopheles, Culex, Aedes.*

moss See **Bryopsida**.

mother cell A cell which divides to give daughter cells; the term is applied particularly to cells which divide to give spores, pollen grains, gametes and blood corpuscles.

motivation As a general term, and used mostly with reference to *human behaviour,* it refers to the desire to act in certain ways in order to achieve a goal; these may be transitory impulses or more persistent intentions. Motives may or may not be conscious and/or rational. In *animal behaviour*, it refers to a set of problems characterized by variations in responsiveness to a constant environmental situation, where the explanation assumes some reversible internal change in the individual, excluding changes due to learning, maturation or fatigue.

motoneuron(e) A motor neuron(e).

motor Pertaining to movement; as nerves which convey movement-initiating impulses to the muscles from the central nervous system.

motor areas Nerve centres of the brain concerned with the initiation and co-ordination of movement.

motor cell One of a number of cells which together can expand or contract and so cause movement in a plant member.

motor end plates The special end organ in which a motor nerve terminates in a striated muscle.

motor habits Repetitive, nonfunctional patterns of motor behaviour that seem to occur in response to stress (e.g. nail biting, thumb sucking). See **stereotyping**.

motor system Tissues and structures concerned in the movements of plant members, as in pulvini.

mould A fungus, esp. one that produces a visible mycelium on the surface of its host or substrate.

moult See **ecdysis**.

mouth parts In Arthropoda, the appendages associated with the mouth.

mRNA, messenger RNA That class of RNA whose sequence of nucleotide triplets determines the sequence of a polypeptide.

MTOC Abbrev. for *MicroTubule-Organizing Centre.*

mucigen A substance occurring as granules or globules in chalice cells and later extruded as mucin.

mucilaginous Pertaining to, containing or resembling mucilage or mucin.

mucinogen A substance producing or being the precursor of mucin, e.g. granules of mucous gland cells.

mucins A group of glycoproteins occurring in mucus, saliva and other secretions. Widely distributed in nature, the polypeptide chains are densely glycosylated.

mucopolysaccharides Heteropolysaccharides each containing a hexosamine in its characteristic repeating disaccharide unit.

mucoproteins Conjugated proteins with carbohydrate side chains, which may include hexoses, hexosamines or glucuronic acids.

mucosa See **mucous membrane.**

mucous glands Glands secreting or producing mucus.

mucous membrane A tissue layer found lining various tubular cavities of the body (as the gut, uterus, trachea etc.). It is composed of a layer of epithelium containing numerous unicellular mucous glands and an underlying layer of areolar and lymphoid tissue, separated by a basement membrane. Also *mucosa.*

mucro A short, sharp, terminal point.

mucronate Terminated by a short point or *mucro.*

mucus The viscous slimy fluid secreted by the mucous glands. adjs. *mucous, mucoid, muciform.*

mulch In horticulture, matter placed on the soil surface in order e.g. to suppress the growth of weed seedlings, or to reduce temperature fluctuations at the soil surface; often organic, e.g. peat, shredded bark, sometimes inorganic, e.g. pebbles.

mull Loose crumbly layer of soil occurring in some deciduous woodlands, in which leaf litter is being broken down rapidly at near-neutral pH in the presence of a rich fauna including earthworms. Cf. **mor, moder.**

Müllerian duct A duct which arises close beside the oviduct, or which, by the actual longitudinal division of the archinephric duct, in many female Vertebrates becomes the oviduct.

Müllerian mimicry Resemblance in colour between two animals, both benefiting by the resemblance. Cf. **Batesian mimicry.**

Müller's muscle The circular ciliary muscle of the Vertebrate eye.

multiarticulate Many-jointed.

multiaxial Having a main axis consisting of several more or less equal files of cells with or without subordinate branches. Cf. **uniaxial.**

multicellular Consisting of a number of cells.

multicipital Many-headed.

multicuspidate Teeth with many cusps.

multifactorial Determined by many genes and non-genetic factors. See **polygenic.**

multigravida A woman who has been pregnant more than once.

multilayered structure Single broad band of microtubules forming a *flagellar root* in the motile cells and gametes of the Charophyceae, bryophytes and vascular plants.

multilocular Having a number of compartments or loculi.

multinet growth Steady state pattern postulated for the growth of the cell walls of some elongating cells in which the cellulose microfibrils, deposited more or less transversely on the cytoplasmic surface of the wall, become passively reorientated by the growth of the cell so as to lie more nearly longitudinally in the older, outer parts of the wall.

multinucleate Having many nuclei. Also *polynucleate.*

multiparous Bearing many offspring at a birth.

multiple fission A method of multiplication found in Protozoa, in which the nucleus divides repeatedly without corresponding division of the cytoplasm, which subsequently divides into an equal number of parts leaving usually a residium of cytoplasm. Cf. **binary fission.**

multiple fruit A fruit formed from the maturing ovaries of a group of flowers, often with their receptacles and their floral parts, e.g. pineapple, fig.

multiple isomorphous replacement Method of solving the *phase problem* in the **X-ray crystallography** of protein crystals. Such a crystal consists of an array of geometrically identical unit cells arranged in three dimensions in which each unit cell contains one or more identical asymmetric units each containing one or a small number of protein molecules. For the method to work, the isomorphous crystals must have identical geometry and molecular structure, the only difference being the substitution of a heavy atom (platinum or mercury) at a small number of sites in each molecule. When diffraction patterns can be obtained from the unlabelled protein *and* two or more isomorphous derivatives, then the *phases* of the unlabelled crystal can be calculated and, together with the *amplitude* data, the molecular image obtained. Abbrev. *MIP.*

multiple personality disorder An extreme form of dissociative reaction in which

---

mutant, mutation and mutation rate

A *mutant* is a change, spontaneous or induced, that converts one allele into another (*point mutation*). More generally, any change of a gene or of chromosomal structure or number. A *somatic mutation* is one occurring in a somatic cell and not in the germ line. A *mutant individual* is one which displays the result of a *mutation*. A mutant gene is one which has undergone *mutation*.

The *mutation rate* is the frequency, per gamete, of mutations of a particular gene or a class of genes. Sometimes, esp. in micro-organisms, the number per unit of time (esp. generation time).

---

2 or more complete personalities, each well developed and distinct, are found in one individual.

multipolar Said of nerve cells having many axons. Cf. **bipolar**, **unipolar**.

multiseriate In several rows; vascular ray several cells wide.

multituberculate Said of tuberculate teeth with many cusps; having many small projections.

multivalent A group of three or more partly homologous chromosomes held together during the prophase of the first division of meiosis. Cf. **bivalent**, **univalent**.

multivariate analysis Statistical analysis of several measurements of different characteristics on each unit of observation.

multivoltine Having more than one brood in a year. Cf. **univoltine**, **bivoltine**.

Mummery's plexus A network of fine nerve fibrils lying between the odontoblasts and the dentine in a tooth.

muramyl dipeptide *N-acetyl-muramyl-L-alanyl-D-isoglutamine.* This is the simplest structural unit of bacterial peptidoglycans, which can mimic the adjuvant effects of mycobacteria in complete Freund's adjuvant. It causes release of IL-1 from macrophages, and is sometimes incorporated in a vaccine to increase its immunogenicity.

muricate Having a surface roughened by short, sharp points.

Musci The mosses; Bryopsida.

muscle A tissue whose cells can contract; a definitive mass of such tissue. See **cardiac-**, **striated-**, **unstriated-**, **voluntary-**.

musculature The disposition and arrangement of the muscles in the body of an animal.

musculocutaneous Pertaining to the muscles and the skin.

mushroom Common name for the fruiting body of an **agaric**, esp. of species of *Agaricus*.

mushroom bodies Paired nerve centres of the protocerebrum in Insects, regarded as the principal association areas. Also known as *corpora pedunculata*.

muskeg Mossy growth found in high latitudes, usually as deep and treacherous swamp, bog or marsh. Unstable peat formation.

musk glands In some Vertebrates, glands, associated with the genitalia, the secretion of which has an odour of musk.

mutagen A substance which causes *mutation*.

mutant See panel.

mutation See panel.

mutation rate See panel.

muticate, muticous Without a point, mucro or awn. Unarmed.

mutualism Any association between 2 organisms which is beneficial to both and injurious to neither. See **symbiosis**.

muzzle See **rhinarium**.

myarian Based on musculature, as a system of classification; pertaining to the musculature.

myasthenia gravis Disease characterized by increasing muscular weakness on exercise caused by faulty transmission at the neuromuscular junction. Autoantibodies are present in the blood against antigens of the acetylcholine receptor on the post synaptic membrane, and are the putative cause of the symptoms by interfering with or damaging this receptor. One form is associated with the presence of a thymoma, and thymectomy may reverse the condition.

mycelium A mass of branching hyphae; the vegetative body (thallus) of most true fungi. adj. *mycelial.*

mycetocytes Cells containing symbiotic micro-organisms occurring in the **pseudovitellus**.

mycetome *Pseudovitellus.* A special organ in some species of Insects, Ticks and Mites, inhabited by intracellular symbionts, e.g. Bacteria, Fungi, Rickettsias.

mycetophagous Fungus-eating.

mycobiont The fungal partner in a symbiosis, e.g. in a lichen.

mycology The study of fungi.

myco-, myc-, myceto- Prefixes from Gk. *mykes*, gen. *myketos*, fungus.

mycophthorous Said of a fungus parasitic on another fungus.

mycoplasma The smallest free-living organism known. *Mycoplasma pneumoniae* is an important cause of atypical pneumonia.

Mycoplasmatales Organisms of this group are associated with pleuropneumonia in cattle and contagious agalactia in sheep; very variable in form, consisting of cocci, rods etc. and a cytoplasmic matrix; very similar to the L-forms of true bacteria; nonpathogenic morphologically similar organisms have also been found in animals and man.

mycorrhiza A symbiotic association between a fungus and the roots, or other structures of a plant. See **endotrophic mycorrhiza, ectotrophic mycorrhiza, vesicular-arbuscular mycorrhiza**.

mycotoxin A substance produced by a fungus and toxic to other organisms, esp. to man and animals, e.g. the **aflatoxins**.

mycotrophic plant A plant which lives in symbiosis with a fungus and thus has mycorrhizas.

mydriasis Extreme dilation of the pupil of the eye.

myelencephalon The part of the hind brain in Vertebrates, forming part of the *medulla oblongata*. Cf. **metencephalon**.

myelin A white fatty substance which forms the medullary sheath of nerve fibres.

myelination Formation of a myelin sheath.

myelin sheath See **medullary sheath**.

myel-, myelo- Prefixes from Gk. *myelos*, marrow.

myelocoel The central canal of the spinal cord.

myelocyte A bone-marrow cell; a large amoeboid cell which is found in the marrow of the long bones of some higher Vertebrates, which give rise, by division, to leucocytes.

myeloid cell Cells derived from stem cells in the bone marrow which mature to form the granular leucocytes (granulocytes) of blood. They differentiate into different forms (neutrophils, eosinophils or basophils) under the influence of **colony stimulating factors**.

myeloma Syn. for **plasmacytoma**. Hence myeloma protein.

myeloplast A leucocyte of bone marrow.

myeloplax A giant cell of bone marrow and other blood-forming organs, sometimes multinucleate, which gives rise to the blood platelets. Also called *megakaryocyte*.

myenteric Pertaining to the muscles of the gut; as a sympathetic nerve plexus controlling their movements.

myoblast An embryonic muscle cell which will develop into a muscle fibre.

myocardium In Vertebrates, the muscular wall of the heart.

myocoel The coelomic space within a myotome.

myocomma A partition of connective tissue between 2 adjacent myomeres. Also *myoseptum*.

myocyte A muscle cell; a deep contractile layer of the ectoplasm of certain Protozoa.

myo-epithelial A term used to describe the epithelial cells of Coelenterata which are provided with taillike contractile outgrowths at the base.

myofibril The contractile filament consisting of actin, myosin and associated proteins within muscle cells.

myogenic Said of contraction arising in a muscle independent of nervous stimuli. Cf. **neurogenic**.

myoglobin A haem protein related to haemoglobin but consisting of only one polypeptide chain. It is present in large amounts in the muscles of mammals such as whales and seals, where it acts as an oxygen store during diving. It was the first protein to have its 3-dimensional structure determined by **X-ray crystallography**.

myolemma See **sarcolemma**.

myology The study of muscles.

myoma A tumour composed of unstriped (leiomyoma) or striped (rhabdomyoma) muscle fibres.

myomere In metameric animals, the voluntary muscles of a single somite.

myometrium The muscular coat of the uterus.

myo-, my- Prefixes from Gk. *mys*, gen. *myos*, muscle.

myoneme In Protozoa, a contractile fibril of the ectoplasm.

myoneural Pertaining to muscle and nerve, as the junction of a muscle and a nerve.

myophily, myiophily Pollination by flies or other Diptera.

myoseptum See **myocomma**.

myosin Large highly asymmetric protein found originally in association with actin in muscle but now known to have a wide cellular distribution. The myosin molecules aggregate into fibres which, in muscle, interdigitate between the actin fibres. The relative movement of the two sets of fibres provides the molecular basis of *muscular contraction*.

myotome A muscle merome; one of the metameric series of muscle masses in a developing segmented animal.

myriapod A general term denoting arthropods with many similar segments and comprising the classes Chilopoda and Myriapoda (Centipedes and Millepedes).

myrmecochory Distribution of seeds or other reproductive bodies by ants.

myrmecophagous Feeding on ants.

myrmecophily (1) Symbiosis with ants, e.g. Acacia. (2) Pollination by ants.

Myrtaceae Family of ca 3000 spp. of dicotyledonous flowering plants (superorder Rosidae). Woody plants, tropical and subtropical and esp. Australian. Includes some sources of spices, e.g. cloves, allspice, and the genus Eucalyptus, trees important in the hardwood forests of Australia.

mysophobia Morbid fear of being contaminated.

myxamoeba The amoeboid stage of a slime mould. See **Myxomycota**.

myxo- Prefix from Gk. *myxa*, mucus, slime.

Myxobacteriales A group of bacteria characterized by the absence of a rigid cell wall, gliding movements and the production of slime. Some genera produce characteristic fruiting bodies.

myxomatosis A highly contagious and fatal viral disease of rabbits, characterized by tumourlike proliferation of myxomatous tissue, esp. beneath the skin of the head and body. Virus lethality has progressively decreased. Vaccine available.

Myxomycetes The true (or acellular) slime moulds. Class of slime moulds (Myxomycota) first feeding phagotrophically as individual **myxamoebae** which can interconvert with flagellated swarm cells, either of which fuse in pairs. The zygote develops as a multinucleate **plasmodium**, which is phagotrophic and eventually forms a fruiting body, liberating spores, e.g. *Physarum*.

Myxomycota *Gymnomycota*. The slime moulds. Wall-less heterotrophic organisms, usually classified with the fungi, phagotrophic on bacteria or parasitic within plant cells. Dispersed by zoospores and/or small, walled, wind-blown spores. Includes Acrasiomycetes, Mysomycetes and Plasmodiophoromycetes.

Myxophyceae Old name for blue-green algae; Cyanophyceae.

myxoviruses RNA containing viruses pathological to Vertebrates, including the influenza virus and related species. Many have considerable genetic variability which results in continual changes in their antigenic status and consequent difficulty in producing an effective vaccine.

N

NAA Abbrev. for *NaphthaleneAcetic Acid*.

nacre, nacreous layer The iridescent calcareous substance, mostly calcium carbonate, composing the inner layer of a Molluscan shell, which is formed by the cells of the whole of the mantle. Mother of pearl.

NAD See **nicotinamide adenine dinucleotide**.

NADP See **nicotinamide adenine dinucleotide phosphate**.

naevus Birthmark; mole. (1) A pigmented tumour in the skin. (2) A patch or swelling in the skin composed of small dilated blood vessels.

nail In higher Mammals, a horny plate of epidermal origin taking the place of a claw at the end of a digit.

naked Lacking a structure or organ.

nanism The condition of being a dwarf (L. *nanus*); dwarfism.

nano- Prefix for 10^{-9}, i.e. equivalent to millimicro or one thousand millionth. Symbol *n*.

nanophanerophyte Phanerophyte, with buds 25 cm to 2 m above soil level.

nanoplankton Plankton of microscopic size from 0.2 to 20 μm.

naphthalene acetic acid A synthetic plant growth regulator with **auxin**-like activity, used esp. in **rooting compounds**.

narcissism (1) An erotic preoccupation with one's own body. (2) In psychoanalytic theory, investment of libidinal energy in the self, the developmental basis of self regard and not necessarily pathological.

narcolepsy A sleep disturbance characterized by uncontrollable sleepiness and the tendency to fall asleep in inappropriate or dangerous circumstances.

narcosis A state of unconsciousness produced by a drug; the production of a narcotic state.

narcotic Tending to induce sleep or unconsciousness; a drug which does this.

nares Nostrils; nasal openings; as the internal or posterior *nares* to the pharynx, the external or anterior *nares* to the exterior. adj. *narial, nariform*.

nasal Pertaining to the nose; a paired dorsal membrane bone covering the olfactory region of the Vertebrate skull.

nasolacrimal canal A passage through the skull of Mammals, passing from the orbit to the nasal cavity, and through which the tear duct passes.

nasopalatine duct In some Reptiles and Mammals, a duct piercing the secondary palate and connecting the vomeronasal organs with the mouth.

nasopharyngeal duct The posterior part of the original vault of the Vertebrate mouth which, in Mammals, due to the development of a secondary palate, carries air from the nasal cavity to the pharynx.

nasoturbinal In Vertebrates, a paired bone or cartilage of the nose which supports the folds of the olfactory mucous membrane.

nastic movement, nasty A plant movement in response to but not orientated by an external stimulus, e.g. the opening or closing of some flowers in response to increasing or decreasing temperature. See **epinasty, hyponasty, photonasty, thermonasty**. Cf. **tropism, taxis**.

natal (1) Pertaining to birth. (2) Pertaining to the buttocks.

natality The inherent ability of a population to increase. *Maximum* or *absolute natality* is the theoretical maximum production of new individuals under ideal conditions, while *Ecological* or *realized natality* is the population increase occurring under specific environmental conditions. Cf. **mortality**.

natatory, natatorial Adapted for swimming.

nates The buttocks.

National Nature Reserve In Britain, extensive area set aside for nature conservation, under the auspices of the *Nature Conservancy Council*.

nativism In philosophy, the position that humans are born with some innate knowledge; in the study of perception, the view that some important abilities are innate.

natural antibody Antibody present in the blood of normal individuals not known to have been immunized against the relevant antigen, e.g. in humans, antibodies against antigens of the **ABO blood group system** or Forssman antibodies. They are generally induced by organisms in the gut owing to shared antigenic determinants. Animals which are reared under germ-free conditions do not develop such antibodies, but do so if the gut is colonized by enterobacteria.

natural classification One based on many characters, and likely to have a predictive value.

natural immunity Immunity conferred before birth and not *acquired* subsequently by exposure to antigens from the environment.

naturalized Introduced from another region but growing, reproducing and maintaining itself in competition with the native vegetation.

natural killer cell *NK cell*. A lymphoid cell which kills a range of tumour cell targets in the absence of prior immunization and without evident antigen specificity.

Morphologically it is a lymphocyte with large granules visible under the light microscope or after staining with azure dyes. NK cells are activated by interferon and may also be important in killing some virally infected cells *in vivo*. Killing results from NK cells binding to their target and inserting granules which contain proteases and **perforin**. Killer cells appear to use a similar mechanism. The relationship of these cells to one another is ill-understood and their origin is also uncertain, but may be common with T lymphocytes, since T cells during prolonged culture *in vitro* sometimes acquire NK cell characteristics.

natural selection An evolutionary theory which postulates the survival of the best adapted forms, with the inheritance of those characteristics wherein its fitness lies, and which arise as random variations due to mutation; it was first propounded by Charles Darwin, and is often referred to as *Darwinism* or the *Darwinian Theory*.

nauplius The typical first larval form of Crustacea; egg-shaped, unsegmented, and having three pairs of appendages and a median eye; found in some members of every class of Crustacea, but often passed over, becoming an entirely embryonic stage.

Nautiloidea A sub-class of the Cephalopoda, having a wide central siphuncle and a planospiral chambered shell. Abundant from early Cambrian to late Cretaceous, but now represented by one genus, *Nautilius*, which lives in tropical seas. All chambers except the terminal living chamber contain gas which buoys up the heavy shell.

navel In Mammals, the point of attachment of the umbilical cord to the body of the foetus.

navicular bone In Mammals, one of the tarsal bones, also known as the *centrale* or *scaphoid*.

navigation In biology, refers to complex forms of long distance orientation by animals.

N-bands See **banding techniques**.

neanic Said of the adolescent period in the life history of an individual.

Nearctic region One of the subrealms into which the Holarctic region is divided; it includes N. America and Greenland.

nearest neighbour analysis Method for determining the frequency of pairs of adjacent bases in DNA. It has shown that there is a deficiency of the pair CG in most higher organisms.

neck (1) The upper tubular part of an archegonium, and of a perithecium. (2) The lower part of the capsule of a moss, just above the junction with the seta.

neck canal cell One of the cells in the central canal in the neck of an archegonium.

Also *neck cell*.

necro- Prefix from Gk. *neckros*, dead body.

necrobiosis The gradual death, through stages of degeneration and disintegration, of a cell in the living body.

necrophagous Feeding on the bodies of dead animals.

necrophorous Carrying away the bodies of dead animals; as certain Beetles, which usually afterwards bury the bodies.

necrosis Death of a cell (or of groups of cells) while still part of the living body. adj. *necrotic*. v. *necrose*.

necrotroph An organism which feeds off dead cells and tissues. A necrotrophic parasite kills host cells and feeds on them once they are dead, e.g. the **damping-off** fungi. Cf. **biotroph**, **facultative parasite**.

nectar A sugary fluid exuded by plants, usually from some part of the flower, occasionally from somewhere else on the plant; it attracts insects, which assist in pollination.

nectar guide Same as **honey guide**.

nectarivorous Nectar-eating.

nectary A glandular organ or surface from which nectar is secreted.

necto- Prefix from Gk. *nektos*, swimming.

necton See **nekton**.

nectopod An appendage adapted for swimming.

need Refers to specific physical or psychological conditions necessary for an individual's welfare and/or sense of wellbeing.

needle A long, narrow, stiffly-constructed leaf, characteristic of pines and similar plants.

negative reaction A tactism or tropism in which the organism moves, or the member grows, from a region where the stimulus is stronger to one where it is weaker.

negative reinforcement In conditioning situations, a stimulus, usually aversive, that increases the probability of escape or avoidance behaviour. Cf. **punishment**.

negative staining Important technique in electron microscopy, in which heavy metals which scatter electrons are deposited around the specimen. This is then seen in negative contrast.

Neisseriaceae A family of bacteria belonging to the order Eubacteriales. Gramnegative, characteristically occurring as paired spheres, parasitic in mammals, e.g. *Neisseria gonorrhoeae* (gonorrhoea).

nekton Actively swimming aquatic organisms, as opposed to the passively drifting organisms or *plankton*. Also *necton*.

nematoblast A cell which will develop a **nematocyst**.

nematocyst An independent effector found in most Coelenterata (and a few Protozoa), consisting of a fluid filled sac and produced at one end into a long narrow pointed hollow thread, which normally lies inverted and coiled up within the sac, but can be everted when the *cnidocil* is stimulated. Used for prey capture and defence.

Nematoda A class of phylum, Aschelminthes, comprising unsegmented worms with an elongate rounded body pointed at both ends; marked by lateral lines and covered by a heavy cuticle composed of protein; have a mouth and alimentary canal; have only longitudinal muscles; nervous system consists of a circum-pharyngeal ring and a number of longitudinal cords; perivisceral cavity, a pseudocoel; cilia absent; sexes separate; development direct, the larvae resembling the adults; many species are of economic importance; mostly free living but some are parasitic. Round Worms, Thread Worms, Eel Worms.

Nemertea A phylum of apparently nonmetameric acoelomate worms with an elongate flattened body, a ciliated ectoderm and a dorsal eversible proboscis not connected with the alimentary canal. Most are marine, but some are fresh water or terrestrial. Ribbon Worms. Also *Nemertini*.

neoblasts In many of the lower animals (Annelida, Ascidia etc.), large amoeboid cells widely distributed through the body which play an important part in the phenomena of regeneration.

neo-Darwinism The modern version of Darwin's theory of evolution by natural selection, incorporating the discoveries of Mendelian and population genetics.

neo-Freudians Schools of thought and therapy based on modification of Freud's theories; usually stress social rather than biological factors as important determinants of unconscious conflict.

neologism Verbal construction such as occurs in schizophrenia, manic depressive psychosis and some aphasias, in which the patient uses coined words, which may have meaning for him but not for others, or else gives inappropriate meanings to ordinary words.

neonychium A pad of soft tissue enclosing a claw of the foetus during the development of many Mammals, to eliminate the risk of ripping the foetal membranes.

neopallium In Mammals, that part of the cerebrum occupied with impressions from senses other than the sense of smell.

neoplasm A new formation of tissue in the body; a *tumour*. adj. *neoplastic*.

neossoptiles The down feathers found on a newly-hatched Bird.

neotenin In Insects, the **juvenile hormone** produced by the corpora allata, which suppresses the development of adult characteristics at each moult except the last, when the corpora allata become inactive and

metamorphosis occurs.

neoteny Retention of some juvenile characteristics by the sexually mature adult, e.g. some Amphibians which have the appearance of tadpoles. Cf. **paedogenesis**.

neotropical region One of the primary faunal regions into which the surface of the globe is divided. It comprises South America, the West Indian islands, and Central America south of the Mexican plateau.

neovitalism The theory which postulates that a complete causal explanation of vital phenomena cannot be reached without invoking some extra-material concept.

nephric Pertaining to the kidney.

nephridiopore The external opening of a nephridium or nephromixium.

nephridium In Invertebrates and lower Chordata, a segmental excretory organ consisting of an intercellular duct of ectodermal origin leading from the coelom to the exterior; more generally, an excretory tubule. adj. *nephridial*. Cf. **coelomoduct**.

nephr-, nephro- Prefixes from Gk. *nephros*, kidney.

nephrodinic Using the same duct for the discharge of both excretory and genital products.

nephrogenic tissue In the embryonic development of Vertebrates, a relatively small intermediate region of the mesoderm lateral and ventral to the somites, from which derive the kidney tubules, their ducts, and the deeper tissues of the gonads. May be segmented, forming nephrotomes, or a continuous band of tissue.

nephrogonoduct Esp. in Invertebrates, a common duct for genital and excretory products.

nephropore See **nephridiopore**.

nephros A kidney. adj. *nephric*.

nephrostome The ciliated funnel by which some types of nephridia and nephromixia open into the coelom.

nephrotoxin A poison or toxin which specifically affects the cells of the kidney.

nepionic Said of the embryonic period in the life history of an individual.

nervation, nervature See **venation**.

nerve (1) A collection of axons leading to or from the central nervous system; also a nerve bundle or tract. (2) A strand of conducting tissue and/or strengthening tissue in a leaf or leaf-like organ; a *vein*. adjs. *nervous, neural*.

nerve cell See **neuron**.

nerve centre An aggregation of nerve cells associated with a particular sense or function.

nerve ending The distal end of a nerve axon, normally a synapse.

nerve fibre An axon.

nerve impulse A regenerative electrical

---

network theory

A theory postulated by N. K. Jerne in 1974 that the immune system is controlled by a network of interaction between antigen binding sites (paratopes) which may be on immunoglobulin molecules or lymphocyte receptors. Each paratope is capable of binding an epitope on an external antigen and also an idiotope with a shape resembling the epitope present on another immunoglobulin molecule (the 'internal image'). An immunoglobulin which is an anti-idiotope will in turn be recognized by another molecule which is an anti-anti-idiotope, and so on.

In an individual the concentration of immunoglobulin molecules bearing any particular idiotope is likely to be so low that no stimulation of anti-idiotope results. However when an external antigen is administered this stimulates a large increase in the concentration of immunoglobulin molecules with complementary paratopes, sufficient to stimulate anti-idiotope production and the network is thereby disturbed. There are many examples showing that this theory is in principle correct, although there is sufficient degeneracy in the system to limit the extent of the network of interactions.

---

potential which travels along an **axon**. See **action potential**.

nerve net The primitive type of nervous system found in Coelenterata, consisting of numerous multipolar neurons which form a net underlying and connecting the various cells of the body wall.

nerve plexus A network of interlacing nerve fibres.

nerve root The origin of a nerve in the central nervous system.

nerve trunk A bundle of nerve fibres united within a connective tissue coat.

nervous system The whole system of nerves, ganglia, and nerve endings of the body of an animal, considered collectively.

nervure One of the chitinous struts which support and strengthen the wings of an Insect.

nest An artefact built to provide temporary shelter as in some primates, or protection for the young and eggs in most birds, or for housing the colony in social Insects.

nest epiphyte An **epiphyte** in which the leaves and/or a tangle of stems and roots form a structure in which leaf litter collects, humifies and is used by the epiphyte to root into as a source of mineral nutrients, e.g. the bird's nest fern, *Asplenium nidus*.

net assimilation rate *Unit leaf rate*. Abbrev. *NAR*. The net photosynthetic rate (i.e. total photosynthesis minus respiration for the plant) per unit leaf area. *NAR = (1/A).(dW/dt)* where *A* is leaf area, *W* is dry weight and *t* is time).

net production See **production**.

network theory See panel.

neural See **nerve**.

neural arch The skeletal structure arising dorsally from a vertebral centrum, formed by the neurapophyses and enclosing the spinal cord.

neural canal The space enclosed by the centrum and the neural arch of a vertebra, through which passes the spinal cord.

neural crest In a Vertebrate embryo, a band of cells lying parallel and close to the nerve cord which will later give rise to the ganglia of the dorsal roots of the spinal nerves.

neural spine The median dorsal vertebral spine, formed by the fusion of the neurapophyses above the neural canal.

neural tube A tube formed dorsally in the embryonic development of Vertebrates by the joining of the 2 upturned neural folds formed by the edges of the ectodermal neural plate, giving rise to the brain and spinal nerve cord.

neuraminidase An enzyme produced by viruses of the myxovirus and paramyxovirus groups and by some bacteria which splits the glycosidic link between neuraminic acid or sialic acid and other sugars. Neuraminic acid is an important structural component of the surface glycoproteins of many cells and contributes largely to the net negative charge of cells. After treatment with neuraminidase they show an increased tendency to agglutinate, and this is used to increase the sensitivity of some agglutination reactions.

neurapophyses A pair of plates arising dorsally from the vertebral centrum, and meeting above the spinal cord to form the neural arch and spine. sing. *neurapophysis*.

neurilemma See **neurolemma**.

neurine $CH_2=CH.N(CH_3)_3OH$, *trimethylvinylammonium hydroxide*, obtainable from brain substance and from putrid meat; related to *choline*, into which it can be transformed. It is a ptomaine base.

neur-, neuro- Prefixes from Gk. *neuron*, nerve.

neuroblasts Cells of ectodermal origin which give rise to neurons.

neurocranium The brain case and sense capsules of a vertebrate skull.

neurocrine Neurosecretory; secretory property of nervous tissue.

neurocyte See **neuron**.

neurogenesis The development and formation of nerves.

neurogenic Activity of a muscle or gland which is dependent on continued nervous stimuli. Cf. **myogenic**.

neuroglia The supporting tissue of the brain and spinal cord of Vertebrates, composed of much branched fibrous cells which occur among the nerve cells and fibres. Also *glia*.

neurohaemal organs Organs which serve as a gateway for the escape of the products of neurosecretory cells from the neurons into the circulating blood, e.g. the *corpora cardiaca* of Insects.

neurohypophysis See **pars nervosa**.

neurolemma, neurilemma A thin homogeneous sheath investing the medullary sheath of a medullated nerve fibre; sheath of Schwann.

neurology The study of the nervous system.

neuromasts Sensory hair cells embedded in a gelatinous cupola found in the lateral line system of lower vertebrates and concerned with **mechanoreception**.

neuromuscular Pertaining to nerve and muscle, as a myoneural junction.

neuron, neurone A nerve cell and its processes. Also *neurocyte*.

neuropil In Vertebrates, a network of axons, dendrites and synapses within the central nervous system.

neuropile In Arthropods, regions within the brain and the central portion of segmental ganglia consisting of dendrites and synapses.

neuropore The anterior opening by which the cavity of the central nervous system communicates with the exterior.

neurosecretory cell A special type of neurone in which the axon terminates against the wall of a blood vessel or sinus into which it secretes a hormone or other factor.

neuroses A loose term for mental disorders in which the individual experiences anxiety, or engages in behaviour to avoid experiencing it; there is no evidence of an organic component, and the individual remains in contact with reality. See **obsessional neurosis**.

neurula In the embryonic development of Vertebrates, the stage after the gastrula in which the processes of organ formation begin, with the formation of the neural tube, mesodermal somites, notochord and archenteron.

neuter (1) Sexless. (2) Lacking functional sexual organs; having neither functional stamens nor functional carpels; sterile. Also *neutral*.

neutral pump A pump which transports only uncharged molecules or appropriately balanced pairs of ions so that there is no net transfer of charge.

neutron therapy Use of neutrons for medical treatment.

niacin *Vitamin B$_3$*; *nicotinic acid*. Deficiency results in **pellagra**.

niche See **ecological niche**.

nicitating Winking; said of the third eyelid of the Vertebrates, which by its movements keeps clean the surface of the eye.

nick (1) A cut between adjacent nucleotides in one strand of a duplex DNA molecule. (2) A particular combination of male and female parents giving desirable offspring, esp. in the breeding of **F1 hybrids**.

nick translation An inexact phrase for a method of radioactively labelling a DNA molecule. The DNA is first nicked in one strand by a short DNase treatment. This allows DNA polymerase I to both remove nucleotides from the exposed end and replace them by highly radioactive nucleotides.

nicotinamide adenine dinucleotide A co-enzyme which serves as an electron acceptor for many dehydrogenases. The reduced form (NADH) subsequently donates its electrons to the *electron transport chain*. Abbrev. *NAD*.

nicotinamide adenine dinucleotide phosphate Phosphorylated derivative of NAD. It also serves as an electroncarrier but the electrons are primarily used for *reductive biosynthesis*. Abbrev. *NADP*.

nicotine An alkaloid of the pyridine series, $C_{10}H_{14}N_2$. It occurs in tobacco leaves, is extremely poisonous and, in small quantities, highly addictive. It is a colourless oil, of nauseous odour; bp 246°C at 97 kN/m^2.

nicotinic acid See **vitamin B**.

nidamental Said of glands which secrete material for the formation of an egg covering.

nidation In the oestrous cycle of Mammals, the process of renewal of the lining of the uterus between the menstrual periods.

nidicolous Said of Birds which remain in the parental nest for some time after hatching. Cf. *nidifugous*.

nidification, nidulation The process of building or making a nest.

nidifugous Said of those Birds which leave the parental nest soon after hatching. Cf. **nidicolous**.

nidus A nest; a small hollow resembling a

nest; a nucleus.

night blindness Abnormal difficulty in seeing objects in the dark; due often to deficiency of vitamin A (retinol) in the diet. Also *nyctalopia*.

night terror A particularly harrowing variety of bad dream experienced by a child.

nigrescent Becoming blackish.

nipple The mamma or protuberant part of the mammary gland in female Mammals, bearing the openings of the milk forming glands.

Nissl bodies Aggregations of **ribosomes** found within nerve cells.

nitrate-reducing bacteria Facultative aerobes able to reduce nitrates to nitrites, nitrous oxide, or nitrogen under anaerobic conditions, e.g. *Micrococcus denitrificans*. This process is termed *denitrification*. A few bacteria use such reduction processes as hydrogen acceptor reactions and hence as a source of energy; in this case the end product is ammonia.

nitrification The oxidation of ammonia to nitrite and nitrate by chemoautotrophic bacteria whose energy requirements come from these exergonic reactions.

Nitrobacteriaceae Family of bacteria belonging to the order *Pseudomonadales*. Important in nitrification processes in the soil and fresh water. Autotrophic bacteria which derive energy from oxidation processes; *Nitrosomonas* from the oxidation of ammonia to nitrites; *Nitrobacter* from the oxidation of nitrites to nitrates. Also *nitrifying bacteria*.

nitrogenase The enzyme system that catalyses the reduction of gaseous nitrogen (dinitrogen, N_2) to ammonia in biological **nitrogen fixation**. It is inactivated by oxygen.

nitrogen balance The state of equilibrium of the body in terms of intake and output of nitrogen; positive nitrogen balance indicates intake exceeds output, negative nitrogen balance denotes output exceeds intake.

nitrogen cycle The sum total of the transformations undergone by nitrogen and nitrogenous compounds in nature in relation to living organisms.

nitrogen fixation, dinitrogen fixation See panel.

nitrophilous Growing characteristically in places where there is a good supply of fixed nitrogen.

nitrozation The conversion of ammonia into nitrites by the action of soil bacteria, *Nitrosomonas*, being the second stage in the nitrification in the soil.

NK cell Abbrev. for *Natural Killer cell*.

nm Abbrev. for *nanometre* = 10 Ångstroms, = 10^{-9} m.

nociceptive Sensitive to pain.

noctilucent Phosphorescent; producing light in the dark.

node The position on a stem at which one or more leaves are attached. Cf. **internode**.

node of Ranvier See **Ranvier's node**.

nodose, nodular Bearing localized swellings or nodules.

nodule Any small rounded structure on a plant, esp. a swelling on a root inhabited by symbiotic, nitrogen-fixing bacteria or actinomycetes.

nomadism The habit of some animals of roaming irregularly without regularly returning to a particular place. Cf. **migration**.

Nomarski microsope A type of **differential interference contrast microscope**.

nomeristic Of metameric animals, having a definite number of somites.

non-caducous See **indeciduate**.

non-disjunction Failure of one or more chromosomes to move with the rest of the set towards the appropriate pole at anaphase.

non-essential organs The sepals and petals of flowers.

non-homologous pairing Pairing between regions of non-homologous chromosomes. In some cases short stretches of similar sequences, possibly repetitive may be involved.

non-medullated See **amyelinate**.

nonsense mutation A base change which causes an amino acid-specifying sequence to be changed into one which causes termination by polypeptide chain synthesis.

nonsense syllable A series of letters, which usually consist of two consonants with a vowel between them that does not constitute a word; used in studies of learning and memory.

non-specific immunity Mechanisms whereby the body is protected against microbial invasion which do not depend upon the mounting of a specific immune response. They include physical barriers to infection (skin, mucous membranes); enzyme inhibitors naturally present in the blood; activation by 'rough' variants of Gram-negative bacteria of complement via the alternative pathway; interferon; lysozyme; phagocytosis etc. These mechanisms are sufficient to protect against microbes which are regarded as non-pathogenic, although they would be capable of multiplication within the dead body.

nonviable Incapable of surviving.

nopaline One sort of **opine**.

NOR See **nucleolar-organizing region**.

norm Shared expectations about how individuals should or do behave.

norm Of a vector in a finite-dimensional real vector space; the square root of the scalar product of the vector with itself; the magnitude of the vector. A norm on a vector space is a consistent definition of the norms of its

nitrogen fixation

The formation of free gaseous nitrogen (dinitrogen, N_2) compounds with other elements. Nitrogen is fixed: (1) into oxides, in the atmosphere by lightning and UV radiation, estimated at $7×10^6$ tonnes of N per annum, globally; (2) industrially by man to make nitrogenous fertilizers, e.g. nitrates, ammonia, ammonia salts and urea ($50×10^6$ tonnes); (3) biologically by free-living or symbiotic prokaryotic organisms into ammonia ($150×10^6$ tonnes).

Free-living nitrogen-fixers include many blue-green algae (notably those with heterocysts, e.g. *Anabaena*) and photosynthetic bacteria, and a variety of heterotrophic bacteria which can be aerobic (e.g. *Azotobacter*), facultatively aerobic (e.g. *Klebsiella*) or anaerobic (e.g. *Clostridium*).

Some *heterotrophic bacteria* are especially associated with the rhizosphere of plant roots to the mutual benefit of bacteria and plant. Such symbiotic fixers include species of the bacterium *Rhizobium*, which inhabit root nodules of the pea and bean family (Leguminosae), of the actinomycete *Frankia* which inhabit root nodules in various plants and of various blue-green algae which form associations with a variety of other plants, including a few angiosperms and some lichens. At least two sorts of animal, termites and shipworms are reported to harbour nitrogen-fixing symbiotic bacteria.

The most important nitrogen-fixing systems agriculturally are the leguminous plants; nitrogen fixing by the *Azolla/Anabaena* symbiosis can improve rice yields when the aquatic fern *Azolla* is grown and used as a green manure. Biological nitrogen fixation is typically inhibited in the presence of adequate nitrate or ammonia in the soil and fixed nitrogen does not accumulate in ecosystems because of leaching and denitrification.

Nitrogen is also fixed inadvertently by man in automobile engines and other combusting devices to produce oxides of nitrogen, NO_X, which are locally significant as aerial pollutants. See **acid rain**, **nitrogenase**.

vectors, or equivalently of a scalar product.

normal distribution A distribution widely used in statistics, to model the variation in a set of observations, as an approximation to other distributions, or as the asymptotic distribution of statistics from large samples. The normal distribution is indexed by two parameters, the mean and variance. See **standard normal distribution**.

normoblast A stage in the development of an erythrocyte from an erythroblast when the nucleus has become reduced in size and the cytoplasm contains much haemoglobin.

nosology Systematic classification of diseases; the branch of medical science which deals with this.

nostrils The external nares.

notochord In Chordata, skeletal rod formed of turgid vacuolated cells. adj. *notochordal*.

notum The tergum of Insects.

NPK Nitrogen, Phosphorus and Potassium as fertilizer.

nucellus Parenchymatous tissue in the ovule of a seed plant, more or less surrounded by the integuments and containing the embryo sac; equivalent to the megasporangium. See **perisperm**.

nuchal Pertaining to, or situated on, the back of the neck.

nuchal crest A transverse bony ridge forming across the posterior margin of the roof of the vertebrate skull for attachment of muscles and ligaments supporting the head.

nuchal flexure In developing Vertebrates, the flexure of the brain occurring in the hinder part of the medulla oblongata, which bends in the same direction as the primary flexure.

nucivorous Nut-eating.

nuclear budding Production of 2 daughter nuclei of unequal size by constriction of the parent nucleus.

nuclear envelope See **nuclear membrane**.

nuclear family In human genetics, a family providing data on the two parents and their children.

nuclear fragmentation The formation of two or more portions from a cell nucleus by direct break-up and not by mitosis.

nuclear matrix The nuclear residue left after removal of chromatin, including pore complexes, lamina, nucleolar residues and ribonucleoprotein fibrils.

nuclear medicine The application of radionuclides in the diagnosis or treatment of disease.

nuclear membrane The double membrane punctuated by **nuclear pore** complexes which surrounds the interphase nucleus, the outer membrane being continuous with the membrane of the **endoplasmic reticulum**. Also *nuclear envelope*.

nuclear pore complex Sites at which the two layers of the nuclear membrane are joined forming pores which connect the nucleoplasm and the cytoplasm, and which are surrounded on either side by symmetrical arrays of granules called *annuli*.

nuclear sap See **nucleoplasm**.

nuclear spindle See **spindle**.

nuclear winter theory Theory, based on model calculations, that nuclear war would be followed by a period of cold resulting from the attenuation of solar energy by dust and smoke in the atmosphere.

nuclease An enzyme which specifically cleaves the 'backbone' of nucleic acids. Their specificity ranges from those like DNase and RNase which cut the phosphodiester bonds in any DNA or RNA, to **restriction enzymes** which only cut a particular 4 or 6 base-pair sequence.

nucleic acid General term for natural polymers in which *bases* (purines or pyrimidines) are attached to a sugar phosphate backbone. Can be single- or double-stranded. Short molecules are called *oligonucleotides*. syn. *polynucleotide*.

nucleolar organizer Region in some chromosomes, recognized as a constriction, at which a nucleolus forms.

nucleolar-organizing region Chromosomal region containing ribosomal genes; there is often a visible constriction at this site on metaphase chromosomes, and the region can be differentially stained. Abbrev. *NOR*.

nucleolus Round body occurring within a cell nucleus, consisting of ribosomal genes and associated polymerases, nascent RNA transcripts and proteins involved in ribosome assembly. adj. *nucleolar*.

nucleoplasm The protoplasm in the nucleus surrounding the chromatin. Cf. **cytoplasm**.

nucleoplasmic ratio The ratio between the volume of the nucleus and of the cytoplasm in any given cell.

nucleoside A desoxyribose or ribose sugar molecule to which a purine or pyrimidine is covalently bound. See **nucleotide**.

nucleosome Chromosome of eukaryotes have a complex 'quaternary' structure, consisting of chromatin beads into which about 145 base pairs of DNA are folded, each bead being separated by less folded chromatin.

nucleotide A **nucleoside** to which a phosphate group is attached at the 5′ position on the sugar. The individual components of a **nucleic acid**.

nucleus The *cell nucleus*. A compartment within the interphase **eukaryotic** cell bounded by a double membrane and containing the genomic DNA, with its associated functions of **transcription** and processing. Additionally in animals (1) any nut-shaped structure; (2) a nerve centre in the brain; (3) a collection of nerve cells on the course of a nerve or tract of nerve fibres.

nudicaudate Having the tail uncovered by fur or hair, as Rats.

nullisomic Cell or organism having one particular chromosome of the normal complement not represented at all in an otherwise diploid (or, more generally, euploid) complement. See **aneuploidy**, of which nullisomy is one sort.

numbers, pyramid of The relative decrease in numbers at each stage in a food chain, characteristic of animal communities.

numerical taxonomy A series of methods, based on the analysis of numerical data, for generating a classification (often in the form of dendrograms) of a group or groups of organisms. See **operational taxonomic units**.

nu nu (nude) Applies to mice or rats with congenital absence of the thymus, and in which the blood and the thymus dependent areas of lymphoid tissues are very severely depleted of T lymphocytes. These animals are homozygous for the gene 'nude', hence *nu nu*, and have no hair. (Note that other hairless strains exist which have normal thymuses.) Such animals are unable to mount thymus-dependent immune responses but can make normal antibody responses to a wide variety of antigens which are 'thymus independent'.

nuptial flight The flight of a virgin Queen Bee, during which she is followed by a number of males, copulation and fertilization taking place in mid air.

nurse cells Cells surrounding, or attached to, an ovum, probably to perform a nutritive function.

nut A hard, dry, indehiscent fruit formed from a syncarpous gynaeceum, and usually containing one seed. The term is used loosely for any fairly large to large hard, dry, one-seeded fruit.

nutation Autonomic swaying movement of e.g. a growing shoot tip, esp. *circumnutation*.

nutlet A small, one-seeded portion of a fruit which divides up as it matures. See **schizocarp**.

nutrient Conveying, serving as, or providing nourishment; nourishing food.

nutrient film technique Method of growing plants without soil, their roots in a gutter-like channel with a nutrient solution trickling over them; used commercially for growing, e.g. tomatoes in greenhouses.

nutrient solution An artificially prepared solution containing some or all of the mineral substances used by a plant in its nutrition. Also *culture medium*.

nutrition The process of feeding and the subsequent digestion and assimilation of food material. adj. *nutritive*.

nyctanthous Flowering at night.

nyctinastic movement, nyctinasty *Sleep movement*. A **nastic movement** in which plant parts, esp. flowers and leaves, take up different position by night and day.

nyctipelagic Found in the surface waters of the sea at night only.

nymph In Acarina, the immature stage intervening between the period of acquisition of four pairs of legs and the attainment of full maturity; in Insecta, a young stage of Exopterygota intervening between the egg and the adult, and differing from the latter only in the rudimentary condition of the wings and genitalia.

NZB, NZW mice Inbred strains of mice which develop spontaneous auto-immune diseases, including anaemia, glomerulonephritis and a condition resembling systemic lupus erythematosus. There is evidence that there is an underlying viral aetiology.

O

O antigen The somatic antigen of Gram-negative bacteria, e.g. genera *Escherichia, Shigella, Salmonella* etc. Species specific cell wall antigen with an outer side chain of repeating units of oligosaccharides which confers strain specificity. The polysaccharide is internally linked to lipid A forming **lipopolysaccharide**.

ob- Prefix meaning *reversed, turned about.* Thus *obclavate* is reversed *clavate*, i.e. attached by the broad and not the narrow end.

obconic, obconical Cone-shaped but attached by the point.

obdiplostemonous Having stamens in two whorls, those in the outer opposite to the petals, e.g. *Geranium*. Cf. **diplostemonous**.

obesity An excessive proportion of fat on the body; cultural norms vary. It is produced by a large range of factors including faulty metabolism and abnormal behaviour.

object permanence, constancy Knowledge of the continued existence of an object, even when the object is not accessible to direct sensory awareness. According to Piaget this ability does not develop until infants are approximately 8 months or more.

oblate Globose, but noticeably wider than long.

obligate Obliged to function in the way specified; e.g. an obligate anaerobe cannot grow (and may not survive) in the presence of free oxygen and an obligate parasite cannot live outside its host. Cf. **facultative**.

obligate parasite A parasite capable of living naturally only as a parasite. Cf. **facultative parasite**. A more useful distinction is between **biotrophs** and **necrotrophs**.

obligate saprophyte An organism which lives on dead organic material and cannot attack a living host.

obliquus An asymmetrical or obliquely placed muscle.

oblongata See **medulla oblongata**.

obovate Having the general shape of the longitudinal section of an egg, not exceeding twice as long as broad, and with the greatest width slightly above the middle; hence, attached by the narrower end.

obovoid Solid, egg-shaped and attached by the narrower end.

observational learning Learning through the observation of another individual (model) which is accomplished without practice or direct experience.

obsession The morbid persistence of an idea in the mind, against the wish of the obsessed person.

obsessive-compulsive disorder A disorder whose main symptoms are *obsessions* and **compulsions**.

obturator Any structure which closes off a cavity, e.g. all the structures which close the large oval foramen formed by the ischiopubic fenestra; the foramen itself.

obtuse Of a leaf etc. having a blunt tip with the sides forming an angle of more than 90°. Cf. **acute**.

obvolvent Folded downwards and inwards, as the wings in some Insects.

occipital condyle In Insects, a projection from the posterior margin of the head which articulates with one of the lateral cervical sclerites. In Craniata, one or two projections from the skull which articulate with the first vertebra.

occipitalia A set of cartilage bones forming the posterior part of the brain case in the Vertebrate skull.

occiput The occipital region of the Vertebrate skull forming the back of the head. adj. *occipital.*

occlusion Closure of a duct or aperture, e.g. the upper and lower teeth of a Vertebrate.

occlusor A muscle which by its contraction closes an operculum or other movable lidlike structure.

ocellus A simple eye or eyespot in Invertebrates; an eye-shaped spot or blotch of colour. adj. *ocellate.*

ochrea, ocrea A sheath around the base of an internode formed from united stipules or leaf bases, e.g. dock (Rumex) and other Polygonaceae.

ochroleucous Yellowish-white.

ochrophore See **xanthophore**.

ochrosporous Having yellow or yellow-brown spores.

octopine An **opine**.

octopod Having 8 feet, arms or tentacles.

ocular Pertaining to the eye; capable of being perceived by the eyes.

ocular micrometer US for *eyepiece graticule.*

oculate Possessing eyes; having markings which resemble eyes.

oculomotor Pertaining to, or causing movements of, the eye; the third cranial nerve of Vertebrates, running to some of the muscles of the eyeball.

odds The ratio of the probability that an event occurs to the probability that it does not occur.

odds ratio The ratio of two odds, used particularly in comparing and modelling conditional probabilities.

Odonata Order of primitive insects with 2 pairs of similar membranous, many-veined wings. Large-eyed diurnal forms with both adults and larvae predatious. Dragonflies, Damselflies.

odontoblast A dentine-forming cell, one of the columnar cells lining the pulp cavity of a tooth.

odontoclast A dentine-destroying cell, one of the large multinucleate cells which absorb the roots of the milk teeth in Mammals.

odontogeny The origin and development of teeth.

odontoid Toothlike.

odontoid process A process of the anterior face of the centrum of the axis vertebra which forms a pivot on which the atlas vertebra can turn.

odontophore In Molluscs, a feeding organ comprising the radula and radula sac, with muscles and cartilages.

odontostomatous Having jaws which bear teeth.

oedema Pathological accumulation of fluid in the tissue spaces and serous sacs of the body; sometimes the term is restricted to such accumulation in tissue spaces only; *pulmonary oedema*, fluid in the lung; *sacral oedema*, fluid at base of spine.

oedematous Affected by oedema.

Oedipus and **Electra complexes** In psychoanalytic theory, refers to unconscious conflicts, occurring during the **phallic stage** of psychosexual development, that centre around the relations a child forms with his parents. A fantasized form of sexual love for the opposite sex develops and a resentment of the same sex parent. In boys, these possessive feelings and the associated conflicts of guilt, are called the **Oedipus complex**; in girls, the **Electra complex**.

oesophagus, esophagus In Vertebrates, the section of the alimentary canal leading from the pharynx to the stomach; usually lacking a serous coat and digestive glands; the corresponding portion of the alimentary canal in Invertebrates. adj. *oesophageal*.

oestrous cycle In female Mammals, the succession of changes in the genitalia commencing with one oestrous period and finishing with the next.

oestrus, oestrum In female Mammals, the period of sexual excitement and acceptance of the male occurring between pro-oestrum and metoestrum; more generally, the period of sexual excitement. adj. *oestral*.

official, officinal Of plants, used in medicine.

offset Short shoot arising from an axillary bud near the base of a shoot and producing a daughter plant at its apex in e.g. the houseleek, *Sempervivum*. Also, a bulbil or cormlet formed near base of parent bulb or corm.

-oid Suffix. after Gk. *oides*, from *eidos*, form.

oil gland The preen gland or uropygial gland of Birds, a cutaneous gland forming an oil secretion used in preening the feathers.

oil-immersion objective The use of a thin film of oil, with the same refractive index as glass, between the objective of a microscope and the specimen. This permits the numerical aperture to exceed 1 and thus extends the resolving power and gives the maximum magnification obtainable with a light microscope.

olecranon In land Vertebrates, a process at the upper end of the ulna which forms the point of the elbow.

olfactory Pertaining to the sense of smell; the first cranial nerve of Vertebrates, running to the olfactory organ.

olfactory lobes Part of the forebrain in Vertebrates, which is concerned with the sense of smell, and from which the olfactory nerves originate.

oligaemia, oligemia Diminution in the volume of the blood.

Oligochaeta A class of Annelida with relatively few chaetae, not situated on parapodia; have a definite prostomium which usually has no appendages; always hermaphrodite; have only 1 or 2 pairs of male and female gonads in fixed segments of the anterior part of the body; reproduction involves copulation and cross fertilization; the eggs are laid in a cocoon and develop directly; terrestrial and aquatic. Earthworms.

oligodendroglia Cells within the CNS which deposit the myelin sheath. Also *oligodendrocyte*.

oligolecithal A type of egg in which there is little yolk, what there is being somewhat more concentrated in one hemisphere. Found in *Amphioxus* (a genus of Cephalochordata) and Mammals. Cf. **mesolecithal, telolecithal**.

olig-, oligo- Prefixes from Gk. *oligos*, a few, small.

oligomerous Consisting of only few parts.

oligonucleotide A nucleic acid with few nucleotides. Cf. **polynucleotide**.

oligopeptides Short polymers of amino acids, less than about 100 peptides long. May be synthesized, but also occur naturally with often powerful biological effects.

oligopod (1) Having few legs or feet. (2) Said of a phase in the development of larval Insects in which the thoracic limbs are large while the evanescent abdominal appendages of the polypod phase have disappeared.

oligotokous Bearing few offspring. Cf. **polytokous**.

oligotrophic Said of lakes, mires or soils which are relatively poor in available plant mineral nutrients. Cf. **eutrophic**.

oligotrophophyte A plant growing in a soil poor in soluble mineral salts.

oliphagous Feeding on few different kinds of food; as phytophagous Insects which are limited to a few related food plants. Cf.

polyphagous.

olivary nucleus A wavy band of grey matter within the medulla oblongata of higher Vertebrates.

omasum See **psalterium**.

ombrogenous Said of a mire or bog receiving water only as precipitation and hence is extremely oligotrophic. Cf. **soligenous**.

ombrophile A plant adapted to rainy places. Also *ombrophyte*.

omental bursa In some Mammals, a pouch formed ventrally and dorsally to the stomach by the mesentery supporting the stomach.

omentum In Vertebrates, a portion of the serosa connecting two or more folds of the alimentary canal. adj. *omental*.

ommatidium One of the visual elements composing the compound eyes of *Arthropoda*.

ommatophore An eyestalk.

omnivore An animal which eats both plants and animals. adj. *omnivorous*.

omnivorous Said of a parasitic fungus which attacks several hosts.

omphalic Pertaining to the umbilicus.

omphaloid Navel-shaped.

oncogene Genetic locus originally identified in RNA tumour viruses which is capable of the **transformation** of the host cell. Implicated as cause of certain cancers. See **proto-oncogene**.

oncogenic virus Generally a virus able to cause cancer, but more specifically one carrying an **oncogene**.

ontogeny, ontogenesis The history of the development of an individual. Cf. **phylogeny**. adj. *ontogenetic*.

onychogenic Nail-forming, nail-producing, as a substance similar to eleidin, occurring in the superficial cells of the nail bed.

onych-, onycho- Prefixes from Gk. *onyx*, gen. *onychos*, a nail or claw.

Onychophora A subphylum of Arthropoda having trachea; a soft thin cuticle; body wall consisting of layers of circular and longitudinal muscles; head not marked off from the body, and consisting of 3 segments, 1 pre-oral, bearing pre-antennae and 2 postoral, bearing jaws and oral papillae respectively; a pair of simple vesicular eyes; all body segments similar, each bearing a pair of parapodia-like limbs which end in claws and containing a pair of excretory tubules; spiracles scattered irregularly over the body; cilia in the genital region; development direct, e.g. the genus *Peripatus*.

oö- Prefix from Gk. *oon*, egg.

oöblastema A fertilized egg.

oöcium A brood pouch.

oöcyst In certain *Protozoa*, the cyst formed around two conjugating gametes; in *Sporozoa*, the passive phase into which an oökinete changes in the host.

oöcyte An ovum prior to the formation of the first polar body; a female gametocyte.

oögamy (1) The union of gametes of dissimilar size, usually of a relatively large non-motile egg and a small active sperm. (2) In *Protozoa*, **anisogamy** in which the female gamete is a hologamete. Cf. **isogamy**.

oögenesis The origin and development of ova.

oögonium An egg-mother-cell or oöcyte. In many algae and the Oömycetes, a unicellular female gametangium containing one or more eggs or oöspheres.

oölemma See **vitelline membrane**.

oölogy The study of ova.

Oomycetes A group of fungus-like, non-photosynthetic organisms, having a non-septate mycelium, reproducing asexually by zoospores or dispersed sporangia that may germinate to give zoospores or a hypha, and sexually by oöspores. They have heterokont flagellation and mitochondrial microvilli suggesting affinity with the Heterokontophyta. Include some 'water fungi' and many plant parasites, e.g. *Pythium* (*damping-off*), the 'downy mildews' and the historically very important potato blight (*Phytophthora infestans*).

oösperm See **oöblastema**.

oöspore A fertilized ovum; in *Protozoa*, an encysted zygote; a thick walled zygote with food reserves formed from fertilized oösphere in some algae and the Oomycetes.

oötheca An egg case, as in the Cockroach.

oötocoid Bringing forth the young in an immature condition and allowing them to complete their early development in a marsupium.

oötocous Oviparous.

open aestivation Aestivation in which the leaves or perianth parts neither overlap nor meet by their edges.

open community A plant community which does not occupy the ground completely, so that bare spaces are visible.

open-field test An experimental procedure in which an animal is released into an open area with no obstacles, and features of its behaviour, such as defaecation, urination and locomotion are observed, these sometimes being related to emotionality.

open vascular bundle A bundle including cambium.

operant chain A complex chain of behaviour built up by the operant technique of **shaping**.

operant conditioning A learning procedure in which the reinforcement follows a particular response on a proportion of occasions. See **operant response, reinforcement**.

operant response A response which acts

---

operculum

A lid or cover, composed of part of a cell wall or, of from one to many cells, which opens to allow the escape of contents from some sort of container. Esp. in *plants*, the lid of an antheridium, a moss capsule, an ascus or other sporangium or the germ pore of á pollen grain.

In the eggs of some *Insects*, a differentiated area of the chorion which lifts up when the larva emerges from the egg. In some tubicolous Polychaeta, an enlarged branch of a tentacle closing the mouth of the tube when the animal is retracted; in Spiders, a small plate partially covering the opening of a lung book; in some Cirripedia, plates of the carapace which can be closed over the retracted thorax; in some Gastropoda, a plate of chitinoid material, strengthened by calcareous deposits, which fits across the opening of the shell.

In the higher *Fish*, a fold which articulates with the hyoid arch in front of the first gill slit, and extends backwards, covering the branchial clefts. A similar structure occurs in the larvae of Amphibians.

---

on the environment to produce an event which affects the subsequent probability of that response.

operational taxonomic units The entities of any taxonomic rank, such as those individuals, species, genera etc. whose relationships are studied in **numerical taxonomy**. Abbrev. *OTU*.

operator Sequence of DNA to which a **repressor** or **activator** can bind. Situated before the coding sequence of a gene and close to the **promoter**.

opercular apparatus In Fish, the operculum, together with the branchiostegal membrane and rays.

operculate (1) Possessing a lid. (2) Opening by means of a lid.

operculum See panel.

operon In bacteria the set of functionally related genes, which have a common promoter and mRNA, thus securing co-ordinated transcription.

Ophiuroidea A class of Echinodermata with a dorsoventrally flattened star-shaped body; arms sharply differentiated from the disk and not containing caecae of the alimentary canal; tube feet lack ampullae and suckers, and lie on the lower surface, although not in grooves; no anus; madreporite aboral; well-developed skeleton; no pedicellariae; free-living. Brittle stars.

ophthalmic Pertaining to or situated near the eye, as the *ophthalmic nerve*, which passes along the back of the orbit in lower Vertebrates.

Opiliones A subclass of Arachnida with rounded bodies, the prosoma and opistosoma broadly jointed and usually with long legs. Mostly predaceous. Harvestmen.

opine Guanidoamino acids (either octopine or nopaline) synthesized and released by plant cells after infection with a Ti plasmid

and used by *Agrobacterium tumefasciens* as carbon and nitrogen source. See **crown gall**.

opisthocoelous Concave posteriorly and convex anteriorly; said of vertebral centra.

opisthoglossal Having the tongue attached anteriorly, free posteriorly, as in Frogs.

opisthomere A postoral somite.

opisthosoma In Chelicerata the segments posterior to those bearing the legs; the abdomen. Cf. **prosoma**.

opportunistic infection An infection to which normal people are resistant or from which they recover quickly, but which occurs in those whose immune system has been compromised by illness. Such bacterial, fungal, protozoal and viral infections occur in patients with terminal cancer or with AIDS, and may be the immediate cause of death.

opportunistic species A species adapted to colonize temporary or local conditions.

opposite Said of two organs, esp. leaves, which arise at the same level but on opposite sides of a stem. Cf. **alternate**, **decussate**, **whorled**. See **phyllotaxis**.

opsonin Factors present in blood and other body fluids which bind to particles and increase their susceptibility to phagocytosis. They may be antibody, or products of complement activation (esp. C3b), or some other substances which bind to particles such as fibronectin.

optic Pertaining to the sense of sight; the second cranial nerve of Vertebrates.

optic lobes In Vertebrates, part of the mid brain, which is concerned with the sense of sight, and from which the optic nerves originate.

optimal proportions Describes the relative proportions of antibody and a soluble antigen which when mixed together produce the maximum degree of cross linkage, such that all the antibody and all the antigen are included in the precipitate which forms. See **lattice hypothesis**.

oral Pertaining to the mouth.

oral characters In psychoanalytic theory, refers to fixation at, or regression to, the oral stage of development; Freud considered that many oral habits reflected this (e.g. smoking).

oral contraception The use of synthetic hormones (oestrogen and progestogen steroids in varying proportions), taken orally in pill form, to prevent conception by reacting on the natural **luteinizing** and **follicle-stimulating hormones** and so inhibiting ovulation and/or fertilization. Colloq. *the pill*.

oral stage In psychoanalytic theory, the first stage of psychosexual development in which stimulation of the mouth and lips is the primary focus of bodily (libidinal) pleasure; occurs in the first year of life.

ora serrata The edge of the retina.

orbicular Flat, with a circular or almost circular outline.

orbiculares Muscles which surround an aperture; as the muscles which close the lips and eyelids in Mammals.

orbit A space lodging an eye; in Vertebrates, the depression in the skull containing the eye; in Arthropoda, the hollow which receives the eye or the base of the eyestalk; in Birds, the skin surrounding the eye.

orbitosphenoid A paired cartilage bone of the Vertebrate skull, forming the side wall of the brain case in the region of the presphenoid.

orchic, orchitic Pertaining to the testis.

Orchidaceae The orchid family, ca 18 000 spp. of monocotyledonous flowering plants (superorder Liliidae). The largest monocot family. Herbs, terrestrial and epiphytic; cosmopolitan. Some are CAM plants. The flowers are zygomorphic, usually showy. The seeds are minute and early stages of growth depend on symbiosis with a mycorrhizal fungus. Of no economic importance except for vanillin (the seed pod of *Vanilla*) and as ornamental and florists' plants.

order Taxonomic rank below **class** and above **family**; for plants, the names usually end in -oles.

ordination A family of multi-variate statistical techniques commonly used to plot ecological data sets collected from a large number of sites, on geometric axes so that similarity is represented by proximity.

organ A part of the body of an animal or a plant adapted and specialized for the performance of a particular function.

organ culture See **plant cell culture**.

organelle A defined structure within a cell,

e.g. nucleus, mitochondrion, lysosome.

organ genus See **form taxon**.

organic mental (brain) disorders Behavioural disorders stemming demonstrably from damage to the brain tissue or to chemical imbalances in the nervous system.

organisms Animals, plants, fungi and micro-organisms.

organized Showing the characteristics of an organism; having the tissues and organs formed into a unified whole.

organogeny, organogenesis The study of the formation and development of organs.

organography A descriptive study of the external form of plants, with relation to function.

orgasm Culmination of sexual excitement. adj. *orgastic*.

oriental region One of the primary faunal regions into which the land surface of the globe is divided. It includes the southern coast of Asia east of the Persian Gulf, the Indian subcontinent south of the Himalayas, southern China and Malaysia, and the islands of the Malay Archipelago north and west of Wallace's line.

orientation The position, or change of position, of a part or organ with relation to the whole; change of position of an organism under stimulus.

orientation behaviour The positioning of the body or of a behavioural sequence, with respect to some aspect of the external environment; it includes simple postural preferences as well as complex navigational behaviours.

orienting reflex First described by Pavlov, an animal's response to the sudden presentation of a novel stimulus. Includes turning the body and head so that the animal's attention can be focused on the source of stimulation (the 'what is it?' reaction).

origin That end of a skeletal muscle which is attached to a portion of the skeleton which remains, or is held, rigid when the muscle contracts, the other end, which is attached to a part of the skeleton which moves as a result of the contraction, being known as the *insertion*.

ornis A Bird fauna. adj. *ornithic*.

ornithine *2-6-Diaminovaleric acid*. It is concerned in urea formation in animals (see **arginine**), and a derivative, *ornithuric acid*, is found in the excrement of Birds.

ornithology The study of Birds.

ornithophily Pollination by Birds.

oro- Prefixes from (1) L. *os*, gen. *oris*, mouth or (2) Gk. *oros*, mountain.

oro-anal Connecting, pertaining to, or serving as, mouth and anus.

oronasal Pertaining to or connecting the mouth and the nose.

ortho- Prefix from Gk. *orthos*, straight.

orthognathous With the long axis of the head at right angles to that of the body, and the mouth directed downwards. Cf. **prognathous**.

Orthoptera An order of the *Insecta*. Large Insects with biting mouthparts; posterior legs often with enlarged femora for jumping; forewings toughened (tegmina) and overlapping when folded; unjointed cerci; well developed ovipositor; possess a variety of stridulatory organs. Grasshoppers, Locusts, Crickets, Cockroaches.

orthotropism A *tropism* in which a plant part becomes aligned directly towards (positive-) or away from (negative-) the source of the orientating stimulus, such as most seedling shoots which are negatively *orthogravitropic* and positively *orthophototropic*. Cf. **diatropism, plagiotropism**.

orthotropous *Atropous*. Said of an ovule which is straight and on a straight stalk, so that the micropyle points away from the stalk.

os (1) An opening, as the *os uteri*. (L. *os*, gen. *oris*, mouth.) pl. *ora*. (2) A bone, as the *os coccygis*. (L. *os*, gen. *ossis*, bone.) pl. *ossa*.

osculum In Porifera, an exhalant aperture by which water escapes from the canal system. adjs. *oscular, osculiferous*.

osmeterium In the larvae of certain Papilionidae (Lepidoptera), a bifurcate sac exhaling a disagreeable odour which can be protruded through a slitlike aperture in the first thoracic segment.

osmole The amount of a solute that when dissolved in water gives a solution of the same osmotic pressure as that expected from one mole of an ideal non-ionized solute. The total osmotic concentration or *osmolarity* of complex solutions is usually estimated by measuring the vapour pressure or the freezing point depression of the solution. Ordinary sea water is approximately 1000 milliosmolar (1000 osmol m^{-3}), mammalian isotonic saline is about 290 milliosmolar. *Osmolality* (abbrev. *Osm*) expresses equivalent figures per kilogram of solvent.

osmoreceptors Cells specialized to react to osmotic changes in their environment, e.g. cells which react to osmotic changes in the blood or tissue fluid and which are involved in the regulation of secretion of antidiuretic hormone by the neurohypophysis.

osmoregulation The process by which animals regulate the amount of water in their bodies, and the concentration of various solutes and ions in their body fluids.

osmosis Diffusion of a solvent through a semipermeable membrane into a more concentrated solution, tending to equalize the concentrations on both sides of the membrane.

osmotic potential ψ_π. That component of the **water potential** due to the presence of solutes; equal to minus the osmotic pressure. Also called solute potential, ψ_s.

osmotic pressure π. The hydrostatic pressure which would have to be applied to the solution in order to make the **chemical potential** of the water in the solution equal to that of pure, free water at the same temperature, or to prevent osmotic water movement through a semipermeable membrane between the solution and pure water.

osmotrophy Nutrition based on the uptake of soluble materials.

os penis A bone developed in the middle line of the penis in some Mammals, as Bats, Whales, some Rodents, Carnivores and Primates.

osphradium A sense organ of certain aquatic Mollusca, consisting usually of a patch of columnar ciliated epithelium and concerned in the assessment of suspended silt in the water entering the mantle chamber. adj. *osphradial*.

ossa See os (2).

osseous Bony; resembling bone.

ossicle A small bone; in Echinodermata, one of the skeletal plates; in Crustacea, one of the calcified toothed plates of the gastric mill.

ossification The formation of bone; transformation of cartilage or mesenchymatous tissue into bone. v. *ossify*.

osteoblast A bone-forming cell.

osteoclast A bone-destroying cell, esp. one which breaks down any preceding matrix, chondrified or calcified during bone formation.

osteocranium The bony brain case which replaces the chondrocranium in higher Vertebrates.

osteocyte A bone cell derived from an osteoblast.

osteodermis An ossified or partially ossified dermis; membrane bones formed by ossification of the dermis. adj. *osteodermal*.

osteogenesis See ossification.

osteology The study of bones.

osteoporosis Decrease in bone density and mass, often occurring in old age.

osteosclereid Sclereid having a columnar middle and enlarged ends like a stylized thigh bone.

oste-, osteo- Prefixes from Gk. *osteon*, bone.

ostiolate Having an opening.

ostiole A pore, esp. one by which spores or gametes escape.

ostium A mouthlike aperture; in Porifera, an inhalant opening on the surface; in Arthropoda, an aperture in the wall of the heart by which blood enters the heart from the pericardial cavity; in Mammals, the internal

oxidative phosphorylation

Glucose is oxidized to carbon dioxide and water with the concomitant synthesis of ATP by **aerobic respiration**. The energy for the synthesis of nearly all the ATP is derived from a series of oxidations which occur, when the electrons liberated from glucose during respiration, pass down the **electron transfer chain** to be accepted ultimately by oxygen, reducing it to water. A stoichiometric ratio of 3 moles of ATP per mole of oxygen consumed was established during the 1950s but the search for the metabolic intermediates was fruitless until P. Mitchell made the then revolutionary suggestion that there were no intermediates in the normally accepted sense and that the synthesis depended upon the integrity of the *inner mitochondrial membrane* and on the disposition of its components across it by a process which he termed *chemiosmosis*.

During chemiosmosis the energy derived from the sequential oxidation of the components of the electron transfer chain is released in three packets which energize proton pumps driving protons across the inner membrane from the mitochondrial matrix. The resulting pH differential and membrane potential generate a 'proton motive force' across the inner mitochondrial membrane which causes a flow of protons back into the matrix through the proton translocating ATPase of the membrane where the ATP is synthesized. Respiratory poisons, such as dinitrophenol, which 'uncouple' the electron flow from ATP synthesis exert their effect by themselves carrying protons through the membrane, thus short circuiting the ATP generating mechanism.

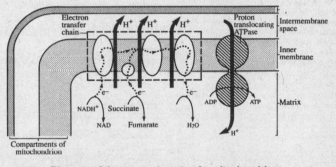

Diagram of the compartments of a mitochondrion.

H^+ = hydrogen ions, e^- = electrons, ATP and ADP = adenine triposphate and diphosphate, $NADH^+$ = oxidized nicotinamide adenine dinucleotide phosphate, NAD = nicotinamide adenine dinucleotide.

aperture of a Fallopian tube. adj. *ostiate*.

Ostracoda A subclass of the Crustacea, with or without compound eyes; having a bivalve shell with adductor muscle; cephalic appendages well-developed and complex; not more than 2 pairs of trunk limbs, often parthenogenetic, e.g. *Cypris*.

otocyst In many aquatic Invertebrates, a sac lined by sensory hairlets, filled with fluid, and containing a calcareous or siliceous concretion (*otolith*) which subserves the equilibristic sense; in Vertebrates, part of the

internal ear which is similarly constructed (as the utriculus).

otolith The calcareous concretion which occurs in an otocyst.

ot-, oto- Prefixes from Gk. *ous*, gen. *otos*, ear.

OTU See **operational taxonomic units**.

Ouchterlony test A precipitin test in which antigen and antibody are allowed to diffuse towards one another in a gel medium.

Oudin test A precipitin test in which antigen diffuses into antibody incorporated in a gel medium.

outbreeding Sexual reproduction occurring between unrelated individuals, thus increasing heterozygosity. Cf. **inbreeding**. See **allogamy**.

outcross A cross to a strain with a different genotype.

outgroup See **ingroup/outgroup**.

ova Pl. of *ovum*.

oval window See **fenestra ovalis**.

ovariole In Insects, one of the egg tubes of which the ovary is composed.

ovary A female gonad; a reproductive gland producing ova. In plants, the hollow structure, the basal part of a carpel or of a syncarpous gynoecium, which contains the ovules. Also *pistil*. adj. *ovarian*.

ovate Egg-shaped with the broadest part nearer to the point of attachment.

overlapping genes Some small DNA viruses exploit the degeneracy of the genetic code by making different proteins from overlapping sequences of DNA. Achieved by displacing the **reading frame** by one or two bases.

overlearning A learning procedure where training or practice on what is being learned continues beyond the point where learning can be said to be adequate (learning to criterion). Overlearning often results in improvements in efficiency and in changes in the organization of performance (e.g. from conscious to *automatic* control).

oviduct The tube which leads from the ovary to the exterior and by which the ova

are discharged. adj. *oviducal*.

oviferous, ovigerous Used to carry eggs, as the *ovigerous* legs of Pycnogonida.

oviparous Egg-laying. Cf. **viviparous**.

oviposition The act of depositing eggs.

ovipositor In some Fish (as the Bitterling), a flexible tube formed by the extension of the edges of the genital aperture in the female; in female Insects, the egg-laying organ.

ovisac A brood pouch; an egg receptacle.

ovotestis A genital gland which produces both ova and spermatozoa, as the gonad of some Gastropoda.

ovoviviparous Producing eggs which hatch out within the uterus of the mother.

ovulation The formation of ova; in Mammals, the process of escape of the ovum from the ovary.

ovule The structure in a seed plant, consisting of embryo sac, nucellus and integuments, which after fertilization develops into the seed.

ovule culture See **plant cell culture**.

ovum A non-motile female gamete. An egg or egg cell.

oxidase One of a group of enzymes occurring in plant and animal cells and promoting oxidation.

oxidative phosphorylation See panel on previous page.

oxydactylous Having narrow-pointed digits.

oxyntic Acid-secreting.

oxyphobic Unable to withstand soil acidity.

P

p See **chromosome mapping**. Also symbol for pico-.

P-680 A (chlorophyll a)-protein complex (absorption peak at $\lambda = 684$ nm) that acts as the light trap in **photosystem II**.

P-700 A (chlorophyll a)-protein complex (absorption peak at $\lambda = 700$ nm) that acts as the light trap in **photosystem I**.

pachydermatous Thick-skinned.

pachyphyllous Having thick leaves.

pachytene The third stage (*bouquet stage*) of meiotic prophase, intervening between zygotene and diplotene, in which condensation of chromosomes commences.

Pacinian corpuscles In Vertebrates, skin receptors in which the nerve ending is surrounded by many concentric layers of connective tissue. Sensitive to pressure. Also *Vater's corpuscles*.

packing ratio of DNA The ratio of the calculated length of a helical DNA molecule to its length after organization into more compact form in chromosomes as **nucleosomes** and higher order coiling or folding. The packing ratio of metaphase chromosome DNA is about 10 000 to 1.

paedogenesis Sexual reproduction by larval or immature forms.

paedomorphosis See neoteny.

paedophilia Committing sexual offences against children; sexual gratification through sexual activity with children.

PAF Abbrev. for *Platelet Activating Factor*.

PAGE See **polyacrylamide gel electrophoresis**.

paired-associate learning A procedure in which a list of pairs is presented in which one item serves as stimulus and the other as response; the subject must learn to respond with the second item when the first item of a pair is presented.

pairing The process by which *homologous chromosomes* are brought together during meiosis preparatory to being distributed one to each gamete.

PAL Abbrev. for *Phenylalanine Ammonia-Lyase*.

Palaearctic region One of the subrealms into which the Holarctic region is divided; it includes Europe and northern Asia, together with Africa north of the Sahara.

palaeo-ecology The study of fossil organisms in terms of their mode of life, their interrelationships, their environment, their manner of death and their eventual burial.

palae-, palaeo-, pale-, paleo- Prefixes from Gk. *palaios*, ancient.

palama The webbing of the feet in Birds of aquatic habit.

palate In Vertebrates, the roof of the mouth; in Insects, the epipharynx. adjs. *palatal, palatine*.

palatine Pertaining to the palate; a paired membrane bone of the Vertebrate skull which forms part of the roof of the mouth.

palea, pale, palet *Valvule*. The usually thin and membranous, upper or inner of the two bracts (lemma and palea) which enclose a grass floret. Sometimes, synonymous with glume.

paleogenetic Originating in the past.

palingenesis The reproduction of truly ancestral characters during ontogeny. adj. *palingenetic*.

palisade Chlorenchyma in which the cells are elongated at right angles to the surface of the organ. Palisade mesophyll is characteristically present towards the upper surface of dorsiventral leaves of mesophytic dicotyledons. Palisade layers also occur in the outer cortex of many photosynthetic stems. See **spongy mesophyll**.

pallescent Becoming lighter in colour with age.

palliative Affording temporary relief from pain or discomfort; a medicinal remedy which does this.

palli-, pallio- Prefixes from L. *pallium*, mantle.

pallium The mantle in Brachiopoda or Mollusca, a fold of integument which secretes the shell. In the Vertebrate brain, that part of the wall of the cerebral hemispheres excluding the corpus striatum and rhinencephalon. adjs. *pallial, palliate*.

Palmae *Arecaceae*. The palm family, ca 2800 spp. of monocotyledonous flowering plants (superorder Arecidae). Trees, mostly tropical, typically with a single trunk (which does not undergo secondary thickening) bearing a crown of pinnate or palmate leaves. The fruit is a one-seeded berry or a drupe. The most important economic products are coconuts, palm oil, dates, copra and fibres.

palmar Pertaining to the palm of the hand.

palmate (1) Having 4 or more equal divisions, lobes or veins radiating from a common point rather in the manner of the fingers of a hand; as in a palmately compound, lobed or veined leaf respectively. (2) Having webbed feet.

palmatifid Cut about half way into lobes in a palmate fashion.

palmelloid form Condition in those algae in which non-motile cells divide within a mucilaginous matrix to give large gelatinous masses containing many cell generations.

palmisect Cut almost to the centre in a palmate fashion.

palm oil A reddish-yellow fatty mass from the fruit of *Elaeis guineensis*, mp 27°–43°C, rel.d. 0.90–0.95.

palp See **palpus**. adj. *palpal*.

palpation Physical examination by touch.

palpebra An eyelid.

palpus In Crustacea and Insects, a jointed sensory appendage associated with the mouthparts; in Polychaeta, sensory appendage of prostomium. Also *palp*.

palynology The study of fossil spores and pollen. They are very resistant to destruction and in many sedimentary rocks are the only fossils that can be used for stratigraphical correlation.

pan (1) A compact layer of soil particles, lying some distance beneath the surface, cemented together by organic material, or by iron and other compounds, and relatively impermeable to water. (2) A depression in the surface of a salt marsh, in which salt water stands for lengthy periods.

pancreas A moderately compact though somewhat amorphous structure found in Vertebrates, the larger part of which consists of exocrine glandular tissue with one or more ducts opening to the small intestine, and also containing scattered islets of endocrine tissue. The former produces six or more enzymes involved in the digestion of proteins, carbohydrates, and fats, while the latter secrete the hormones insulin and glucagon.

pancreozymin A polypeptide hormone secreted by the intestinal wall which stimulates the pancreas to secrete digestive enzymes.

pandemic Of an epidemic, occurring over a wide area such as a country or a continent; an epidemic so widespread.

pandurate, panduriform Shaped like the body of a fiddle.

pangamic Of indiscriminate mating.

panic attack An attack of intense terror and anxiety, usually lasting several minutes, though possibly continuing for hours; apprehension often persists for long periods after the panic attack.

panic disorder A **panic attack** occurring in the absence of any phobic stimulus.

panicle Strictly, a branched raceme with each branch bearing a raceme of flowers, e.g. oat. Loosely, any branched inflorescence of some degree of complexity. adj. *paniculate*.

panmixis, panmixia Random mating within a population, esp. a model system. adj. *panmictic*.

panniculus carnosus In some Mammals, an extensive system of dermal musculature covering the trunk and part of the limbs, by means of which the animal can shake itself.

pannose Felted.

pantophagous Omnivorous.

pantothenic acid See **vitamin B**.

papain A protein-digesting enzyme present in the juice from the fruits and leaves of the papaya tree (*Carica papaya*); commercially produced as a meat tenderizer.

paper chromatography Chromatography using a sheet of special grade filter paper as the adsorbent. Advantages include sensitivity to microgram quantities, the bands can be formed in two dimensions and cut out with scissors.

papilionaceous Having some likeness to a butterfly; esp., the flowers of the Papilionaceae, including the pea.

papilla In plant cell walls, a small nipple-shaped projection; a small conical projection of soft tissue, esp. on the skin or lining of the alimentary canal; the conical mass of soft tissue or pulp projecting into the base of a developing feather or tooth. adjs. *papillary, papillate*.

papillae foliatae In some Mammals, two small oval areas at the back of the tongue, marked by a series of alternating transverse ridges and richly provided with taste buds.

pappus Modified calyx in the Compositae, consisting of a ring of feathery hairs or the like around the top of the fruit, as in the dandelion, where it aids in wind dispersal.

papulae The dermal gills of Echinodermata, small finger-shaped, thin-walled respiratory projections of the body wall.

PAR Abbrev. for *Photosynthetically Active Radiation*.

para- Prefix from Gk. *para*, beside.

paracentesis *Tapping*. The puncture of body cavities with a hollow needle, for the removal of inflammatory or other fluids. See **amniocentesis**.

paradoxical sleep Stage of sleep when dreaming is assumed to occur. It is characterized by rapid eye movements (REM), loss of muscle tone, and an *electroencephalogram* very similar to the waking state (hence paradoxical). Also *REM sleep*.

paraeiopod In Crustacea, a walking leg.

paraesthesia An abnormal sensation, such as tingling, tickling and formication.

paraganglia In higher Vertebrates, small glandular bodies, occurring in the posterior part of the abdomen, which show a chromaphil reaction and are believed to secrete adrenaline.

paragnathous Animals having jaws of equal length; as Birds which have upper and lower beak of equal length.

paralalia A form of speech disturbance, particularly that in which a different sound or syllable is produced from the one which is intended.

paralimnion The zone of a lake floor between the water's edge or shoreline, and the lakeward margin of rooted vegetation.

parallel descent The appearance of similar characteristics in groups of animals or plants which are not directly related in evolutionary descent. Also *parallel evolution*, *parallelism*.

parallelism See **parallel descent**.

parallelodromous Leaves having parallel veins.

paramere Half of a bilaterally symmetrical structure; one of the inner pair of gonapophyses in a male Insect.

parameter The population value of a particular characteristic descriptive of the distribution of a random variable.

paramorph A general term for any taxonomic variant within a species, usually used when more accurate definition is not possible.

paramylon, paramylum Reserve polysaccharide, a linear β,1→3 glucan present as highly refractive solid bodies in the cytoplasm of the Euglenophyceae, Xanthophyceae and Haptophyceae.

paranephric Situated beside the kidney.

paranephros See **suprarenal body**.

paranoia A delusion of grievance beyond all bounds of reality. Occurs in a variety of mental diseases including **schizophrenia**.

paranoid disorder A personality disorder characterized by extreme suspiciousness in all situations, and with almost all people; delusions of persecution or grandeur may occur, but without the serious disorganization associated with schizophrenia.

paranoid schizophrenic One of Bleuter's four subtypes of schizophrenia; characterized by delusions of persecution or grandness, with hallucinations and a loss of contact with reality.

paraphasia A defect of speech in which words are misplaced and wrong words substituted for right ones; due to a lesion in the brain.

paraphilias Sexual patterns in which arousal is caused by something other than what is considered a normal sexual object or activity.

paraphyletic group, paraphyly In cladistics, a group that includes a common ancestor and some, but not all, of its descendants. Cf. **polyphyletic group**, **monophyletic group**.

paraphysis (1) A sterile filament borne among the reproductive structures of many algae, fungi and bryophytes. (2) A thin-walled sac developed as an outgrowth from the non-nervous roof of the telencephalon, represented in Mammals by the pineal organ. pl. *paraphyses*. adj. *paraphysate*.

parapineal organ See **parietal organ**.

paraplegia Paralysis of the lower part of the body and of the legs.

parapodium In Mollusca, a lateral expansion of the foot; in Polychaeta, a paired fleshy projection of the body wall of each somite used in locomotion. adj. *parapodial*.

parapophyses A pair of ventrolateral processes of a vertebra arising from the sides of the centrum.

paraprotein Immunoglobulin derived from an abnormally proliferating clone of neoplastic plasma cells. The immunoglobulin and the cells making it will all have the same Ig class, subclass and light chain determinants.

parapsid In the skull of Reptiles, the condition when there is one temporal vacuity, this being high behind the eye, usually with the post-frontal and supratemporal meeting below. Found in Mesosaurs and Ichthyosaurs. See **temporal vacuities** for other types.

parapsychology The study of certain alleged phenomena, the *paranormal*, that is beyond the scope of ordinary psychology, e.g. ESP, psychokineses etc.

paraquat *1,1'*-dimethyl-4,4'-dipyridylium salts, used as a weedkiller. When taken orally causes severe and often irreversible damage to lungs, liver and kidneys.

parasexual cycle Genetic system in some fungi which allows limited **recombination** as a result of the doubling of the chromosomes in a nucleus, followed by crossing over and the gradual return to the haploid state by progressive chromosome loss, *haploidization*.

parasite An organism which lives in or on another organism and derives subsistence from it without rendering it any service in return. See **parasitism**.

parasitic castration Castration brought about by the presence of a parasite, as in the case of a Crab parasitized by *Sacculina*.

parasitic male A dwarf male in which all but the sexual organs are reduced, and which is entirely dependent on the female for nourishment, as in some deep-sea Angler fish (Ceratioids).

parasitism A close internal or external partnership between two organisms which is detrimental to one partner (the *host*) and beneficial to the other partner (the *parasite*); the latter often obtains its nourishment at the expense of the nutritive fluids of the host. The term usually refers to such a feeding relationship but other forms exist. See **social parasitism**.

parasitoid An animal which is parasitic in one stage of the life history and subsequently free-living in the adult stage, as the parasitic Hymenoptera.

parasitology The study of, usually animal, parasites and their habits.

parasphenoid In some of the lower Vertebrates, a membrane bone of the skull, which forms part of the cranial floor.

parasymbiosis The condition when two organisms grow together but neither assist nor harm one another.

parasympathetic nervous system In Vertebrates, a subdivision of the autonomic nervous system, also known as the *craniosacral system*. The action of these nerves tends to slow down activity in the glands and smooth muscles which they supply, but promotes digestion, and acts antagonistically to that of the **sympathetic nervous system**. Parasympathetic nerves are cholinergic.

parathormone The hormone secreted by the **parathyroid** which controls the metabolism of calcium and phosphorus.

parathyroid An endocrine gland of Vertebrates found near the thyroid or embedded in it. Two pairs are usually present, deriving from the third and fourth pairs of gill pouches. Probably only secretes one hormone (*parathormone*).

paratonic movement A plant movement in response to an external stimulus, e.g. taxis, tropism. See **autonomic movement**.

paratope Same as *antigen binding site*. See **network theory**.

paratyphoid Enteric fever due to infection by *Salmonella spp.* other than *S. typhi*; similar to, but milder than, typhoid fever.

paraxonic foot A foot in which the skeletal axis passes between the third and fourth digits, as in Artiodactyla. Cf. **mesaxonic foot**.

Parazoa A subkingdom of the animals. Multicellular organisms though their cells are less specialized and interdependent than in the Metazoa. Contains the single phylum **Porifera**. Cf. **Protozoa, Metazoa**.

parencephalon A cerebral hemisphere.

parenchyma Soft spongy tissue of indeterminate form, consisting usually of cells separated by spaces filled with fluid or by a gelatinous matrix, and generally of mesodermal origin. In plants, typically blunt ended, somewhat elongated cells with thin or evenly thickened cell walls, they are not adapted for water transport but are sometimes photosynthetic, as *chlorenchyma*, or able to act as a store. In animals, the specific tissue of a gland or organ as distinct from the interstitial (connective) tissue or stroma. adj. *parenchymatous*.

parenteral Said of the administration of therapeutic agents by any way other than through the alimentary tract.

paresis Slight or incomplete paralysis. See **general paresis** (general paralysis of the insane, GPI).

parietal *Peripheral*, as the paired dorsal membrane bone of the Vertebrate skull, situated between the auditory capsules; pertain-ing to, or forming part of, the wall of a structure.

parietal foramen Small rounded aperture in the middle of the united parietals of the skull. Site of the pineal eye.

parietal organ The anterior diverticula of the pineal apparatus. When present may be developed as an eyelike organ, the pineal eye.

parietal placentation Placentas which develop along the fused margins of the carpels of a unilocular ovary, e.g. violet.

parietes The walls of an organ or a cavity. sing. *paries*.

paronychia A felon or whitlow. Purulent inflammation of the tissues in the immediate region of the finger nail.

parosmia Abnormality of the sense of smell.

parotid gland In some Anura, an aggregation of poison-producing skin glands on the neck; in Mammals, a salivary gland situated at the angle of the lower jaw.

pars A part of an organ. pl. *partes*.

pars anterior See **pars distalis**.

pars distalis The anterior part of the adenohypophysis of the pituitary.

pars intermedia In higher Vertebrates, part of the posterior lobe of the pituitary body, which is derived from the hypophysis at first but tends to become spread over the surface of the **pars nervosa** as development proceeds.

pars nervosa In higher Vertebrates, part of the posterior lobe of the pituitary body, developed from the infundibulum.

parthenocarpy The production of fruit without seeds either spontaneously or by artificial induction by **auxins**.

parthenogenesis The development of a new individual from a single, unfertilized gamete, often an egg. adj. *parthenogenetic*.

parthenospore A spore formed without previous sexual fusion; an azygospore.

parthen-, partheno- Prefixes from Gk. *parthenos*, virgin.

partial parasite (1) A plant capable of photosynthesis but dependent on another plant for water and mineral nutrients, e.g. mistletoe. (2) A plant capable of living independently but able to become parasitic in suitable circumstances, *facultatively parasitic*.

partial reinforcement Refers to conditions in which a response is reinforced only some of the time; such responses are more resistent to extinction than responses acquired through **continuous reinforcement**. Also *intermittent reinforcement*.

partial umbel One of the smaller group of flowers which altogether make up a compound umbel.

partial veil In some basidiomycete fruiting

bodies, e.g. mushrooms, a membrane joining the edge of the cap to the stalk, rupturing to leave an annulus. Cf. **universal veil.**

partite Split almost to the base.

parturition In viviparous animals, the act of bringing forth young. adj. *parturient.*

parvifoliate Having leaves which are small in relation to the size of the stem.

passage cell An endodermal cell in a root, usually opposite the protoxylem, which retains unthickened cell walls in an endodermis in which most of the cells have developed secondary walls.

Passeriformes An order of Birds containing those perching which comprise about half the known species of bird. Mostly small, living near the ground and having 4 toes arranged to allow gripping of the perch. Young helpless at hatching. Rooks, Finches, Sparrows, Tits, Warblers, Robins, Wrens, Swallows and many others.

passive-aggressive behaviour Indirectly expressed resistance to the demands of others, e.g. forgetting appointments, losing important objects.

passive cutaneous anaphylaxis A test *in vivo* to reveal the presence of mast cell-sensitizing antibody. Antibody is injected intracutaneously into the skin of an animal. After an interval sufficient for the antibody to become attached, antigen is injected intravenously together with a blue dye which binds to serum albumin. Where the antigen reacts with cell-fixed antibody, histamine and other substances are released which increase vascular permeability, and the dye leaks out to give a blue spot the size of which is proportional to the amount of antibody attached.

passive immunization Use of antibody from an immune individual to provide temporary immunity in a non-immune individual, e.g. with diphtheria antitoxin, tetanus antitoxin or serum from persons convalescent from measles.

passive permeability The flux of solutes across a cell membrane by simple diffusion at a rate proportional to the difference in concentration of the solute across the membrane.

patagium A lobelike structure at the side of the prothorax in some Lepidoptera; in Bats and some other flying Mammals, a stretch of webbing between the forelimb and the hindlimb; in Birds, a membranous expansion of the wing. adj. *patagial.*

patella In higher Vertebrates, a sesamoid bone of the knee joint or elbow joint.

patent Said of leaves and branches which spread out widely from the stem.

pathetic muscle The superior oblique muscle of the Vertebrate eye.

pathetic nerve The 4th cranial nerve of

Vertebrates, running to the superior oblique muscle.

pathogen An organism, e.g. parasite, bacterium, virus, which causes disease.

pathology That part of medical science which deals with the causes and nature of disease, and with the bodily changes wrought by disease. See **phytopathology.**

patristic similarity Similarity due to common ancestry.

patroclinous Exhibiting the characteristics of the male parent more prominently than those of the female parent. Cf. **matroclinous.**

Paul-Bunnell test Test used in the diagnosis of infectious mononucleosis. Serum from persons with infectious mononucleosis contains a particular heterophil antibody which will agglutinate both sheep and horse erythrocytes, and is not found in other conditions. The antibody is distinct from antibodies against Epstein-Barr virus, the actual cause of the disease.

paunch See **rumen.**

pavement epithelium A variety of epithelium consisting of layers of flattened cells.

PCO cycle Abbrev. for *Photorespiratory Carbon Oxidation cycle.* See **photorespiration.**

PCR Abbrev. for *Polymerase Chain Reaction.*

PCR cycle Abbrev. for *Photosynthetic Carbon Reduction cycle.* See **Calvin cycle.**

peak dose Maximum absorbed radiation dose at any point in an irradiated body, usually at a small depth below the surface, due to secondary radiation effects.

pearl An abnormal concretion of nacre formed inside a mollusc shell round a foreign body such as a sand particle or a parasite.

peat The name given to the layers of dead vegetation, in varying degrees of alteration, resulting from the accumulation of the remains of marsh vegetation in swampy hollows in cold and temperate regions. Geologically, peat may be regarded as the youngest member of the series of coals of different rank, including brown coal, lignite and bituminous coal, which link peat with anthracite.Peat is very widely used as a fuel, after being air-dried, in districts where other fuels are scarce and in some areas, e.g. in Russia and Ireland, it is used to fire power stations. It is low in ash, but contains a high percentage of moisture, and is bulky; specific energy content about 16 MJ/kg or 7000 Btu/lb.

peck order The classic example of **social hierarchy** in farmyard hens in which animals within a group form some consistent relationship most apparent in their aggressive interactions; the same individuals dominate or are dominated by particular animals within the group.

pecten Any comblike structure; in some Vertebrates (Reptiles and Birds), a vascular process of the inner surface of the retina; in Scorpionidea, tactile sensory organs under the mesosoma.

pectineal, pectinate Comblike. Said (1) of a process of the pubis in Birds; (2) of a ridge on the femur to which is attached the *pectineus muscle*, one of the protractors of the hindlimb.

pectines Comblike chitinous structures of mechanoreceptive function attached to the ventral surface of the second somite of the mesosoma in *Scorpionidea*.

pectins Calcium-magnesium salts of polygalacturonic acid, partially joined to methanol residues by ether linkage. They occur in the cell walls, esp. in the *middle lamellas* and *primary walls* of vascular plants. They are soluble in water and can be precipitated from aqueous solutions by excess alcohol. Acid solutions gel with 65–70% of sucrose, the basis of their use in jam making.

pectization The formation of a jelly.

pectorales In Vertebrates, muscles connecting the upper part of the forelimb with the ventral part of the pectoral girdle. sing. *pectoralis*.

pectoral fins In Fish, the anterior pair of fins.

pectoral girdle In Vertebrates, the skeletal framework with which the anterior pair of locomotor appendages articulate.

pedate leaf A leaf with three divisions of which the 2 laterals are forked once or twice.

pedes See pes.

pedicel A stalk bearing one flower. Cf. peduncle.

pedicel The second joint of the antennae in Insects; more generally, the stalk of a sedentary organism; the stalk of a free organ, as the *optic pedicel* in some Crustacea.

pedicellaria In Echinodermata, a small pincerlike calcareous structure on the body surface with 2 or 3 jaws provided with special muscles and capable of executing snapping movements; it may be stalked or sessile.

pedicellate Having, or borne on, a pedicel.

pedicle In the vertebrae of the Frog, a pillarlike process springing from the centrum and extending vertically upwards to join the flat, nearly horizontal lamina, which forms the roof of the neural canal. Intervertebral foramina, for the passage of the spinal nerves, occur between successive pedicles. Generally any pillarlike process supporting an organ.

pediculosis Infestation of the body with lice.

pedipalp In Chelicerata, the appendage, borne by the first postoral somite, of which

the gnathobases function as jaws; it may be a tactile organ or chelate weapon.

pedology The study of soil.

peduncle (1) A stalk bearing several flowers. Cf. **pedicel**. (2) In Brachiopoda and Cirripedia, the stalk by which the body of the animal is attached to the substratum; in some Arthropoda, the narrow portion joining the thorax and abdomen or the prosoma and opisthosoma; in Vertebrates, a tract of white fibres in the brain.

pedunculate Having, or borne on, a peduncle.

pelagic Living in the middle depths and surface waters of the sea.

Pelecaniformes A varied order of Birds, mainly fish-eating and colonial nesters, with 4-toed webbed feet, bodies adapted for diving, and long beaks with wide gapes and sometimes with a pouch. Pelicans, Cormorants, Gannets.

Pelecypoda Syn. for **Bivalvia**.

pelleted seed Seed coated with a layer of inert material, esp. to make smaller, angular seeds into larger, rounded bodies that can be drilled more precisely, sometimes also to incorporate pesticides etc.

pellicle (1) Layer of interlocking, helically wound, proteinaceous strips just below the plasmalemma of *Euglenophyceae*. (2) A thin cuticular investment, as in some Protozoa. adj. *pelliculate*.

pelma See **planta**.

peltate A flattened rounded plant organ attached to its stalk at about the middle of its lower surface, e.g. the leaf of the nasturtium (*Tropaeolum maius*).

pelvic fins In Fish, the posterior pair of fins.

pelvic girdle In Vertebrates, the skeletal framework with which the posterior pair of locomotor appendages articulate.

pelvis The pelvic girdle or posterior limb girdle of Vertebrates, a skeletal frame with which the hind limbs or fins articulate; in Mammals, a cavity, just inside the hilum of the kidney, into which the uriniferous tubules discharge and which is drained by the ureter. adj. *pelvic*.

pen In Cephalopoda, the shell or cuttle bone.

pendulous placentation See **apical placentation**.

penetrance The proportion of individuals of a specified genotype in which a particular gene exhibits its effect.

penis The male copulatory organ in most higher Vertebrates; a form of male copulatory organ in various Invertebrates, as Platyhelminthes, Gastropoda. adj. *penial*.

Penman-Monteith equation Describes the dependence of evapotranspiration or transpiration rates on climatological variables and surface properties of the vegetation.

pennae See **plumae**.

pennate (1) Generally, winged. (2) Said of diatoms which are bilaterally symmetrical or approximately so.

pentadactyl Having five digits.

pentadactyl limb The characteristic free appendage of Tetrapoda with five digits.

pentamerous Having five members in a whorl.

pentarch Said of a stele having five strands of protoxylem.

Pentastomida A subphylum of Arthropoda. Elongate vermiform parasites of carnivorous mammals, living in the nasal sinuses, which have 2 pairs of claws at the sides of the mouth, and no respiratory or circulatory system.

pentose shunt A series of metabolic reactions which converts glucose-6-phosphate into ribose-5-phosphate with concomitant generation of NADPH.

pent-, penta- Prefixes from Gk. *pente*, five, used in the construction of compound terms, e.g. *pentactinal*, 5-rayed.

PEP carboxylase *Phosphoenopyruvate carboxylase*. The enzyme that catalyses the reaction of phosphoenopyruvate and CO_2 to give oxaloacetic acid, esp. as the first step in both the **Hatch-Slack** pathway and **crassulacean acid metabolism** in both of which the oxaloacetic acid is then reduced to malic acid.

pepsin Protease of the gastric juice which is able to function optimally under the acidic conditions of the stomach.

peptidase An enzyme which degrades peptides into their constituent amino acids.

peptide A sequence of amino acids held together by peptide bonds. With rare exceptions peptides are unbranched chains joined by peptide bonds between their α-amino and α-carboxyl groups. Peptides can vary in length from dipeptides with 2 to polypeptides with several hundred amino acids.

peptide bond The bond formed by the condensation of the amino group and carboxyl group of a pair of amino acids.

percentile See **quantile**.

perception Refers to the individual's apprehension of the world and of the body through the action of various sensory sytems.

perceptual defence The tendency to identify anxiety-provoking stimuli less readily than neutral stimuli, esp. if presented for only brief periods.

perceptual learning Refers to learning about aspects of the stimulus in situations where there is no external reinforcement for doing so; it is usually demonstrated by an increased ability on a subsequent task needing discrimination.

Perciformes The largest order of bony fish

with about 7000 species. An advanced and successful group inhabiting marine and fresh water habitats. Basses, Perches, Tuna.

percurrent Extending throughout the entire length.

pereiopods In higher *Crustacea*, the thoracic appendages modified as walking legs. Cf. *pleopod*.

perennation Survival from one growing season to the next with, usually, a period of reduced activity between.

perennial A plant that lives for more than 2 years. Most flower in most years after the first or second but some are **hapaxanthic**.

perfect (1) A flower having functional stamens and carpels. (2) Reproducing sexually.

perfoliate Said of a leaf, the basal part of which encircles the stem completely, so that the stem appears to pass through the leaf.

perforate (1) Having holes, as many shells. (2) Containing small rounded transparent dots which give the appearance of holes. (3) Those gastropod shells which have a hollow columella.

perforation A hole in the common wall between two consecutive elements of a *vessel*, resulting from the dissolution of the wall. Cf. **pit**.

perforation plate The part of the common wall between two consecutive elements of a xylem *vessel* which has one or more perforations.

perforin A protein present in K and NK cells which resembles the C9 component of complement, and forms ring-like tubular structures which become inserted into the cell membranes of target cells and cause leakage of their contents.

performance test Mental tests consisting primarily of motor or perceptual items and not requiring verbal ability. Cf. **verbal test**.

peri- Prefix from Gk. *peri*, around.

perianth The set of sterile structures which typically surrounds the stamens and carpels of a flower and which may be differentiated into an outer, often green and protective, *calyx* of sepals and an inner, often coloured, *corolla* of petals. See **perianth segment**.

perianth segment *Tepal*. A member of the perianth, esp. if there is no differentiation into sepals and petals.

periblast In meroblastic eggs, the margin of the blastoderm merging with the surrounding yolk. See **periplasm**.

periblastic Of cleavage, superficial.

periblem A *histogen*, the precursor of the cortex.

peribranchial Surrounding a gill or gills.

pericardium Space surrounding the heart; the membrane enveloping the heart. adj. *pericardial*.

pericarp Fruit wall derived from the ovary wall.

pericellular Surrounding a cell.

perichaetium (1) A cuplike sheath surrounding the archegonia in some liverworts. (2) The group of involucral leaves around the archegonia of a moss.

perichondrium The envelope of areolar connective tissue surrounding cartilage.

perichordal Encircling or ensheathing the notochord.

periclinal Parallel to the nearest surface. Opposite of anticlinal.

periclinal chimera A plant or shoot in which tissues of one genetic constitution form a complete layer throughout, the remaining tissues being of different genetic constitution. See **chimera, tunica-corpus concept.**

pericranium The fibrous tissue layer which surrounds the bony or cartilaginous cranium in Vertebrates.

pericycle A layer, one or more cells thick, of ground tissue at the periphery of the stele next to the endodermis; present in roots, rare in stems.

periderm Secondary protective tissue often replacing epidermis in longer-lived stems and roots and consisting of cork, cork cambium and phelloderm. See **perisarc.**

peridesmium The coat of connective tissue which ensheathes a ligament.

perididymis The fibrous coat which encapsulates the testis in higher Vertebrates.

peridinin Major carotenoid accessory pigment in the Dinophyceae.

peridium A general term for the outer wall of the fruit body of a fungus, when the wall is organized as a distinct layer or envelope surrounding the spore-bearing organs partially or completely.

perigynium See **perichaetium.**

perigynous Said of a flower having the perianth and stamens inserted on a flat or cup-shaped structure which arises below the ovary but is not fused to it, e.g. raspberry. Cf. **epigynous, hypogynous, superior.**

perilymph The fluid which fills the space between the membranous labyrinth and the bony labyrinth of the internal ear in Vertebrates. adj. *perilymphatic.* Cf. **endolymph.**

perimedullary zone Same as **medullary sheath.**

perimysium The connective tissue which binds muscle fibres into bundles and thus muscles.

perineal glands In some Mammals, a pair of small glands beside the anus which secrete a substance with a characteristic odour.

perineum, perinaeum The tissue wall between the rectum and the urinogenital ducts in Mammals. adj. *perineal, perinaeal.*

perineurium The coat of connective tissue which ensheathes a tract of nerve fibres.

periodicity Rhythmic activity.

periosteum The covering of areolar connective tissue on bone.

periostracum The horny outer layer of a Molluscan shell.

periotic In higher Vertebrates, a bone enclosing the inner ear and formed by the fusion of the otic bones. Also *petrosal.*

peripheral Situated or produced around the edge.

periplasm (1) The space between the plasma membrane and outer membrane of *Gram-negative* bacteria. It contains proteins secreted by the cell and a rigid peptideoligosaccharide complex, the *peptidoglycan.* (2) A bounding layer of protoplasm surrounding an egg just beneath the vitelline membrane, as in Insects.

periplasmic space A space between the cell wall and the plasmalemma.

periproct The area surrounding the anus.

perisarc In some Hydrozoa, the chitinous layer covering the polyps etc. Cf. **coenosarc.**

perisperm Storage tissue, derived from nucellus (and hence wholly maternal), present in some seeds in which the endosperm does not develop, e.g. many Caryophyllaceae.

perissodactyl Having an odd number of digits. Cf. **artiodactyl.**

Perissodactyla An order of Mammals containing the 'odd-toed' hooved animals, i.e. those with a mesaxonic foot with the skeletal axis passing down the third digit. Horses, Tapirs, Rhinoceros and extinct forms.

peristaltic Compressive; contracting in successive circles; said of waves of contraction passing from mouth to anus along the alimentary canal. Cf. **antiperistaltic, systaltic.** n. *peristalsis.*

peristome (1) The margin of the aperture of a gastropod shell. (2) In some Ciliophora, a specialized food-collecting, frequently funnel-shaped, structure surrounding the cell mouth. (3) More generally, the area surrounding the mouth. adj. *peristomial.*

peristomium In *Annelida,* the somite in which the mouth is situated; in some forms (as *Nereis*), 2 somites have been fused to form the apparent peristomium.

perisystole The period between diastole and systole in cardiac contraction.

perithecium A more or less flask-shaped **ascocarp** with a pore or ostiole at the top through which the asci are discharged.

peritoneal cavity In Vertebrates, that part of the coelom containing the viscera; the abdominal body cavity.

peritoneum In Vertebrates, a serous membrane which lines the peritoneal cavity and extends over the mesenteries and viscera. adj. *peritoneal.*

peritonitis Inflammation of the peritoneum.

peritrichous Said of bacteria which have flagella distributed all over their surface.

peritrophic Surrounding the gut; as the *peritrophic membrane* of Insects, a membranous tube lining the stomach and partially separated from the stomach epithelium by the peritrophic space.

perivascular sheath A sheath of connective tissue around a blood or lymph vessel.

perivitelline Surrounding an egg yolk.

permanent dentition In Mammals, the second set of teeth, which replaces the milk dentition.

permanent wilting point The water content of a soil at which a plant will wilt and not recover without additional water, even if shaded or left overnight. For most crop plants this corresponds to a soil **matric potential** of about −15 bars (−1.5 MPa). Cf. **hydraulic capacity**.

permissible dose See **maximum permissible dose**.

pernicious anaemia *Addison's anaemia*. Disease characterized by atrophic gastritis with achlorhydria and lack of gastric **intrinsic factor** which leads to failure of absorption of dietary vitamin B_{12} and consequent megaloblastic anaemia and perhaps peripheral neuritis. Gastric secretions and the blood of affected persons contain auto-antibodies against intrinsic factor and against a microsomal antigen present in gastric parietal cells.

peroral Surrounding the mouth, as the *peroral membrane* of *Ciliophora*, which surrounds the cytopharynx.

peroxidases Enzymes which activate hydrogen peroxide and induce reactions which hydrogen peroxide alone would not effect.

peroxisome Small membrane-bound cytoplasmic organelle containing oxidizing enzymes and catalase; common in leaf cells where they contain some of the enzymes of the glycollate pathway.

persistent Continuing to grow or develop after the normal period for the cessation of growth or development, as teeth (cf. **deciduous**); said also of structures present in the adult which normally disappear in the young stages.

person An individual organism, particularly of colonial forms.

personal dosimeter Sensitive tubular electroscope using a metallized quartz fibre unit which is viewed against a calibrated scale. It is charged by a generator and ionizing radiation discharges it. Used by workers in places of potential radiation hazard.

personality The integrated organization of all the psychological, intellectual, emotional and physical characteristics of an individual which determines the unique adjustment he or she makes to the world.

personality disorders An inflexible and well-established behaviour pattern that is maladaptive for the individual in terms of social or occupational functioning; usually recognizable by adolescence.

personal space The space immediately surrounding one's body that an individual considers private; it varies with cultural and other factors.

personnel monitoring Monitoring for radioactive contamination of any part of an individual, his breath or excretions, or any part of his clothing.

pertusate Perforated; pierced by slits.

pes The podium of the hind limb in land Vertebrates. pl. *pedes*.

petal A member of the **corolla**, often brightly coloured and conspicuous.

petalody The transformation of stamens, or carpels, into petals. See **double**.

petaloid Looking like a petal; petal-shaped, as the dorsal parts of the ambulacra in certain Echinoidea.

petiolate Leaf having a stalk or petiole. Cf. **sessile**.

petiole The stalk of a leaf.

petiolule The stalk of a leaflet of a compound leaf.

petrosal See **periotic**.

petrous Stony, hard (as a portion of the temporal bone in higher Vertebrates); situated in the region of the petrous portion of the temporal bone.

Peyer's patches Nodules of lymphoid tissue in the submucosa of the small intestine, which are important for the development of immunity (or of immunological unresponsiveness) to antigens present in the gut. B Lymphocyte precursors of IgA producing cells are more plentiful than in other lymphoid tissues.

P_{fr} The physiologically active form of the plant pigment **phytochrome**, having a peak of absorption at $\lambda = 730$ nm which converts it to P_r (to which it also changes slowly in the dark). See **P_r**.

PHA Abbrev. for *PhytoHaemAgglutinin*.

phaeic, phaeochrous Dusky. n. *phaeism*.

phaeochrous See **phaeic**.

phaeo-, pheo- Prefixes from Gk. *phaios*, dusky.

Phaeophyseae The brown algae, a class of the eukaryotic algae in the division Heterokontophyta. Branching filamentous and parenchymatous types. Mostly marine, littoral and sublittoral, the brown sea weeds. Source of alginic acid. Isogamous, anisogamous or oögamous; alternation of generations (isomorphic or heteromorphic), or almost diplontic. Includes kelps and wracks.

phagocyte A cell which exhibits amoeboid movement, in particular it throws out **pseudopods** to engulf foreign bodies, such as bacteria.

phagocytosis The ingestion of cells or particles which first attach to and then become surrounded by the cell membrane. This is invaginated to form an intracellular vesicle (phagosome), towards which lysosomes move and fuse to release their content of hydrolytic enzymes into the vesicle, which becomes a phagolysosome. Cells of the mononuclear phagocyte system and polymorphonuclear leucocytes are the main phagocytic cells in mammals.

phagotrophy Heterotrophic nutrition in which cells ingest solid food particles.

phag-, phago-, -phage, -phagy Prefixes and suffixes from Gk. *phagein*, to eat.

phalanges In Vertebrates, the bones supporting the segments of the digits; fiddle-shaped rings composing the reticular lamina of the organ of Corti. sing. *phalanx*.

Phalangida See **Opiliones**.

phalanx See **phalanges**.

phallic stage In psychoanalytic theory, the third stage of psychosexual development in which the child is preoccupied with his or her genitals; from the third to fifth or sixth year of life. See **Oedipus and Electra complexes**.

phallus The penis of Mammals; the primordium of the penis or clitoris of Mammals. adj. *phallic*.

phanero- Prefix from Gk. *phaneros*, visible.

phanerophyte Woody plant with perennating buds more than 25 cm above the soil surface. See **Raunkiaer system**.

pharate In Insects, refers to a phase of development when the old cuticle of one stage is separate from the hypodermis of the next stage, but has not yet been ruptured and cast off. The so-called *pupa* of many Insects actually represents a pharate adult, and the so-called *prepupa* is a pharate pupa.

pharmacology The scientific study of the action of chemical substances on living systems.

pharynx In Vertebrates, that portion of the alimentary canal which intervenes between the mouth cavity and the oesophagus and serves both for the passage of food and the performance of respiratory functions; in Invertebrates, the corresponding portion of the alimentary canal lying immediately posterior to the buccal cavity, usually having a highly muscular wall. adj. *pharyngeal*.

phase-contrast microscopy See panel.

Phe Symbol for phenylalanine.

phellem See **cork**.

phelloderm Parenchymatous tissue formed centripetally by the cork cambium, or phellogen, as part of the periderm.

phellogen See **cork cambium**.

phenetic classification That based on overall similarity in as many characters as possible, usually without weighting.

phenocopy An environmentally caused copy of a genetic abnormality. It is non-heritable.

phenogram A branching diagram, *dendrogram*, reflecting the overall similarities of groups of organisms.

phenolics Large and diverse group of plant secondary metabolites. The phenol group includes *flavonoids*, *lignin* and *tannins*.

phenology The study of plant development in relation to the seasons.

phenomenology In philosophy, the study of the psychic awareness that accompanies experience and that is the source of all meaning for the individual. In psychiatry, it refers to the description and classification of an individual's mental activity, including subjective experience and perceptions, mental performance (e.g. memory) and the somatic accompaniments of mental events (e.g. heart rate).

phenotype The observable characteristics of an organism as determined by the interaction of its **genotype** and its environment. adj. *phenotypic*.

phenotypic variance Measure of the amount of variation among individuals in the observed values of a quantitative character.

phenoxyacetic acids Synthetic compounds many with auxin-like activity including the important selective weedkillers, 2,4-D, 2,4,5-T and MCPA.

phenylalanine *2-amino-3-phenylpropanoic acid*. The L-isomer is a polarizable amino acid. Symbol Phe, short form F.

Side chain: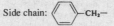

See **amino acid**.

phenylalanine ammonia-lyase Enzyme catalysing the elimination of ammonia from L-phenylalanine in the synthesis of many *secondary* plant metabolites, e.g. phenolic acids, flavonoids, alkaloids, lignins, tannins etc. Abbrev. *PAL*.

phenylmethylsulphonyl fluoride Widely used as a protease inhibitor.

pheromone A chemical substance secreted by an animal, which influences the behaviour of other animals (cf. **hormone**) of the same species, e.g. the *queen substance* of honey bees. See **semiochemical**.

Philadelphia chromosome Distinctive, small chromosome found in patients suffering from chronic myelocytic leukaemia, corresponding to chromosome 22 after a reciprocal **translocation** with chromosome 9.

phleb-, phlebo- Prefixes from Gk. *phleps*, gen. *phlebos*, vein.

phloem Tissue with the major function of

phase-contrast microscopy

A method of particular value in examining living unstained cells, whereby phase differences are converted to differences of light intensity by small modifications to a microscope. Living cells are not usually coloured, i.e. they are pure phase objects, but the light transmitted by their different structures will have phase differences caused both by variations in refractive index arising from changes in protoplasmic concentration and by differences in thickness.

In addition to the image formed by the microscope objective which can be viewed magnified by the eye through the eyepiece, the condenser has a *plane* below it which is focused at a *conjugate plane* just above the objective (in the conventional vertical representation of a microscope) so that an annular (or other shaped) stop (*phase stop*) placed below the condenser is focussed at the conjugate plane. A *phase plate* is placed at this upper conjugate plane. It consists of glass with a deposit of dielectric material, retarding the light by a B wavelength, and of a shape to correspond with the phase stop. When the phase plate is aligned with the image of this stop, any rays displaced by refractive-index gradients in the specimen will pass inside or outside the phase plate (see the diagram). They will then, because of their phase relationship, interfere with the unrefracted rays from the specimen to give changes in luminosity at the image plane.

For optimum contrast the phase plate must also absorb a fraction of the unrefracted light. Further, an auxiliary telescope is needed to aid the alignment of the phase plate with the image of the phase stop. Analogous and newer methods for viewing living material include the **interference microscope** and the **differential interference contrast microscope** (*Nomarski microscope*).

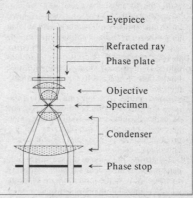

Eyepiece

Refracted ray

Phase plate

Objective

Specimen

Condenser

Phase stop

---

transporting metabolites, esp. sugars, from sources to sinks. See **mass flow hypothesis**. The tissue consists of **sieve elements**, and companion cells and/or parenchyma cells, often with fibres or sclereids.

phloem ray That part of a vascular *ray* which traverses the secondary phloem.

phloroglucinol A stain for lignin when acidified.

phobia An excessive anxiety in some specific situation or in the presence of some object, which presents no apparent threat.

phobic disorder A disorder characterized by intense anxiety to some external object or situation, and avoidance of the phobic stimulus.

Pholidota An order of old-world Mammals having imbricating horny scales, interspersed with hairs, over the head, body and tail. They have a long snout, no teeth and a long thin tongue. The hindfeet are plantigrade, and the forefeet have long curved claws. Nocturnal animals, feeding on ants or termites. Pangolins.

phonation Sound production.

-phore Suffix from Gk, *pherein*, carry.

phoresis Electrical passage of ions through a membrane.

phoresy Transport or dispersal achieved by clinging to another animal, e.g. certain Mites which achieve dispersal by attaching themselves to various Insects.

Phoronidea A small phylum of hermaphrodite unsegmented coelomate animals of tubicolous habit, having a U-shaped gut, a dorsal anus, and a lophophore in the form of a double horizontal spiral; marine forms occurring in the sand and mud of the sea bottom.

phosphatase An enzyme which dephosphorylates its substrate by hydrolysis of the phosphate ester bond.

phosphate fixation The reaction of orthophosphate ions with Ca, Al or Fe ions to give low-solubility hydroxyphosphates which renders much of the phosphate unavailable to plants. Occurs in most mineralized or artificially fertilized soils.

phosphatidyl choline A phosphatide in which the choline forms the organic base. Previously termed *lecithin*. Important component of biological membranes.

phosphatidyl inositol A phosphatide containing inositol. They are important in the response of cells to external ligands.

phosphoenolpyruvate The phosphate ester of the enol form of pyruvic acid, $CH_2=COPO_3H_2–COOH$. An important metabolite in glycolysis and the substrate for **PEP carboxylase**.

phospholipase An enzyme which degrades a phospholipid. There are 3 main types, phospholipase A, C and D: A removes a hydrocarbon chain from the phospholipid, C cleaves the glycerol/phosphate ester bond and D the phosphate/base linkage.

phosphoproteins Proteins which have been enzymically phosphorylated so that they contain phosphate groups. An important functional modifier.

phosphorescence Luminosity; production of light, usually (in animals) with little production of heat, as in Glow worms. adj. *phosphorescent*.

phosphorylase An enzyme which catalyses the cleavage of a bond by the addition of orthophosphate (cf. **hydrolysis**). Thus glycogen phosphorylase liberates the terminal glucose of glycogen as glucose-1-phosphate.

photic zone The zone of the sea where light penetration is sufficient for photosynthesis, corresponding to the limnetic zone of freshwater habitats. Cf. **aphotic zone**.

photobiology The study of light as it affects living organisms.

photogenic Emitting light, light-producing, e.g. *photogenic bacteria*.

photogenin See **luciferase**.

photolysis of water The notional splitting of water molecules into oxygen, electrons and protons in the light reactions of photosynthesis. The term emphasizes the fact that the released oxygen comes from water not carbon dioxide.

photomorphogenesis Control of plant morphogenesis by the duration and nature of the light. Cf. **etiolation**, **photoperiodism**, **phytochrome**.

photonastic movement, photonasty Nastic movement resulting from change in illumination.

photopeak The energy of the predominant photons released during the decay of a radionuclide.

photoperiodicity The controlling effects of the length of day on such phenomena as

the flowering of plants, the reproductive cycles of Mammals, migration and diapause of Insects, and seasonal changes in the feathers of Birds and the hair of Mammals.

photoperiodism Response by an organism to day length. See **short-day plant**, **long-day plant**, **day-neutral plant**, **phytochrome**.

photophilous Light-seeking, light-loving; said of plants which inhabit sunny places.

photophobia Intolerance of the eye to light with spasm of the eyelids.

photophore A luminous organ of Fish.

photophosphorylation The production in photosynthesis of ATP from ADP and inorganic phosphate using energy from light. In cyclic photophosphorylation only **photosystem I** is involved and NADP is not reduced; in non-cyclic photophosphoryation *photosystem I* and *II* are both involved and NADP is reduced.

photophygous Shunning strong light.

photoreceptor (1) A structure such as that associated with an *eyespot* or a molecule (e.g. **phytochrome**) which absorbs the stimulating light in phototaxis etc. (2) A sensory cell specialized for the reception of light, e.g. rods and cones of the vertebrate eye.

photorespiration Light-stimulated respiration (O_2 uptake and CO_2 production) resulting from the production of phosphoglycolic acid by the oxygenase reaction of **ribulose 1,5-bisphosphate carboxylase oxygenase** and its subsequent metabolism, in the so-called photorespiratory carbon oxidation cycle, to phosphoglyceric acid and CO_2.

photosensitizing dye Intercalating dye which binds to DNA and makes it susceptible to breakage when exposed to ultraviolet light.

photosynthate The substances, esp. sugars produced in photosynthesis.

photosynthesis See panel.

photosynthetically active radiation See panel.

photosynthetic carbon reduction cycle Same as **Calvin cycle**.

photosynthetic pigment See panel.

photosynthetic quotient The ratio of carbon dioxide absorbed to oxygen released in a photosynthesizing structure or organism. Abbrev. *PQ*.

photosystem I Reaction centre in photosynthesis comprising chlorophyll a, as *P700*, and other pigments and molecules, in which energy absorbed as light is used to transfer electrons from a weak oxidant to a strong reductant. The latter either reduces **ferredoxin** and ultimately NADP or is used in cyclic photophosphorylation. Abbrev. *PSI*.

photosystem II Reaction centre in photo-

---

photosynthesis

The use of energy from light to drive chemical reactions most notably the reduction of carbon dioxide to carbohydrates coupled with the oxidation of water by plants and algae. Energy from light is also used to drive the reduction of nitrate and sulphate in plants and algae, and in **nitrogen fixation** in blue-green algae and photosynthetic bacteria.

Globally, about 5×10^{11} tonnes of organic matter is produced by photosynthesis annually, 60% on land (almost all by seed plants) and the remainder in the sea, mostly by diatoms. Photosynthesis is the commonest form of autotrophy and photo-autotrophs form the basis of virtually all food chains with the exception of those deep-sea communities which are based on chemo-autotrophic bacteria oxidizing sulphur. See **photosystems I and II**, **light reactions**, **dark reactions**, **Calvin cycle**, C_3, C_4, **CAM plants**, **photophosphorylation**, **photorespiration**, **Z scheme**. adj. *photosynthetic*.

Photosynthetically active radiation is that with a wave-length $\lambda = 400–700$ nm, active in photosynthesis. Abbrev. *PAR*.

Photosynthetic pigments are those involved in the absorption of light in photosynthesis. In plants and algae, including blue-green algae, they are *chlorophyll a* and the *accessory pigments*. In photosynthetic bacteria, other than the blue-green algae, they are the various baterial chlorophylls; photosynthetic bacteria are unable to oxidize water to free oxygen but rather oxidize some reducing agent, e.g. hydrogen, hydrogen sulphide, sulphur or some organic compounds.

---

synthesis comprising chlorophyll a, as *P680*, and other pigments and molecules, in which energy absorbed as light is used to transfer electrons from a strong oxidant to a weak reductant, the former oxidizing water to oxygen. Abbrev. *PS II*.

phototaxis Locomotory response or reaction of an organism or cell to the stimulus of light. adj. *phototactic*.

phototrophic Said of organisms obtaining their energy from sunlight.

phototropism A *tropism* in which a plant part becomes orientated with respect to the direction of light.

phragma A septum or partition; an apodeme of the endothorax formed by the infolding of a portion of the tergal region of a somite; an endotergite.

phragmoplast Complex of interdigitating microtubules aligned more or less parallel to the earlier spindle microtubules, developing at the end of mitosis and spreading like an expanding doughnut to the side walls while the *cell plate* forms in the centre. Characteristic of vascular plants, bryophytes and some algae.

phrenology Historically, the notion that a person's skull formations reflect mental abilities and personality characteristics.

pH stat Metabolic reactions which collectively maintain constant pH in the cytoplasm.

phycobilin One of a number of red (phycoerythrin) or blue-green (phycocyanin) *ac-*

cessory photosynthetic pigments, found in the Rhodophyceae, Cyanophyceae and Cryptophyceae.

phycobiont The algal partner in a lichen or other symbiotic association.

phycology The study of the algae.

Phycomycetes A probably unnatural group of fungi and fungus-like organisms comprising the Mastigomycotina (including the Oomycetes) and the Zygomycotina.

phyc-, phyco- Prefixes from Gk. *phykos*, seaweed and, by extension, algae.

phyletic classification A scheme of classification based on presumed evolutionary descent.

phyllid The leaf of a moss or liverwort.

phyllobranchia A gill composed of numerous thin plate-like lamellae.

phylloclade A flattened leaf-like stem, functioning as a photosynthetic organ. Cf. **cladode**.

phyllode A flat, more or less expanded, leaf-like petiole functioning as a photosynthetic organ, usually in the absence of a leaf blade.

phyllody The abnormal, often pathological, development of leaves in place of the normal parts of a flower.

phylloplane, phyllosphere The leaf surface, esp. as a habitat for micro-organisms.

phyllopodium The thin leaflike swimming foot characteristic of some crustaceans. Cf. **stenopodium**.

phyllotaxis, phyllotaxy The arrangement of leaves on a stem.

phylo- Prefix from Gk. *phylon*, race.

phylogeny, phylogenesis The evolutionary development or history of groups of organisms. adj. *phylogenetic*. Cf. **ontogeny**.

phylum A category or group of related forms constituting one of the major subdivisions of the animal kingdom. A *division* in plants.

physical containment The construction of laboratory or work station so as to prevent the contamination of the worker or the environ by harmful organisms.

physio- Prefix from Gk. *physis*, nature.

physiography The science of the surface of the earth and the inter-relations of air, water and land.

physiological Relating to the functions of plant or animal as a living organism.

physiological anatomy The study of the relation between structure and function.

physiological drought The condition in which a plant is unable to take in water because of low temperature, or because the water available to it holds substances in solution which hinder absorption by the plant.

physiological psychology The study of anatomy and physiology in relation to psychological phenomena.

physiological race *Biological race, forma specialis*. A group of individuals within the morphological limits of a species but differing from other members of the species in habits (as host, larval food etc.), e.g. the several races of a parasitic fungus each confined to a different host.

physoclistous Of fish, having no pneumatic duct connecting the air bladder with the alimentary canal. Cf. **physostomous**.

physostomous Of Fish, not **physoclistous**, i.e. having a pneumatic duct.

phytic acid Inositol hexaphosphate, present in e.g. seeds as its calcium magnesium salt, phytin, perhaps as a phosphorus-storage compound.

phytoalexin A substance produced by a plant in response to some stimulus. Often phenolic or terpenoid, they inhibit the growth of some micro-organisms, esp. pathogenic fungi.

phytochemistry The study of the chemical constituents and esp. the *secondary* metabolites of plants.

phytochrome A blue-green, phycobilin-like pigment reported from seed plants and a wide variety of other photosynthetic eukaryotes, acting as the light receptor molecule in a number of morphogenetic processes including photoperiodism, reversal of etiolation and the germination of some seeds and spores. See P_{fr} and P_r.

phytoferritin An iron-protein complex, similar to the ferritin of animals, apparently a store of iron.

phytohaemagglutinins *PHA*. Lectins extracted from the beans of *Phaseolus vulgaris*. They bind to N-acetyl-beta-D-galactosamine residues and can agglutinate certain erythrocytes which bear these on their surface. The purified lectins are potent mitogens of T lymphocytes.

phytohormone A plant hormone.

phytology Syn. for *botany*.

phytopathology *Plant pathology*. The study of plant diseases, esp. of plants in relation to their parasites.

phytophagous, 'phytophilous Feeding on plants.

phyto-, -phyte Prefix and suffix from Gk. *phyton*, plant.

phytoplankton The photosynthetic members of the plankton.

phytoplankton blooms The very high densities of plankton which occur quickly, and persist for short times, usually at regular times of the year.

phytosanitary certificate A certificate of health for plants or parts of plants for export.

phytosociology The study of the association of plant species.

phytotoxic substance A substance toxic to plants; sometimes also refers to animals.

phytotoxin Plant substance toxic to animals or other organisms.

phytotron A large and elaborate **growth room** for plants, or a collection of such.

P$_i$ In biochemical equations, inorganic phosphate e.g. $H_2PO_4^{2-}$.

pia mater In Vertebrates, the innermost of the three membranes surrounding the brain and spinal cord, a thin vascular layer.

pica Unnatural craving for unusual food.

Piciformes An order of Birds, containing climbing, insectivorous and wood-boring species. The beak is hard and powerful, the tongue long and protrusible, and the feet zygodactylous. Woodpeckers, Toucans.

pico- SI prefix for 1-millionth-millionth, or 10^{-12}. Formerly, *micromicro-*. Abbrev. *p*.

pigeon's milk In Pigeons and Doves, a white slimy secretion of the epithelium of the crop in both sexes. It contains protein and fat, and is produced during the breeding season under the influence of **prolactin**, being regurgitated to feed the young.

pigment Substances which impart colour to the tissues or cells of animals and plants.

pigmentary colours Colours produced by the presence of drops or granules of pigment in the integument, as in most Fish. Cf. **structural colours**.

pigment cell See **chromatophore**.

pileate Crested.

pileus The cap of the fruiting body of some fungi, e.g. mushrooms or toadstools (Basidiomycotina).

piliferous layer The epidermis, bearing

root hairs, of a young root.

pilose Hairy, with long, soft hairs.

pin The long-styled form of such heterostyled flowers as the primrose, *Primula vulgaris*, with the pinhead-like stigma conspicuous at the top of the corolla tube. Cf. **thrum**. See **heterostyly**.

pinacocytes The flattened epithelial cells forming the outer part of the dermal layer in Sponges.

pincers Claws adapted for grasping, as *chelae, chelicerae*.

pineal apparatus In some Vertebrates, two median outgrowths from the roof of the diencephalon, one (originally the left) giving rise to the *parietal organ* and the other (originally the right) to the *pineal organ*, (also *body* or *gland*). In Cyclostomata the *pineal organ* forms an eyelike functional photosensitive structure involved in a diurnal rhythm of colour change, and may be sensitive to light in some Fish. It persists in higher Vertebrates and may function as an endocrine gland.

pineal eye An anatomically imprecise term referring to eyelike structures formed by one or other of the two outgrowths of the pineal apparatus. In Cyclostomata it derives from the pineal organ, but in Lizards it derives from the parietal organ.

pinna A leaflet that is part of a pinnate leaf; in Fish, a fin; in Mammals, the outer ear; in Birds, a feather or wing.

pinnate (1) Feather-like; bearing lateral processes. (2) In plants, having 4 or more regular divisions, lobes, veins etc. arranged in 2 rows along a common midrib, rachis or stalk, rather in the manner of the barbs of a feather, as in a pinnate leaf. Many ferns have leaves that are 2-, 3- or more pinnate.

pinnatifid Cut about half way into lobes in a **pinnate** fashion.

pinnatiped, pinniped Having the digits of the feet united by flesh or membrane. Cf. **fissiped**.

pinnule One of the lobes or divisions of a leaf that is 2- or more-*pinnate*.

pinocytosis 'Cell drinking'. Ingestion by cells of vesicles containing fluid from the environment. There are two forms: *micropinocytosis*, whereby small droplets or microvesicles are pinched off from the cell membrane and interiorized, carrying with them any materials selectively adsorbed at the cell surface. This process is common to many cells. Pinocytosis of ligand-receptor complexes into clathrin-coated vesicles is an important mechanism whereby ligands can be carried into the interior; *macropinocytosis*, or the ingestion of large vesicles or vacuoles, similar to phagocytosis is carried out by cells of the mononuclear phagocyte system.

pioneer species A species of which the members tend to be among the first to occupy bared ground; these plants are often intolerant of competition and esp. of shading, and may be crowded out as the community develops.

piscivorous Fish-eating.

pisiform Pea-shaped; as one of the carpal bones of Man.

pistil Ovary, style and stigma; either of a single carpel in an apocarpous flower, or of the whole gynoecium in a syncarpous flower.

pistillate Said of a flower with functional carpels but not functional stamens.

pit A localized, thin area of a cell wall, typically where the primary wall is not covered by a secondary wall. Plasmodesmata may be present. A pit is usually one half of a pit pair and the term is sometimes loosely used to mean a pit pair. See **bordered pit**, **simple pit**. Cf. **perforation**.

pit cavity The space within a pit, from pit membrane to cell lumen.

pitcher An urn-shaped or vase-shaped modification of a leaf, or part of a leaf, developed by certain plants; it serves as a means of trapping insects and other small animals which are killed and digested.

pithed Having the central nervous system (spinal cord and brain) destroyed.

pith medulla Ground tissue in the centre of a shoot or root.

pith ray See **medullary ray**.

pit membrane That part of the middle lamella and primary wall that lies across the distal end of a pit cavity.

pitted Having a secondary wall interrupted by pits.

pituitary gland The major endocrine gland of Vertebrates, formed by the fusion of a downgrowth from the floor of the diencephalon (the *infundibulum*) and an upgrowth of ectoderm from the roof of the mouth (*hypophyseal pouch*). The former becomes the *neurohypophysis*, comprising the neural lobe and infundibulum, and the latter becomes the *adenohypophysis*, comprising the *pars intermedia*. The *pars intermedia* and the neural lobe together form the posterior lobe. The *adenohypophysis* produces several hormones affecting growth, adrenal cortex activity, thyroid activity, reproduction (*gonadotropic hormones*) and melanophore cells. The *neurohypophysis* secretes two hormones, **vasopressin** and **oxytocin**.

pivot joint An articulation permitting rotary movements only.

placebo A pharmacologically inactive substance which is administered as a drug either in the treatment of psychological illness or in the course of drug trials.

placenta (1) In Eutheria, a flattened cake-like structure formed by the intimate union of the allantois and chorion with the uterine wall of the mother; it serves for the respiration and nutrition of the growing young. (2) The part of the ovary of a flowering plant where the ovules form and remain attached while they mature as seeds. (3) Any mass of tissue to which sporangia or spores are attached. adjs. *placental, placentate, placentiferous, placentigerous*.

Placentalia See **Eutheria**.

placentation In plants, the arrangement of the placentas in an ovary, and of the ovules on the placenta. In animals, the method of union of the foetal and maternal tissues in the placenta.

placenta vera A deciduate placenta in which both maternal and foetal parts are thrown off at birth. Cf. **semiplacenta**.

placode Any platelike structure; in Vertebrate embryos, an ectodermal thickening giving rise to an organ primordium.

Placodermi A class of Gnathostomata comprising the earliest jawed Vertebrates, known from fossils from the Silurian to Permian. All had a heavy defensive armour of bony plates and hyoid gill slits with no spiracle. The hyoid arch did not support the jaw. Most possessed paired fins.

placoid Plate-shaped, as the scales and teeth of Selachii.

plagio- Prefix from Gk. *plagios*, slanting, oblique.

plagiotropism A **tropism** in which a plant part becomes aligned at an angle to the source of the orientating stimulus. Cf. **orthotropism, diatropism**.

plague A disease of rodents due to infection with the *Yersinia pestis*, transmitted to Man by rat fleas, epizootics in rats invariably preceding epidemics. In Man the disease is characterized by enlargement of lymphatic glands (*bubonic plague*), severe prostration, a tendency to septicaemia, and occasional involvement of the lungs.

plain muscle See **unstriated muscle**.

planetary boundary layer That zone of the atmosphere affected by the biological and physical properties of the surface beneath it. It extends about 1 km above the surface.

plankton Animals and plants floating in the waters of seas, rivers, ponds and lakes, as distinct from animals which are attached to, or crawl upon, the bottom; esp. minute organisms and forms, possessing weak locomotor powers.

planogamete A motile or wandering gamete; a zoogamete.

planospore See **zoospore**.

planozygote A motile zygote.

plant A photosynthetic organism or one re-

lated to it. For classification purposes, it will always include the seed plants, almost always the pteridophytes and bryophytes, usually the algae and the fungi and sometimes the bacteria also.

planta The sole of the foot in land Vertebrates; the flat apex of a proleg in Insects. adj. *plantar*.

Plantae The plant kingdom.

plant cell culture See panel

plant genetic manipulation See panel.

plantigrade Walking on the soles of the feet, as Man. Cf. **digitigrade, unguligrade**.

plant pathology See **phytopathology**.

plantula A larval form of some Invertebrates, esp. Coelenterata; it consists of an outer layer of ciliated ectoderm and an inner mass of endoderm cells.

plaque Areas of cell destruction in a bacterial colony grown on a solid medium, or tissue culture monolayer preparation due to infection with a virus.

plasm Protoplasm, esp. in compound terms, as *germ plasm*.

plasma cells See panel.

plasmacytoma *Myeloma*. See panel.

plasmalemma The boundary membrane of the cell which regulates the passage of molecules between the cell and its surroundings. The plant *cell wall* is outside the plasmalemma. *Plasma membrane* is the commoner term for animal cells.

plasmalogen Phosphatide in which a hydrocarbon chain is bound to a glycerol carbon by an unsaturated ether bond rather than an ester link.

plasma membrane The bounding membrane of cells which controls the entry of molecules and the interaction of cells with their environment. Like most cell membranes it consists of a lipid bilayer traversed by proteins. *Plasmalemma* is the commoner term in botany.

plasma-, plasmo-, -plasm Prefixes and suffix from Gk. *plasma*, gen. *plasmatos*, anything moulded.

plasmid A genetic element containing nucleic acid and able to replicate independently of its host's chromosome. Often carries genes determining **antibiotic resistance**. Much used in recombinant DNA procedures.

plasmocyte See **leucocyte**.

plasmodesma A fine tube of protoplasm which connects the protoplasts of two adjacent cells through the intervening wall. See **primary pit field, symplast**.

plasmodium A multinucleate mass of naked (wall-less) protoplasm, i.e. a *syncytium* formed by the union of uninucleate individuals without fusion of their nuclei. It moves in an amoeboid fashion and constitutes the thallus in the Myxomycetes. Cf.

plant cell culture

Including *tissue* and *organ culture*. The culture of explants, i.e. cells, tissues or organs from plants, under aseptic conditions in or on a sterile growth medium typically containing sugar(s), as an energy and carbon source, mineral salts and growth substances, sometimes solidified with e.g. agar.

Cells in an explant will grow and divide indefinitely in an appropriate medium. Depending on the conditions, cell growth and division may be more or less disorganized, forming individual cells or clumps of mostly parenchyma cells with little differentiation (callus cultures on solid media and suspension cultures in liquid media). Alternatively, organized growth may persist or develop: isolated shoots or roots may grow more or less indefinitely and, given appropriate levels of growth substances, organized meristems may develop in callus cultures to give roots, shoots or small plants (*plantlets*). Sometimes plantlets arise through a developmental process resembling that of normal embryos and are then termed *embryoids*. Techniques exist for the 'weaning' of plantlets and their continued growth in ordinary soil.

Cell culture, because of the possibilities it offers for the control of conditions, is important in research into e.g. the cell cycle, cell growth and development and the **totipotency** of plant cells. It also has important practical applications.

In **meristem culture**, shoot apical meristems, which even in virus-infected plants may be free of virus particles esp. after heat treating the parent plant, can be excised and cultured to give virus-free material from many vegetatively-propagated crop plants.

In *micropropagation*, shoot-tips may be cultured under conditions which promote the growth of a mass of proliferating lateral shoots which, after 6–8 weeks, may be separated and subcultured. Rooting may be induced by transfer to a suitable medium. This method is especially useful for the rapid multiplication of pathogen-free plants and new cultivars and is used with e.g. strawberries, potatoes and some ornamental plants. Alternatively, a callus or suspension culture may be established, subcultured and eventually induced to form plantlets. Such cultures are however prone to *somaclonal variation*, e.g. chromosomal abnormality and other genetic change.

Embryo culture may be used to 'rescue' embryos that might otherwise abort. *Ovule culture* may be used to overcome some kinds of incompatibility barriers. *Protoplast culture* is used in **plant genetic manipulation**.

In *anther culture*, anthers are excised and transferred to a suitable medium at an appropriate stage in development and may then produce haploid cultures, embryoids and plantlets. Such cultures or plantlets may, e.g. if treated with **colchicine**, develop into diploid plants, *autodiploids* which are completely homozygous, a condition which is otherwise only approached after many generations of inbreeding.

Plant cell cultures have potential as sources for pharmaceuticals, flavourings etc. which are presently extracted from plants and a few are in commercial production from cultures. The antiseptic dye, *shikonin*, was the first. **Immobilized culture** is a promising method.

---

pseudoplasmodium. adj. *plasmodial*.

plasmogamy The fusion of cytoplasm as distinct from fusion of nucleoplasm; *plastogamy*. In plants, the fusion of protoplasts, e.g. of gametes in a sexual reproductive cycle. In most organisms it is followed more or less immediately by karyogamy (fusion of nuclei); in some fungi it may result in a hete-rokaryon. See **dikaryophase**.

plasmolysis Process in which the protoplast of a plant cell shrinks away from the wall following water loss due to exposure to a solution of higher osmotic pressure, the wall being permeable to the solute but the plasmalemma not. Cf. **cytorrhysis**.

plastic Said of a genotype which can de-

plant genetic manipulation

The use of various techniques, except in most usages those of conventional breeding, to produce plants containing foreign DNA as part of their genomes. The techniques include both the isolation of DNA from another organism and its transfer to a plant and also protoplast fusion and culture. *Genetic engineering* and *genetic manipulation* usually refer to the former rather than the latter techniques. Both however make it possible to transfer genes from much less closely related organisms than does conventional breeding, even from bacteria with DNA transfer. Plants in which foreign DNA is incorporated and expressed are called *transformed* or *transgenic* plants. See **plant cell culture** for additional methods.

DNA containing a desired gene may be prepared as **cDNA** or be selected from a **DNA library**. One method of transferring the DNA to the plant is to use a natural vector, most successfully the **Ti plasmid** of *Agrobacterium tumefaciens*, which inserts DNA into the nuclear genome; plants may then be regenerated from the transformed cells by plant cell culture. In a second method, **protoplasts** are stimulated to take up the DNA directly by polyethylene glycol or by **electroporation**, the plants being regenerated from protoplasts by **protoplast culture**. Both methods have been more successful with dicotyledons than with cereals because *Agrobacterium* naturally infects dicotyledons only and current methods have rarely achieved regeneration from the protoplasts of cereals. In a third method, successful with at least one cereal, DNA is injected directly into the bud which will give rise to the ear.

In *protoplast fusion*, protoplasts are prepared from cells of two plants and induced to fuse by polyethylene glycol or electric shock (*electrofusion*). The fused protoplast initially contains the complete nuclear, chloroplast and mitochondrial genomes of both plants. During protoplast culture, individual chromosomes, chloroplasts or mitochondria may be lost which, together with recombination in both mitochondria and chloroplasts, may result in cells with almost any combination of parental genomes. Some of these cells may be capable of regenerating hybrid or **cybrid** plants,

Genetic manipulation is important in research into the functioning of the genome and has considerable promise in the production of new cultivars. Useful traits which may be introduced into crop plants include: herbicide resistance (to increase the range of herbicides that can be used with a given crop), disease resistance (plants able to make the coat protein of a virus may be resistant to that virus), pest resistance (plants making the toxin of *Bacillus thuringiensis*, toxic to caterpillars) and male sterility (to facilitate the production of F_1 hybrids). Protoplast fusion has made possible the transfer of **cytoplasmic male sterility** and some cytoplasmically-inherited herbicide resistance between crops. Once made, transgenic plants may be used in conventional breeding programs. See also **genetic manipulation**.

velop into substantially different phenotypes under different conditions.

plastid One of a class of cytoplasmic organelles in plants and eukaryotic algae, comprising the chloroplasts and related organelles, surrounded by 2 membranes and containing DNA, e.g. *chloroplast, etioplast, leucoplast, amyloplast, chromoplast*.

plastochron, plastochrone The interval of time between the appearance of successive leaf primordia at the shoot apex, or between other similar successive events.

plastocyanin Copper-containing protein, involved in the transfer of electrons from photosystem II to photosystem I in photosynthesis.

plastoquinone Terpenoid involved in the transfer of electrons from photosystem II to photosystem I in photosynthesis.

plastron (1) The ventral part of the bony exoskeleton in Chelonia; any similar structure. (2) In some aquatic Insects (e.g. Coleoptera and Hemiptera) a thin air film over certain parts of the body held by minute

---

plasma cells and plasmacytoma

Name given to the end stage of differentiation of B lymphocytes into cells wholly devoted to synthesis and secretion of immunoglobulins. They have a very highly developed endoplasmic reticulum and a prominent Golgi apparatus, and in stained preparations are recognizable by a basophilic cytoplasm with a juxta-nuclear 'vacuole', and an eccentrically placed nucleus with a 'clock face' appearance. Plasma cells are not known to revert to resting B lymphocytes, but rather to become exhausted and die. They are prominent in sites of intensive antibody synthesis.

A *plasmacytoma* or *myeloma* is a tumour of plasma cells, which is often preferentially localized in the bone marrow and where it produces typical erosion of the local bone (hence the term *myeloma*). Plasmacytomas continue to secrete an immunoglobulin product, although this may sometimes have sections of the normal amino acid sequence missing. Such tumours arise spontaneously but rarely in several species; in certain strains of mice they can be caused to appear regularly following intraperitoneal administration of mineral oil.

Mouse **myelomas** are widely used to study immunoglobulin synthesis and gene structure, and for the production of **hybridomas**.

---

hydrofuge hairs. This serves as a physical gill rather than as a store of air. adj. *plastral*.

plate See **plax**.

platelet activating factor *PAF*. A lipid released by various cells, including basophil leucocytes and monocytes, in the presence of antigen. Induces platelet aggregation and degranulation. Activity is transient since PAF is inactivated by phospholipase A which is also released.

platydactyl Having the tips of the digits flattened.

Platyhelminthes A phylum of bilaterally symmetrical, triploblastic Metazoa; usually dorsoventrally flattened; the space between the gut and the integument is occluded by parenchyma; the execretory system consists of ramified canals containing flame cells; there is no anus, coelom or haemocoele; the genitalia are usually complex and hermaphrodite. Flat worms.

platysma A broad sheet of dermal musculature in the neck region of Mammals.

platyspermic Having seeds which are flattened in transverse section, as in the Cordaitales and the conifers. Cf. **radiospermic**.

plax A flat platelike structure, as a lamella or scale.

play A concept used by both students of animal behaviour and developmental psychologists without rigid definition. It occurs largely but not solely in the young of warm-blooded mammals and can involve almost any behaviour; play behaviours tend to occur in isolation from their normal function, are given voluntarily rather than in conditions of necessity, and often merge behaviours from different functional systems. Exaggerated movements often occur and may be preceded

or accompanied by a signal that the activity is a playful one, (e.g. play fights).

play therapy A form of child psychotherapy which uses play as the means of communication to reveal unconscious conflicts, and which encourages children to vent their feelings through symbolic play.

pleasure principle In Freudian theory, the motive to seek immediate pleasure and gratification, without regard to consequence, which governs the **id**.

Plectomycetes A class of fungi in the Ascomycotina in which the fruiting body (ascocarp) is a cleistothecium. Includes *Eurotium* and allied genera which are the perfect stages of many spp. of *Aspergillus* and *Penicillium*.

plectostele Type of protostele in which the xylem and phloem are arranged in alternating planes across the stele.

pleiomerous Having a large number of parts.

pleiomorphic, pleomorphic, pliomorphic Having more than one shape.

pleiotropy The condition when one gene affects more than one character. adj. *pleiotropic*.

plei-, pleio-, pleo-, plio- Prefixes from Gk. *pleion*, more.

pleochromatic Presenting different colours according to changes in the environment or with different physiological conditions.

pleomorphism, pleomorphous Same as *polymorphism, polymorphic*.

pleopod In Arthropoda (esp. Crustacea) an abdominal appendage adapted for swimming.

plerome A **histogen**, the precursor of the stele.

pleura The serous membrane lining the pulmonary cavity in Mammals and Birds.

pleurapophysis A lateral vertebral process; usually applied to the true ribs.

pleurocarp A moss bearing the archegonia, and therefore the capsule and its stalk, on a short side branch, not at the tip of a main stem or branch. Cf. **apocarp.**

pleurodont Having the teeth fastened to the side of the bone which bears them, as in some Lizards.

pleurogenous Borne on a lateral position.

pleuron In some Crustacea, a lateral expansion of the tergite; more generally, in Arthropoda. The lateral wall of a somite. pl. *pleura.* adj. *pleural.*

pleuropneumonia - like organisms A group of organisms, closely resembling the pleuropneumonia organisms, isolated from the throat and vagina. See **Mycoplasmatales.** Abbrev. *PPLO.*

pleur-, pleuro- Prefixes from Gk. *pleura,* side.

plexus A network; a mass of interwoven fibres, as a *nerve plexus.* adj. *plexiform.*

plica A fold of tissue; a foldlike structure. adjs. *plicate, pliciform.*

plicate Folded.

plio- Prefix. See **plei-.**

pliomorphic See **pleiomorphic.**

ploidy Pertaining to chromosome number, e.g. *haploid, diploid, polyploid.*

plumae Feathers having a stiff shaft and a firm vexillum, and usually possessing hamuli; they appear on the surface of the plumage and determine the contours of the body in addition to forming the remiges and rectrices. adjs. *plumate, plumous, plumose, plumigerous.*

plume (1) A feather; any featherlike structure. (2) A light, hairy or feathery appendage on a fruit or seed serving in wind dispersal.

plumose Hairy; feathered.

plumulae Feathers having a soft shaft and vane and lacking hamuli; in some cases the shaft is entirely lacking; they form the deep layer of the plumage. adjs. *plumulate, plumulaceous.*

plumule (1) Down feather. Form covering of nestling, sometimes persisting in adult, between the contour feathers. Barbules and hooks little developed. (2) Embryonic shoot above the cotyledon or cotyledons in a seed. See **epicotyl.**

plurilocular Divided into several compartments by septa. Cf. **unilocular.**

plus sign (+) Symbol used before the generic or specific epithet to denote, respectively, an intergeneric or interspecific **graft chimaera.**

plus strain, (+) strain One of two arbitrarily designated, mating types of a heterothallic species. See **heterothallism.**

PMSF See **phenylmethylsulphonyl fluoride.**

pneumathode A more of less open outlet of the ventilating system of a plant, usually some loosely packed cells on the surface of the plant; through it exchange of gases between the air and the interior of the plant is facilitated.

pneumatic Containing air, as in physostomous Fish, the *pneumatic duct* leading from the gullet to the air bladder, and in Birds, those bones which contain air cavities.

pneumato- See **pneum-.**

pneumatocyst (1) Any air cavity used as a float. (2) In Fish, the **air bladder.** (3) The cavity of a pneumatophore.

pneumatophores Specialized roots of swamp plants and mangroves which grow vertically upwards into the air from the root system in the mud and which, containing much aerenchyma, apparently facilitate gas exchange for the submerged roots.

pneumococcal polysaccharide Type specific polysaccharide present in the capsules of *Streptococcus pneumoniae.* The capsules are related to the virulence of the organism, since they prevent ingestion by neutrophils unless coated with antibody. Each type of pneumococcus has different oligosaccharide repeating units in its capsule, so protection is type specific. The pneumococcal polysaccharides are **thymus-independent antigens.** Pneumococcal vaccines are composed of a mixture of polysaccharides derived from the types prevailing in the community.

pneumococcus A Gram-positive diplococcus, a causative agent of pneumonia, though it may occur normally in throat and mouth secretions.

pneumon-, pneumono- Prefixes from Gk. *pneumon,* gen. *pneumonos,* lung.

pneumostome In Arachnida, the opening to the exterior of the lung books. In Pulmonata, the opening to the exterior of the lung formed by the mantle cavity.

pneum-, pneumo-, pneumat-, pneumato- Prefixes from Gk, *pneuma,* gen. *pneumatos,* breath.

Poaceae See **Gramineae.**

pod See **legume.**

podal Pedal.

podex The anal region. adj. *podical.*

Podicipitiformes An order of Birds, containing compact-bodied species of cosmopolitan distribution. Almost completely aquatic, and build floating nests. Toes lobate and feet placed far back. Grebes.

podite A walking leg of Crustacea.

podium (1) In land Vertebrates, the third or distal region of the limb; manus or pes; hand or foot. (2) Any footlike structure, as the locomotor processes or tubefeet of Echino-

---

poison

Any substance or matter which, introduced into the body in any way, is capable of destroying or seriously impairing life. Poisons include products of decomposition or of bacterial organisms, and in some cases the organisms themselves; also very numerous chemical substances forming the residue of industrial, agricultural and other processes (see **pollution**).

Poisons are generally classified as *irritants* (e.g. cantharides, arsenic) and *corrosives* (e.g. strong mineral acids, caustic alkalis); *systemics* and *narcotics* (e.g. prussic acid, opium, barbiturates, henbane); *narcotic-irritants* (e.g. nux vomica, hemlock).

Among gases, carbon monoxide, carbon dioxide, sulphuretted hydrogen, sulphur dioxide, sulphide of ammonium and numerous others (e.g. fumes of leaded petrol) are of significance in industry and daily life.

---

dermata. pl. *podia.* adj. *podial.*

podomere In Arthropoda, a limb segment.

pod-, podo-, -pod Prefixes and suffix from Gk. *pous,* gen. *podos,* foot.

podsol, podzol A common soil type developed on siliceous mineral soil in areas of very high rainfall and low evaporation. The soil has an ash coloured layer below the surface, from which minerals and clay particles have been washed, and an orange-brown deeper layer where some of these have accumulated. Sometimes there is an impermeable layer of oxidized iron. The overall reaction is acid and the soil usually supports only a calcifuge flora.

Pogonophora A phylum of coelomate animals possibly related to the Hemichordata, without an alimentary canal or bony tissue, and having a simple nervous system. There is a muscular heart, and the blood contains haemoglobin. The extremely long bodies are divided into three sections, the posterior one serving to anchor the animals in the chitinous tubes in which they live, and the anterior part bearing well-developed tentacles which probably serve to gather food, digest it and absorb the products of digestion.

poikilohydric Lacking structures or mechanisms to regulate water loss and, hence, having water content determined rapidly by the water potential of the environment, as in algae, lichens, bryophytes, submerged vascular plants. Cf. **homoiohydric.**

poikilosmotic Of an aquatic animal, being in osmotic equilibrium with its environment, the concentration of its body fluids changing if the environment becomes more dilute or more concentrated. Marine Invertebrates are frequently poikilosmotic. Cf. **homoiosmotic.**

poikilothermal See **cold-blooded.**

poikil-, poikilo- Prefixes from Gk. *poikilos,* variegated, variable, many-coloured.

point quadrat Device for measuring the canopy cover by lowering a thin pointed shaft into the vegetation many times.

poison See panel.

Poisson distribution A probability distribution often applied to the number of occurrences of particular (esp. rare) events in a particular time period.

pokeweed mitogen Mitogenic lectins obtained from *Phytolacca americana.* There are five different substances of which one is active for both T and B lymphocytes, but the others act only on T lymphocytes.

polar body During the maturation divisions (*meiosis*) of the ovum, one daughter set of chromosomes passes to a small cell, the polar body, and takes no further part in gametogenesis.

polarity Existence of a definite axis.

polar nuclei The two (typically haploid) nuclei which migrate from the poles to the centre of the embryo sac of an angiosperm and there fuse with the second male gamete to form the, typically triploid, primary **endosperm** nucleus.

pole Point; apex; an opposite point (as *aboral pole*); axis. Specifically, one end of the achromatic spindle in cell division, where the spindle fibres come together.

pollo- Prefix from Gk. *polios,* grey.

poliomyelitis Inflammation of the grey matter of the spinal cord. Due to infection, chiefly of the motor cells of the spinal cord, with a virus; characterized by fever and by variable paralysis and wasting of muscles. Often used as a synonym for *acute anterior poliomyelitis,* or *infantile paralysis.*

pollen The microspores of seed plants. The individual pollen grains may contain an immature or mature male gametophyte.

pollen analysis A method of investigating the past occurrence and abundance of plant species by a study of pollen grains and other spores preserved in peat and sedimentary deposits. Also *palynology.* See **sporopollenin.**

pollen chamber A cavity in the micropylar end of the nucellus of the ovule of some gymnosperms in which pollen grains lodge as a result of pollination and where they develop further and germinate.

pollen count A graded assessment of the level of plant pollen in the atmosphere; of importance to sufferers from **hay fever**.

pollen flower A flower which produces no nectar but has abundant pollen which attracts pollinating insects.

pollen mother cell A cell, typically diploid, in a pollen sac, which divides by meiosis to form a **tetrad** of pollen grains; the **microsporocyte** of a seed plant.

pollen sac A chamber in which pollen grains (microspores) are formed in seed plants, e.g. such chambers in the **anther**.

pollen tube A tubular outgrowth of the intine, produced on the germination of a pollen grain. In angiosperms it grows to the embryo sac and there delivers the male gamete(s), *siphonogamy*.

pollex The innermost digit of the anterior limb in Tetrapoda.

pollination The transfer of pollen from the pollen sac to the micropyle in gymnosperms (see **pollination drop**) or to the stigma in angiosperms, usually by means of wind, insects, birds, bats or water.

pollination drop A drop of sugary fluid that is secreted into the micropyle at the time of pollination in many gymnosperms, and which traps pollen grains that may then float up to the nucellus or be drawn there by the reabsorption of the drop.

pollinium A mass of pollen grains, held together by a sticky secretion or retained within the pollen sac wall, which is transported usually by an insect, in the pollination of e.g. orchids.

pollution Modification of the environment by release of noxious materials, rendering it harmful or unpleasant to life.

poly- Prefix from Gk, *polys*, many.

polyacrylamide gel electrophoresis Technique used for separating nucleic acids or proteins on the basis of charge, shape and size. The highly cross-linked polymer of acrylamide forms a transparent gel matrix through which macromolecules move under the influence of an electric field. Proteins are frequently electrophoresed in the presence of the detergent, *sodium dodecylsulphate*, which forms polyanionic complexes whose mobility is a simple function of size. Abbrev. *PAGE*.

polyadelphous Said of a flower having the stamens joined by their filaments into several separate bundles.

polyandrous Said of a flower having a large, indefinite number of stamens.

polyandry The practice of a female animal mating with more than one male.

polyarch Said of a stele having many protoxylem strands.

polycarpellary Consisting of many carpels.

polycarpic Potentially able to fruit many times, as in most perennial plants. Cf. **monocarpic**. See **hapaxanthic**.

polycarpous See **apocarpous**.

Polychaeta A class of Annelida of marine habit, having locomotor appendages (parapodia) bearing numerous setae; there is usually a distinct head; the perivisceral cavity is subdivided by septa; the sexes are generally separate, with numerous gonads; and development after external fertilization involves metamorphosis, with a free swimming trochosphere larva. Marine Bristle worms.

polychasium A cymose inflorescence in which the branches arise in sets of three or more at each node.

polyclimax theory Idea, proposed by A.G. Tansley in 1939, that not one but several climax vegetation types are possible in one climatic region because the environment is influenced by local factors such as soil and the activity of animals. See **climax**.

polyclonal activators General term for substances that activate many clones of lymphocytes as opposed to an antigen which only activates clones which have receptors which recognize it. See **lectin, mitogen**. Unless cloned antigen-reactive cell lines are available, the only way in which to study the behaviour of normal B and T lymphocytes has been to stimulate them with polyclonal activators.

polycormic Having several strong vertical trunks.

polycotyledonous Having more than two cotyledons, e.g. pine.

polycyclic Said of vascular tissues of a stem having two or more concentric solenosteles, dictyosteles or rings of vascular bundles.

polydactylism, polydactyly Having more than the normal number of digits. adj. *polydactylous*.

polyembryony The development of more than one embryo in one ovule, seed or fertilized ovum.

polygamous (1) Mating with more than one of the opposite sex during the same breeding season. n. *polygamy*. (2) Having staminate, pistillate and hermaphrodite flowers on the same and on distinct individual plants.

polygenes The genes that control *quantitative characters* and whose individual effects are too small to be detected.

polygenic Of a character whose genetic component is determined by many genes with individually small effects.

---

polymerase chain reaction

Method of amplifying DNA sequences a few hundred bases long by over a million-fold without using methods of **genetic manipulation** which need biological vectors. It requires two oligonucleotide **primers** which flank the sequence to be amplified but which bind to opposite strands.

A cycle in which the DNA, and later the newly synthesized polynucleotide, is first denatured at high temperature (95°), then annealed to the primers at a lower temperature (about 50°) followed by the polymerase reaction at an intermediate temperature (about 70°), is repeated some 20 times. Although different DNA **polymerases** can be used, the highest specificity is obtained (with the temperatures mentioned) when a thermostable polymerase like that isolated from the bacterium, *Thermus aquaticus*, is used. This enzyme survives the high temperature of the denaturation step and need not be replaced at each cycle. Results suggest that the method can amplify a single target sequence in a million cells.

---

polygoneutic Having several broods in a year.

polygraph *Lie detector.* A physiological recording device which can pick up physiological changes in the form of electrical impulses, and record them onto a moving roll of paper.

polygynous Said of a male animal which mates with more than one female.

polymerase Enzymes producing a polynucleotide sequence, complementary to a pre-existing *template* polynucleotide. DNA polymerase requires a **primer** from which to start polymerization whereas RNA polymerase does not.

polymerase chain reaction See panel.

polymerous Having many members in a whorl, e.g. many petals.

polymorphism (1) Condition of a population in which variant alleles are more common than can be explained simply by mutation, suggesting a selective advantage for heterozygotes or for uncommon phenotypes. (2) The occurrence of different structural forms at different stages of the life cycle of the individual. adjs. *polymorphic*, *polymorphous*.

polymorphonuclear leucocyte Cell of the myeloid series which in its mature form has a multilobed nucleus and cytoplasm containing granules. They are present in inflammatory exudates and in the blood, where unlike erythrocytes, they have only a short residence time. Term includes neutrophil, eosinophil and basophil leucocytes.

polynucleotide Syn. for long chain of nucleic acid.

polyoestrous Exhibiting several oestrous cycles during the breeding season. Cf. **monoestrous**.

polyoma A small DNA virus oncogenic for mice. Similar to **SV 40**.

polyoma virus A virus that can induce a wide variety of cancers in mice, hamsters, and also in other species.

polyp An individual of a colonial animal.

polypeptide A linear condensation of *amino acids* which, alone or associated with others, forms a *protein* molecule.

polypetalous Having a corolla of separate petals which are not fused with each other. Cf. **gamopetalous**.

polyphagous Feeding on many different kinds of food, as *Sporozoa* which exist in several different cells during one life cycle, or phytophagous Insects with many food plants.

polyphyletic group, polyphyly (1) Any group of species whose members derive from two or more independent ancestral lines. (2) In cladistics, a group that does not include the most recent ancestor of its members. Cf. **monophyletic** and **paraphyletic groups**.

polyphyllous Having a perianth of separate members, not fused with each other. Cf. **gamophyllous**.

polyphyodont Having more than 2 successive dentitions. Cf. **diphyodont, monophyodont**.

polyploid Having more than twice the normal haploid number of chromosomes. The condition is known as *polyploidy*. *Artificial polyploidy*, which can be induced by the use of chemicals (notably **colchicine**), is of economic importance in producing hybrids with desired characteristics.

polyprotodont Having numerous pairs of small subequal incisor teeth; pertaining to the Polyprotodontia. Cf. **diprotodont**.

polysepalous Having a calyx of separate sepals which are not fused with each other. Cf. **gamosepalous**.

polysome, polyribosome An assembly of ribosomes held together by their association with a molecule of **messenger RNA**.

polysomy A chromosome complement in which some of the chromosomes are present in more than the normal diploid number, e.g. a *trisomic*.

polyspermy Penetration of an ovum by several sperms.

polyspondyly The condition of having more than two vertebral centra corresponding to a single myotome. adjs. *polyspondylic, polyspondylous*.

polystely The condition in which the vascular tissue in a stem exists as two or more separate steles interconnecting only at intervals such as where the stem branches, e.g. some *Selaginella* spp.

polystichous Arranged in several rows.

polyteny Special case of **polyploidy** in which chromatids remain very closely paired after duplication, through all or part of their length. In some cases homologues are also closely paired, e.g. in **salivary glands** of dipterans.

polytokous Bringing forth many young at a birth; prolific; fecund. n. *polytoky*.

polytrophic In Insects, said of ovarioles in which nutritive cells alternate with the oöcytes; more generally, obtaining food from several sources. Cf. **acrotrophic**.

Polyzoa Another name for the Ectoprocta or Bryozoa.

pome False fruit, of the subfamily Pomoideae of the Rosaceae, in which the true fruit forms the core, containing the seeds, surrounded by a greatly expanded, fleshy receptacle, e.g. apple.

pons A bridgelike or connecting structure; a junction. pl. *pontes*. adj. *pontal*.

pons Varolii In Mammals, a mass of transversely coursing fibres joining the cerebellar hemispheres.

pontal flexure The flexure of the brain occurring in the same plane as the cerebellum; it bends in the reverse direction to the primary and nuchal flexures and tends to counteract them.

population Any specified reproducing group of individuals.

pore A small more or less rounded aperture, esp. (1) the aperture of a **stoma**, (2) a rounded aperture (rather than a slit) in a dehiscing anther or capsule, (3) a vessel seen in a transverse section of wood, (4) a more or less circular germinal aperture in the wall of a pollen grain. Also *porus*. See **aperturate**. adjs. *poriferous, porous, poriform*.

poricidal Opening or dehiscing by pores.

Porifera A phylum of sessile, aquatic, filter feeding animals. They lack sense organ systems but possess tissues, e.g. epithelial cells (pinacocytes) and flagellated cells (choanocytes). They have skeletons of silica or collagen. Sponges.

porogamy The entry of the pollen tube into the ovule, in an angiosperm, by way of the micropyle (the commoner route). Cf. **chalazogamy**.

porometer An instrument for investigating the *stomal aperture* by measuring the rate at which air (or other gas) flows through the leaf (viscous flow porometry) or the rate at which water vapour diffuses through the leaf (diffusion porometry).

porous dehiscence The liberation of pollen from anthers, and of seeds from fruits through pores in the wall of the containing structure.

porrect Directed outwards and forwards.

porta Any gatelike structure. adj. *portal*.

portal system, circulation A vein which breaks up at both ends into sinusoids or capillaries; as (in Vertebrates) the *hepatic portal system*, in which the hepatic portal vein collects from the capillaries of the alimentary canal and passes the blood into the sinusoids and capillaries of the liver, and the *renal portal system*.

porus Same as **pore** sense (4).

position effect Differential effect on transcription and repression of the same gene in different chromosomal locations.

positive reaction A tactism or tropism in which the plant moves, or the plant member grows, from a region where the stimulus is weaker to one where it is stronger.

positive reinforcement A situation in which a response is followed by a positive event or stimulus (the positive reinforcer or reward) which increases the likelihood that the response will be repeated.

positive taxis See taxis.

post- Prefix from L. *post*, after.

post-capillary venules Small vessels through which blood flows after leaving the capillaries and before reaching the veins. The main route by which leucocytes migrate into inflammatory sites is between the endothelial cells of post-capillary venules. Specialized venules with high, rather than flattened endothelial lining are present in the thymus dependent area of lymph nodes. Through these, lymphocytes recirculate from blood to lymph.

postcardinal Posterior to the heart, as the *postcardinal* sinus of Selachii.

postcaval vein In higher Vertebrates, the posterior vena cava conveying blood from the hind parts of the body and viscera to the heart. Called *inferior vena cava* in Man.

postclimax A relict of a former vegetational climax held under edaphic control in an area where the climate is no longer favourable for its development.

posterior (1) The side nearer the axis of a bud, flower or other lateral structure. (2) In a bilaterally symmetrical animal, further away from the head region; behind. Cf. **anterior**.

post-fertilization stages The developmental processes which go on between the union of the gametic nuclei in the embryo sac and the maturity of the seed.

post-hypnotic suggestion Refers to suggestions made to an individual during a hypnotic trance, which determine a behavioural or experiential response occurring after the individual has returned to ordinary consciousness.

postical Relating to or belonging to the back or lower part of a leaf or stem.

post-trematic Posterior to an aperture; as, in Selachii, that branch of the ninth cranial nerve which passes posterior to the first gill cleft. Cf. **pretrematic**.

postventitious Delayed in development.

postzygapophysis A facet or process on the posterior face of the neurapophysis of a vertebra, for articulation with the vertebra next behind. Cf. **prezygapophysis**.

potamous Living in rivers and streams.

potash Potassium, usually expressed notionally as K_2O, in a fertilizer.

potential See **latent**.

potometer An apparatus to measure the rate of uptake of water by a plant or a detached shoot, etc. and often, thus, indirectly to estimate transpiration.

pouch Any saclike or pouchlike structure; as the abdominal *brood pouch* of marsupial Mammals.

powdery mildew One of several plant diseases of e.g. cereals and apples, caused by biotrophic fungi of the order Erysiphales.

pox viruses A group of fairly large viruses, brick-shaped in shadow-cast electron micrographs, usually characterized by the formation of cytoplasmic inclusion bodies in the cells they invade; usually cause skin lesions, e.g. smallpox virus.

PPQ bar See **pterygopalatoquadrate bar**.

P-protein *Phloemprotein*. Present in phloem cells esp. in sieve elements.

P$_r$ Form of the plant pigment **phytochrome** having a peak of absorption at $\lambda \cong 660$ nm which converts it to P$_{fr}$.

prae- Prefix. See **pre-**.

praecoces Birds which when hatched have a complete covering of down and are able at once to follow the mother on land or into water to seek their own food. Cf. **altrices**.

Prausnitz-Kustner reaction Skin reaction for detection and measurement of human reaginic (IgE) antibodies. Serum is injected intracutaneously into a volunteer. Reaginic antibody fixes to skin cells whereas other immunoglobulins diffuse away. Antigen is injected into the same site after 48 hours. A **weal and flare response** develops immediately and its intensity can be graded. The test is not now used because of the danger of transferring hepatitis, and has

been replaced by *in vitro* assays.

pre-adaptation Change of structure preceding appropriate change of habit.

prebiotic The very different conditions existing on earth before the appearance of life, which provided an environment in which the first living organisms could evolve from nonliving molecules.

precaval vein The anterior vena cava conveying blood from the head and neck to the right auricle. Called *superior vena cava* in Man.

prechordal Anterior to the notochord or to the spinal cord.

precipitation In immunology describes the formation of a visible aggregate when antigen and antibody are mixed in aqueous solution so as to form large macromolecular complexes (see **optimal proportions**). These form flocculent precipitates in a tube and appear as a white line in a gel medium where the antigen and antibody interact after diffusing toward one another. Precipitation only occurs when the proportions of the reagents are suitable, even though antigen-antibody combination has taken place.

precipitin test A serological test in which the reaction between soluble antigen and antibody results in the formation of a visible precipitate.

preclimax A *seral* stage which just precedes the climax.

precoracoid An anterior ventral bone of the pectoral girdle in Amphibians and Reptiles, corresponding to the epicoracoid of Monotremes.

predation A form of species interaction in which an individual of one species of animal (the *predator*) directly attacks, kills and eats, one of another species (the *prey*). The predator is usually larger than its prey. Cf. **parasite**.

predator See **predation**.

predentin(e) The soft primitive dentin, composed of reticular Korff's fibres, which later becomes calcified to form dentin(e).

preen gland See **oil gland**.

preening A form of grooming behaviour performed by birds as part of feather maintenance.

preferendum That part of the range of a species in which it lives and functions most successfully.

preferential mating See **sexual selection**.

prefloration Aestivation.

prefoliation Vernation.

pre-frontal lobotomy A form of psychosurgery in which the nerve fibres connecting the thalamus and frontal lobes are cut, in order to reduce the effects of emotions on intellectual processes (the premises for this supposition are dubious).

pregnancy *Gestation.* The state of being with child.

prehallux In Amphibians and Mammals, a rudimentary additional digit of the hind limb.

prehensile Adapted for grasping.

prelacteal In Mammals, said of teeth developed prior to the formation of the milk dentition.

premature ejaculation Inability of the male to postpone ejaculation long enough to satisfy the female.

premaxillary A paired membrane bone of the Vertebrate skull which forms the anterior part of the upper jaw; anterior to the maxilla. Also *premaxilla.*

premeiotic mitosis The nuclear division immediately preceding the organization of nuclei which will divide by meiosis.

premolars In Mammals, the anterior grinding or cheek teeth, which are represented in the milk dentition.

premorse Looking as if the end had been bitten off.

pre-operational thinking In Piaget's theory, the period from about two to six years when children's thinking is characterized by an ability to represent objects and events internally, but an inability to manipulate these representations in a logical way. According to Piaget, this accounts for the errors children make when asked to perform certain tasks. Cf. the periods of **sensorimotor development, concrete operations, formal operations.**

pre-operculum In Fish, an anterior membrane bone forming part of the gill cover.

prepollex In some Vertebrates (Amphibians, Mammals) a rudimentary extra digit of the forelimb.

pre-prophase band A band of microtubules 2–3 μm wide which forms and moves to the plasmalemma shortly before prophase begins, predicting the position where the new cell wall will join the old.

prepubic Pertaining to the anterior part of the pubis; in front of the pubis, as bony processes in some Marsupials and Rodents.

prepuce In Mammals, the loose flap of skin which protects the glans penis. adj. *preputial.*

presbyopia Long sightedness and impairment of vision due to loss of accommodation of the eye in advancing years.

pressure bomb Thick-walled metal vessel used in investigations of plant water relations, e.g. to apply pressure by compressed air to an excised leaf placed within the vessel with its petiole emerging. Water is forced from the leaf when the air pressure equals the **water potential** of the leaf cells.

pressure probe Device for measuring the **turgor pressure** within a plant cell by inserting into the cell a fine, fluid-filled capillary connected to a pressure transducer.

presternum In Anura, an anterior element of the sternum, of paired origin and doubtful homologies; the reduced sternum of whalebone whales; the anterior part of the sternum.

pretrematic Anterior to an aperture, as (in Selachii) that branch of the 9th cranial nerve which passes anterior to the first gill cleft. Cf. **post-trematic.**

prezygapophysis A facet or process on the anterior face of the neurapophysis of a vertebra, for articulation with the vertebra next in front. Cf. **postzygapophysis.**

Priapulida A phylum of coelomate, superficially segmented wormlike animals, living in mud. They have a straight gut with an anterior mouth and a posterior anus. The nervous system is not separated from the epidermis and the urinogenital system is simple, with solenocytes.

prickle A hard, sharp-pointed outgrowth of the epidermis, a multicellular trichome which is not vascularised, the 'thorn' of the rose.

primacy effect (1) In impression formation, the fact that attributes noted early on carry a greater weight than attributes noted at a later time. (2) In memory, the tendency for the first items on a list to be remembered better than other items on the list.

primacy process thinking In psychoanalytic theory, the mode of thinking characteristic of unconscious mental activity; it is governed by the **pleasure principle** rather than the **reality principle.** Cf. **secondary process thinking.**

primaries In Birds, the remiges attached to the manus.

primary Original, first formed, as *primary meristem, primary body cavity*; principal, most important, as *primary feathers, primary axis.*

primary body That part of the plant body formed directly from cells cut off from the apical meristems.

primary body cavity The blastocoel or segmentation cavity formed during cleavage, or that part of it which is not subsequently obliterated by mesenchyme.

primary cell wall See **primary wall.**

primary constriction Region at which two chromatids are joined in metaphase chromosomes, and to which spindle microtubules attach. It appears narrower and more condensed than the arms.

primary flexure The flexure of the mid brain by which, in Vertebrates, the forebrain and its derivatives are bent at a right angle to the axis of the rest of the brain.

primary growth The growth that results from division and expansion of those cells which are produced at apical meristems. Cf.

secondary growth.

primary immune response The response made by an animal to an antigen on the first occasion that it encounters it. Characteristically low levels of antibody are produced after several days and these gradually decline. However the immune system has been 'primed' so that a secondary response can be evoked on subsequent challenge with the same antigen. Responses of cell mediated immunity follow a similar pattern.

primary meristem Any of the three meristematic tissues derived in a pattern appropriate to the organ from the apical meristem, the *protoderm, ground meristem* and *procambium*. Cf. **secondary meristem**.

primary node The node at which the cotyledons are inserted.

primary phloem Phloem tissue that differentiates from the procambium during the primary growth of a vascular plant, consisting of protophloem and metaphloem. Cf. **secondary phloem**.

primary pit field A thin area in the primary wall of a plant cell, often penetrated by plasmodesmata, within which one or more pits may develop if a secondary wall is formed.

primary production See **production**.

primary ray See **medullary ray**.

primary reinforcer A stimulus that increases the probability of preceding responses even if the stimulus has never been experienced before.

primary sere A sere starting from a new, bare surface not previously occupied by plants. Land newly exposed on a rising coast or by a retreating glacier.

primary succession A succession beginning on an area not previously occupied by a community, e.g. a newly exposed rock or sand surface. Cf. **secondary succession**.

primary tissue Tissue formed from cells derived from primary meristems.

primary wall The earlier-formed part of the **cell wall** characteristically laid down while the cell is expanding and typically richer in pectins than the **secondary wall**.

primary xylem Xylem tissue that differentiates from the procambium during primary growth in a vascular plant, consisting of *protoxylem* and *metaxylem*. Cf. **secondary xylem**.

Primates An order of Mammals; dentition complete, but unspecialized; brain, esp. the neopallium, large and complex; pentadactyl; eyes well developed and directed forwards, the orbit being closed behind by the union of the frontal and jugal bones; basically arboreal animals; uterus a single chamber; few young produced, parental care lasting a long time after birth. Lemurs, Tarsiers, Monkeys, Apes and Man.

primed When referring to a whole animal it describes the condition in which prior contact with an antigen will result in a secondary response on subsequent challenge with the same antigen. When referring to a cell population describes the presence of cells which have already been activated by an antigen.

primer A nucleotide sequence with a free 3′ –OH group, needed to initiate synthesis by DNA polymerase. A short single chain primer base-paired to a specific site on a longer DNA strand can initiate polymerization from that site. See **polymerase chain reaction**.

primitive Original, first-formed, of early origin, the ancestral condition, as the *primitive streak*.

primitive streak In developing Birds and Reptiles, a thickening of the upper layer of the blastoderm along the axis of the future embryo; represents the fused lateral lips of the blastopore.

primordial germ cells In the early embryo, cells which will later give rise to the gonads.

primordium An organ, cell or other structure in the earliest stage of development or differentiation. Also *anlage*. adj. *primordial*.

principle of reinforcement Skinner's term for the **law of effect**.

prisere Same as **primary sere**.

prismatic Prism-shaped; composed of prisms.

prismatic layer In the shell of *Mollusca*, a layer consisting of calcite or aragonite lying between the periostracum and the nacreous layer. In the shell of *Brachiopoda*, the inner layer of the shell, composed mainly of calcareous, but partly of organic material.

Pro Symbol for **proline**.

pro- Prefix from Gk. and L. *pro*, before in time or place, used in the sense of 'earlier', 'more primitive' or 'placed before'.

probability density function The first derivative of the cumulative distribution function of a continuous random variable, often identified as the probability that a random variable takes a value in an infinitesmal interval divided by the length of the interval.

proband In human genetics, the affected individual who brings a family to the notice of the investigator.

probe A radioactive or otherwise labelled single-stranded sequence of nucleic acid which hybridizes to another single stranded nucleic acid, usually separated into discrete spots on a nitrocellulose sheet. The label then detects the presence of a sequence complementary to the probe sequence. See **Southern blot**.

problem solving behaviour Diverse strategies used by animals and humans to overcome difficulties in attaining a desired goal; it implies a higher order of intelligent behaviour than simple **trial and error** strategies.

Proboscidea An order of Mammals having a long prehensile proboscis with the nostrils at the tip, large lophodont molars, and a pair of incisors of the upper jaw enormously developed as tusks; semiplantigrade; forest-gliving herbivorous forms of Africa and India. It includes Elephants and the extinct Mammoths.

proboscis An anterior trunklike process; in Turbellaria and Polychaeta, the protrusible pharynx; in Nemertinea, a long protrusible muscular organ lying above the mouth; in some Insects, the suctorial mouthparts; in Hemichorda, a hollow club-shaped or shield shaped structure in front of the mouth; in Proboscidea, the long flexible prehensile nose.

procambium *Provascular tissue.* The cells of *primary meristem* which are typically longer than broad, which differentiates into primary xylem and primary phloem and, in some cases, cambium.

procartilage An early stage in the formation of cartilage in which the cells are still angular in form and undergoing constant division; embryonic cartilage.

procaryote See **Prokaryote.**

Procellariiformes An order of Birds. Wandering ocean species often very large, with long narrow wings. Lay one white egg, often in a burrow. Petrels, Shearwaters, Albatrosses.

process An extension or projection.

Prochlorophyceae Prokaryotic algae with the pigmentation of green algae (chlorophyll a and b, and no phycobilins) rather than blue green algae. Two or three species so far identified. Of interest because they possibly represent the group from which green plant chloroplasts may have evolved. See **endosymbiotic hypothesis.**

procoelous Concave anteriorly and convex posteriorly; said of vertebral centra.

proctal Anal.

proctodaeum That part of the alimentary canal which arises in the embryo as a posterior invagination of ectoderm. Cf. **stomodaeum, midgut.** adj. *proctodaeal.*

proct-, procto- Prefixes from Gk. *proktos,* anus.

procumbent Lying loosely on the ground surface.

procuticle In the cuticle of Insects, a multilaminar layer initially present below the *epicuticle* and pierced by pore canals running perpendicularly to it. In soft transparent areas this undergoes no apparent change after formation, but in other cases the outer part becomes hard, dark sclerotized *exocuticle,* the inner unchanged part then being called the *endocuticle.* See **tanning,** sclerotin.

producers In an ecosystem, autotrophic organisms, largely green plants, which are able to manufacture complex organic substances from simple inorganic compounds. Cf. **consumers, decomposers.**

production Biomass, or heat of combustion of the biomass, expressed on an area basis, (units: $g\,m^{-2}$ or $MJ\,m^{-2}$). Primary production is production by green plants. Secondary production is the biomass produced by heterotrophic organisms. The rate of production is called the *productivity* with units of $g\,m^{-2}\,year^{-1}$. Gross primary productivity is the rate of community photosynthesis, whereas net primary productivity is the rate of community photosynthesis minus the rate of community respiration.

productivity See **production.**

pro-ecdysis The period of preparation for a moult in Arthropoda during which the new cuticle is laid down and the old one ultimately detached from it.

pro-embryo The structure formed by the first few cell divisions of the zygote of seed plants, before differentiation into suspensor and the embryo proper.

progeria Premature old age. Occurring in children, the condition is characterized by dwarfism, falling out of hair, wrinkling of the skin and senile appearance.

proglottis One of the reproductive segments forming the body in *Cestoda;* produced by strobilization from the back of the scolex. pl. *proglottides.*

prognathous Having protruding jaws; having the mouth parts directed forwards. Cf. **orthognathous.**

programmed learning, instruction The process of learning from a systematic presentation of data constructed so that each step leads to the next. The learner does not need to refer to material other than the programme and progresses by answering questions at each stage which are necessary for the understanding of the rest. Such programmes can be presented on a *teaching machine.*

projection In psychoanalytic theory, a **defence mechanism,** in which we ascribe to others various feelings, thoughts and motives, esp. of an undesirable sort, which really belong to the self, in order to ward off the anxiety associated with them.

projective technique, test A method of psychological testing, used in personality assessment, in which relatively unstructured stimuli are presented (e.g. an inkblot or a picture) and elicit subjective responses from the subject; these are presumed to involve the *projection* of the subject's personality on to the test material.

prokaryon The nucleus of a blue-green alga or of a bacterium. See **eukaryote, prokaryote**.

prokaryote, procaryote Major division of living organisms, which have no defined nucleus and their genetic material is usually a circular duplex of DNA. They have no endoplasmic reticulum. Bacteria are the prime example but it also includes the blue-green algae, the Actinomycetes and the Mycoplasmata. Cf. **eukaryote**. adj. *prokaryotic*.

prolamellar body Three-dimensional lattice composed of tubules, formed within an **etioplast** and rapidly converting into **thylakoids** on illumination.

prolan Former general name for gonadotrophic hormones of mammals found in various tissues and body fluids during pregnancy; prolan A is equivalent to **follicle-stimulating hormone** and prolan B to **luteinizing hormone**.

proleg One of several pairs of fleshy conical retractile projections borne by the abdomen in larvae of most Lepidoptera, Sawflies and Scorpion-flies; used in locomotion.

proliferation Growth or extension by the multiplication of cells. adj. *proliferous*.

prolification Development of buds in the axils of sepals and petals.

proline *Pyrrolidine-2-carboxylic acid*. The L-isomer is a non-polarizable amino acid and constituent of proteins, particularly **collagen**. Symbol Pro, short form P.

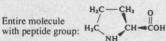

Entire molecule with peptide group:

See **amino acid**.

promeristem The initial cells and their immediate derivatives in an apical meristem.

prometaphase The stage in cell division, mitosis and meiosis, between prophase and metaphase.

promontory A projecting structure; a small ridge or eminence.

promoter A DNA region in front of the coding sequence of a gene which binds RNA polymerase and therefore signals the start of a gene.

pronation In some higher Vertebrates, movement of the hand and forearm by which the palm of the hand is turned downwards and the radius and ulna brought into a crossed position. Cf. **supination**. adj. *pronate*.

pronator A muscle effecting pronation.

pronephros *Archinephros*; in Craniata, the anterior portion of the kidney, functional in the embryo but functionless and often absent in the adult. Also called *fore-kidney, head kidney*. adj. *pronephric*. Cf. **mesonephros; metanephros**.

pronotum The notum of the prothorax in Insects. adj. *pronotal*.

pronucleus The nucleus of a germ cell after the maturation divisions.

pro-oestrus In Mammals, the coming on of heat in the oestrus cycle.

pro-otic An anterior bone of the auditory capsule of the Vertebrate skull.

propagation The reproduction of a plant by asexual or sexual means, esp. in horticulture. See **plant cell culture**.

propagule Any structure, sexual or asexual and independent from the parent which serves as a means of reproduction. Also *disseminule*.

properdin A component of the alternative pathway of complement activation present in small amounts in the blood, which is not an immunoglobulin, although at one time considered to be so. It complexes with C3b and stabilizes the alternative pathway C3 convertase.

prophage A phage genome which replicates in synchrony with its host. May be integrated into the host genome.

prophase First stage of mitosis or meiosis during which chromosomes condense and become recognizably discrete.

prophyll The first leaf in most monocots or either of the first two leaves in most dicots, on a shoot.

proplastid Small undifferentiated plastid.

proprioceptor A sensory nerve ending receptive to internal stimuli, particularly signalling the relative positions of body parts. Also *interoceptor*. adj. *proprioceptive*.

prop root Adventitious root arising on a stem above soil level, growing into the soil and serving as additional support for the stem, as in maize or some mangroves. Also *stilt root*.

proscapula See **clavicle**.

proscolex See **cysticercus**.

prosencephalon In Craniata, the part of the forebrain which gives rise to the cerebral hemispheres and the olfactory lobes.

prosocoele In Craniata, the cavity of the forebrain or first brain vesicle in the embryo; fore ventricle.

prosoma In Arachnida, the region of the body comprising all the segments in front of the segment bearing the genital pore; in *Acarina*, the gnathosoma together with the podosoma.

prostaglandins Biologically active lipids generated by the action of cyclooxygenase enzymes on arachidonic acid (c.f. **leukotrienes** which are generated by lipoxygenase action). A variety of prostaglandins are produced during anaphylactic reactions, some of which have antagonistic actions on platelet activation and vascular permeability, so that the net outcome is difficult to predict.

---

protein

Protein molecules consist of one, or a small number, of *polypeptide* chains each of which is a linear polymer of several hundred amino acids linked through their amino and carboxylate groups by **peptide bonds**. The amino acid side chains can have a positive or negative charge, a short aliphatic chain or an aromatic residue. Because the 20 amino acids can be arranged in nearly any sequence, the potential diversity of structure and function is enormous.

The properties of each polypeptide depends on the amino acid sequence (its *primary structure*) which itself determines the correct folding of the chain in three dimensions in its *native conformation* and thus give its specific biological activity. See **secondary structure, tertiary structure, alpha helix, beta pleated sheet.** Because the three dimensional structure is largely dependent on weak forces, it is usually readily disrupted by extremes of pH or heat with a resulting loss of biological activity. See **denaturation.** Covalent cross links, particularly disulphide bonds, when present provide a more stable component of higher orders of structure but, in most proteins, are not themselves sufficient to stabilize the native conformation.

If the polypeptide sequence contains many hydrophilic amino acids the resulting proteins are water soluble, e.g. most enzymes, but they may also be assembled into extensive polymeric complexes in structures such as the cytoskeleton, cilia, mitotic spindles etc. If hydrophobic amino acids predominate, the protein is water insoluble and fibrous, e.g. hair, silk and **collagen.** These fibrous proteins, in contrast to those which are water soluble, usually have extremely stable secondary and tertiary structure and are less prone to denature.

---

prostate In Cephalopoda, said of a gland of the male genital system associated with the formation of spermatophores; in eutherian Mammals (except Edentata and Cetacea), including Man, said of a gland associated with the male urogenital canal.

prosthetic group Non-proteinaceous entity essential for an enzyme's activity. It is functionally equivalent to a *co-enzyme* and differs only in being tightly bound to its protein.

prostomium In annelid Worms, that part of the head region anterior to the mouth.

protamines Family of short, basic proteins which are bound to sperm DNA in place of histones.

protandry The maturation of the male organs before the female are receptive. A form of dichogamy. Cf. **protogyny.**

protease Enzyme which hydrolyses the peptide bonds of a protein. They hydrolyse different sites according to the amino acids adjacent to the peptide bond under attack. e.g. **chymotrypsin, trypsin.** syn. *peptidase.*

protective layer In an **abscission zone,** a layer of cells lying immediately proximal to the abscission layer and protecting the underlying tissues from desiccation and invasion by parasites after abscission has occurred.

protein See panel.

protein A A protein in the cell walls or ex-tracts made from certain strains of *Staphylococcus aureus* which binds to the Fc fragment of IgG from a variety of species. This property has made protein A a useful reagent for isolating IgG and for detecting it in complexes. Biologically it has an antiphagocytic effect.

protein structure The three dimensional structure of a protein. See **protein.** Determined primarily by **X-ray crystallography** of protein crystals down to a resolution of 0.2 nm.

proteolysis The degradation of proteins into peptides and amino acids by cleavage of their peptide bonds.

proteolytic, proteoclastic Referring to enzymes which catalyse the breakdown of proteins into simpler substances, e.g. trypsin.

proterandrous See **protandrous.**

proterokont A bacterial flagellum; it is not homologous with the flagella found in higher organisms.

proter-, protero-. Prefixes from Gk. *proteros,* before, former.

prothallus (1) The gametophyte of the pteridophytes, growing photosynthetically at the soil surface, when it may resemble a thallose liverwort, or heterotrophically in association with a fungus underground, and bearing, when mature, archegonia and/or antheridia. The embryo and young sporophyte, developing from a fertilized egg in the archegonium are at first dependent on the prothallus. (2) Sometimes the gametophyte of gymnosperms.

prothorax The first or most anterior of the three thoracic somites in Insects. Cf. **mesothorax, metathorax.**

Protista A paraphyletic group in some classifications of mostly unicellular organisms including, usually the protozoa, Euglenophyceae, Crytophyceae, Dinophyceae and slime moulds, sometimes the flagellate members of the Chlorophyta and Heterokontophyta, and in older usages the bacteria and blue-green algae.

proto- Prefix form Gk. *protos*, first.

protocercal See **diphycercal**

Protochordata A division of Chordata comprising the subphyla Hemichordata, Urochordata and Cephalochordata, which are distinguished by the absence of a cranium, vertebral column and of specialized anterior sense organs. Cf. **Vertebrata.**

protoderm Primary **meristem** which gives rise to the epidermis and which may arise from independent initials in the apical meristem.

protogyny The maturation of the female organs before the male organs liberate their contents. See **dichogamy.** Cf. **protandry.**

protomorphic Primordial; primitive.

protonema Juvenile stage of the gametophyte of mosses and liverworts.

protonephridial system The excretory system of Platyhelminthes, consisting of flame cells and ducts.

protonephridium A larval nephridium, usually of the flame cell type.

proton motive force The electrochemical gradient which is derived from a membrane potential together with a proton gradient across the membrane. Such gradients operate across the inner mitochondrial membrane and the **thylakoid** membrane. Essential for the generation of ATP during oxidative phosphorylation and photosynthesis.

proton-translocating ATPase *H⁺-ATP-ase.* Primary electrogenic **active transport** system, powered by the hydrolosis of ATP, pumping protons out of a plant cell across the plasmalemma or into the vacuole across the tonoplast. The resulting pH and electrical potential gradients drive a number of secondary active transport processes coupled to the return movements of the protons.

proto-oncogene A gene which is required for normal function of the organism, but which when it has mutated can become an **oncogene.**

protophloem The first-formed **primary phloem**; characteristically maturing while the organ is elongating.

protoplasm The living material within a cell divided into discrete structures, e.g. the mitochondria, ribosomes, nuclei, chromosomes and nucleoli in **eukaryotes** and the chromosome and ribosomes in **prokaryotes.** adj. *protoplasmic.*

protoplasmic circulation The streaming motion that may be seen in the protoplasm of a living cell.

protoplast (1) The living part of a plant cell including the nucleus, cytoplasm and organelles, all bounded by the plasmalemma, but excluding any cell wall. (2) The above structure isolated from its cell wall usually by treatment of a tissue with wall-degrading enzymes or mechanically.

protoplast culture The use of methods of **plant cell culture** with isolated protoplasts, esp. for the regeneration of plants from protoplasts that have been induced to take up foreign DNA or that result from protoplast fusion. See **plant genetic manipulation.**

protoplast fusion See **plant genetic manipulation.**

protostele A stele without leaf gaps or, sometimes, with a solid core of xylem.

Prototheria A subclass of primitive Mammals, which probably left the main stock in the Mesozoic, and are found only in Australasia. The adults have no teeth, the cervical vertebrae bear ribs, the limbs are held laterally, the shoulder girdle has precoracoids and an inter clavicle, and large yolky eggs are laid. The young are fed after hatching on milk produced by specialized sweat glands, whose ducts do not unite to open on nipples. One order, the Monotremata. Duck-billed Platypus, Spiny Anteater.

prototroph A **wild-type** organism able to grow in its unsupplemented medium. Cf. **auxotroph.**

prototype An ancestral form; an original type or specimen.

protoxylem The first-formed primary xylem; typically maturing while the organ elongates, having narrow tracheary elements with annular or helical thickening and parenchyma only, and becoming stretched and crushed as the organ elongates.

Protozoa A phylum of unicellular or acellular animals. Nutrition holophytic, holozoic or saprophytic; reproduction by fission or conjugation; locomotion by cilia, flagella or pseudopodia; free-living or parasitic. sing. *protozoon.*

protractor A muscle which by its contraction draws a limb or a part of the body forward or away from the body. Cf. **retractor.**

provascular tissue See **procambium.**

proventriculus (1) In Birds, the anterior thin-walled part of the stomach, containing the gastric glands. (2) In Oligochaeta and Insects, the gizzard, a muscular thick-walled chamber of the gut posterior to the crop. (3) In Crustacea, the stomach or gastric mill.

provitamin A vitamin precursor, such as β-carotene which gives vitamin A.

proxemics The study of the spatial features of human social interaction, e.g. **personal space.**

proximal Pertaining to or situated at the inner end, nearest to the point of attachment. Cf. **distal.**

pruinose Having a bloom on the surface, esp. a whitish bloom like hoar frost.

pruniform Shaped like a plum.

psalterium In ruminant Mammals, the third division of the stomach. Also *omasum, manyplies.*

psammophyte A plant adapted to growing on sand or sandy soils.

pseudautostyly A type of jaw suspension in which the upper jaw is fused with the ethmoidal, orbital and otic regions of the cranium. Cf. **autostyly.** adj. *pseudautostylic.*

pseudo-aposematic Warning or aposematic coloration borne by animals which are not dangerous or distasteful, but show **Batesian mimicry** of animals which are.

pseudobrachium In some Fish, an appendage used for propulsion along a substratum or on dry land; formed by modification of the pectoral fin.

pseudobulb Swollen, solid, above-ground stem of some orchids, acting as a storage organ.

pseudocarp See **false fruit.**

pseudocoele (1) In higher Vertebrates, a space enclosed by the inner walls of the closely opposed cerebral hemispheres; the 5th ventricle. (2) A body formed from a persisting blastocoel. Also *pseudocoelom.* Cf. **coelom.**

pseudocopulation Attempts of a male insect to mate with a flower that resembles a female of its species, as in the pollination mechanisms of many orchids.

pseudocyesis In some Mammals, uterine changes following oestrus and resembling those characteristic of pregnancy.

pseudodementia Refers to mental conditions in which there are symptoms which suggest dementia, but which are caused by other factors (e.g. drug use or depression).

pseudodont Having horny pads or ridges in place of true teeth, as Monotremes.

pseudogamy A form of apomixis in which, although fertilization does not occur, the stimulus of pollination is necessary for seed production.

pseudogene Defective copy of a gene and therefore not transcribed.

pseudoheart In Oligochaeta, one of a number of paired contractile anterior vessels by which blood is pumped from the dorsal to the ventral vessel; in Echino-

dermata, the axial organ.

pseudometamerism The condition of repetition of parts, found in some Cestoda, which bears a superficial resemblance to metamerism.

Pseudomonadaceae A family of bacteria belonging to the order Pseudomonadales. Gram-negative rods, occurring in water and soil and including also animal and plant pathogens, e.g. *Pseudomonas aeruginosa* (blue pus); *Xanthomonas hyacinthi* (yellow rot of hyacinth). Acetobacter species are used in production of vinegar.

Pseudomonadales One of the two main orders of true bacteria, distinguished by the polar flagella of motile forms. They are Gram-negative, spiral, spherical or rod-shaped cells. See **Eubacteriales.**

pseudoparenchyma A *false tissue* made of interwoven fungal hyphae.

pseudoperianth Cylindrical sheath growing up around the archegonium and young sporophyte of some liverworts.

pseudopod (1) A broad finger-like protrusion of the cell surface which may be used in amoeboid cells for locomotion. Also *pseudopodium.* (2) A footlike process of the body wall, characteristic of some Insect larvae.

pseudopregnancy See **pseudocyesis.**

Pseudoscorpionidea An order of Arachnida, resembling Scorpionidea, but with no tail; the pedipalps are large, chelate and contain poison glands; small carnivorous forms found under stones, leaves, bark and moss; occasionally found in houses. Also called *Chelonethida, False Scorpions.*

pseudovilli Projections from the surface of the trophoblast in some Mammals, as distinct from the true villi, which are definite outgrowths.

pseudovitellus In some hemipterous Insects, an abdominal mass of cells which contains symbiotic micro-organisms. See **mycetocytes, mycetome.**

pseud-, pseudo- Prefixes from Gk. *pseudes,* false.

ψ$_p$ Symbol for *pressure potential.*

Psittaciformes An order of Birds containing one family. Mainly vegetarian, with powerful hooked beaks; feet typically zygodactylous. The birds are often vividly coloured, and capable of mimicry. Parrots, Cockatoos.

psittacosis *Parrot disease, ornithosis.* An acute or chronic contagious disease of wild and domestic birds which is transmittable to other animals and Man, and caused by *Chlamydia psittaci.* Results in respiratory and systemic infections including, in man, a disease resembling pneumonia.

psittacosis-lymphogranuloma viruses A group of Rickettsiae-like organisms, sometimes classified with the animal viruses, but which may be more closely related to bacteria. Large antigenically-related organisms, which contain RNA and DNA, muramic acid in the cell wall and are susceptible to sulphonamides and certain antibiotics. Includes causative agents of trachoma, inclusion conjunctivitis, lymphogranuloma venereum and psittacosis.

psychiatry That branch of medical science which deals with the study, diagnosis, treatment and prevention of mental disorders.

psychism The doctrine, difficult to sustain, that living matter possesses attributes not recognized in nonliving matter.

psychoanalysis (1) A theory of personality developed by Freud in which the ideas of unconscious conflict and **psychosexual development** are central. (2) A method of therapy based on (1) above which attempts to help the individual gain insight into his or her unconscious conflicts using a variety of psychoanalytic techniques. See **free association, transference, Freud's theory of dreams**.

psychodynamics A theory of the working of an individual's mind.

psychogalvanic reflex The decrease in the electrical resistance of the skin under the stimulation of various emotional states.

psychogenic Having a mental origin.

psychogenic disorders Disorders whose origins are psychological rather than organic.

psychokinesis The alleged ability of some people to alter physical reality in the absence of any known mechanism for accomplishing it (e.g. bending metal objects without touching them).

psychometrics The application of mathematical and statistical concepts to psychological data, particularly in the areas of mental testing and experimental data.

psychopath A medical-legal term referring to a behaviour disorder characterized by repetitive, antisocial behaviour with emotional indifference and without guilt, where the individual does not learn from experience or punishment. It is a category not recognized in Scottish law. Also *anti-social personality*.

psychopathology The study of psychological disorders.

psychopharmacology Refers to the study and use of drugs that influence behaviour, emotions, perception and thought, by acting on the central nervous system.

psychophily Pollination by butterflies.

psychophysics The branch of psychology that studies the relationship between characteristics of the physical stimulus and the psychological experience they produce.

psychophysiological disorders Those physical disorders which are thought to be due to emotional factors but which involve genuine organic disorders (e.g. high blood pressure). Formerly called *psychosomatic disorders*.

psychophysiology The measurement of physiological processes such as heart rate and blood pressure in relation to various mental and emotional states, See **psychophysiological disorders**.

psych-, psycho- Prefixes from Gk. *psyche*, soul, mind.

psychosexual development (1) In psychoanalytic theory, a progressive series of stages in which the source of bodily pleasure changes during development, defined by the zone of the body thought which this pleasure is derived. See **oral stage, anal stage, phallic stage, genital stage**. (2) The term *psychosexual* also refers to mental aspects of sexual phenomena.

psychosexual disorders Seriously impaired sexual performance of various kinds, or unusual methods of sexual arousal.

psychosis A very general term used to describe mental illnesses which result in a severe loss of mental and emotional function, in contrast to **neurosis**, where the individual remains competent to cope with reality.

psychosomatic See **psychophysiological disorders**.

psychosurgery Biological intervention which involves surgery to remove or destroy brain tissue, aimed at changing undesirable behaviour.

psychotherapy The treatment of mental and emotional disturbance by psychological means, often in an extended series of therapist-client sessions; refers to several different forms of treatment and techniques. The term is usually restricted to treatments supervized or conducted by trained psychologists or psychiatrists.

psychrophilic Growing best at a relatively low temperature, esp. (of a micro organism) having a temperature optimum below 20°C. Cf. **mesophilic, thermophilic**.

Pteridophyta (1) Division of the plant kingdom containing all the vascular plants which do not bear seeds, i.e. the ferns, clubmosses, horsetails etc. There is an alternation of generations of, typically, a smaller, more or less thalloid, independent gametophyte and larger, longer-lived sporophyte usually with roots, stems and leaves. Usually divided into 8 classes; Rhyniopsida, Psilotopsida, Zosterophyllopsida, Lycopsida, Trimerophytopsida, Sphenopsida, Filicopsida and Progymnospermopsida. (2) Sometimes, confusingly, the ferns alone.

Pteridospermopsida The seed ferns and allies. Class of extinct gymnosperms, mostly Carboniferous to Jurassic.

pterygial In Fish, an element of the fin skeleton; pertaining to a fin; pertaining to a wing.

pterygium In Vertebrates, a limb.

pterygoid A paired cartilage bone of the Vertebrate skull, formed by the ossification of the front part of the PPQ bar; a membrane bone which replaces the original pterygoid in some Vertebrates; more generally, wing-shaped.

pterygopalatoquadrate bar In Fish with a cartilaginous skeleton, the rod of cartilage forming the upper jaw and known as the *PPQ bar*.

pterylosis In Birds, the arrangement of the feathers in distinct feather tracts or pterylae, whose form and arrangement are important in classification.

ptilinum In certain Diptera (Cyclorrhapha), an expansible membranous cephalic sac by which the anterior end of the puparium is thrust off at emergence.

ptyxis Manner in which an individual unexpanded leaf, sepal or petal is folded, rolled or coiled in the bud. Also called *vernation*.

puberulent Minutely pubescent.

pubescence A covering of fine hairs or down. adj. *pubescent*.

pubis In Craniata, an element of the pelvic girdle (contr. of *os pubis*). adj. *pubic*.

puff ball Fruiting body of some fungi, esp. of the order Lycoperdiales in the Gasteromycetes.

puffs Visibly decondensed bands of **polytene chromosomes** in which active transcription of RNA is occurring.

pulmonary In land Vertebrates, pertaining to the lungs; in pulmonate *Mollusca*, pertaining to the respiratory cavity.

Pulmonata An order of the Gastropoda. Hermaphrodite; exhibit torsion; have a shell but no operculum; mantle cavity forming a lung with no ctenidium but a vascular roof and a small aperture (the pneumostome). Land and freshwater snails, land slugs.

pulmonate Possessing lungs or lung-books; air-breathing.

pulmones Lungs. sing. *pulmo* adj. *pulmonary*.

pulp A mass of soft spongy tissue situated in the interior of an organ, as *spleen pulp, dental pulp*.

pulse The periodic expansion and elongation of the arterial walls caused by the pressure wave which follows each contraction of the heart.

pulsed-field gel electrophoresis Variant of **agarose gel electrophoresis** which allows the fractionation of very large DNA fragments (up to 2 million base pairs) by applying the electric field in pulses from different angles.

pulse labelling Adding a pulse of radioactive material to a cell and then studying the subsequent metabolic stages.

pulvinule The small **pulvinus** of a leaflet.

pulvinus (1) Swollen base of petiole or pinna, containing motor tissue responsible for sleep movements etc. as in many Leguminosae. (2) Thickened region of grasses at a node of the stem capable, by growth, of re-erecting a lodged culm.

pump Molecular mechanism in a membrane which brings about the active transport (electrogenic or neutral) of a solute, e.g. a proton-translocating ATPase. See **active transport**. Cf. **carrier**.

punctate With translucent or coloured dots, or shallow pits.

punctuated equilibrium A concept of the process of evolution in which the fossil record is interpreted as long periods of stasis interrupted by relatively short periods of rapid change and speciation.

punctum A minute aperture; a dot or spot in marking. adj. *punctate*.

pungent Ending in a point stiff and sharp enough to prick. Also, acrid to taste.

punishment In conditioning situations, the weakening of a response which is followed by an aversive or noxious stimulus, or by the withdrawal of a pleasant one. Cf. **negative reinforcement**.

pupa An inactive stage in the life history of an Insect during which it does not feed and reorganization is taking place to transform the larval body into that of the imago. adj. *pupal*.

puparium The hardened and separated last larval skin which is retained to form a covering for the pupa in some Diptera.

pupil The central opening of the iris of the eye. adj. *pupillary*.

pupilometer An instrument for measuring the size and shape of the pupil of the eye and its position with respect to the iris.

pupiparous Giving birth to offspring which have already reached the pupa stage, as some two-winged Flies, e.g. *Glossina*, the Tsetse Fly.

pure culture See **axenic culture**.

pure line *Inbred line*. A group of individuals, with their ancestors and descendants, usually the product of continued *inbreeding*, which breed true among themselves and which are, therefore, presumably homozygous at most loci.

purposive behaviour Behaviour that is carried out with the design of achieving a desired end; it may be conscious or unconscious in its nature.

pustule (1) A blister-like spot, on a leaf, stem, fruit etc. from which erupts a fruiting structure of a fungus. (2) A small elevation of the skin containing pus. adjs. *pustular, pustulous*.

putamen In Birds, the shell membrane of the egg; in higher Vertebrates, the lateral part of the lentiform nucleus of the cerebrum.

putrefaction The chemical breaking down or decomposition of plants and animals after death. This is caused by the action of anaerobic bacteria and results in the production of obnoxious or offensive substances.

puzzle box A box in which an animal is confined, and from which it can escape only by performing a particular series of manipulations which it must discover by trial and error or problem-solving behaviour.

p-value The probability of observing an outcome as, or more extreme than, that actually arising from a particular experiment or sample when a particular hypothesis is true. A low p-value is taken to indicate evidence against the particular hypothesis.

pycnidiospore A spore formed within a pycnidium.

Pycnogonida An order of the Chelicerata. Marine animals with long legs containing diverticulae of the digestive system and with reduced opisthosoma. Sea Spiders.

pycnosis The shrinkage of the stainable material of a nucleus into a deeply staining knot, usually a feature of cell degeneration.

pycnoxylic wood This is the secondary xylem in gymnosperms composed mainly of tracheids, with relatively narrow rays as in e.g. conifers. Cf. **manoxylic wood**.

pycn-, pycno-, pykn-, pykno- Prefixes from Gk. *pyknos*, compact, dense.

pygal Pertaining to the posterior dorsal extremity of an animal; in *Chelonia*, a posterior median plate of the carapace. (Gk. *pyge*, rump.)

pygostyle In Birds, a bone at the end of the vertebral column formed by the fusion of some of the caudal vertebrae.

pylorus In Vertebrates, the point at which the stomach passes into the intestine. adj. *pyloric*.

py-, pyo- Prefixes from Gk. *pyon*, pus.

pyramid A conical structure, as part of the medulla oblongata in Vertebrates. adj. *pyramidal*.

pyramidal tract In the brain of Mammals, a large bundle of motor axons carrying voluntary impulses from particular areas of the cerebral cortex.

pyramid of numbers   See **numbers, pyramid of**.

pyrenocarp See **perithecium**.

pyrenoid Small, dense, rounded, refractile, proteinaceous body within or associated with the chloroplast, in some members of at least most classes of eukaryotic algae and often surrounded by the appropriate storage carbohydrate.

Pyrenomycetes A class of fungi in the Ascomycotina in which the fruiting body or ascocarp is usually a perithecium. Includes the **powdery mildews**, *Claviceps* (ergot) and *Neurospora* (used in genetic research) etc.

pyrethrins Active constituents of pyrethrum flowers used as standard contact insecticide in fly sprays etc; remarkable for the very rapid paralysis ('knock-down' effect) produced on flies, mosquitoes etc. Pyrethrum root is the source of a similar substance used as a sialagogue. Chemically modified pyrethrins, which have greater persistence and other desirable properties are now available.

pyrexia *Fever*. An increase above normal of the temperature of the body. adj. *pyrexial*.

pyridine alkaloids A group of **alkaloids** based on the pyridine ring, including coniine from hemlock.

pyridoxal See **vitamin B6**.

pyriform Pear-shaped, as the *pyriform organ* of a Cyphonautes larva.

pyroninophilic cells Cells stained with methyl green pyronin stain which have bright red cytoplasm. This indicates the presence of large amounts of RNA, and implies very active protein synthesis. It is characteristic of plasma cells.

pyrophilous Growing on ground which has been recently burnt over.

pyxidium, pyxis A capsule dehiscing by means of a transverse circular split, the top coming off like a lid.

Q

q See **banding techniques.**

Q₁₀ See **temperature coefficient.**

Q-bands See **banding techniques.**

quadrant A section of a segmenting ovum originating from 1 of the 4 primary blastomeres.

quadrat A small area (say 0.1 to 10 m²) of vegetation marked out for ecological study; a device of laths or strings to mark out such an area.

quadrate (1) Square to squarish in cross-section or in face view. (2) A paired cartilage bone of the Vertebrate skull formed by ossification of the posterior part of the PPQ bar, or the corresponding cartilage element prior to its ossification; except in Mammals, it forms part of the jaw-articulation.

quadratus A muscle of rectangular appearance, e.g. *quadratus femoris.*

quadriceps A muscle having 4 insertions, as one of the thigh muscles of Primates.

quadrivalent A group of four at least partly homologous chromosomes held together by chiasmata during the prophase of the first division of meiosis, commonly found during meiosis in tetraploids.

quadr-, quadri- Prefixes from L. *quattuor,* four.

quadrumanous Of Vertebrates, having all 4 podia constructed like hands, as in Apes and Monkeys.

quadruped Of Vertebrates, having all four podia constructed like feet, as Cattle.

quality In radiography, it indicates approximate penetrating power. Higher voltages produce higher quality X-rays of shorter wavelength and greater penetration. (This term dates from a period before the nature of X-rays was completely understood).

quality factor See **relative biological effectiveness.**

quantile The argument of the cumulative distribution function corresponding to a specified probability; (of a sample) the value below which occur a specified proportion of the observations in the ordered set of observations.

quantitative character A character displaying *continuous variation.* Cf. **unit character.**

quantitative genetics The genetics of *quantitative characters.*

quantity of radiation Product of intensity and time of X-ray radiation. Not measured by energy, but by energy density and a coefficient depending on ability to cause ionization.

quartet, quartette A set of 4 related cells in a segmenting ovum.

queen In social Insects, a sexually reproducing female.

queen substance *Queen bee substance.* A pheromone produced by queen honeybees (*Apis mellifera*; Hymenoptera) consisting of 9-ketodecanoic acid. Its effects include the suppression of egg-laying and of the building of queen cells by workers.

quiescent centre Region, within the apical meristem of many roots, in which the cells either do not divide, or divide very much more slowly than the cells around it.

quill See **calamus.**

quill feathers In Birds, the remiges and rectrices.

quillwort Common name of *Isoetes* spp.

quinacrine fluorescence See **banding techniques.**

quincuncial aestivation A common type of **imbricate** aestivation of a five-membered calyx or corolla in which 2 members overlap their neighbours by both edges, 2 are overlapped on both edges and 1 overlaps 1 neighbour and is overlapped by the other, as in e.g. calyx of roses, corolla of Caryophyllaceae.

quinine $C_{20}H_{24}O_2N_2.3H_2O$, an alkaloid of the quinoline group, present in Cinchona bark. It crystallizes in prisms or silky needles, mp 177°C. It is a diacid of very bitter taste and alkaline reaction. The hydrochloride and sulphate were once widely used as a febrifuge but have been largely superseded as a remedy for malaria but is still used as a treatment for leg cramps.

R

r See **r-strategist**.

r The instantaneous population growth rate defined as

$$r = \frac{1}{N}\frac{dN}{dt}.$$

where N is the population size and t is time. For any organism there is a maximum r, achieved in ideal conditions, called the innate capacity for increase r_{max}.

R Symbol for *röntgen* unit in X-ray dosage.

rabies An acute disease of dogs, wolves, and other carnivores, due to infection with a virus, and communicable to Man by the bite of the infected animals. In Man the disease is characterized by intense restlessness, mental excitement, muscular spasms (esp. of the mouth and throat), convulsions and paralysis. Also *hydrophobia*.

Rabl configuration Spatial arrangement of interphase chromosomes with centromeres clustering at one side of the nucleus and telomeres at the other.

race A population, within a species, that is genetically distinct in some way, often geographically separate; a breed of domesticated animals. See **physiological race**.

raceme A simple (unbranched) **racemose inflorescence** in which the flowers are visibly stalked. Cf. **spike**.

racemose Shaped like a bunch of grapes; said esp. of glands.

racemose inflorescence One in which the main axis (and, in a compound raceme, each of its main branches) does not end in a flower but continues to grow bearing flowers in acropetal succession on its lateral branches, e.g. raceme, spike, panicle. Also called *indefinite* or *indeterminate inflorescence*. Cf. **cymose inflorescence, mixed inflorescence**.

rachilla (1) The axis of the spikelet of a grass, on which are borne the glumes and the florets. (2) A secondary (or tertiary etc.) axis of a pinnately compound leaf.

rachiodont Having some of the anterior thoracic vertebrae with the hypapophysis enlarged, forwardly directed, and capped with enamel to act as an egg-breaking tooth, as certain egg-eating Snakes.

rachi-, rachio- Prefixes from Gk. *rhachis*, spine. Also *rhachi-, rhachio-*.

rachis, rhachis (1) The main axis of an inflorescence or of a pinnately compound leaf etc. (2) The shaft or axis; the shaft of a feather; the vertebral column. adj. *rachidial*.

rad Former unit of radiation dose which is absorbed, equal to 0.01 J/kg of the absorbing (often tissue) medium. See **gray**.

radial longitudinal section A section cut longitudinally along a diameter of a more or less cylindrical organ. Cf. **tangential longitudinal section**. Abbrev. *RLS*.

radial symmetry The condition in which an organ or the whole of an organism can be divided into two similar halves by any one of several planes which include the centre line. Cf. **bilateral symmetry**.

radiate Said of a capitulum which has conspicuous **ray florets**.

radiation See panel.

radiation danger zone A zone within which the **maximum permissible dose rate** or **concentration** is exceeded.

radiation hazard The danger to health arising from exposure to ionizing radiation, either due to external irradiation or to radiation from radioactive materials within the body.

radiation sickness Illness, characterized by nausea, vomiting and loss of appetite after excessive exposure to radiation either from radiation therapy or accidentally. If the exposure has been great it will cause bone marrow suppression with loss of blood cells, leading to anaemia, inability to overcome infection and internal bleeding.

radiation therapy The use of any form of radiation, e.g. electromagnetic, electron or neutron beam, or ultrasonic, for treating disease.

radical (1) Pertaining to the root. (2) Said of leaves, flowers etc. arising at soil level from a root stock, rhizome or the base of a stem, as in rosette plants. Cf. **cauline**.

radicivorous Root-eating.

radicle The primary root of an embryo, normally the first organ to emerge when a seed germinates.

radio-allergosorbent test *RAST*. Method for measuring extremely small amounts of IgE antibody specific for various allergens. Blood serum is reacted with allergen-coated particles which are then washed to remove non-reacting proteins. Radiolabelled anti-human IgE is then added and this binds to the IgE antibody, bound to the particles via the allergen. Provided that the amount of allergen supplied and the anti-IgE are present in excess, the radioactivity on the particles after washing is proportional to the amount of allergen-specific antibody in the serum sample.

radiobiology Branch of science involving study of effect of radiation and radioactive materials on living matter.

radiocarbon dating A method of determining the age in years of fossil organic material or water bicarbonate, based on the known decay rate of ^{14}C to ^{14}N. See **radiocarbon**.

radiation

The energy disseminated from a source which falls off as the inverse square of the distance from the source in the absence of absorption. This definition includes acoustic waves but it mainly refers to energy in the electromagnetic spectrum which requires no supporting medium and to the particles emitted during radioactive decay.

The *electromagnetic spectrum* extends from the longest radio waves of wavelengths up to 10^5 m through infrared radiation, the visible spectrum and ultraviolet waves down to X-rays and γ-rays with a wavelength of 10^{-14} m. *Non-ionizing* radiations which extend from radio to ultraviolet waves are generally thought to be less damaging to biological systems although short wave ultraviolet, because it is strongly absorbed by nucleic acids and causes the formation of thymidine dimers, is highly mutagenic if it reaches the chromosomes. *Ionizing* radiation which includes X-rays and the products of radioactive decay, α- and β-particles and γ-rays, are much more hazardous because they can penetrate living tissue and release considerable energy when they collide with biological molecules.

An average person living in the UK receives about 2.2 milliSieverts of ionizing radiation per year. Of this 36% is due to radioactivity in the air, principally due to radon gas, a natural decay product of radioactive elements in rocks and soil, 15% comes from outer space as cosmic radiation and 35% directly from buildings, land, food and water. The remainder, about 14%, is man made and is nearly all the result of medical diagnosis and treatment, mainly X-rays. Only about 0.5% is due to earlier nuclear weapons testing, compared to 0.5% due to air travel and 0.1% to nuclear power generation.

There is still considerable controversy about how hazardous such small excess doses are and it relates to whether the dose-response curve measured in experimental animals at high doses can be linearly extrapolated to zero or whether there is some threshold below which radiation becomes disproportionately ineffective. The question is important because, despite the small size of any expected effect, it could be spread across a very large population. *Repair enzymes* which locate and repair damaged DNA strands, using the undamaged strand as a template, are widespread and relevant to this discussion because they would be expected to be most effective when dealing with small amounts of damage. The main problem however is the difficulty of measuring the effects of a small excess of radiation over background in experimental animals. It has been estimated to require the observation of perhaps a million mice throughout their natural lives in order to give a significant result. Other approaches have been to study disease incidence in people who live in situations of higher than average background radiation due, for example, to granite in buildings or at high altitudes where the cosmic radiation is much increased but even in these situations it is difficult to detect significant effects.

radiography Process of image production using X-rays.

radio-immunosorbent test Method for measuring IgE immunoglobulin in samples of serum. The serum is mixed with a standard amount of purified radiolabelled IgE, and the mixture is exposed to particles coated with antibody specific for the IgE heavy chain. After appropriate incubation and washing, the amount of radioactivity bound to the particles is determined. Since IgE in the serum will compete with the radiolabelled IgE for binding to the anti-IgE, the amount of radioactivity bound will be less than is bound in the absence of such competition. The reduction of binding provides a measure of the amount of IgE in the test sample. Abbrev. *RIST*.

Radiolaria An order of marine planktonic Sarcordina, the members of which have numerous fine radial pseudopodia which do not anastomose; there is usually a skeleton of siliceous spicules.

radiolarian ooze A variety of non-calcareous deep-sea ooze, deposited at such depth that the minute calcareous skeletons of such organisms as Foraminifera pass into solution, causing a preponderance of the less soluble siliceous skeletons of Radiolaria. Confined to the Indian and Pacific Oceans, and passes laterally into red clay.

radiology The science and application of X-rays, gamma-rays and other penetrating ionizing or non-ionizing radiations.

radiomimetic Said of drugs which imitate the physiological action of X-rays, notably in suppressing new cell growth, particularly those used in treating cancer.

radionuclide Any nuclide (isotope of an element) which is unstable and undergoes natural radioactive decay.

radionuclide imaging The use of radionuclide substances to image the normal or abnormal physiology or anatomy of the body. ^{99}Technetium is an important radionucluide used for diagnostic imaging in medicine.

radiopaque Opaque to radiation, esp. X-rays.

radioresistant Able to withstand considerable radiation doses without injury.

radiosensitive Quickly injured or changed by irradiation. The gonads, the bloodforming organs and the cornea of the eye are the most radiosensitive in man.

radiospermic Having seeds which are rounded in cross section, as in the Cycadopsida. Cf. **platyspermic**.

radiotherapy Theory and practice of medical treatment of disease, particularly any of the forms of cancer, with large doses of X-rays or other ionizing radiations.

radium needle A container in form of needle, usually platinum-iridium or gold alloy, designed primarily for insertion into tissue.

radius In land Vertebrates, the pre-axial bone of the antebranchium; in Insects one of the veins of the wing; in Echinodermata and Coelenterata, one of the primary axes of symmetry. adj. *radial*.

radix The root or point of origin of a structure, as the *radix aortae*.

radon seeds Short lengths of gold capillary tubing containing radon used in treatment of malignant and nonmalignant neoplasms.

radula In Mollusca, mechanism for rasping consisting of a strip of epithelium bearing numerous rows of horny or chitinous teeth. adjs. *radular, radulate, raduliform*.

rain forest The natural forest of the humid tropics. Developing where the rainfall is heavy (> 2500 mm/year) and characterized by a great richness of species, very tall trees (> 30 m), lianes and epiphytes.

rain shadow A dry area, often a desert, on the sides of mountains away from the sea, due to the deposition of most of the moisture from the winds blowing off the ocean on the slopes facing the ocean. The higher the mountain, the greater the effect.

raised bog Type of *Sphagnum* bog, originating from a valley bog or a fen by the upward growth of the vegetation and the failure of the dead plant material to decompose. The consequent raised bog is convex.

ramentum Thin, chaffy, brownish scale, esp. on the stem, petiole or leaf of a fern.

ramet Any physically and physiologically independent individual plant, whether grown from a sexually-produced seed or derived by vegetative reproduction. See **clone**. Cf. **genet**.

ramiform Branching.

ramus The barb of a feather; in Vertebrates, one lateral half of the lower jaw, the mandible; in Rotifera, part of the trophi; any branchlike structure; a ramification.

ranalian complex Group of families, including the Ranunculaceae, containing what are thought to be the most primitive extant flowering plants. See **Magnoliidae**.

random coil A section of a polypeptide chain which is not folded into any specific secondary structure.

random mating Occurs when any individual in a population has an equal chance of mating with any other of the opposite sex.

random searching A process of completely unorganized 'search' by which some ecologists suggest that some animal populations find food, mates and suitable places to live.

random variable The mathematical representation of a variate associated with a stochastic phenomenon.

range The difference between the largest and smallest values in a set of observations.

ranine Pertaining to, or situated on, the under surface of the tongue.

rank The number in serial order corresponding to a given data value when all values are placed in ascending order of magnitude; to place in ascending order of magnitude.

rank test A statistical procedure carried out on the ranks rather than the values of the observations.

Ranunculaceae The buttercup family, ca 1800 spp. of dicotyledonous flowering plants (superorder Magnoliidae). Mostly herbs, mostly north temperate; the floral parts are free, the stamens numerous and hypogynous. Many are poisonous and have little economic importance except for use as ornamentals. *Anemone, Helleborus, Delphinium* etc.

Ranvier's nodes Constrictions of the neurolemma occurring at regular intervals along medullated nerve fibres.

raphe, rhaphe (1) Ridge on an ovule or seed representing that part of the stalk that is fused to the ovule. (2) A broad junction, as between the halves of the Vertebrate brain.

raphide A needle-shaped **crystal**, usually calcium oxalate, usually one of a bundle in the vacuole of a cell, esp. in a leaf.

raptatory, raptorial Adapted for snatching or robbing, as birds of prey.

rasorial Adapted for scratching.

RAST Abbrev. for *RadioAllergoSorbent Test*.

Rathke's pouch In developing Vertebrates, the diverticulum formed from the dorsal aspect of the buccal cavity ectoderm which gives rise to the adenohypophysis. Also *craniobuccal pouch*.

rationalization In psychoanalytic theory, a mechanism of defence by means of which unacceptable thoughts or actions are given acceptable reasons which justify it, and also hide its true motivation.

ratio schedule of reinforcement A program of reinforcement in which a certain number of responses are necessary in order to produce the reward. On **fixed ratio schedules** the number of responses is always the same; on **variable ratio schedules** the number of responses varies from trial to trial.

rattle the series of horny rings representing the modified tail-tip scale in Rattlesnakes (*Colubridae*).

Rauber's cells In Mammals, cells of the trophoblast situated immediately over the embryonic plate.

Raunkiaer system A classification of the vegetative or life forms of plants according to the positions of the perennating (resting) buds and the protection they receive during an unfavourable season of cold or drought.

ray A skeletal element supporting a fin; a sector of a radially symmetrical animal. In plants, a panel of tissue, usually mostly parenchyma, one to several cells wide and a few to many cells high, produced by ray initials in the cambium and extending radially into the secondary xylem (*xylem ray*) and secondary phloem (*phloem ray*), and with the functions of radial transport and storage. See **medullary ray, ray tracheid.**

ray floret (1) Usually a peripheral floret of a capitulum, regardless of its morphology. (2) Sometimes a **ligulate floret.** Cf. **disk floret.**

ray initial A more or less isodiametric cell, one of a group of such in the vascular **cambium**, each giving rise to one of the radial files of cells making up a ray. Cf. **fusiform initial.**

ray tracheid Tracheids shorter than the ordinary axial tracheids. They are found at the top and bottom margins of the ray with their long axes in the radial direction. Occur in the wood of some conifers, e.g. pines.

R-bands See **banding techniques.**

rDNA, ribosomal DNA Genes specifying the several kinds of ribosomal RNA molecules.

reaction Any change in behaviour of an organism in response to a stimulus.

reaction formation In psychoanalytic theory, a **defence mechanism** by which a forbidden impulse is mastered by exaggeration of the opposing tendency (hate is converted to oversolicitous love).

reaction time The interval between the presentation of a signal and the subject's response to it.

reaction wood Wood of distinctive anatomical structure formed on branches and leaning trunks, producing as it matures tensile or compressive forces that tend to maintain a growing branch at its appropriate orientation (in spite of increasing mass) or to correct misorientation. Reaction wood makes unsatisfactory timber and pulp. See **compression wood, tension wood.**

reactive depression A type of depression clearly linked to environmental events (e.g. after a death).

reactive schizophrenia Those cases of schizophrenia in which onset is sudden and linked to some precipitating event in the environment.

reading frame See panel.

reagin, reaginic antibody Antibody which fixes to tissue cells of the same species so that, in the presence of antigen, histamine and other vasoactive agents are released. The term was used to describe such antibodies in humans before IgE was identified, and is often still used. In some species antibodies of immunoglobulin classes other than IgE also have similar properties.

reality principle In psychoanalytic theory, the mental activity that leads to instinctual gratification by accommodating to the demands of the real world; it is acquired during development. Cf. **pleasure principle.**

recall A method of measuring retention in which some material must be produced from memory. Cf. **recognition.**

recapitulation theory States that stages in the evolution of the species are reproduced during the developmental stages of the individual, i.e. ontogeny tends to recapitulate phylogeny. Superficially apparent in some instances. Also called *biogenetic law, Haeckel's law.*

recency effect In recall, the tendency to recall items from the end of the list more readily than those in the middle. See **primacy effect.**

receptacle Structure on which reproductive organs are borne, esp. (1) swollen tip of a thallus with conceptacles in brown algae, (2) area bearing archegonia or antheridia in liverworts, or sporangia in ferns, (3) the end of a stalk on which is borne either the parts of a single flower or the involucre and florets of a head or capitulum.

---

reading frame

In the genetic code, three bases are required to specify one amino acid. An mRNA molecule can therefore be read off the DNA in three different *reading frames*, depending on the starting base as in the table below. Usually alternative reading frames contain many **stop codons,** e.g. 'amber', and are not *open*. Certain small viruses, notably SV40 and polyoma, exploit this possibility however and specify different polypeptides from a single sequence of DNA. They have therefore *overlapping genes*. The table gives an example.

mRNA	G A A G G C U U U A C U U C A A G U A G A U G C ....
Frame 1	Glu Ala Phe Thr Ser Ser Arg Cys ....
Frame 2	Lys Ala Leu Leu Gln Val Asp ....
Frame 3	Arg Leu Tyr Phe Lys AMBER

---

receptaculum (1) A receptacle; a sac or cavity used for storage. (2) A sac in which ova are stored, as in some Oligochaeta.

receptaculum seminis A sac in which spermatozoa are stored, as in many Invertebrates; a spermotheca.

receptive Capable or being effectively pollinated or fertilized.

receptor See panel.

receptor mediated endocytosis The internalization of ligands bound to certain receptors on the cell surface which become clustered into *coated pits* and enter the cell *via* **coated vesicles** and **endosomes**.

recess A small cleft or depression, as the *optic recess*.

recessive Describes a gene (allele) which shows its effect only in individuals that received it from both parents, i.e. in homozygotes. Also describes a character due to a recessive gene. Cf. **dominant**.

reciprocal cross A cross made both ways with respect to sex, i.e. Af×Bg and Ag×Bf. Consistent differences between the offspring of such crosses suggests **cytoplasmic inheritance.**

reciprocal hybrids A pair of hybrids obtained by crossing the same two species, in which the male parent of one belongs to the same species as the female parent of the other, e.g. mule and hinny.

reciprocal translocation Mutual exch-, ange of non-homologous portions between two chromosomes.

recognition A method of measuring retention in which a stimulus has to be identified as having occurred before.

recombinant inbred strains *RI strains.* Inbred strains, mostly of mice, which have been made by crossing two different inbred parental strains to yield an F1 generation, and from this an F2 generation. Pairs of F2 mice are then crossed and their progeny inbred until they are homozygous at most loci. After prolonged inbreeding all members of the RI strain tend to complete genetic identity and homozygosity. However genes of the parental strains have become reassorted. Such RI strains provide a means of assessing the functions of gene products associated with one another, as occurs in the formation of complex receptor molecules. A single genetic character can be bred from one strain into another, by forming the F1 generation followed by repeated back-crossing over 20 or more generations, at each of which the offspring have been selected for the presence of a particular genetic character from one parent on the genetic background of the other parent.

recombinant, recombinant DNA (1) An organism which contains a combination of alleles differing from either of its parents. (2) DNA which contains sequences from different sources, made usually as the result of laboratory procedures *in vitro* (3) Individual, gamete or chromosome resulting from *recombination*.

recombination Reassortment of genes or characters in combinations different from what they were in the parents, in the case of *linked* genes by *crossing over*.

recovery rate That at which recovery occurs after radiation injury. It may proceed at different rates for different tissues.

rectal gills In the larvae of some Odonata, tracheal gills in the form of an elaborate system of folds in the wall of the rectum, used in respiration.

recti-, recto- Prefixes from L. *rectus*, straight; recto- often in terms pertaining to the rectum.

rectirostral Having a straight beak.

rectrices In Birds, the stiff tail feathers used in steering. sing. *rectrix*. adj. *rectricial*.

rectum (L. *rectum intestinum*, straight intestine.) The posterior terminal portion of the alimentary canal leading to the anus. adj. *rectal*.

---

receptor

A chemical grouping on a macromolecule or a cell which can combine selectively with other complementary molecules or cells, e.g. enzyme receptors for substrate, cell surface receptors for hormones or growth factors.

In immunology refers to cell surface sites such as Fc receptors, or antigenbinding molecules on the membranes of B or T lymphocytes. Binding to a receptor on a cell membrane is often followed by transduction of a signal across the membrane and a response on the part of the cell, e.g. RNA transcription. See **steroid regulated genes**.

An element of the nervous system specially which is adapted for the reception of stimuli; for example, a sense-organ or a sensory nerve-ending.

---

rectus A name used for various muscles which are of equal width or depth over their length, e.g. the *rectus abdominis* in Vertebrates.

recurrent Returning towards point of origin.

recurvirostral Having the beak bent upwards.

red algae Rhodophyceaea.

red blood corpuscle See **erythrocyte**.

red body See **red gland**.

red corpuscle See **erythrocyte**.

Red Data Book Catalogue of rare and endangered species prepared by the *International Union for Conservation of Nature and Natural Resources*, started in 1966, and covering the whole world.

red gland, body In some Fish, a structure found in the wall of the air bladder, responsible for secretion or absorption of gas. Also *rete mirabile*.

redia The secondary larval stage of Trematoda, possessing a pair of locomotor papillae and a rudimentary pharynx and intestine, and capable of paedogenetic reproduction.

redirected behaviour Behaviour directed at inappropriate or irrelevant objects, often as a result of **frustration** or **conflict**.

red light For plant responses mediated by phytochrome, light of wavelength around 630 nm.

red muscles In Vertebrates, phasic muscles which are therefore rich in sarcoplasm and myoglobin, and are of red colour.

red snow Lying snow coloured by the growth near the surface of algae, esp. *Chlamydomonas nivalis*, containing haematochrome.

red tide Water containing sufficient dinophytes or other organisms to colour it red. Called a *bloom*. Some blooming dinophytes contain sufficient toxin to make shellfish feeding on them fatally poisonous to man.

reduction division See **meiosis**.

reed See **abomasum**.

refection In Rabbits, Hares and probably other herbivores, the habit of eating freshly-passed faeces. Also *autocoprophagy*.

reflected Said of a structure, esp. a membrane, which is folded back on itself.

reflex A simple, automatic, involuntary and stereotyped response to some stimulus (e.g. an eye blink).

reflex action An automatic or involuntary response to stimulus.

reflex arc The simplest functional unit of the nervous system, consisting of an afferent sensory neuron conveying nerve impulses from a receptor to the CNS, generally the spinal cord, where they are passed, either directly or via an internuncial or association neuron, to a motor neuron, which conveys them to a peripheral effector, such as a muscle.

reflexed Bent back abruptly.

refractory period For an organism or an excitable tissue, the period of zero response following a previous response.

refugium An area where species have survived the great changes undergone by the region as a whole, because local conditions are favourable. Examples of refugia are the area escaping glaciation in the Ice Ages, and hedgerows (where woodland species escape the influence of cultivation).

regeneration Regrowth of tissues or organs, such as amphibian limbs, after injury; the formation of new plants from cultured tissues. See **tissue culture**.

regression (1) In psychoanalytic theory, a defence mechanism which involves a reversion to an earlier and less threatening mode of functioning. (2) A tendency to return from an extreme to an average condition, as when a tall parent gives rise to plants of average stature. (3) In statistics, a model of the relationship between the expected value of a random variable and the values of one or more possibly related variables.

regular Said of a flower which is radially symmetrical or **actinomorphic**.

regulator gene A gene whose product controls the rate at which the product of another gene is synthesized.

regurgitation (1) The flowing of blood in reverse direction to the circulation in the heart as a result of valvular disease, e.g. *aortic regurgitation*. (2) The reverse movement of the gastric contents.

reinforcement Refers to situations when a response is predictably followed by an event, the *reinforcer*, and the event can be shown to increase or alter the future probability of the response. Cf. **positive reinforcement**, **negative reinforcement**.

reiterated, repeated sequences DNA sequences repeated many times within a genome; common in higher organisms.

rejection The process by which the body rejects tissue transplanted into it.

rejuvenescence Renewal of growth from old or injured parts.

relative abundance A rough measure of population density, relative, e.g. to time (as the number of birds seen per hour) or percentage of sample plots occupied by a species of plant.

relative growth rate Abbrev *R*. Mathematical expression of growth

$$R = \frac{1}{W}\frac{dW}{dt}$$

where *W* is weight and *t* time.

relaxation time In excitable tissues, the period during which activation subsides after cessation of a stimulus.

relearning A method of measuring retention on a learning/memory task; the material is relearned again some time after the original learning. The difference between the original learning and relearning (the *savings*) is a measure of the original learning.

releaser *Social releaser*. A term originating in classical ethology; it refers to aspects of stimulus (the sign stimulus) which are esp. effective in releasing a specific response in all individuals of a species. It also implies that both the relevant stimulus features and the response to them have become mutually adapted through evolution.

relevé List of the plant species at any site with visual estimation of canopy cover.

relict A species, whether terrestrial, marine or freshwater, which occurs at the present time in circumstances different from those in which it originated.

REM *Rapid Eye Movement*. Occurs at certain stages during dreaming sleep and is believed necessary for brain repair.

rem See **röntgen equivalent man**.

remiges In Birds, the large contour feathers of the wing. sing. *remex*.

remiped Having the feet adapted for paddling, as many aquatic birds.

remission An abatement (often temporary)

of the severity of a disease; the period of such abatement.

REM sleep See **paradoxical sleep**.

renal Pertaining to kidneys.

renal portal system In some lower Vertebrates, that part of the venous system which brings blood from the capillaries of the posterior part of the body and passes it into capillaries of the kidneys.

renaturation The converse of **denaturation**. The complementary strands of DNA or DNA and RNA will reform duplex molecules. The kinetics of the process depend on the number of copies of each sequence and the concentration of the molecules. Usually achieved by heating single strands at about 20°C below the melting temperature (Tm). The basis of the specificity of molecular **probes**.

rendzina Shallow, dark, intrazonal soil, rich in calcium carbonate, developed on limestone, esp. chalk.

reniform Kidney-shaped.

repeated DNA A sequence which occurs more than once in the haploid genome. Such sequences are often short and occur many times.

repetition compulsion In psychoanalytic theory, the factor in mental life which compels early patterns of behaviour to be repeated, irrespective of pleasure/displeasure thereby experienced by the individual.

replica plating Typically, transferring the *pattern* of bacterial colonies on an agar plate by impressing velvet, stretched over a holder, onto the agar and then placing the velvet in turn onto a number of further sterile plates. If the latter contain, say, different antibiotics, a sensitive strain can be selected by noting colonies which fail to grow and then picking them from the master plate.

replication Duplication of genetic material, usually prior to cell division.

replication fork The fork where duplex DNA becomes split into two double strands as replication moves from the *origin of replication*.

replicon A part of a DNA molecule which is replicated from a single origin. In prokaryotes there is usually one origin per genome, but in eukaryotes there are many spaced along the chromosome.

replum *False septum*. Partition across ovary or fruit formed by ingrowth from placentas, not by the walls of the carpels. Cf. **septum**.

repression In psychoanalytic theory, the process by which an unacceptable thought, impulse or memory is rendered unconscious.

repressor A protein which binds to an **operator** site and prevents transcription of the associated gene.

reproduction The generation of new individuals in the perpetuation of the species.

---

respiration

The term can be applied to events which occur at the level of the whole organism or its constituent cells. The former usage relates to the exchange of oxygen and carbon dioxide between the organism and its environment, 'breathing'. Small organisms can exchange gases across their body surface but larger animals require a richly vascularized respiratory surface, such as gills or lungs, and mechanisms for the movement of water or air over the respiratory surface.

Respiration at the cellular level consists of the metabolic processes which degrade foodstuffs with the synthesis of ATP. It is of two major types: aerobic, which requires oxygen as a terminal electron receptor and anaerobic, where some other terminal acceptor is used.

Aerobic respiration of glucose consists of its total oxidative degradation by **glycolysis** and the **tricarboxylic acid cycle**, to carbon dioxide and water with the generation of 36 molecules of ATP per glucose molecule, made up of 2 molecules by **substrate level posphorylation** and 34 molecules by **oxidative phosphorylation.**

During anaerobic respiration in higher organisms the oxidation of glucose is incomplete and consists essentially of those stages of aerobic respiration which precede the tricarboxylic acid cycle, producing ATP only by substrate level phosphorylation, and degrading the glucose to pyruvate. The electrons arising from this limited oxidation are accepted by a metabolite (see **fermentation**) or alternatively, in micro-organisms, by some other molecule, e.g. sulphur with the production of hydrogen sulphide. In some micro-organisms an extensive oxidative anaerobic respiration is possible, analogous to aerobic respiration with an **electron transfer chain**, including cytochromes and oxidative phosphorylation, but using terminal electron acceptors such as *nitrate* which is reduced to gaseous nitrogen, *sulphate* which is reduced to hydrogen sulphide or *carbonate* reduced to methane.

---

adj. *reproductive.*

reproductive behaviour The varied activities that lead to production and rearing of offspring. Includes **agonistic behaviour**, courtship and other mate interactions, as well as maternal behaviour.

Reptilia A class of Craniata. They are pentadactyl and have shelled amniote eggs. The vertebrae are gastrocentrous, the kidney metanephric and the skin completely covered by epidermal scales or, sometimes, by bony plates. They are poikilothermous, breathe by lungs and retain aortic arches. Known fossils from late Carboniferous, they were dominant and various in the Mesozoic *(Dinosaurs)* but became less numerous in the Cretaceous. Living forms include Lizards, Snakes, Turtles, Tortoises, Crocodiles and Alligators.

repugnatorial glands In Arthropoda, glands, usually abdominal in position, which produce a repellent secretion of odoriferous pungent or corrosive nature which can be used in self defence.

repulsion When two specified *non-allelic* genes are on different but homologous chromosomes, having come from different parents, they are in *repulsion*. Cf. **coupling**.

resilience The capacity of ecosystems and populations to return to a previous state after they have been disturbed.

resin canal, resin duct Duct of schizogenous origin lined with resin-secreting cells which contain resin, as in the leaves and sometimes wood of conifers.

resistance In psychoanalytic theory, the opposition encountered during psychoanalytic treatment to the process of making unconscious memories and impulses, conscious.

resistant Not readily attacked by a parasite, disease or drug. See **antibiotic resistance**.

resolving power of the eye The angle subtended by a small object which can just be determined visually.

respiration See panel.

respiratory centre In Vertebrates, a nerve centre of the hindbrain which regulates the respiratory movements.

respiratory failure Occurs when the respiratory system is no longer able to maintain normal tensions of oxygen or carbon dioxide in the body.

respiratory movement The muscular movements associated with the supply of air or water to the respiratory organs.

respiratory organs The specialized structures like lungs and gills which enable oxygen to be transferred to the body fluids.

respiratory pigment In the blood of many animals, a coloured compound formed by a metal-containing prosthetic group bound to a protein, the whole forming a complex with a high affinity for oxygen. Concerned with oxygen transport, e.g. haemoglobin, haemocyanin.

respiratory quotient The ratio of moles CO_2 evolved to moles O_2 absorbed in respiration; unity when the substrate is carbohydrate, lower when protein or fat. Abbrev. *RQ*.

respiratory substrate Any chemical compound broken down during respiration to release the chemical energy stored in its bonds.

respiratory system See **respiratory organs**.

respiratory valve In some Fish, e.g. Trout, a pair of transverse membranous folds, one attached to the floor, the other to the roof of the mouth, which prevent water from escaping through the mouth during expiration.

respondant In classical conditioning, a response that is elicited by a known stimulus (e.g. a knee jerk).

response The effect of stimulation; it may, as in muscular and glandular responses, be easily observable and measurable, but it may also be an inferred response which is not immediately apparent in behaviour.

response latency The time elapsing between the onset of a stimulus and the beginning of an animal's response to it.

restiform Ropelike.

resting nucleus Nucleus of a cell which is not undergoing active growth and division.

resting spore A thick-walled spore able to endure drought and other unfavourable conditions, and normally remaining quiescent for some time before it germinates.

restitution nucleus A single nucleus formed following failure of the chromosomes to separate properly at anaphase and hence containing, say, twice the expected chromosome number.

restriction and modification See panel.

restriction endonuclease, enzyme See panel.

restriction fragment Because of their sequence specificity, **restriction endonucleases** will cleave any DNA into defined polynucleotide fragments, which can be separated on an agarose gel.

restriction fragment length polymorphism A restriction fragment identified by blotting whose length is variable in the population. Those which map close to sites of genetic diseases are a useful aid to antenatal diagnosis in families at risk. Abbrev. *RFLP*.

resupinate *Inverted*, e.g. the flowers of orchids in which, because of a 180° twist in the stalk, what appears to be the lower petal is, in fact, morphologically the upper petal.

rete A netlike structure. pl. *retia*.

rete Malpighii, mucosum See **Malpighian layer**.

rete mirabile A network of small blood vessels, as in the so-called **red gland** of Fish.

reticular Resembling a net; of or pertaining to the reticuloendothelial system.

reticular tissue A form of connective tissue in which the intercellular matrix is replaced by lymph; it derives its name from the network of collagenous fibres which it shows. Also *retiform tissue*.

reticulate thickening Secondary wall deposition in the form of an irregular network in tracheids or vessel elements of **metaxylem**.

reticulodromous Having a network of veins.

reticuloendothelial system A term formerly used to describe the system of cells that have the ability to take up certain dyes and particles (such as carbon in the form of India ink) when injected into the living animal. It has been replaced by **mononuclear phagocyte system**.

reticul-, reticulo- Prefixes from L. *reticulum*, net.

reticulum In ruminant Mammals, the second division of the stomach, or *honeycomb bag*; any netlike structure. adj. *reticular*.

retiform See **reticular tissue**.

retina The light-sensitive layer of the eye in all animals. The human retina contains two types of sensitive element: **rod** and **cone**. adj. *retinal*.

retinene Vitamin A_1-aldehyde, a component of rhodopsin.

retinulae In Arthropoda, the visual cells of the compound eye, forming the base of each ommatidium.

retractile Capable of being withdrawn, as the claws of most Felidae.

retractor A muscle which, by its contraction, draws a limb or a part of the body towards the body. Cf. **protractor**.

retrices In Birds, the stiff tail feathers used in steering. sing. *retrix*. adj. *retricial*.

retrieval *Memory*. The process of searching for and bringing stored information into consciousness.

retrieval cue Environmental or internal stimuli that help the retrieval of an experience.

retro- Prefix from L. *retro*, backwards, behind.

retrocerebral glands In Insects, a collective name applied to a number of endocrine glands in the head, behind the brain, which are concerned with the co-ordination of

restriction and modification

Some bacteria are able to restrict their susceptibility to lysis by phage or other genetic elements by cleaving the invading DNA with restriction enzymes. Their own DNA is made immune by methylating the susceptible sites (*modification*). Such bacteria are said to have a *restriction-modification* system.

Restriction enzymes are a class of endonucleases able to cleave DNA at a specific nucleotide sequence, although there is no evidence that all organisms containing restriction enzymes use a restriction-modification system. Different enzymes, as obtained from a wide range of organisms, have different specificities, often recognising 4 or 6 base pairs. Because of this specificity, restriction enzymes will cleave a sample of DNA into defined polynucleotide fragments which can then be separated according to their length. The pattern of fragments will depend on both the source and the enzyme used.

This procedure is sufficiently sensitive to be able to detect a difference of one base pair in certain circumstances. The enzymes are an essential prerequisite for the procedures of **genetic manipulation.** Where the cleavage sites are not directly opposite each other as in most of the examples shown, the cleaved DNA has *sticky ends* which facilitate the joining of one piece of DNA to another cut by the same enzyme, as in the insertion of a eukaryote DNA into that of a bacterial **vector**. The AluI enzyme gives *blunt ends*.

These enzymes are denoted by a symbol representing genus and species followed by a roman numeral if a number of different enzymes have been isolated from the one species. In the accompanying table, the four bases are shown as A G C or T (N standing for any of them) and the cleavage position by an arrow. The upper sequence reads in the 5′ to 3′ direction and an overscore indicates a methylation site.

Examples of restriction enzymes

EcoRI
G↓AATT C
C TTAA↑G
BamHI
G↓GATC C
C CTAG↑G
AluI
AG↓CT
TC↑GA
HpaII
C↓CG G
G GC↑C
HinfI
G↓ANT C
C TNA↑G

Examples of specific methylases

HindI
CAC
BbvSI
GA(A or T)GC

---

postembryonic development and metamorphosis. See **corpora cardiaca,corpora allata**.

retrograde amnesia A type of amnesia that often occurs after a head injury, or from electrical shock; there is loss of memory for events leading up to the injury, although the period of time that is lost to memory varies with the conditions of injury.

retrovirus A virus of higher organisms whose genome is RNA, but which can insert a DNA copy of its genome into the host's chromosome. Important because they include the **oncogenic viruses**, and because they can be used as **vectors** for the introduction of DNA sequences into eukaryotic cells.

retuse Having a slight notch at a more or less obtuse apex.

revehent Carrying back.

reverse genetics The process of removing a gene from an organism, altering it in a known way and reinserting it. The organism is then tested for any altered function.

reverse transcriptase An enzyme which makes a double-stranded DNA copy of an RNA virus genome, much used in the laboratory to make DNA copies of mRNA.

reversion The process by which a *mutant* phenotype is restored to normal by another mutation of the same gene, i.e. a *back-mutation*. But sometimes used in the sense of *suppression*.

revolute Leaves with margins rolled outwards or downwards (i.e. towards abaxial surface). Cf. **involute**. See **vernation**.

rexigenous, rhexigenous Said of a space formed by the rupture of cells. Cf. **lysigenous, schizogenous**.

---

rheumatoid arthritis, rheumatoid factor

Rheumatoid arthritis is a chronic inflammatory polyarthritis, which may be accompanied by systemic disturbances such as fever, anaemia and enlargement of lymph nodes. The synovia of joints are infiltrated with granulomata containing plasma cells, lymphocytes, macrophages and germinal centres, causing inflammation and swelling of particularly the small joints of the extremities. B lymphocytes are present in the inflamed tissue and in lymphoid tissues elsewhere which make *rheumatoid factor*. This is present in the blood as well as locally. It can combine with other immunoglobulins to activate complement, and this is thought to be the immediate cause of the inflammation. In this sense rheumatoid arthritis is an autoimmune disease, but the initiating cause or causes are not known.

Rheumatoid factor is an antibody reactive with determinants present on the heavy chain constant region of immunoglobulins of many species, including that in which the antibody is made. Causes the inflammation characteristic of rheumatoid arthritis. The determinants are revealed when the immunoglobulin molecules are slightly distorted, as by combination with an antigen or by mild denaturation, and this property is used in various tests to detect the factors. Rheumatoid factors are usually polyclonal, and may be of the IgM, IgG or IgA class, although IgM is much the commoner. In some subjects monoclonal rheumatoid factors are made, and if the amount is sufficient these will combine with other immunoglobulins to form **cryoglobulins**.

RFLP See **restriction fragment length polymorphism**.

rhabdom In the compound eyes of Arthropoda, the structure containing the visual pigment and concerned with phototransduction.

rhabdomeres One of the constituent portions of the rhabdom, secreted by a single visual cell.

rhachi-, rhachio- See **rachi-**.

rhachis See **rachis**.

rhamphotheca In Birds, the horny coverings ensheathing the upper and lower jaws.

rhaphe See **raphe**.

Rheiformes An order of Birds, containing two species of large running bird, found on the South American pampas, occupying an ecological niche approximately similar to that of the Emu. Have 3 toes. Rheas.

rheo- Prefix from Gk. *rheos*, current, flow.

rheoreceptors Receptors of Fish and certain Amphibians which respond to stimulus of water current, e.g. lateral line system.

rhesus blood group system Human blood group system, so called because the antigen involved is also present on rhesus monkey red cells and was first detected when these were used to immunize rabbits. The rhesus blood group system is genetically complex and there are several alleles. The most important is that which is known as the D-antigen. Antibodies against rhesus antigens do not occur naturally in the blood but may be produced after transfusion into a rhesus(D)-negative person of rhesus(D)-positive blood or in a rhesus(D)-negative mother

who bears a rhesus(D)-positive child. In the former case a subsequent transfusion of positive blood may cause a **transfusion reaction**, and in the latter give rise to **erythroblastosis fetalis** in the child.

rhesus factor Blood group **antigens** possessed by 85% of the population (Rhesuspositive). Of importance in blood transfusion during pregnancy. A rhesuspositive baby born to a rhesus-negative mother with rhesus antibodies may develop **haemolytic disease of the newborn**.

rhesus monkey One of the species, *Macaca mulatta*, of the macaque monkeys, native to S.E. Asia. Robust and intelligent, they have been widely used in medical research. See **Rhesus factor**.

rheumatic fever An acute inflammatory disease involving the heart and the joints which generally follows a few weeks after an infection by *Streptococcus pyogenes* of Lancefield Group A. The characteristic lesions are degeneration and necrosis of fibrous tissue and nodules of necrotic fibrous tissue surrounded by macrophages, lymphocytes and plasma cells. These are probably some form of hypersensitivity reaction to Group A streptococci. Antibodies are present against an antigen of the streptococci which shares antigenic determinants with the sarcolemma of heart muscle.

rheumatoid arthritis See panel.

rheumatoid factor See panel.

rhinal Pertaining to the nose.

rhinarium In Mammals, the moist skin

around the nostrils, also known as the muzzle, which is lacking in Anthropoids.

rhinencephalon The olfactory lobes of the brain in Vertebrates.

rhinocoele The cavity of the rhinencephalon; olfactory ventricle of the Craniate brain.

rhin-, rhino- Prefixes from Gk. *rhis*, gen. *rhinos*, nose.

rhizo- Prefix from Gk. *rhiza*, root or root-like.

Rhizobaceae A family of bacteria belonging to the order Eubacteriales (Bergey classification). Includes the symbiont *Rhizobium* which is important in nitrogen fixation by leguminous plants.

rhizodermis The outermost layer of cells of a root in its primary state.

rhizoid Outgrowth from an alga, fungus, bryophyte or pteridophyte gametophyte, attaching to or growing into the substrate and serving in anchorage and, possibly, absorption.

rhizome Stem, usually underground, often horizontal, typically non-green and root-like in appearance but bearing scale leaves and/or foliage leaves, e.g. nettle, many *Iris* spp. Cf. **stolon**.

rhizomorph A strand, like a length of thin string, composed of densely packed hyphae, by means of which some fungi spread, e.g. boot-lace or honey fungus, *Armillaria mellea*, dry-rot fungus *Serpula (Merulius) lacrymans*.

rhizophagous Root-eating.

Rhizopoda See **Sarcodina**.

rhizopodium Long, very fine, sometimes branched, cytoplasmic process from an algal cell, esp. in the Chrysophyceae.

rhizosphere Zone of soil, in the immediate vicinity of an active root, influenced by the uptake and output of substances by the root and characterized by a microbial flora different from the bulk soil.

rhodamine A fluorochrome commonly conjugated with antibodies for use in **indirect immunofluorescence**.

rhodophane A coloured oily substance, globules of which are found in the cones of Birds and in parts of the retina in some forms.

Rhodophyceae Red algae, eukaryotic, chloroplasts with chlorophyll a and phycobilins, thylakoids single. No flagellate stages. Reserve carbohydrate floridian starch (α–1→4 glucan) in the cytoplasm. Life cycle often complex, e.g. triphasic alternation of generations, sexual reproduction oögamous. Unicellular, filamentous or parenchymatous. Mostly marine, littoral and sublittoral (the red seaweeds); some in fresh water or soil; some parasitic on other red algae. Source of agar and carragheen; a few are eaten, e.g. *Porphyra*, laver-bread.

rhodopsin The light sensitive protein present in the eye. Its light sensitivity is due to the prosthetis group of 11-cis-retinol.

rhod-, rhodo- Prefixes from Gk. *rhodon*, rose.

rhombencephalon See **hindbrain**.

rhyncho- Prefix from Gk. *rhynchos*, beak, snout, proboscis.

Rhynchocephalia An order of the Lepidosauria with two temporal vacuities, a large parietal foramen which, in the one living form, *Sphenodon*, contains a non-functional median eye. Known as fossils from Middle Trias, *Sphenodon* survives in coastal islands off New Zealand.

rhynchodont Having a toothed beak.

rhynchophorous Having a beak.

rhytidome Dead, outer bark, consisting of layers of periderm with some cortex and/or secondary phloem.

rib A small ridge or rib-like structure; in Vertebrates, an element of the skeleton in the form of a curved rod connected at one end with a vertebra; it serves to support the body walls enclosing the viscera.

riboflavin See **vitamin B**.

ribose $C_5H_{10}O_5$, a pentose, a stereoisomer of arabinose. D-Ribose occurs in ribose nucleic acids (RNA).

ribosome Consisting of three subunits of RNA and protein, this complex bead-like structure can associate with mRNA and is the site of synthesis in the cytoplasm of polypeptides encoded by the mRNA.

ribulose A 5-carbon ketose sugar. See **ribulose bisphosphate**.

ribulose 1,5-bisphosphate carboxylase oxygenase See panel.

ribulose bisphosphate See panel.

richness The number of species in a defined area.

rictus Of Birds, the mouth aperture adj. *rictal*.

righting reflex A reflexive response to falling which ensures that the animal lands upright.

rigor A state of rigidity and irresponsiveness into which some animals pass on being subjected to a sudden shock as a defensive mechanism; 'shamming dead'.

rigor mortis Stiffening of the body following upon death.

rima A narrow cleft. adjs. *rimate, rimose, rimiform*.

ring See **annulus**.

ring culture A system for growing e.g. tomatoes in greenhouses, using bottomless containers filled with fresh compost (through which the mineral elements are supplied) resting on a bed of sand or other inert material (through which water is supplied). The system makes it easier to avoid the infection of the roots by fungi that is

ribulose 1,5-bisphosphate carboxylase oxygenase, RUBISCO

The enzyme that catalyses both the reaction of ribulose bisphosphate (RUBP, 5-carbons) and CO_2 to give two molecules of phosphoglyceric acid (PGA, 3-carbons) as the first step of the **Calvin cycle** in photosynthesis, and also the apparently wasteful reaction of RUBP with oxygen to give one molecule each of PGA and phosphoglycollic acid (2-carbons). The former, carboxylation, reaction results in the net fixation of carbon (1C per RUBP); the phosphoglycolate produced in the latter, oxygenation, reaction is mostly metabolised to PGA and CO_2 ($\frac{1}{2}$ molecule each) resulting in a net loss of CO_2 ($\frac{1}{2}$ C per RUBP) (photorespiration). Oxygen and CO_2 compete, and oxygenation becomes more important at higher temperatures. Under temperate conditions in an ordinary (C_3) plant the ratio of carboxylation to oxygenation might be 4 : 1, representing the loss of 30% of the potential fixation.

RUBISCO in higher plants is made of eight large (55kD) subunits, coded by **chloroplast DNA** and made in the chloroplast, and eight small (15kD) subunits, coded by nuclear DNA and made in the cytoplasm. There is one reaction site. RUBISCO has a low reaction rate; in a typical (C_3) leaf about half the soluble protein is RUBISCO, making it the most abundant protein on earth.

Although some of the metabolites produced from phosphoglycolate are used in other reactions, and photorespiration during CO_2 shortage may serve to dissipate energy from light which would otherwise damage the photochemical system, the oxygenation reaction is usually interpreted as an unfortunate or inevitable side effect of an active centre selected for carboxylation, photorespiration being a method of recovering some of the carbon which would otherwise be lost through oxygenation. RUBISCO presumably originated when the atmosphere was rich in CO_2 and relatively anaerobic. Perhaps because of its central role in assimilation it appears to have undergone relatively little evolutionary change despite its low reaction rate and an oxygenation reaction which became increasingly important as atmospheric CO_2 concentration fell and O_2 rose. Conditions in which photosynthesis is limited by CO_2 appear to have favoured CO_2-concentrating mechanisms. These include the C_4 metabolism of some plants in hot sunny places (favouring oxygenation) and the possible active uptake of bicarbonate by some submerged aquatics (CO_2 diffusion being slow in water). Attempts to select (C_3) crop plants lacking the oxygenation reaction have so far failed.

Alternative names include *RuBPc/o, RuBP carboxylase, carboxydismutase* and *fraction I protein.*

Ribulose bisphosphate is the 1,5-bisphosphate ester of the 5-carbon ketose sugar, ribulose, the substrate of RUBISCO. Its alternative names include *RuBP, ribulose diphosphate, RuDP.*

common in greenhouse crops grown in the soil.

ring gland See **Weismann's ring.**

ringing The removal of the outer tissues from a strip encircling or partly encircling a stem or trunk, e.g. experimentally to interrupt phloem transport while leaving xylem transport more or less undisturbed, the phloem being peripheral to the xylem in most stems, horticulturally to encourage flowering and fruiting in over-vigorous fruit trees or by e.g. rabbits or deer and then, if it is extensive, often fatal.

ring-porous Said of wood, having much larger and/or more vessels in the early wood in each annual ring than in the late wood, so that the early wood may appear in cross sections of stems as a ring of small holes. Characteristic of some mainly north-temperate deciduous trees, e.g. chestnut, elm and deciduous oaks. Cf. **diffuse-porous.**

ring-spot An area, e.g. on a leaf, surrounded by a ring or rings of chlorotic, necrotic or abnormally dark green tissue; a characteristic symptom in some virus diseases.

RIST Abbrev. for *RadioImmunoSorbent Test.*

RI strains Abbrev. for *Recombinant Inbred strains.*

ritualization The evolutionary process by which a behaviour pattern is modified to enhance its communication value, usually through exaggeration or repetition of some of its elements.

RNA, ribonucleic acid Polynucleotide containing ribose sugar and uracil instead of thymine. Can hold genetic information as in viruses, but is also the primary agent for transferring information from the genome to the protein synthetic machinery of the cell. See **mRNA**.

Robertsonian translocation Balanced translocation in which the breakpoints in the translocated chromosomes cannot be identified.

rod One of the noncolour-sensitive light perceptive elements of the vertebrate retina. Rods respond to lower illumination levels than **cones**.

Rodentia An order of Mammals. Generally small animals with never more than a single pair of chisel-shaped upper incisors which have open roots. Lower incisors can move like scissors as there is no anterior symphysis between the mandibles. Canines are never present, but a wide diastema between the gnawing incisors and the grinding cheek teeth which vary in numbers and frequently have persistently open roots. The glenoid cavities are elongated antero-posteriorly, the lower jaw being moved forwards for gnawing and backwards for grinding, the jaw muscles being greatly enlarged; herbivorous with a large caecum. They are almost universally distributed, and are of considerable economic and medical importance as pests of stored food and carriers of plague fleas. Adaptive, with a wide radiation, terrestrial, amphibian, burrowing, arboreal, gliding, saltatorial. Squirrels, Beavers, Gophers, Voles, Rats, Mice, Hamsters, Porcupines, Guinea Pigs and Agoutis.

rogue (1) A plant that is not true to type. (2) To remove such plants from a crop, esp. one grown for the production of seed.

role A pattern of behaviour, and the expectancy of it, that is associated with individuals who hold a particular position in a society.

röntgen Unit of X-ray or gamma dose, for which the resulting ionization liberates a charge of each sign of $2.58{\times}10^{-4}$ coulombs per kilogram of air. Symbol R.

röntgen equivalent man Former unit of biological dose given by the product of the absorbed dose in R and the relative biological efficiency of the radiation. Abbrev. *rem*. Now replaced by the **effective dose equivalent**, unit the sievert.

röntgenology Same as **radiology**, US.

root (1) The typically descending axis of a plant and other axes that are anatomically similar and/or clearly homologous. In most vascular plants, roots may be recognized by their endogenous origin, lack of leaves and possession of a root cap. Roots typically function in anchorage and the absorption of water and mineral salts in the soil. See **aerial root**, **root tuber**. Cf. **stem**. (2) See **flagellar root**.

root cap *Calyptra*. A hollow cone of cells protecting the apical meristem of a growing root, which is renewed from within as it wears.

root hair A tubular outgrowth from a cell in the epidermis of a young root, possibly important in the uptake of the more slowly diffusing mineral ions, e.g. phosphate.

rooting compound Usually powders containing **auxins** in which a cutting is dipped to promote rootgrowth.

root-mean-square Square root of sum of squares of individual observations divided by total number of observations. Abbrev. *rms*.

root pressure The positive pressure that may develop in the xylem when water uptake by osmosis follows ion uptake in the root, and transpiration is low. It may result in guttation or bleeding.

rootstock (1) A rhizome, esp. a short, erect one. (2) A stock for grafting, esp. one from a clone selected for desirable effects on the scion, e.g. dwarfing, early fruiting.

root tuber Swollen adventitious root acting as a storage organ, e.g. Dahlia. Cf. **taproot**, **tuber**.

Rorschach inkblot test A **projective test** which requires the subjects to look at inkblots and report on what they see; the answers are used to interpret his or her fantasy life, personality, intelligence, and also as an aid to psychiatric diagnosis.

Rosaceae The rose family, ca 3300 spp. of dicotyledonous flowering plants (superorder Rosidae). Trees, shrubs and herbs; cosmopolitan, esp. N. temperate regions. The flowers are polypetalous, and perigynous or epigynous. Includes many important tree and bush fruit of temperate regions, e.g. apple, pear, plum, cherry, almond, raspberry, strawberry and also many ornamentals including the rose.

Rosenmüllers organ See **epididymis**.

rosette Any rosette-shaped structure; in some Oligochaeta, a large ciliated funnel by which the contents of the vesiculae seminales pass to the exterior; in some Crinoidea, a thin calcareous plate (*rosette plate*, *rosette ossicle*) formed by the coalescence of the basal plates.

rosette plant A plant in which the leaves radiate out at about soil level and which has a more or less leafless flowering stem. See

rubber

The main source of natural rubber or caoutchouc is the tree, *Hevea brasiliensis*, originally a native of Central and South America, but since the end of the 19th century widely grown in plantations in S.E. Asia. Commercial rubber consists of caoutchouc, a polymerization product of isoprene, of resinlike substances, nitrogenous substances, inorganic matter and carbohydrates. The caoutchouc portion is soluble in CS_2, CCl_4, trichloromethane or benzene, forming a viscous colloidal solution. When heated, rubber softens at 160°C, and melts at about 220°C.

Rubber easily absorbs a large quantity of sulphur either by heating or in the cold by contacting with S_2Cl_2 etc. This process is called **vulcanization**. Carbon black, in a fine state of division, is used as a reinforcing filler; other substances, produced by the condensation of aldehydes with amines, retard the oxidation of vulcanized rubber. The uses of rubber are innumerable.

hemicryptophyte.

Rose-Waaler test A test for the presence of rheumatoid factor in blood. Serum dilutions are mixed with sheep erythrocytes coated with an amount of antibody insufficient to cause agglutination. If rheumatoid factor is present this combines with Fc exposed on the red cell surface and causes agglutination.

Rosidae Subclass or superorder of dicotyledons. Trees, shrubs and herbs, flowers mostly polypetalous, stamens (if numerous) developing centripetally, rarely with parietal placentation. Contains ca 60 000 spp. in 108 families including Leguminosae, Rosaceae, Crassulaceae, Myrtaceae, Melastomaceae, Euphorbiaceae and Umbelliferae.

rostellum A small beak-like outgrowth, esp. one from the column of the flower of some orchids.

rostrum In Birds, the beak; a beak-shaped process; in Cirripedia, a ventral plate of the carapace; in some Crustacea, a median anterior projection of the carapace. adjs. *rostral*. *rostrate*.

rot The disintegration of tissue resulting from the activity of invading fungi or bacteria.

rotate Said of a corolla which is wheelshaped, with the petals or lobes spreading out at right angles to the axis of the flower.

Rotifera A class of small, unsegmented, pseudocoelomate animals, phylum Aschelminthes. A distinctive anterior ciliary apparatus is used for locomotion and food gathering. Aquatic. *Wheel animalcules.*

rotor A muscle which by its contraction turns a limb or a part of the body on its axis.

rotula In higher Vertebrates, the kneecap.

rough colony A bacterial colony produced by mutation from a smooth colony. The morphological change is frequently accompanied by physiological changes, e.g. altered virulence.

rough endoplasmic reticulum Cisternal form of **endoplasmic reticulum** bearing ribosomes on the cytoplasmic surface; functioning in the synthesis of protein for export from the cell, sometimes at least, through the golgi apparatus.

round dance Circular movements of a worker bee on returning from a foraging trip which communicates information that a food source is less than 150 yards from the hive. See **waggle dance**.

roundworm Name applied to a number of parasitic Nematodes, esp. those of the genus *Ascaris*, including the large intestinal roundworm of Man (*A. lumbricoides*).

Rous' sarcoma A tumour, occurring in fowls, which is transmitted by an oncogenic RNA virus. The first demonstration that viruses can cause some cancers.

r-strategist Organism which assigns much of its resources to reproduction; usually opportunistic and colonizing species (weeds) with high fecundity and low competitive ability. See **logistic equation, K-strategist**.

rubber See panel.

Rubiaceae Family of ca 7000 spp. of dicotyledonous flowering plants (superorder Asteridae). Cosmopolitan; most tropical species are trees and shrubs, all the (fewer) temperate species are herbs. The leaves are opposite and have stipules, the ovary is inferior and usually of two carpels. Includes coffee and *Cinchona*, the source of quinine.

RUBISCO See **Ribulose 1,5-bisphosphate carboxylase oxygenase**.

RuBP carboxylase, RuBPCase See **Ribulose 1,5-bisphosphate carboxylase oxygenase**.

ruderal A plant which grows usually on rubbish heaps or waste places.

rudiment The earliest recognizable stage of a member or organ.

RuDP Ribulose 1,5-bisphosphate, formerly called *ribulose diphosphate*.

Ruffini's organs In Vertebrates, a type of

cutaneous sensory nerve ending concerned with the perception of heat.

rufous Red-brown.

rugose Wrinkled. dim. *rugulose*.

rumen The first division of the stomach in Ruminants and Cetacea, being an expansion of the lower end of the oesophagus used for storage of food; the paunch.

rumination, ruminant The regurgitation of food that has already been swallowed, and its further mastication before reswallowing. *adj.* and *n.* **ruminant**.

runner Stem growing more or less flat on the ground, with long internodes, rooting at the nodes and/or the tip and there producing new plantlet(s) from axillary or terminal bud(s), as in strawberry.

runt disease Disease which develops after injection of allogeneic lymphocytes into immunologically immature experimental animals. Characterized by loss of weight, failure to thrive, diarrhoea, splenomegaly and often death. This is an example of a **graft-versus-host reaction**.

rupicolous Living or growing on or among rocks.

russet A brownish, roughened, corky layer or patch on the surface of a fruit (or other organ) as a varietal characteristic, or as the result of disease or of injury from insects or spraying.

rust One of a number of plant diseases, some economically very important, caused by biotrophic fungi of the order Uredinales and often recognizable by the rounded or elongated pustules of rust coloured spores on stems or leaves, e.g. black rust of cereals caused by *Puccinia graminis*.

rut The noise made by certain animals as Deer, when sexually excited; oestrus; to be sexually excited, i.e. to be in the oestrous period; to copulate.

rutilant Brightly coloured in red, orange or yellow.

S

S Abbrev. for Svedberg unit, referring to the sedimentation coefficient of proteins analysed in an ultracentrifuge.

7S, 19S antibody Immunoglobulins with sedimentation coefficients about 7 S or 19 S respectively. Terms often used as synonyms for IgG and IgM.

sabulose, sabuline Growing in sandy places.

sac Any sort of bag-like structure or pouch.

saccadic eye movements The rapid, ballistic movements of the eyes used in scanning a scene; these involuntary eye movements occur about every quarter of a second even when the eyes are fixated on an object.

saccate Bag-like or pouch-like. Said of pollen grains with a **saccus** or two.

Saccharomyces cerevisiae The yeast used widely in bread and alcohol manufacture. It can also be used as an eukaryotic *host* for growing and *expressing* DNA sequences.

saccule, sacculus A small sac; the lower chamber of the auditory vesicle in Vertebrates. adj. *sacculate*.

sacculiform Shaped like a little bag.

saccus A large, hollow, pouch-like projection of the outer part of the wall of a pollen grain.

sacral ribs Bony processes uniting the sacral vertebrae to the pelvis, distinct in Reptiles but fused to the transverse processes in other Tetrapoda.

sacral vertebrae In higher Craniata, those vertebrae which articulate with the ilia of the pelvis via sacral ribs, there being one in the Frog and two in the Lizard, coming between the lumbar vertebrae and the caudal vertebrae (if any). In Birds and Mammals they are fused with other vertebrae to form the **sacrum**.

sacroiliac joint In some Craniata, the almost immovable joints between the **sacrum** and the two ilia of the pelvis. The articular surfaces of the bones are partly covered with cartilage and partly roughened for the attachment of the *sacroiliac ligament*.

sacrum In the skeleton of some Craniata, part of the vertebral column which articulates immovably with the ilium of the pelvis at the **sacroiliac joint**, being composed of several fused vertebrae, including the **sacral vertebrae**. In Birds it consists of one thoracic vertebra, five or six lumbar vertebrae, the two sacral vertebrae and the anterior five caudal vertebrae. Sometimes called the *synsacrum*. In Mammals it comprises varying numbers of vertebrae in different orders (e.g. four in the Rabbit and five in Man), the first one or two being regarded as sacral, and the others as caudal. The former have low spines

and expanded ventral surfaces for the attachment of muscles.

sadism Sexual gratification through the infliction of pain on others; pleasure in cruel behaviour.

sado-masochism The pairing of a **sadist** and a **masochist** to satisfy their mutually complementary sexual needs.

sagittal Elongate in the median vertical longitudinal plane of an animal, as the *sagittal* suture between the parietals, the *sagittal* crest of the skull; used also of sections.

sagittate Shaped like an arrowhead with the barbs pointing backwards.

Sahelian drought The pattern of drought in the northern regions of West Africa, where several years of below-average rainfall often occur in succession.

Salientia A superorder of Amphibia. Adults four-legged, the hind limbs being esp. well-developed, and short-bodied, with no tail. Toads and Frogs. Also known as *Anura, Batrachia*.

saliva The watery secretion produced by the salivary glands, whose function is to lubricate the passage of food and, sometimes, to carry out part of its digestion. In the Insects saliva may contain amylase, invertase, protease and lipase, depending on diet, and in some blood-sucking insects it contains anticoagulants. In Mammals it contains water, mucin and, in Man and some herbivores, the amylase ptyalin, which catalyses the breakdown of starch to maltose.

salivary gland chromosome See panel.

salivary glands Glands present in many land animals, the ducts of which open into or near the mouth.

Salmonella A group of Gram-negative, carbohydrate fermenting, nonsporing, bacilli, all pathogenic to animals. Associated with food poisoning in Man. Includes *S. typhi* and *S. paratyphi*.

Salmoniformes Order of freshwater and **anadromous** *Osteichthyes* of great commercial and sporting importance. Salmon, Trout, Charr, Pike.

salping- Prefix from Gk. *salpinx*, gen. *salpingos*, trumpet, referring esp. to the Fallopian tubes. See **salpinx**.

salpinx *Lagenostome*. Structure, adapted for the reception of pollen, at the distal end of the nucellus of ovules of many seed ferns. In animals, the eustachian tube; the fallopian tube; a trumpet-shaped structure. adj. *salpingian*.

salsuginous Growing on a salt marsh.

saltation A sudden heritable variation in a species. Now applied more often to large morphological changes which occur during

salivary gland chromosome

Polytene chromosome found in the salivary glands of larval dipterans, e.g. *Drosophila melanogaster*. Conspicuously banded, whether stained or not, and used for gene mapping and other studies of chromosome organization.

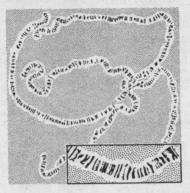

The illustration shows the five chromosomes which are joined at the chromocentre (middle towards the top). The inset is an enlarged view of part of a chromosome, each of which contains about 1000 strands of a normal chromosome aligned side by side. The pattern of bands has long been carefully described and it is possible to see inversions, duplications and deletions of regions down sometimes to just part of a band which can be correlated with mutations and nowadays detailed molecular alterations. This almost unique situation has ensured the importance of the fruit fly (*Drosophila*) in genetic, evolutionary and molecular studies. See **bands**.

evolution over a time period shorter than that required by similar changes earlier or later. Said to be difficult to explain by **natural selection**.

saltatorial, saltatory Used in, or adapted for, jumping, as the third pair of legs in Grasshoppers.

saltatory conduction The process of nervous conduction along a myelinated nerve axon where the impulse jumps from one **node of Ranvier** to the next.

salt gland Structure at the leaf surface which actively secretes sodium chloride in many salt marsh and mangrove species.

saltigrade Progressing by jumps, as Grasshoppers.

salt marsh A marsh characterized by saline soil, most often in estuaries and subject to marine inundation. See **halosere**.

samara A dry, indehiscent fruit of which part of the wall forms a flattened wing, e.g. ash key.

samariform Winged, like an ash key.

sampling The survey of a small but representative part of a population or stand of vegetation in order to obtain an estimate of some characteristic of the whole, e.g. age distribution or species present.

sampling distribution The probability distribution describing the variation of a statistic in repeated sampling, or the hypothetical repetitions of the same experiment.

sampling error Variation due to a sample from a population necessarily giving only incomplete information about the population.

sandwich technique A technique for the detection of antibody or antibody-producing cells in histological preparations. A first layer of antigen is applied and allowed to react with the antibody in the section. After washing, this is followed by a second layer of fluorochrome-labelled antibody specific for the antigen. The antigen is 'sandwiched' between the two layers of antibody.

sap An aqueous solution present in xylem,

phloem, cell or vacuole, released on wounding. Cf. **phloem** sap, **xylem** sap.

saprobe An organism such as a bacterium or fungus which obtains its nourishment osmotrophically from dead organic matter. Cf. **saprophyte**.

saprogenous Growing on decaying matter.

saprophilous Saprogenous.

saprophyte An organism which lives heterotrophically and osmotrophically on dead organic matter. adj. *saprophytic*. Cf. **saprobe**.

saprotrophy Heterotrophic nutrition based on non-living (dead) organic matter.

sapr-, sapro- Prefixes from Gk. *sapros*, rotten, rancid.

sap wood *Alburnum*. The outer, younger part of the wood of a tree or shrub, pale in colour, with living parenchyma cells and still conducting sap. Cf. **heart wood**.

sarcodic, sarcodous, sarcoid Pertaining to or resembling flesh.

Sarcodina Class of Protozoa with pseudopodia containing both irregular, amoeboid forms and others possessing regular calcareous or siliceous tests, e.g. *Radiolaria, Foraminifera*.

sarcolemma The extensible sheath of a muscle fibre enclosing the contractile substance.

sarcoma A malignant tumour of connective tissue origin (e.g. of fibrous tissue, bone, cartilage); the tumour invades adjacent tissue and organs, and metastases are formed via the blood stream. pl. *sarcomata* or *sarcomas*. Cf. **carcinoma**

sarcomere The basic contractile unit of the *myofibril*.

sarcophagous Feeding on flesh.

sarcoplasmic reticulum A network of tubules associated with muscle fibrils which acts as the source of calcium ions that stimulate contraction.

sarc-, sarco- Prefixes from Gk. *sarx*, gen. *sarkos*, flesh.

sartorius A thigh muscle of Tetrapoda which by its contraction causes the leg to bend inwards.

satellite The part of a chromosome distal to the **secondary constriction**.

satellite DNA That from eukaryotic chromosomes which separates from the remainder of the genome on **buoyant density** centrifugation. Contains closely related repeated sequences with a base composition different from the rest of the DNA. See **simple sequence DNA**.

savanna (1) Extensive area in which grasses are an important part of the vegetation. (2) Common vegetation type in dry parts of Africa, consisting of trees and grasses, usually burnt every year.

saxicole, saxicolous Growing on rocks;

living on or among rocks.

scab A discrete localized superficial lesion characterized by roughening, abnormal thickening and, esp. cork formation; a disease in which such scabs are formed, e.g. potato and apple scab.

scabellum In Diptera, the dilated basal portion of a haltere.

scabrid Rough to the touch; scaly.

scabrous Having a surface roughened by small wartlike upgrowths. dim. *scaberulous*.

scaffold Protein core of histone-depleted metaphase chromosomes left after nuclease treatment.

scaffold/radial loop model Model of metaphase chromosome structure which postulates a non-histone protein core to which the linear DNA molecule has an ordered series of attachment points. These occur every 30 000 to 90 000 bases, with the intervening DNA forming a loop packed by supercoiling or folding.

scale (1) A small exo-skeletal outgrowth of tegumentary origin, of chitin, bone or some horny material, usually flat and platelike. (2) A thin, flat, semitransparent plant member, usually of small size, and green only when very young, if then. See **bud scale, scale leaf**.

scale bark, scaly bark (1) Bark which becomes detached in irregular patches. (2) **Rhytidome**.

scale leaf Leaf, often thin, more or less flattened against the stem or other leaves, and either photosynthetic, protective (bud scales) or holding reserves (bulb scales).

scandent Climbing.

scanning electron microscope, SEM See panel.

scanning transmission electron microscope *STEM*. One which uses field emission from a very fine tungsten point as the source of electrons. The electrons transmitted through the sample are either unscattered, elastically scattered or inelastically scattered; they are collected, separated and analysed to produce an image. See **scanning electron microscope** in panel.

scansorial Adapted for climbing trees.

scape (1) The flowering stem, nearly or quite leafless, arising from a rosette of leaves and bearing a flower, several flowers or a crowded inflorescence; e.g. the dandelion. adj. *scapigenous*. (2) The basal joint of the antenna in Insects.

scaphoid See **navicular bone**.

Scaphopoda A class of Mollusca, being bilaterally symmetrical with a tubular shell open at both ends, a reduced foot used for burrowing, a head with many prehensile processes, a radula, separate cerebral and pleural ganglia, no ctenidia, circulatory system rudimentary, larva a trochosphere.

scanning electron microscope, SEM

A form of **electron microscope** in which a very fine beam of electrons at 3–30 kV is made to scan a chosen area of specimen as a *raster* of parallel contiguous lines. Usually the specimen is a solid object and secondary electrons, which are emitted from the surface or from near the surface in numbers depending on its nature and topography, are collected and after analysis form a signal which modulates the beam of a cathode ray tube scanned in synchrony with the scanning beam (see diagram).

The images resemble those seen in a hand lens but have much finer resolution (say 5 nm), can be usefully magnified about 100 000 times and have, for comparable magnifications, much greater depth of focus than a light microscope. Many SEMs are also capable of **X-ray analysis.**

In the *scanning transmission electron microscope* or *STEM*, the specimen is scanned much as in the SEM but transmitted electrons are collected and used to form the picture on the screen.

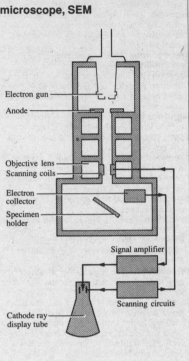

Electron gun
Anode
Objective lens
Scanning coils
Electron collector
Specimen holder
Signal amplifier
Scanning circuits
Cathode ray display tube

scaph-, scapho- Prefixes from Gk. *skaphe*. boat.

scapula In Vertebrates, the dorsal portion of the pectoral girdle, the shoulder blade; any structure resembling the shoulder blade. adj. *scapular*.

scapulars In Birds, small feathers attached to the humerus, and lying along the side of the back.

scarification Any treatment, e.g. with sulphuric acid or by mechanical abrasion, which makes the coat of a seed more permeable to water and promotes imbibition and germination.

scent-marking A form of communication between individuals of a species involving the deposition of glandular secretion onto ground or some other surface; the scent dissipates at a slow rate to form a relatively long-lived signal.

schedule of reinforcement A rule that determines the frequency and manner in which behaviour is reinforced.

schema A mental pattern, or body of knowledge, that provides a framework within which to place newly acquired knowledge.

Schick test A test to assess the degree of susceptibility or immunity of individuals to diphtheria. A standard quantity of diphtheria toxin is injected intracutaneously into one forearm. In non-immune persons an area of redness and swelling appears in 1–2 days, reaching a maximum at 4–7 days followed by pigmentation and scaling. This is a positive Schick reaction and indicates that diphtheria immunization is required.

schizo- Prefix from Gk. *schizein*, to cleave.

schizocarp Fruit, derived from a syncarpous ovary, which when mature becomes divided into separate, one-seeded, indehiscent parts (mericarps).

schizocoel Coelom produced within the mass of mesoderm by splitting or cleavage. Cf. **enterocoel**. adj. *schizocoelic*.

schizogamy In Polychaeta, a method of reproduction in which a sexual form is produced by fission or germination from a sexless form.

schizogenesis Reproduction by fission.

schizogenous Said of a space formed by the separation of cells by splitting of their common wall along the middle lamella. Cf. **lysigenous**, **rexigenous**.

schizogony In Protozoa, vegetative reproduction by fission.

schizont In Protozoa, a mature trophozoite about to reproduce by schizogony.

schizophrenia A group of psychoses marked by severe distortion and disorder of thought, perception, motivation and mood; delusion and hallucination are common, as are bizarre behaviours and social withdrawal. The term was invented by Eugen Bleuler and was previously known as *dementia praecox*.

school, schooling In Fish, refers to groups of individuals who maintain a constant distance and orientation from their neighbours, and who all swim at a constant pace; the primary function of schooling is as an ante-predator strategy.

school phobia Severe reluctance to attend school.

Schwann cell A neuroglial cell which deposits the myelin sheath along myelinated axons.

sciatic Situated in, or pertaining to, the ischial or hip region.

SCID Abbrev. for *Severe Combined Immuno-Deficiency syndrome*.

scintillation camera *Gamma camera*. An imaging device which may have either a single sodium iodide crystal or multiple crystals and which is capable of detecting and recording the spatial distribution of an internally administered radionuclide.

scion A piece of a plant; in horticulture usually a young, often dormant, shoot that is inserted into the **stock** when a **graft** is made.

sciophyte A plant which is adapted to living in shady places.

sclera The tough fibrous outer coat of the Vertebrate eye. adj. *sclerotic*.

sclere A skeletal structure; a sponge spicule.

sclereid A short **sclerenchyma cell**, star-shaped, rod-shaped or rounded; often occurs in tissues of other cell types but in some seed coats as a tissue of sclereids.

sclereide (1) A general term for a cell with a thick, lignified wall, i.e. any sclerenchymatous cell. (2) A thick-walled cell mixed with the photosynthetic cells of a leaf, giving them mechanical support. (3) A stone cell.

sclerenchyma (1) Tissue composed of, or the collective term for, **sclerenchyma cells**. (2) Hard skeletal tissue, as of Corals.

sclerenchyma cell Cell with thick, usually lignified, walls, often dead when mature, and usually having a supporting function in the plant; either a **sclereid** or a **fibre**. Cf. **collenchyma**.

sclerified, sclerosed, sclerotized Hardened, e.g. as a result of secondary wall formation and/or lignification.

sclerite A hard skeletal plate or spicule.

sclerophyll A leaf with well-developed sclerenchyma and hence tough and fibrous or leathery, usually evergreen and typical of trees and shrubs in places with rather warm and dry, esp. Mediterranean, climates, e.g. olive and *Eucalyptus*.

scleroproteins Insoluble proteins forming the skeletal parts of tissues, e.g. keratin from hoofs, nails, hair etc. chondrin and elastin from ligaments.

sclerosis The hardening of cell walls or of tissues by thickening and lignification. In Man, induration or hardening, as of the arteries. See **multiple sclerosis**.

sclerotic The sclera of the eye; pertaining to the sclera.

sclerotic cell See **sclereid**.

sclerotin In the cuticle of Insects and some other Arthropods, a protein which has become strengthened and dark through cross linkage by the action of quinones. See **tanning**.

sclerotium A hard mass of fungal hyphae, often black on the outside, crust-like to globular, and serving as a resting stage from which an active mycelium or spores are formed later.

sclerotization In Insects, the process by which most of the **procuticle** becomes hardened and darkened to form tough, rigid, discrete sclerites of **exocuticle**. See **tanning**.

scler-, sclero- Prefixes from Gk. *skleros*, hard.

scobicular, scobiform Looking like sawdust.

scolex The terminal organ of attachment of a tapeworm (Cestode). pl. *scoleces*, not *scolices*. adj. *scolecid, scoleciform*.

scolophore A subcuticular spindle-shaped nerve ending in Insects, sensitive to mechanical vibrations. Also called *scolopidium*.

scolopidia Campaniform sensillae.

scopa The pollen brush of Bees, consisting of short stiff spines on the posterior metatarsus.

scopophilia Pleasure in looking.

scorch Necrosis, like that of leaf margins looking as if seared by heat, caused by infection, mineral deficiency or weather conditions etc.

Scorpionidea An order of Arachnida with the opisthosoma being divided into a distinct mesosoma and metasoma, and consisting of 12 segments and a telson. The chelicera and pedipalps are chelate, there are 4 pairs of walking legs, the mesosomatic segments carry the genital operculum, the pectines and 4 pairs of lung books, the metasoma forming a flexible tail wielding the terminal sting. Viviparous and terrestrial. Known as fossils from the Silurian. Scorpions.

scoto- Prefix from Gk. *skotos*, darkness, dark (in the sense of not illuminated).

scotoma (1) A blind or partially blind area in the visual field, the result of disease of, or damage to, the retina or optic nerve or visual cortex. (2) The appearance of a black spot in front of the eye, as in choroiditis. pl. *scotomata*.

scotomization A defence mechanism where the individual fails to consciously perceive parts of the environment or of himself/herself. Derived from scotoma, a blind spot in the visual field.

screen To investigate a large number of organisms for the presence of a particular property as in screening for a mutation for *antibiotic resistance*.

scrobiculate Having the surface dotted all over with small rounded depressions; pitted.

scrobiculus A small pit or rounded depression.

Scrophulariaceae Family of ca 3000 spp. of dicotyledonous flowering plants (superorder Asteridae). Almost all herbs and cosmopolitan. The flowers are gamopetalous, the ovary superior and the fruit usually a capsule. Includes the foxglove from which the important cardiac drugs digitalin and digoxin are derived, and some ornamental plants, e.g. *Antirrhinum*.

scrotum In Mammals, a muscular sac forming part of the ventral body wall into which the testes descend. adj. *scrotal*.

scute An exoskeletal scale or plate. adj. *scutate*.

scutellum More or less shield-shaped structure, possibly a modified cotyledon, attached to the side of the embryo in a grass grain, and which at germination secretes hydrolytic enzymes into, and absorbs sugars etc. from, the adjacent endosperm.

Scyphomedusae See **Scyphozoa**.

Scyphozoa A class of Cnidaria in which the polyp stage is inconspicuous and may be completely absent. Where present it is known as a *scyphistoma*, and gives rise to the ephyra larvae. Velum and nerve ring generally absent, gonads endodermal, no skeleton. Jellyfish.

SDS See **sodium dodecyl sulphate**.

SDS gel electrophoresis Polyacrylamide gel electrophoresis in the presence of the denaturant **SDS**.

sea lily See **Crinoidea**.

search image Refers to the perceptual phenomena of an increased accuracy of discrimination for certain objects in the environment, e.g. a predator's improved ability to see camouflaged prey against its background.

seaweed Macroscopic marine alga. Most

seaweeds belong to the Phaeophyceae or Rhodophyceae, some to the Chlorophyceae.

sebaceous *Sebiparous*. Producing or containing fatty material, as the *sebaceous* glands of the scalp in Man.

sebaceous cyst A cyst formed as a result of blockage of the duct of a sebaceous gland, often present on the face, scalp or neck.

sebiferous Conveying fatty material.

sebum The fatty secretion, produced by the sebaceous glands, which protects and lubricates hair and skin.

secodont Having teeth adapted for cutting.

secondary Arising later; of subsidiary importance; in Insects, the hindwing; in Birds, a quill feather attached to the forearm, and also called the *cubital*.

secondary body cavity See **coelom**.

secondary cell wall See **secondary wall**.

secondary constriction Non-centromeric constriction of the chromosomes, often at the site of the **nucleolar organizing region**.

secondary growth See **secondary thickening** (1).

secondary immune response Response of the body to an antigen with which it has already been primed (see **primary immune response**). There is a very rapid production of large amounts of antibody over a few days, followed by a slow exponential fall. The response of **cell mediated immunity** follows a similar pattern.

secondary meristem (1) Meristem producing secondary tissues, (2) Meristem derived by dedifferentiation from differentiated cells, e.g. the interfasicular cambium, cork cambium.

secondary messenger A cytoplasmic component which mediates the action of hormones bound to receptors on the cell surface, e.g. cyclic AMP.

secondary metabolites Applied to those compounds which do not function directly in biochemical activities like photosynthesis, respiration and protein synthesis which support growth. They include alkaloids, terpenoids, flavonoids which may function in defence against insects, fungi and herbivores, in **allelopathy** or as attractants to pollinators or fructivores.

secondary phloem Phloem formed by the activity of a cambium.

secondary-process thinking In psychoanalytic theory, logic-bound thinking, governed by the **reality principle**; it involves the ego functions of remembering, reasoning and evaluation that mediate between instinctual needs and adaptation to the external world.

secondary reinforcement An initially neutral stimulus that acquires reinforcing properties through pairing with another stimulus that is already reinforcing.

secondary sexual characters Features which distinguish between the sexes other than the reproductive organs.

secondary structure The first level of 3-dimensional folding of the backbone of a polymer. Thus a polypeptide chain can be folded into an α-helix or β-pleated sheet and the nucleotide chains of DNA into a double helix.

secondary substances Plant biochemicals which are involved in no known biosynthetic pathways, but which are often detected in high concentrations in leaves and other organs, and so are presumed to be chemical defences.

secondary succession A succession proceeding in an area from which a previous community has been removed, e.g. a ploughed field. Cf. **primary succession**.

secondary thickening (1) The increase in girth of a stem or root that results from the activity of a cambium after elongation has ceased. (2) Confusingly, the formation by a cell of a secondary wall.

secondary wall A later-formed part of the *cell wall* (sometimes the major part) laid down on the cytoplasmic side of the *primary wall* after cell expansion has ceased, typically richer than the primary wall in cellulose.

secondary xylem Xylem formed by the activity of a cambium. Wood.

second ventricle In Vertebrates, the cavity of the right lobe of the cerebrum.

secretion The elimination from the cytoplasm of a cell or from a multicellular structure (a gland) of an organic substance or inorganic ions.

secretor Persons who secrete ABO blood group substances into mucous secretions such as gastric juice, saliva and ovarian cyst fluid. Over 80% of humans are secretors. The status is genetically determined.

secretory Secretion-forming.

secretory duct A duct containing material secreted by its lining of epithelial cells, e.g. resin duct.

secretory piece A large polypeptide attached to dimers of the secreted form of IgA. Structurally unrelated to immunoglobulins and synthesized by epithelial cells in the secretory gland, not by the plasma cells that synthesize the immunoglobulin. Has strong affinity for mucus thus prolonging retention on mucous surfaces, and it may inhibit destruction of IgA enzymes in the gut.

secritin A polypeptide hormone secreted by the intestinal wall which stimulates the pancreas to secrete bicarbonate ions.

section (1) A thin slice of a biological or mineral material sufficiently transparent to be studied with the compound microscope. (2) A taxonomic group, esp. a subdivision of

a genus, but sometimes of a higher rank.

sectorial Adapted for cutting.

sectorial chimera That in which one component forms a longitudinal strip of tissue down a shoot, the strip including more than one layer of the tunica and/or corpus.

secund Having the lateral members, leaves or flowers, all turned to one side.

sedentary Said of Mammals which remain attached to a substratum.

seed The matured ovule of a seed plant containing usually one embryo, with, in some species, endosperm or perisperm, surrounded by the seed coat or testa.

seed bank The total seed content of the soil.

seed leaf Same as **cotyledon**.

seed plant Member of those plant groups that reproduce by seeds, i.e. the gymnosperms and the angiosperms; the division Spermatophyta.

segment One of the joints of an articulate appendage; one of the divisions of body in a metameric animal; a cell or group of cells produced by cleavage of an ovum. adj. *segmental.*

segmental In metameric animals, repeated in each somite; as *segmental arteries*, *segmental papillae.*

segmental interchange The exchange of portions between two chromosomes which are not homologous.

segmentation Meristic repetition of organs or of parts of the body; the early divisions of a fertilized ovum, leading to the formation of a blastula or analogous stage.

segmentation cavity See **blastocoel**.

segregation The separation of the two alleles in a heterozygote when gametes are formed, each carrying one or the other; and, consequentially, the appearance of more than one genotype in the progeny of a heterozygote.

seismonasty A nastic response to a shock, esp. mechanical shock, as in e.g. *Mimosa pudica*.

selection The process by which some individuals come to contribute more offspring than others to form the next generation in *natural selection* through intrinsic differences in survival and fertility, in *artificial selection* through the choice of parents by the breeder.

selective mating See **preferential mating**.

selenodont Having cheek teeth with crescentic ridges on the grinding surface.

self Self-fertilize or self-pollinate. Cf. **cross**.

self-compatible Said of an individual plant or a clone, capable of self fertilization. See **homothallism**.

self cure In animals infested with intestinal nematodes, about 10 days after the initial establishment of the infection, there suddenly begins an expulsion from the intestine of the majority of the population of worms. This is probably due to an immune response by the host, manifested by an immediate hypersensitivity reaction involving release of histamine and other vasoactive substances from mast cells in the gut wall.

self fertilization The fertilization of an egg by a male gamete from the same individual or the same clone (genet). Cf. **cross fertilization**, **self pollination**. See **autogamy**.

self incompatible Said of an individual plant or a clone, incapable of self fertilization. See **heterothallism**.

selfing Self fertilization, self pollination.

selfish DNA A class of DNA sequences thought to have been selected during evolution only by its ability to spread and duplicate itself in the genome of higher organisms, with minimal damage to the 'host'.

self pollination The transfer of pollen to a stigma from an anther of the same flower or individual or clone (genet).

self sterile Not capable of producing viable offspring by self fertilization.

self sterility In a hermaphrodite animal or plant, the condition in which self fertilization is impossible or ineffective.

self-thinning curve Curve describing the survival of individuals in a crowded population with time.

sell A hard outer case or exoskeleton of inorganic material, chitin, lime, silica etc.

SEM See **Scanning Electron Microscope**.

semantic memory Refers to general knowledge, e.g. grammar, principles, theories.

semantide A molecule carrying information, as in a gene or messenger RNA.

sematic Warning; signalling; serving for warning or recognition, as *sematic colours*.

semeiotic Pertaining to, or relating to, the symptoms and signs of disease.

semen The fluid formed by the male reproductive organs in which the spermatozoa are suspended. adj. *seminal*.

semi- Prefix from L. *semi*, half.

semicircular canals Structures forming part of the labyrinth of the inner ear of most Vertebrates, there usually being three at right angles to each other, two being vertical and one horizontal. Movements of the head cause movement of the endolymph in the canals, which moves the gelatinous cupula attached to the sensory hairs of a neuromast sense organ in the swelling (called *ampullae*) at the base of the canals, initiating nerve impulses which travel to the brain via the 8th cranial nerve. They thus serve as organs of dynamic equilibrium.

seminal Pertaining to the seed. Seminal roots are adventitious roots produced at the base of the stem of young seedlings of e.g. cereals.

seminal receptacle See **vesicula seminalis**.

seminal roots Adventitious roots that develop from the hypocotyl, as in grass seedlings.

seminiferous Semen-producing or semen-carrying.

semiochemical A chemical substance produced by an animal and used in communication. See **pheromone**, **allomone**.

semiotics The study of communication.

semi-oviparous Giving birth to imperfectly developed young, as marsupial Mammals.

semipalmate Having the toes partially webbed.

semipermeable membrane A membrane which permits the passage of solvent but is impermeable to dissolved substances.

semiplacenta A nondeciduate placenta in which only the foetal part is thrown off at birth.

semistreptostyly In Vertebrates, the condition of having a slightly movable articulation between the quadrate and the squamosal. Cf. **streptostyly**, **monimostyly**.

sempervirent Evergreen.

senescent Said of that period in the life history of an individual when its powers are declining prior to death.

senile-degenerative disorders Deterioration of intellectual, emotional and motor functioning with advancing age.

senility Condition of exhaustion or degeneration due to old age.

senses Any number of responses to stimulation through the specialized sense organs (i.e. eyes, ears etc.). The responses of these organs translate into neural impulses, often referred to as *sensation*.

sense strand That strand of a double-stranded DNA molecule which is transcribed into messenger or other RNA.

sensiferous, sensigerous Sensitive.

sensillum In Insects, small sense organs of varied function on the integument typically comprising a cuticular and/or hypodermal structure. pl. **sensilla**.

sensitive Capable of receiving stimuli.

sensitive period Periods of time during development when an individual is particularly sensitive to environmental and social experiences and which affect learning in a variety of ways in a wide range of species.

sensitization (1) Administration of an antigen to provoke an immune response so that, on later challenge, a more vigorous secondary response will ensue. This involves the recruitment of **primed** cells. (2) Coating of cells with antibody e.g. for use in complement fixation tests.

sensorimotor development The development of co-ordination between perception and action (e.g. hand-eye co-ordination).

sensorimotor intelligence stage According to Piaget, the first period of intellectual development (0–2 years) in which the infant's interactions with the environment consist of motor responses to classes of sensory stimuli (e.g. looking, grasping etc.). During this period the infant progresses from simple reflex actions to complex ways of playing with and manipulating objects, which leads eventually to internal representations of the world.

sensorium The seat of sensation; the nervous system. adj. *sensorial*.

sensory Directly connected with the sensorium; pertaining to, or serving, the senses.

sensory adaptation A short-term change in the response of a sensory system as a consequence of repeated or protracted stimulation.

sensory deprivation Refers to experimental work, mostly with humans, in which total sensory input is reduced beyond normal conditions through the use of special chambers or devices (e.g. translucent goggles).

sensory store The portion of the memory system that maintains representations of sensory information for very brief intervals; divided into **echoic** and **iconic memory**. See **long term memory, short term memory**.

sepal A member of the **calyx**; typically a green, more or less leaf-like structure, several enclosing the rest of the flower in the bud, but sometimes petaloid.

separation anxiety Anxiety at the prospect of being separated from someone one is strongly attached to and one believes to be necessary for survival; all children experience this fear at between 6–8 months of age to about 2 years but individual differences in its intensity are great.

separation layer See **abscission layer**

septate Divided into cells, compartments or chambers by walls or partitions.

septate fibre A fibre of which the lumen is divided into several compartments by transverse septa.

septicidal Said of a dehiscent fruit, opening by breaking into its component carpels leaving the placental axes standing. Cf. **loculicidal, septifragal**.

septifragal Said of a dehiscent fruit, opening by the breaking away of the outer wall leaving the septa standing. Cf. **loculicidal, septicidal**.

sept-, septi-, septo- Prefixes used in the construction of compound terms. (1) L. *septum*, partition. (2) L. *septem*, seven. (3) Gk. *septos*, rotten.

septum Generally, a partition separating

two cavities. A cell wall or multicellular structure acting as a partition as in a fungal hypha, between cells or between adjacent chambers in an ovary. adj. *septal*.

septum transversum See **diaphragm**.

sequence The linear order of bases in a nucleic acid or of amino acids in a protein.

sequencing Biochemical procedure for determining the sequence of a nucleic acid or protein.

sequestrene, sequestrol, sequestered iron Preparations of chelated mineral elements, esp. iron and some trace elements, used horticulturally to correct such mineral deficiencies as **lime-induced chlorosis** by application to leaves (*foliar feeding*) or to the soil.

Ser Symbol for **serine**.

sere A particular example of plant communities which succeed each other. Hydroseres originate in water, xeroseres occur in dry places, lithoseres develop on rock surfaces. adj. *seral*. See **primary sere**.

serial learning, recall Refers to a learning or memory task in which the subject is required to repeat a list of items in the same order as they were presented.

serial-position effect The observation that in verbal learning, items at the beginning and end of a list are recalled better than those in the middle of the list.

serine *3-hydroxy-2-aminopropanoic acid*. The L-isomer is a polar amino acid and constituent of protein. Symbol Ser, short form S. Side chain: HO—CH$_2$— See **amino acid**.

serological determinants Antigenic determinants on cells that are recognized by and accessible to antibodies, as opposed to determinants which are not recognized by antibodies but only by T lymphocytes.

serological typing A technique that is used for the identification of pathogenic organisms, e.g. bacteria, particularly strains within a species, when morphological differentiation is difficult or impossible. It is based on antibody-antigen reactions in which specific proteins of the organism act as antigens.

serology The study of **serum**.

serophyte Any micro-organism which will grow in the presence of fresh serum exuding into a wound, such as the **streptococcus** and the **staphylococcus**.

serosa See **serous membrane**.

serotaxonomy The use of serological techniques to compare proteins extracted from different plants as an aid in taxonomy.

serotherapy, serum therapy The curative or preventive treatment of disease by the injection into the body of animal or human sera which contain antibodies to the bacteria or toxins causing the disease.

sex determination, chromosomes

In many organisms (including Vertebrates) sex is determined by the possession of a particular combination of chromosomes. In some cases, presence or absence of one special chromosome, known as the *accessory* or *X-chromosome*, is the determining factor, e.g. in the insect, *Pyrrhocoris apterus*, males and females have 13 and 14 chromosomes respectively. In many species there are two sex chromosomes, and the sex of the individual depends on whether it has two identical chromosomes, the *homogametic sex*, or one of each of the two types, the *heterogametic sex*. Where the female is homogametic, as in Mammals, the two chromosomes are designated XX, and the male's chromosomes are known as XY. Where the male is homogametic, as in Birds, the male's chromosomes are called WW and the female's chromosomes are known as ZW. See **chromosome.**

serotonin *5-hydroxytryptamine.* Causes smooth muscle contraction, increased vascular permeability and vasoconstriction of larger vessels. Present in platelets, from which it is released on activation. Also present in mast cells of some species.

serous Watery; pertaining to, producing or containing a watery fluid or serum.

serous membrane, serosa One of the delicate membranes of connective tissue which line the internal cavities of the body in Craniata; the chorion.

serrate Leaf margin toothed like a saw.

serrulate Minutely serrate.

serum (1) The watery fluid which separates from blood or lymph in coagulation. (2) Blood serum containing antibodies, taken from an animal that has been inoculated with bacteria or their toxins, used to immunize people or animals. adj. *serous.*

serum albumin A globular protein obtained from blood and body fluids, having a transport and osmoregulatory function. A crystalline, water-soluble substance, not precipitated by NaCl.

serum sickness A hypersensitivity reaction to the injection of foreign antigens in large quantity esp. those contained in antisera used for passive immunization. Symptoms appear some days after a single dose of the antigen, and consist of local swelling at the injection site, enlarged lymph nodes, fever, joint swellings, urticaria and, more rarely, glomerulonephritis. May have a more prolonged effect on joints, heart and kidney. The symptoms are due to the localization in the tissues of immune complexes formed between antibodies produced during the developing immune response and the large quantities of antigen still present. This is an example of Type III hypersensitivity reaction.

sesamoid A small rounded ossification forming part of a tendon usually at, or near, a joint, as the patella.

sessile (1) Having no stalk. (2) Fixed and stationary.

set See **mental set.**

seta One of the relatively long and thin, or bristle-like, structures, as the stalk supporting the capsule of the sporophytes of mosses (Bryopsida); a *chaeta.* adjs. *setaceous, setiferous, setigerous, setiform, setose, setulose.*

set of chromosomes A **haploid** complement of chromosomes.

severe combined immunodeficiency syndrome *SCID.* The most severe form of congenital immunological deficiency state, in which thymic agenesis with lymphocyte depletion coexists with deficiency of plasma cells and antibody deficiency syndrome. Also known as *Swiss type hypogammaglobulinaemia,* first described in that country. Death in early life from infection is the rule, but the condition may be cured by bone marrow transplantation.

sex (1) The sum total of the characteristics, structural and functional, which distinguish male and female organisms, esp. with regard to the part played in reproduction. (2) As a verb to determine sex. adj. *sexual.*

sex cells See **gametes.**

sex chromosomes See panel.

sex determination See panel.

sex gland See **gonad.**

sex-limited character A character developed only by individuals which belong to a particular sex.

sex-linked See panel.

sex mosaic An individual showing characteristics of both sexes; an *intersex* or *gynandromorph.*

sex reversal The gradual change of the sexual characters of an individual, during its lifetime, from male to female or vice versa.

sex roles A set of attitudes, behaviours, perceptions and feelings which are commonly held to be associated with either being male or being female.

sex transformation See **sex reversal**.

sexual behaviour All behaviour leading to the fertilization of eggs by sperm.

sexual coloration Characteristic colour difference between the sexes, esp. marked at the breeding season. See **epigamic**.

sexual dimorphism Marked differences between the males and females of a species, esp. differences in superficial characters, such as colour, shape, size etc.

sexual organs The gonads and their accessory structures; reproductive system.

sexual reproduction The union of gametes or of gametic nuclei, preceding the formation of a new individual.

sexual selection Selection occurring as a result of mate selection.

Sezary syndrome A disease syndrome characterized by general redness and thickening of the skin. The skin is infiltrated with lymphocytes with an unusual hairy appearance, and large numbers of similar lymphocytes (Sezary cells) are present in the blood. They have been shown to be T lymphocytes, but what causes them to move into the skin is not known.

shade plant A plant adapted to living at low light intensities. Cf. **sun plant**.

shadowing technique A technique for shadow casting used in electron microscopy, in which a very thin nongranular film of a metal, e.g. chromium, gold, uranium, is deposited obliquely on to the surface of the specimen prior to examination. This gives a three-dimensional effect and improves the clarity of the surface contours.

shaft The part of a hair distal to the root; the straight cylindrical part of a long limb bone; the rachis, or distal solid part of the scapus of a feather.

shaping The training of a response by successively reinforcing responses that are increasingly similar to the target behaviour, until that behaviour is reached. Also *method of successive approximations*.

shear To cut the long, stiff DNA duplex by hydrodynamic means.

sheath An enclosing or protective structure. (1) The elytron of some Insects. (2) Any tubular structure surrounding an organ or plant part, e.g. leaf sheath. (3) A tissue layer that surrounds other tissues, e.g. bundle sheath.

sheath of Schwann See **neurolemma**.

shell gland In some Invertebrates, a glandular organ which secretes the materials for the formation of the shell. Also *shell sac*.

shell ligament The dorsal ligament joining the valves of the shell in bivalve Mollusca.

shell sac See **shell gland**.

shell shock See **traumatic neurosis**.

shielding Use of dense matter to attenuate radiation when it might be harmful to oper-

ator or measuring system. The most common material used for large areas is concrete and for smaller areas, lead.

shikimic acid 3,4,5-trihydroxy-1-cyclo-hexene-1-carboxylic acid, an important cyclic intermediate in the synthesis, in plants, of the aromatic amino acids and other aromatic compounds from non-aromatic precursors.

shinbone The *tibia*.

shock (1) *Physiological*, a response to trauma characterized by a state of collapse, pallor, lowered blood pressure etc. (2) *Psychological*, an unexpected and intense experience which compels a total reorientation to life. See **trauma**.

shoot A stem with all its branches and appendages developed from a bud.

shoot-tip culture See **plant cell culture**.

short-day plant A plant which naturally flowers or shows other morphogenetic change only, or better, as the days shorten. It requires the stimulus of dark period(s) longer than some critical length.

short shoot In many, esp. woody, plants, a side shoot with very short internodes, on which are borne most (e.g. larch) or all (e.g. pine) of the foliage leaves and most or all of the fruit (e.g. most apples). See **spur**. Cf. **long shoots**.

short sightedness See **myopia**.

short-term memory According to the 3-store model of memory, a memory system that keeps memories for short periods, has a limited storage capacity and stores items in a relatively unprocessed form. See **long-term memory, sensory store**.

shoulder girdle See **pectoral girdle**.

Shwartzman reaction *Sanarelli Shwartzman phenomenon*. If endotoxin is administered intravenously followed by a second dose by the same route 24 hours later, renal tubular necrosis and adrenal haemorrhages occur. If the first dose is given into the skin the second dose results in destruction of venules and haemorrhagic necrosis at the skin site. Although apparently a hypersensitivity reaction, there is no identifiable immunological basis.

sibs, siblings Brothers and/or sisters. *Full sibs* have both parents in common, *half sibs* have one parent in common.

sieve area See panel.

sieve element See panel.

sieve plate See panel.

sievert A unit of radiation dose, being that delivered in 1 hr at a distance of 1 cm from a point source of 1 mg of radium element enclosed in platinum 0.5 mm in thickness. Numerically equal to ca. 8.4 röntgens or 21.6 C/kg. See **effective dose equivalent**.

sieve tube See panel.

sigmoid flexure An S bend.

sieve element

The cell type in which translocation occurs in the **phloem** of vascular plants, characterized by the presence of sieve areas in the walls and the disappearance of the **tonoplast** at maturity.

A *sieve area* is a small area of the common wall between two adjoining sieve elements; it develops from a primary pit-field and is perforated by many pores (< 1 –10 μm diameter) through which the protoplasts connect and the translocating solutes move.

In angiosperms, the sieve elements are *sieve tube members* (or *elements*) organized into *sieve tubes*, each consisting of a number of such members joined end-to-end and connected by *sieve plates* (highly differentiated sieve areas, typically with relatively large pores) in their common end walls (see diagram). A sieve tube member has one or more **companion cells** and its nucleus disappears as the cell matures .

The sieve elements of pteridophytes and gymnosperms are *sieve cells* which are elongated, have sieve areas with narrow pores rather than sieve plates, lack companion cells (but see **albuminous cell**) and have persistent nuclei.

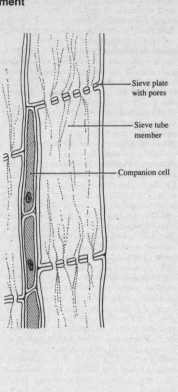

Sieve plate with pores

Sieve tube member

Companion cell

sign Any objective evidence of disease or bodily disorder, as opposed to a **symptom** which is a subjective complaint of a patient.

signal peptides, sequences *Leader peptides.* Short N-terminal peptide sequences of newly synthesized membrane proteins which direct the protein towards the appropriate membrane, facilitate its transfer across it and are usually deleted during the subsequent maturation of the protein.

significance A threshold value of probability at or below which the results of a statistical investigation are held to contradict a particular hypothesis.

sign stimulus Part of a complex stimulus configuration which is relevant to a particular response and evokes the strongest response (e.g. the red breast of the robin).

silicole A plant which grows on soils rich in silica, and usually acid in reaction.

siliqua A capsule with the general charac-

ters of a **silicle**, but at least 4 times as long as it is broad. Also *silique*.

silk A fluid substance secreted by various Arthropoda. It is composed mainly of fibroin, together with sericin and other substances, and hardens on exposure to air in the form of a thread. Used for spinning cocoons, webs, egg cases etc.

Siluriformes Order of mainly freshwater, bottom-living Osteichthyes with barbels used in detecting food and usually without scales. Catfish.

simian virus 40 See **SV 40**.

similarity coefficient Index of similarity between two stands of vegetation, based on their species composition.

simple Consisting of one piece or component; unbranched; not **compound**.

simple eye See **ocellus**.

simple fruit A fruit formed from one pistil.

simple leaf A leaf in which the lamina consists of one piece, which, if lobed, is not cut into separate parts reaching down to the midrib.

simple pit A pit of which the cavity does not become markedly narrower towards the cell lumen. Cf. **bordered pit.**

simple sequence DNA A block of a DNA sequence which consists of many repeats of a short, unit sequence. The repeats are not necessarily identical.

simulation Mimicry; assumption of the external characters of another species in order to facilitate the capture of prey or escape from enemies. v. *simulate.*

simulation by computer The investigation of thought processes by the use of computers programmed to imitate them.

single-cell protein Protein-rich material from cultured algae, fungi (including yeasts) or bacteria, used (potentially) for food or as animal feed. Abbrev. *SCP.*

sinistrorse Helical, twisted or coiled in the sense of left-hand screw thread or of an *S-helix.* Cf. **dextrorse.**

sink Region within a plant (or a cell) where a demand exists for particular metabolites, e.g. growing shoots, roots and developing tubers (sinks for photosynthate); mitochondria (sinks for oxygen).

sinuate A leaf margin with rounded teeth and notches; wavy.

sinus A cavity or depression of irregular shape.

sinusoid In Vertebrates, a sinus-like blood space connected usually with the venous system and lying betwen the cells of the surrounding tissue or organ.

sinus venosus In a Vertebrate embryo, the most posterior chamber of the developing heart; in lower Vertebrates, the tubular chamber into which this develops and which receives blood from the veins or sinuses, and passes it into the auricle.

siphon A tubular organ serving for the intake or output of fluid, as the pallial siphons of many bivalve Mollusca. adj. *siphonate.*

Siphonaptera An order of wingless Insects ectoparasitic on warm-blooded animals. Laterally compressed, mouthparts for piercing and sucking, coxae large, tarsi 5-jointed with prominent claws. Metamorphosis is complete, larvae legless, pupae exarate and enclosed in a cocoon. Serve as vectors for some diseases, e.g. myxomatosis, plague. Fleas. Also *Aphaniptera.*

siphoneous, siphonaceous Having large, tubular, multinucleate cells without cross walls; coenocytic.

siphonogamy Reproductive process in which non-motile male nuclei are carried to the egg cell through a pollen tube, as in conifers and angiosperms. Cf. **zooidogamy.**

siphonostele A stele with a more or less continuous ring of vascular tissue surrounding a pith, i.e. a medullated protostele or a solenostele.

sipho-, siphono- Prefixes from Gk. *siphon,* gen. *siphonos,* tube.

siphuncle In Nautiloidea, a narrow vascular tube extending from the visceral region of the body through all the chambers of the shell to its apex. adj. *siphunculate.*

Sipunculida A phylum of marine worm-like animals without segmentation and an anterior introvert with a mouth surrounded by tentacles. The gut is U-shaped and opens dorsally. Mostly detritus feeding burrowing forms.

Sirenia An order of large aquatic Mammals of herbivorous habit; the fore limbs are fin-like, the hind limbs lacking; there is a horizontally flattened tailfin; the skin is thick with little hair and underlying blubber, there are two pectoral mammae, no external ears, and the neck is very short. Manatees, Dugongs.

sister cell One of the two cells formed by the division of a pre-existing cell.

sister-chromatid exchange Reciprocal exchange of DNA between the chromatids of a single chromosome.

sister nucleus One of the two nuclei formed by the division of a pre-existing nucleus.

Site of Special Scientific Interest In Britain, an area deemed by the Nature Conservancy Council to be important for nature conservation, but not large enough to become a *National Nature Reserve.*

site-specific mutagenesis The possibility of altering a DNA sequence at a defined position by e.g. synthesizing an alternative sequence and reinserting into its host chromosome. A technique for **reverse genetics.**

size-exclusion chromatography Chromatographic separation of particles according to size by a solid phase of material of restricted pore size which functions as a molecular sieve. Such materials include dextran, cross-linked agarose and, for **HPLC,** coated silica.

Sjögren's disease Chronic inflammatory disease of salivary and lacrimal glands often accompanied by inflammation of the conjunctiva. Blood of persons with this condition often contains anti-nuclear antibodies and rheumatoid factor, as well as autoantibody reactive to the duct epithelium of the affected glands.

skeletal muscle See **voluntary muscle.**

skeleton The rigid or elastic, internal or external, framework, usually of inorganic material, which gives support and protection to the soft tissues of the body and provides a basis of attachment for the muscles, forming a system of jointed levers which they can move. adjs. *skeletal, skeletogenous*.

skewness The degree of asymmetry about the central value of a distribution.

skiagram, skiagraph Same as *radiograph*.

skin The protective tissue layers of the body wall of an animal, external to the musculature. See **epidermis**.

skin dose Absorbed or exposure radiation dose received by or at the skin of a person exposed to sources of ionization. Cf. **tissue dose**.

Skinner box A chamber designed for the study of **operant conditioning**; it is provided with mechanisms for an animal to operate and an automatic device for presenting rewards according to **schedules of reinforcement** preset by the experimenter.

skin sensitizing antibody Antibody capable of attachment to skin cells so that, on subsequent combination with an antigen, an immediate type hypersensitivity reaction (Type I) occurs. Mainly IgE.

skin test Any test in which substances are injected into or applied to the skin in order to observe the host's response. Used extensively in the study of hypersensitivity and immunity, e.g. tuberculin test, Schick test.

skull In Vertebrates, the brain case and sense capsules, together with the jaws and the branchial arches.

sleep A state which is characterized by prolonged periods of immobility, often with an associated and species-typical posture, and an increased reluctance to respond to stimulation; most animals are found to sleep during a particular period of the day and for a species-characteristic duration.

sleep movement See **nyctinastic movement**.

slice The cross-sectional portion of the body which is scanned for the production of e.g. a **computed tomography** or **magnetic resonance** image.

sliding growth The sliding of the wall of one cell past that of the next, as has been postulated to occur during the growth of fibres.

slime mould See **Myxomycota**.

slime plug Accumulation of P-protein on a sieve area.

slough The cast-off outer skin of a Snake.

slow-reacting substance A A pharmacologically active material comprising **leukotrienes** C, D and E, released by **mast cells** and other cells in the course of immediate hypersensitivity reactions. Causes contraction of smooth muscle, esp. bronchial muscle, and increased vascular permeability. The term is that applied when it was first discovered. Abbrev. *SRS-A*.

slow virus A transmissable agent which causes disease after very long incubation periods often of many years, e.g. kuru in Man, scrapie in Sheep.

slow-wave sleep *Quiet sleep, NREM sleep.* Sleep characterized by the presence of high amplitude, slow wave changes in potential, as measured on the EEG.

small nuclear RNA Discrete set of RNA molecules found in ribonucleoprotein particles (SnRNPs) which are responsible for processing **HnRNA** to give **mRNA**.

smallpox See panel.

smallpox vaccination See panel.

smegma A thick greasy secretion of the sebaceous glands of the glans penis.

smooth colony A bacterial colony of a typical regular, glistening appearance; sometimes developed from a rough colony by mutation and frequently differing from it physiologically, e.g. altered sensitivity to bacteriophage.

smooth endoplasmic reticulum Tubular form of endoplasmic reticulum without ribosomes, well developed in gland cells which secrete terpenoids, flavonoids etc.

smooth muscle See **unstriated muscle**.

smut One of a number of plant diseases, some economically important, caused by biotrophic fungi of the order Ustilaginales and characterized by the production of masses of usually black spores within the host. See **bunt**.

social facilitation Refers to an increase in the performance of a behaviour as a consequence of its performance by other individuals nearby (e.g. yawning in primates).

socialization The process whereby the child learns the norms of its society and acquires the attitudes and behaviour that conform to them.

social-learning theory A theoretical approach to the development of social behaviour; it stresses the importance of learning by observing others, and of reinforcement; contains elements of both a *stimulus-response* approach and a *cognitive approach* to the issues of learning.

social organization The totality of all social relationships among members of a particular group.

social parasitism A parasitic relationship between members of one species or with individuals of another species which involves exploiting aspects of the host's social behaviour, e.g. their nesting or hunting behaviour. A common example is the female cuckoo which lays its egg in another bird's nest.

social perception Refers to several related areas of study: (1) how individuals perceive others, e.g. the judgment of emotions in others, studies of impression formation (or person perception); (2) the psychological processes that underlie these activities in the perceiver; (3) the study of how social factors

---

smallpox and vaccination

Variola. An acute, highly infectious viral disease characterized by fever, severe headache, pain in the loins, and a rash which is successively macular, papular, vesicular and pustular, affecting chiefly the peripheral parts of the body. Until recent times, one of the major killing diseases of Man but has now been eradicated.

Vaccination is the method of producing active immunity against smallpox (variola), discovered by Jenner in Gloucestershire, UK in 1796. Not practised since smallpox was eliminated on a worldwide scale, but immensely important historically. Vaccinia virus is used, prepared from vesicular lesions of vaccinia in the skin of calves or sheep (more recently from cultures of amnion *in vitro*). This virus shares cross reacting antigens with variola virus and therefore induces protective immunity against smallpox. Vaccination is performed by introducing live vaccinia virus into a site in the superficial layers of the skin. Successful primary vaccination is characterized by development of a vesicular lesion at the site after 6–9 days, reaching a maximum on the 12th day, after which it subsides leaving a scab and later a scar. Protective immunity, which depends primarily on the ability to mount a delayed type reaction against virus infected cells, gradually declines. Re-vaccination is needed at intervals of about 3 years. If some residual immunity is present the local reaction produced by re-vaccination may be accelerated, so that a small vesicle develops at the 4–5th day, and the reaction is at its height on the 7th day.

Rare complications are *generalized vaccinia*, which may occur in subjects with atopic eczema (which interferes with the development of the local delayed hypersensitivity response); *chronic progressive vaccinia* particularly if cell-mediated immunity is deficient due to disease or to drug treatment; and very rarely *post-vaccinial encephalomyelitis*, which is most likely to occur after primary vaccination of young infants.

---

influence perception (the effects of group pressure on the perception of an event).

social phobia The fear of performing certain actions when exposed to the scrutiny of others.

social psychology Refers to the interpersonal relations, face to face interactions, as well as to the social attitudes that are the subject matter of research in social psychology.

social symbiosis Refers to a relationship between members of different species from which one or both derive some advantage; the nature of the benefit for each species can be quite different, e.g. one partner may gain protection from predators while the other gains a nesting site or food supply.

society A minor plant community within an association, dominated by a species which is not a general dominant of the association, e.g. an alder-dominated community on wet ground within an oak wood.

sodium dodecyl sulphate *SDS*. An anionic detergent widely used as a powerful denaturant and solubilizing agent, esp. for the fractionation of proteins and nucleic acids according to size by **acrylamide gel electrophoresis**.

soft commissure In Mammals, the point at which the thickened sides of the thalamencephalon touch one another across the constructed third ventricle.

soft palate In Mammals, the posterior part of the roof of the buccal cavity which is composed of soft tissues only.

soft radiation General term used to describe radiation whose penetrating power is very limited, e.g. low energy X-rays.

soft rot Rot in which tissues (usually parenchyma) soften because of the destruction of middle lamellas and cell walls, as in many bacterial and fungal diseases of stored fruits and vegetables.

softwood The wood from a conifer and coniferous trees generally. Ant. *hardwood*.

soil-acting herbicide A plant poison absorbed from the soil by the roots, such as sodium chlorate.

soil flora Fungi, bacteria and algae living in the soil.

soil structure The nature of a soil in terms of manner in which the particles are aggregated into larger bodies or *peds*.

soil texture The nature of a soil in terms of the proportions of mineral particles of various sizes (sand, silt and clay).

Solanaceae Family of ca 2500 spp. of dicotyledonous flowering plants (superorder Asteridae). Mostly herbs, some shrubs and small trees, cosmopolitan. The petals are fused, the ovary is superior, the fruit is usually a berry, less often a capsule. Includes potato, tomato, aubergine, sweet peppers, chillies, cayenne pepper, tobacco. Many are poisonous and often contain tropane alkaloids of medicinal importance, e.g. belladonna or deadly nightshade (atropine), and henbane (hyocyamine).

solar Having branches or filaments radially arranged.

solar plexus In higher Mammals, a ganglionic centre of the autonomic nervous system; situated in the anterior dorsal part of the abdominal cavity, from which nerves radiate in all directions.

soldier In some social Insects, a form with esp. large head and mandibles adapted for defending the community, for fighting, and for crushing hard food particles.

solenocyte In Invertebrates and lower Chordata, an excretory organ consisting of a hollow cell with branched processes, in the lumen of which occurs a bunch of cilia which by their movements maintain a downward current.

solenoid model The organization of nucleosomes and spacer regions into a *solenoid coil*, containing 6–7 nucleosomes per turn, with a diameter of 30 nm.

solenostele Stele with a central pith and consisting of an annulus of xylem completely enclosed by an endodermis, and with phloem between the endodermis and the xylem throughout.

soligenous Said of a mire or fen, receiving water that has passed through a mineral soil and hence is oligotrophic to relatively eutrophic depending on the nature of the soil. Cf. **ombrogenous**.

solitaria phase One of the two main phases of the Locust (Orthoptera), which occurs when nymphs are reared in isolation. They adjust their colour to match their background and lack the higher activity and gregarious tendencies of the **gregaria phase**.

soluble complex Applied to antigen-antibody complexes in soluble form rather than in precipitates. Occurs *in vivo* or *in vitro* when there is an excess of antigen over antibody so that a large lattice is not formed. (See **lattice hypothesis**). Soluble complexes cause tissue damage *in vivo*, esp. if complement is also activated by them. See **hypersensitivity reactions Type III, Immune complex disease, serum sickness, glomerulonephritis**.

solute potential Same as **osmotic potential**.

soma The body of an animal, as distinct

from the germ cells. Cf. **germen**. pl. *somata*. adj. *somatic*.

somaclonal variation Variability commonly found among plants that have been regenerated from **plant cell cultures**.

somatic Of cells of the body, as distinct from the *germ line*.

somatic cell One of the nonreproductive cells of the parent body, as distinct from the reproductive or germ cells.

somatic cell hybrid Cell formed by fusion of cells from the same or different species, in which there is also nuclear fusion.

somatic doubling A doubling of the number of chromosomes in the nuclei of somatic cells.

somatic hybridization The production of hybrid cells by the fusion of non-gametic nuclei either naturally in the **parasexual cycle** of some fungi or artificially, such as by *protoplast fusion*. See **plant genetic manipulation**.

somatic mutation Mutations occurring in the genes of cells of the body other than the germ cells which give rise to sperm or ova, and therefore not inheritable. Such mutations normally occur rarely in most cells, but the genes controlling synthesis of the variable region of immunoglobulin molecules exhibit a greatly increased rate of mutation. There are three hypervariable sites in the V region genes at which mutations are esp. likely to occur during the rapid proliferation of B lymphocytes which respond to antigenic stimulation. This increases the range of different immunoglobulin molecules made by the cell population involved, and includes some with increased affinity for the stimulating antigen.

somatic pairing Closely paired arrangement of **chromatids** in **polytene** chromosomes.

somatoblast In development, a cell which will give rise to somatic cells.

somatoform disorders Those conditions in which psychological conflicts take on a somatic or physical form, includes the hypochondrias and conversion disorders.

somatogenic Arising as the result of external stimuli; developing from somatic cells, as opposed to germ cells.

somatopleure The outer body wall of coelomate animals; the outer layer of the mesoblast which contributes to the outer body wall. Cf. **splanchnopleure**. adj. *somatopleural*.

somatostatin A peptide of the hypothalamus which controls the secretion of *somatotropin* by the pituitary.

somatotropin A peptide secreted by the anterior pituitary which causes the liver to produce *stomatomedins* which stimulate the growth of bones and muscle.

Southern blot

Method of revealing rare DNA fragments in a complex mixture of DNA. Gel electrophoresis separates the fragments, usually made by cleavage with **restriction enzymes,** into their size classes. The fragments are then denatured by placing the gel in alkali after which a nitrocellulose filter sheet is laid over the gel and the DNA fragments transferred from gel to filter by placing layers of absorbent paper above the latter (*blotting*), the gel being immersed in a high concentration of salt. This causes the solute and DNA fragments to diffuse through the nitrocellulose sheet to which the latter adsorb, maintaining their relative positions. The nitrocellulose sheet is then incubated under renaturing conditions with a solution containing radio-labelled molecules able to base-pair with some sequence in the original DNA (*probing*). Finally, the position and therefore fragment size of the complementary sequence in the original DNA can be located by autoradiography.

Named after its discoverer, E.M.Southern, it is one of the most powerful methods in molecular biology. Analogous methods for proteins and RNA are called western and northern blots.

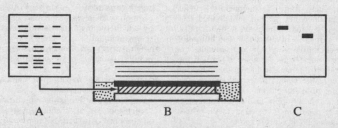

A   B   C

The three main stages in the Southern blot transfer procedure.

A = Plan view of the flat agarose gel. B = The 'sandwich' with the gel (hatched), lying on a support, is at the bottom, the nitrocellulose sheet (solid) in the middle and the layers of absorbent paper on the top (lines). C = The autoradiograph exposed from the nitrocellulose filter sheet after the latter has been hybridized with a radioactive probe.

somatotropism Directed growth movements in plants so that the members come to be placed in a definite position in relation to the substratum.

somatotype theory A theory proposed by Sheldon who suggested that bodily characteristics reflect clusters of personality traits; the theory is no longer held in any serious regard, although the terms **ectomorph, mesomorph** and **endomorph** still occasionally appear.

somite One of the divisions or segments of the body in a metameric animal; a mesoblastic segment in a developing embryo.

somnambulism *Sleepwalking.* A dissociative disorder in which a person walks while asleep.

sorus (1) In plants generally, a cluster of reproductive structures on a thallus. (2) In ferns, a cluster of sporangia on a leaf, often covered by an *indusium* which acts as a protective structure.

source strength Activity of radioactive source expressed in disintegrations per second.

Southern blot See panel.

space parasite A plant which inhabits intercellular spaces in another plant, obtaining shelter but possibly taking nothing else.

spadiceous Shaped like a palm branch. Also *spadiciform, spadicose.*

spadix A spike, the axis of which is fleshy. The characteristic inflorescence of the arum family; commonly associated with a **spathe**.

spasm Involuntary contraction of muscle fibres. adj. *spasmodic*.

spathe A large, sometimes coloured or showy bract which subtends and may enclose a **spadix.**

spatial summation The summational effect of stimuli spread over space (e.g. the brightness of a beam of light has a lower absolute threshold for larger diameters of beam).

spatula Any spoon-shaped structure. adjs. *spatulate, spathulate.*

spawn (1) To deposit eggs or discharge spermatozoa; a collection of eggs, such as that deposited by many Fish. (2) The mycelium of a fungus, esp. the mycelial preparations used to propagate the cultivated mushroom.

specialist Organism with a restricted food source, living in a restricted habitat, often displaying specific behaviour or structural adaptations.

speciation Formation of new biological species including formation of polyploids.

species A group of individuals that (1) actually or potentially interbreed with each other but not with other such groups, (2) show continuous morphological variation within the group but which is distinct from other such groups. Taxonomically, species are grouped into genera and divided into subspecies and varieties or, horticulturally, into cultivars. In the system of **binomial nomenclature** of plants and animals, the second name (i.e. the name by which the species is distinguished from other species of the same genus) is termed the *specific epithet* or *specific name*. The latter, however, correctly refers to the full name, e.g. *Lilium candidum* (Madonna Lily), where *candidum* is the *specific epithet.*

species/area curve Curve relating the number of species found (y axis) to the area over which the observer searched (x axis). The shape of the curve provides information about **diversity** and **richness**.

specific Of a parasite, restricted to a particular host. n. *specificity.*

specific characters The constant characteristics by which a species is distinguished.

specific dynamic action The special calorigenic property of foodstuffs, particularly of proteins, of raising the metabolic rate after ingestion by an amount in excess of their calorific value. May be expressed as the ratio of the calories in excess of the basal to urinary nitrogen in excess of the basal.

specific name See **species.**

spectrin A family of closely related cytoskeletal proteins which consist of an α-β heterodimer of 2 high molecular weight polypeptides. The eponymous member of the spectrin family constitutes a major component of the erythrocyte membrane. Other members are found in brain and intestinal epithelium.

spelaeology, speleology The study of the fauna and flora of caves.

sperm See **spermatozoon.**

spermaceti A glistening white wax from the head of the sperm whale, consisting mainly of cetyl palmitate, $C_{15}H_{31}.COO.C_{16}H_{33}$; mp $41°$–$52°C$, saponification number 120–135, iodine number nil. Used in the manufacture of cosmetics and ointments.

spermagonium See **spermogonium.**

spermary See **testis.**

spermatheca A sac or cavity used for the reception and storage of spermatozoa in many Invertebrates; receptaculum seminis.

spermatic Pertaining to spermatozoa; pertaining to the testis.

spermatid A cell formed by division of a secondary spermatocyte, and developing into a spermatozoon without further division.

spermatoblast A spermatid.

spermatocyte A stage in the development of the male germ cells, arising by growth from a spermatogonium or by division from another spermatocyte, and giving rise to the spermatids.

spermatogenesis Sperm formation; the maturation divisions of the male germ cells by which spermatozoa are produced from spermatogonia.

spermatogonium A sperm mother cell; a primordial male germ cell. adj. *spermatogonial.*

spermatophore A packet of spermatozoa enclosed within a capsule.

Spermatophyta Division containing the seed plants, i.e. the gymnosperms and the angiosperms. Also called *Magnoliophyta.*

spermatozoid Motile, flagellated, male gamete, as found in most algae, some fungi and in bryophytes, pteridophytes, cycads and *Ginkgo.* syn. *antherozoid.*

spermatozoon The male gamete, typically consisting of a head containing the nucleus, a middle piece containing mitochondria, and a tail whose structure is similar to that of a flagellum. pl. *spermatozoa.* Abbrev. *sperm.*

sperm cell Male gamete, motile or not.

spermiducal glands In many Vertebrates, glands opening into or near the spermiducts.

spermiduct *Spermaduct.* A duct by which sperms are carried from the testis to the external genital opening; the vas deferens. adj. *spermiducal.*

spermogonium A flask-shaped structure in which spermatia are formed as in some ascomycetes, rusts and perhaps some lichens.

sperm-, sperma-, spermi-, spermo-, spermato- Prefixes from Gk. *sperma,* gen. *spermatos,* seed.

sphagnicolous Living in peat moss.

Sphagnum The bog mosses. A genus of mosses (*Bryopsida*) of ca 350 spp. with upright, branching gametophytes, the leaves of which are very absorbent because of a regular pattern of dead cells with holes through their walls. *Sphagnum* spp. dominate many bogs. Sphagnum peat is acidic and lacking in mineral nutrients. Cf. **sedge** peat.

S phase The period in the cell cycle during which the nuclear DNA content doubles.

Sphenisciformes An order of Birds in which flight feathers are lacking; the wings are stiff and used as paddles in swimming; the feet are webbed; the bones are solid and there are no air sacs; flightless marine forms, with streamlined bodies, powerful swimmers and divers; confined to the southern hemisphere. Penguins.

Sphenodon See **Rhynchocephalia**.

sphenoidal Wedge-shaped.

Sphenopsida The horsetails and their allies. They are a class of Pteridophyta dating from the Devonian onwards. Sporophytes with roots, stems and whorled leaves (*microphylls*). Sporangia are borne usually reflexed on sporangiophores arranged in whorls often on terminal cones. Mostly homosporous. Spermatozoids multiflagellate. Includes the Sphenophyllales, Calamitales and Equisetales.

sphen-, spheno- Prefixes from Gk. *sphen*, a wedge.

spherosome A small (< 10 μm), spherical, refractile body rich in lipid, in the cytoplasm of plant cells. Probably bounded by a half unit membrane. Cf. **lipid body**.

sphincter A muscle which by its contraction closes or narrows an orifice. Cf. **dilator**.

sphingomyelin A membrane phospholipid derived from **sphingosine** by the addition of a long chain hydrocarbon, phosphate and organic base.

sphingosine A hydrophobic amino alcohol which is a component of the phospholipids known as *sphingomyelins*.

sphygmus The pulse; the beat of the heart and the corresponding beat of the arteries.

spicate Bearing or pertaining to spikes; spike-like.

spicule A small pointed process; one of the small calcareous or siliceous bodies, forming the skeleton in many Porifera and Coelenterata. adjs. *spicular, spiculate, spiculiferous, spiculiform*.

spiculum Any spicule-like structure; in Snails, the dart.

spike A racemose inflorescence with sessile, often crowded, flowers on an elongated axis, e.g. *Plantago*.

spikelet A small or secondary spike. In grasses, the basic unit of the inflorescence usually consisting of a short axis (rachilla) bearing 2 bracts (glumes) and 1 or more florets.

spina A small sharp-pointed process.

spinal Pertaining to the vertebral column or to the spinal cord.

spinal canal The tubular cavity of the ver-

tebral column which houses the spinal cord.

spinal cord In Craniata, that part of the dorsal tubular nerve cord posterior to the brain.

spinal reflex Reflex situated in spinal cord in which higher nerve centres play no part.

spindle (1) In mitosis and meiosis, spindle-shaped structure (i.e. widest in the middle and tapering towards the poles), containing longitudinally-running **microtubules**, formed within the nucleus or the cytoplasm at the end of prophase, with centrioles, if present, usually at the poles. Apparently forms a structural framework for the movements of chromosomes and chromatids. (2) Special sensory receptor in muscle, *muscle spindle*.

spindle fibre One of the microtubules of the mitotic or meiotic spindle.

spine A stiff, straight, sharp-pointed structure, e.g. a pointed process of a vertebra like the neural spine; a fin ray; the scapular ridge. adjs. *spinate, spiniform, spinose, spinous*.

spinneret In Spiders, one of the spinning organs, consisting of a mobile projection bearing at the tip a large number of minute pores by which the silk issues.

spinning glands The silk-producing glands of Arthropoda.

spinose, spinous Covered with spines.

spinous process A process of (1) the proximal end of the tibia, (2) the sphenoid bone; the neural process or spine of a vertebra.

spinule A very small spine or prickle.

spiny vesicle Same as **coated vesicle**.

spiracle In Insects and some Arachnida, one of the external openings of the tracheal system; in Fish, the first visceral cleft, opening from the pharynx to the exterior between the mandibular and the hyoid arches; in amphibian larvae, the external respiratory aperture; in Cetacea, the external nasal opening. adjs. *spiracular, spiraculate, spiraculiform*.

spiral See **helical thickening**.

spiral cleavage A type of segmentation of the ovum occurring in many Turbellaria, most Mollusca, and all Annelida; the early micromeres rotate with respect to the macromeres, so that the micromeres lie opposite to the furrows between the macromeres; the direction of rotation (viewed from above) is normally clockwise (*dexiotropic*) and in 'reversed cleavage' it is anti-clockwise (*laeotropic*). Also *alternating cleavage*.

spiral valve In Lampreys, Elasmobranchii and some Lung fish, part of the intestinal canal, which is provided with an internal spiral fold to increase its absorptive surface.

spirillum A bacterium which is a variation of the rod form, i.e. a curved or cork-screw shaped organism. Varies from 0.2–50 μm in length. One or more flagella may be present.

Also the name of a particular genus, which includes *Sp. minus* (rat-bite fever).

spirochaetes Filamentous flexible bacteria showing helical spirals; without true flagella. Divided into two families: the Spirochaetaceae and the Treponemataceae. Saprophytic and parasitic members. Pathogenic species include the causative agents of syphilis (*Treponema pallidum*), sprirochaetosis (relapsing fever) in Man (*Borrelio recurrentis*), infectious jaundice (*Leptospira icterohaemorrhagiae*) and yaws (*Treponema pertenue*).

splanchnic Visceral.

splanchno- Prefix from Gk. *splanchnon*, inward parts.

splanchnocoel In Vertebrates, the larger posterior portion of the coelom which encloses the viscera, as opposed to the pericardium.

splanchnopleure The wall of the alimentary canal in coelomate animals; the inner layer of the mesoblast which contributes to the wall of the alimentary canal. Cf. **somatopleure**. adj. *splanchnopleural*.

spleen An organ of Vertebrates concerned with the formation and destruction of the red blood cells.

splenomegaly Abnormal enlargement of the spleen.

splicing Specifically refers to the excision of the **introns** from an mRNA and the rejoining of **exons** to form the mature message. Loosely used for joining nucleic acid molecules as in *gene splicing*.

split brain A condition in which the *corpus callosum* and some other fibres are cut so that the two cerebral hemispheres are isolated.

spodogram A preparation of the ash of a plant esp. from a section, used in investigating structure by light or electron microscopy.

spondyl A vertebra. adj. *spondylous*.

Sponge See **Porifera**.

spongin A horny skeletal substance, occurring usually in the form of fibres, in various groups of Porifera.

spongioblasts Columnar cells of the neural canal giving rise to neuroglia cells.

spongy layer, -mesophyll, -parenchyma, -tissue Chlorenchyma in which the cells are irregularly lobed leaving very large continuous intercellular spaces. Spongy **mesophyll** is usually present towards the lower surface of dorsiventral leaves of mesomorphic dicotyledons. See **palisade**.

spontaneous behaviour Behaviour pattern which occurs in the apparent absence of any stimuli.

spontaneous generation The production of living matter or organisms from nonliving matter. Generally believed to occur in micro-organisms before Pasteur.

spontaneous recovery The return of an extinguished response (see **extinction**) following a rest period in which neither the conditioned nor the unconditioned stimulus are presented, when an animal is returned to the original conditioning situation.

spontaneous remission Recovery without treatment.

sporangium A hollow, walled, structure in which spores are produced.

spore One of a variety of reproductive bodies, usually unicellular, often acting as a disseminule or as a resting stage, usually germinating without fusing with another cell. For plants with an alternation of generations. See **sporophyte**. Many algae and fungi produce spores of some sort which reproduce the same stage or the next stage of the life cycle. In Protozoa, it is a minute body formed after multiple fission; more strictly, a seedlike stage in the life cycle of Protozoa arising as a result of sporulation, and contained in a tough resistant envelope. See **conidium**, **aplanospore**, **hypnospore**, **endospore**, **exospore**, **zoospore**, **zygospore**.

spore mother cell A cell which gives rise to a spore, esp. one which divides by meiosis to give four cells which develop into spore(s). Also *sporocyte*.

spore print The marks obtained by placing e.g. the cap of a mushroom or toadstool, gills downward, on a piece of paper and allowing the spores to fall on to the paper.

spori-, sporo- Prefixes from Gk. *sporos*, seed.

sporocarp A hard multicellular structure enclosing sporangia in some fungi and some heterosporous ferns.

sporocyst The tough resistant envelope secreted by, and surrounding, a Protozoan spore.

sporocyte Same as **spore mother cell**.

sporogenesis The process of spore formation.

sporogenous Producing or bearing spores.

sporogenous layer Same as **hymenium**.

sporogonium Same as the **sporophyte** but in bryophytes.

sporogony In Protozoa, propagation, usually involving sexual processes and always ending in the formation of spores.

sporont A stage in the life history of some Protozoa which, as a gametocyte, gives rise to gametes, which in turn, after a process of syngamy, may give rise to spores.

sporophore Any structure that bears spores.

sporophyll A leaf that bears sporangia.

sporophyte The typically diploid generation of the life cycle of a plant showing alternation of generations, which produces, by meiosis, the spores which germinate to give the **gametophyte**.

sporopollenin Polymerized carotenoids, a constituent of the cell walls of pollen grains (the exine) and of many spores, and exceedingly resistant to decay. See **pollen analysis**.

Sporozoa A class of parasitic Protozoa, usually being at some stage intracellular, in the principal phase have no external organs of locomotion or are amoeboid, lack a meganucleus, and form large numbers of spores after syngamy, which constitute the infective stage.

sporozoite In Protozoa, an infective stage developed within a spore.

sport See **bud sport**.

sporulation The production of spores.

sprain A wrenching of a joint associated with tearing or stretching of its ligaments, damage to the synovial membrane, effusion into the joint, occasionally rupture of muscles or tendons attached to the joint, but without dislocation.

spread The establishment of a species in a new area.

spreading agent Substance added to a solution for e.g. spraying on a fungicide, in order to promote even distribution over the target.

spring wood Same as **early wood**.

sp.,spp. Abbrevs. for *species*, sing. and pl.

spur (1) A projection; arising esp. from the base of a petal, sepal or gamopetalous corolla, etc. usually containing nectar, e.g. *Aquilegia, Linaria*. (2) A short shoot, esp. one of the condensed lateral fruiting shoots of many fruit trees. See **calcar**.

spuriae In Birds, the feathers of the bastard wing.

squama A scale; a scalelike structure.

Squamata An order of diapsid Reptiles in which the skull has lost either one or both temporal vacuities, the quadrate is movably articulated with the skull, and there is no inferior temporal arch. Snakes and Lizards.

squamiform Scale-like.

squamous epithelium Epithelium consisting of one or more layers of flattened scalelike cells; pavement epithelium.

squamule A small scale. adj. *squamulose*.

squarrose Said of leaves, hairs, scales etc. which stick out more or less at right angles to the stem or other structure.

squash Spreading of tissue or chromosomes on a microscope slide by application of pressure.

S-R theory The *stimulus response theory* of learning which holds that the basic components of learning are S-R bonds, stimuli and responses which become forged together as learning proceeds.

ssp., sspp. Abbrevs. for *subspecies*, sing. and pl.

stabilate A population of usually a microorganism preserved in viable condition on a unique occasion by e.g. freezing.

stability The ability of an ecosystem to resist change.

stadium An interval in the life history of an animal between two consecutive ecdyses.

stage micrometer A device for measuring the magnification achieved with a given microscope or for calibrating an eyepiece graticule. It usually consists of a small accurate scale mounted on a microscope slide.

stag-headed Of a tree, having the upper branches dead but with regrowth of the crown from new branches.

stagnicolous Living in stagnant water.

staling The accumulation with time of metabolites in a culture medium which results in the slowing down of growth.

stamen The microsporophore (microsporophyll), or male reproductive organ, of flowering plants, i.e. the structure, within the flower, that bears the pollen. See **filament, anther, androecium**.

staminal Pertaining to a stamen; derived from a stamen.

staminate Of flowers, male.

staminode An imperfectly developed, vestigial, anther-less or petaloid stamen.

stand Any living assemblage of land plants.

standard deviation The square root of the mean of the squared deviations from the mean of a set of observations; the square root of the variance.

standard error The estimated standard deviation of an estimate of a parameter.

standard normal distribution The normal distribution with mean zero and variance one, which is extensively tabulated.

standing crop The total dry mass of organisms present in an area, obtained by harvesting sample plots, drying and weighing the harvested biomass and expressing the standing crop in units like $g\,m^{-2}$.

standing-off dose Absorbed dose after which occupationally exposed radiation workers must be temporarily or permanently transferred to duties not involving further exposure. Doses are normally averaged over 13-week periods and standing-off would then continue for the remainder of the corresponding period.

stapes In Amphibians, a small nodule of cartilage in connection with the fenestra ovalis of the ear; in Mammals, the stirrup-shaped innermost auditory ossicle. adj. *stapedial*.

Staphylococcus A Gram-positive coccus of which the individuals tend to form irregular clusters. The commonest types, associated with various acute inflammatory and suppurative conditions, including mastitis in

---

stem cell

An undifferentiated cell which divides, one daughter cell giving rise, usually by a succession of stages, to a mature functional cell such as the erythrocyte. The stem cell's name usually ends in *-blast,* as in *erythroblast.*

They are cells present in bone marrow and other haemopoietic tissues which are capable both of self-replication and generation of progenitor cells for a series of different cell lines. Stem cells are stimulated to proliferate and differentiate under the influence of colony stimulating factors present in the environment. Known also as *CFU–S* (colony forming unit–spleen), referring to the tissue in which it was first studied.

They can give rise to progenitors of lymphocytes or of granulocytes and macrophages. Interleukin-3 can stimulate the growth and differentiation of all the cell types. However there are different colony stimulating factors for granulocytes and macrophages, and for specific kinds of granulocytes.

---

animals, are *S. aureus* (golden yellow colonies) and *S. albus* (white colonies).

starch grain A rounded or irregular mass of starch; within a chloroplast in green algae; within a chloroplast, amyloplast or other plastid in vascular plants and bryophytes.

starch plant A plant in which carbohydrate is stored as starch. Cf. **sugar plant**.

starch sheath (1) A 1-layered cylinder of cells lying on the inner boundary of the cortex of a young stem, with prominent starch grains in the cells. It is homologous with an endodermis. (2) A layer of starch grains around a pyrenoid in an algal cell.

start codon Triplet sequence of nucleotides in **mRNA** which signals the initiation of translation and hence the first amino acid in a polypeptide chain. Also *initiation codon.*

startle colours Bright colours on the body or wings of animals which often resemble vertebrate eyes and which are normally concealed. They are exposed on being disturbed and are anti-predator devices.

stasis Stoppage of growth.

state-dependent learning Learning in which the recall depends on the degree of similarity between the physiological state of the individual at the time of training and at the time of testing.

state-dependent memory Refers to the fact that events learned in one mental state are best remembered when the individual is put back in that state (e.g. a particular mood, a drug-induced state).

statenchyma A tissue consisting of cells containing statoliths.

statistic A numerical quantity calculated from a set of observations.

statocyst, statocyte An organ for the perception of the position of the body in space, consisting usually of a sac lined by sensory cells and containing a free hard body or bodies, either introduced or secreted; an *oto-*

cyst. In plants, a cell containing starch grains or other solid inclusions, acting as statoliths.

statolith (1) A secreted calcareous body contained in a statocyst. (2) A starch grain, or other solid body in a cell, that moves in response to gravity and appears to function as a gravity sensor.

steapsin A fat-digesting enzyme occurring in the digestive juices of various animals, as the pancreatic juice of Vertebrates.

stele The primary vascular system and associated ground tissue of the stems and roots of a vascular plant.

stellate Radiating from a centre, like a star.

stellate hair A hair which has several radiating branches.

stem The above-ground axis of a plant, and other axes (above- or below-ground) that are anatomically similar and/or clearly homologous. In most vascular plants, stems may be recognized by their bearing foliage leaves and/or scale leaves.

stem-and-leaf plot An arrangement of a set of values into rows with common leading digits (stems), entries in each row being the remaining digit of each value, truncated if necessary.

stem cell See panel.

stem succulent A plant with a succulent, photosynthetic stem and with the leaves small or sometimes represented by spines. Many are CAM plants, e.g. cacti. Cf. **leaf succulent**.

steno- Prefix from Gk. *stenos,* narrow.

stenohaline Capable of existence within a narrow range of salinity only. Cf. **euryhaline**.

stenophyllous Having narrow leaves.

stenopodium The typical biramous limb of Crustacea, having slender exopodite and endopodite. Cf. **phyllopodium**.

stereokinesis Movement of an organism in response to contact stimuli.

stereome A general term which is used for the mechanical tissue of the plant.

stereome cylinder A cylinder of strengthening tissue lying in a stem, usually just outside the phloem.

stereospondyly The condition of having the parts of the vertebrae fused to form one solid piece; cf. **temnospondyly**. adj. *stereospondylous*.

stereotaxis Response or reaction of an organism to the stimulus of contact with a solid body, as the tendency of some animals to insert themselves into holes or crannies, or to attach themselves to solid objects. adj. *stereotactic*.

stereotype An oversimplified and very generalized belief about groups of people which is applied to individuals identified as members of the group.

stereotyped behaviour Behaviour patterns which are performed on different occasions with very little variation in their component parts, typical of many animal displays. Animals under stress, e.g. in close confinement, may develop very fixed and idiosyncratic behaviours, e.g. rigid pacing actions.

sterile (1) Unable to breed. (2) Free from living organisms, esp. culture media, foodstuffs, surfaces, medical supplies etc. which are free from micro-organisms that could cause spoilage or infection.

sterile flower (1) A flower with neither functional carpels nor functional stamens. (2) Sometimes a male flower.

sterilization (1) Causing the loss of sexual reproductive function. (2) Making sterile by removing unwanted organisms by heat, radiation, chemicals or by filtration.

sternal See **sternum**.

sternebrae In Mammals, a median ventral series of bones which alternate with the ribs.

sternum The ventral part of a somite in Arthropods; the breast bone of Vertebrates, forming part of the pectoral girdle, to which, in higher forms, are attached the ventral ends of the ribs. adj. *sternal*.

steroid hormones See panel.

steroid regulated genes See panel.

sticky ends When DNA is cut by a **restriction enzyme**, the cut is usually staggered between the two chains, so that one chain is longer by one or two bases than the other. This end is able to base pair with a complementary end on another molecule cut by the same enzyme, causing the two molecules to stick together. Also *cohesive ends*.

stigma (1) See **eye spot**. (2) The part of the carpel of a flowering plant that is adapted for the reception and germination of the pollen. (3) In Arthropoda, one of the external apertures of the tracheal system. (4) In Urochordata, a gill slit. Generally, a spot or mark of

distinctive colour, as on the wings of many Butterflies. pl. *stigmata*.

stilt root See **prop root**.

stimulus Refers to some aspect of the environment, either internal or external to the individual, which produces some response, although this is not always an immediate response nor an easily observable one. More specifically, an agent which will cause propagation of a nerve impulse in a nerve fibre. pl. *stimuli*.

stimulus control An **operant conditioning** technique in which a predictable relationship is established between a given stimulus and a given response by eliminating all other stimuli associated with that response and all other responses associated with that stimulus.

stimulus generalization The principle that when a subject has been conditioned to make a response to a stimulus, other similar stimuli will tend to evoke the same response, although to a lesser degree; the greater the similarity to the original stimulus, the greater this tendency will be.

stimulus threshold The value of a quantified stimulus which elicits a particular response at a definite intensity. See **absolute threshold**, **difference threshold**.

sting A sharp-pointed organ by means of which poison can be injected into an enemy or a victim, as the poisonous fin-spines of some Fishes, the ovipositor of a worker Wasp. See **urticaria**.

stinging hair An epidermal hair which is capable of injecting an irritating fluid into the skin of an animal when its tip is broken by contact, as in nettles.

stipe A stalk, esp. (1) of the fruiting body of a fungus, (2) the part connecting holdfast and lamina of a large algal thallus.

stipes A stalklike structure; an eyestalk. pl. *stipites*. adjs. *stipitate*, *stipiform*.

stipular trace The vascular tissue running into a stipule.

stipule In many dicotyledons one of a pair of appendages which start development as outgrowths of the flank of a leaf primordium, often serve to protect the leaves in the bud and mature as leaf-like photosynthetic structures or as spines, scales etc.

stochastic Developing in accordance with a probabilistic model; random.

stock (1) Usually a rooted stem into which a scion is placed in grafting. (2) The perennial part of a herbaceous perennial. (3) A strain maintained for breeding or propagation.

stolon A cylindrical stemlike structure. (1) See **runner**. (2) Arching stem that forms a new rooted plant at the tip, e.g. blackberry. (3) Slender horizontally-growing underground stem which forms a new plant at the end. (4) A tubular outgrowth in hydroid col-

steroid hormones

A class of lipophilic hormones which are synthesized from **cholesterol** and consist of a four-membered ring system with various substitutions. In mammals there are three major divisions: the adrenal steroids, cortisol and aldosterone, the sex steroids, progesterone, oestrogen and testosterone, which are synthesized by the gonads, and vitamin D3 which is converted into its active form in the liver and kidney. The *adrenal steroids* influence body homeostasis, controlling glycogen and mineral metabolism as well as mediating the stress response. They also affect the immune and nervous sustems. The *sex steroids* determine the control and development of the embryonic reproductive system, control reproduction and reproductive behaviour in the adult and the development of secondary sexual characteristics. *Vitamin D* plays an important role in the regulation of calcium and phosphorus homeostasis and is necessary for normal bone development.

Steroid hormones travel to their target tissues in the blood and, since they are poorly soluble in water, are transported by high specificity carrier proteins such as *corticosteroid binding globulin, sex hormone binding globulin* and *progesterone binding protein.* They also associate with low affinity with α1-acid glycoprotein and serum albumin. Steroids enter cells by diffusing across the membrane since they are readily lipid soluble. They influence the physiology of particular target tissues by regulating the expression of certain genes and these effects are mediated by specific high-affinity intracellular receptor proteins (*steroid receptors*). Binding of a steroid to its cognate receptor increases the affinity of the receptor for specific binding sites on the chromosome – DNA sequences known as *steroid response elements* – which usually occur upstream of **steroid regulated genes**. Binding of the steroid-receptor complex to its response element then results in changes of gene activity. A 90 kd **heat shock protein** has been shown to associate with steroid receptors and this association is disrupted by hormone binding. It has been proposed that the 90 kd protein may regulate the activity of unliganded receptor.

Steroid receptors consist of discrete domains responsible for DNA binding, steroid binding and gene activation and repression. The DNA binding region is rich in basic amino acid residues and cysteines whose position is highly conserved in evolution. The conserved cysteines fold into two finger-like structures which each co-ordinate a zinc atom, and are involved in DNA binding. Target sequence specificity is determined by the DNA binding region. The hormone binding region is relatively large and contains a high proportion of hydrophobic residues. Separable regions which are required for activation or repression of target genes, have been identified in various receptors. Deletion of the *transactivation domain*, for example, impairs the ability of the receptor to stimulate transcription.

onies of Coelenterata and Entoprocta from which new individuals or colonies may arise. adj. *stolonate.*

-stoma Suffix from Gk. *stōma*, mouth, applied esp. in zoological nomenclature. pl. *stomata.*

stoma A small aperture. A pore in the epidermis of a leaf or stem etc. of a vascular plant, of variable aperture, surrounded and controlled by two **guard cells** and providing regulated gas exchange between the tissues and the atmosphere. pl. *stomata, stomates.* Also *stomate.*

stomach In Vertebrates, the saclike portion

of the alimentary canal intervening between the oesophagus and the intestine. The term is loosely applied in Invertebrates to any saclike expansion of the gut behind the oesophagus. adj. *stomachic.*

stomach insecticide One acting on ingestion and applicable only to insects which eat as distinct from sucking insects which draw food in liquid form from host plant or animal; may be used on foliage against leafeating insects, or as poison bait ingredient against locusts etc. e.g. lead arsenate, DDT, Gammexane.

stomatal complex A stoma with its guard

steroid regulated genes

Steroid hormones affect the physiology of particular target tissues by modulating the expression of specific sets of genes (*steroid regulated genes*). The activity of steroid responsive genes may be regulated either by the direct effect on the rate of transcription initiation by RNA polymerase II or by other less-well characterized mechanisms, such as the selective stabilization of certain mRNAs. The rate of transcription of particular steroid regulated genes may be increased or depressed by steroid. In both cases the steroid-receptor complex recognizes specific DNA sequences upstream or within the gene – these are known as *steroid response elements*.

Steroid response elements have the properties of inducible transcriptional **enhancers**: they can act independently of their orientation in relation to the transcription site and they can also act at varying distances upstream or downstream of the mRNA start site. Further, they can stimulate transcription from heterologous promoters. Like other enhancers steroid response elements interact with sequence-specific DNA binding proteins (*steroid receptors*). In this case the activity of the enhancer binding factor is induced by interaction with a *ligand*, the steroid. Comparison of the DNA sequences of various steroid response elements has revealed that they contain one or more copies of a short imperfect inverted repeat sequence and that as little as 15 bp containing one inverted repeat can act as a steroid response element. Three different classes of steroid hormone – the androgens, glucocorticoids and progestins – are known to act through similar 15 bp DNA sequences. The response elements for oestrogen and ecdysone (an insect steroid hormone) are related but distinct.

Upon binding to a response element the steroid-receptor complex is thought to stimulate the rate of transcription by making protein-protein contacts with other **transcription factors** or possibly RNA polymerase II (see diagram). Genes which are repressed by steroid hormones contain response elements similar to those stimulated by steroid. In some instances the negative response elements overlap binding sites for essential transcription factors and binding of a receptor may then interfere with binding or functioning of such factors, resulting in a decrease in gene activity.

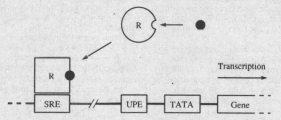

Organization of a steroid regulated gene

Upon interacting with ligand (●), a steroid receptor (R) becomes active and can then bind to a steroid response element (SRE), which may be close to or far upstream of the promotor. This binding leads to an increase in the rate of transcription, which may involve protein-protein interactions between receptor and transcription factors which recognize upstream promoter elements (UPE), the TATA box binding factor or RNA polymerase II.

cells and any subsidiary cells.

stomatal Also *stomate, stomatiferous, stomatose, stomatous*. adjs. from *stoma*, an

aperture.

stomatogastric Pertaining to the mouth and stomach; said esp. of that portion of the autonomic nervous system which controls the anterior part of the alimentary canal.

stomium A part of the wall of a fern sporangium composed of thin-walled cells where splitting begins during dehiscence.

stomodaeum That part of the alimentary canal which arises in the embryo as an anterior invagination of ectoderm. Cf. **proctodaeum, midgut.** adj. *stomodaeal.*

-stomy Suffix from Gk. *stŏma,* mouth, referring esp. to the formation of an opening by surgery.

stone cell *Brachysclereid.* A more or less isodiametric sclereid, e.g. in the fruit of the pear.

stoneworts See **Charales.**

stool A tree or shrub cut back to ground level and allowed to produce a number of new shoots, as in **coppice** management, as a method of managing fruit bushes (cf. **leg**) or to provide shoots for making cuttings etc.

stop codon Specific triplet sequences, which in mRNA do not code for an amino acid but cause protein synthesis to stop. They are UAA, UAG and UGA.

storied cork Protective cork. Protective layer of suberized cells develops around stems of woody monocotyledons (e.g. palms), in which the cells occur in radial files each file of several cells derived from a single precursor.

storied, stratified Said of a vascular cambium and the secondary xylem derived from it if the cells are arranged in more or less horizontal tiers (i.e. with the end walls more or less aligned).

strain A variant group within a species, often breeding true and maintained in culture or cultivation, with more or less distinct morphological, physiological or cultural characteristics. (The term is not used in formal taxonomy.)

strand plant A sea-shore plant growing just above the normal upper limit of the tide.

stranger anxiety A common fear infants have of unfamiliar people, onset usually the end of the first year, until the child is two years or so; related to separation anxiety in that the two tend to co-occur.

strategy, *r* and *K* In evolutionary ecology, a strategy is a suite of genetic traits which confer a selective advantage on an individual in a particular environment and which usually include growth rate, fecundity and longevity. The *r strategists* are those species best adapted to a temporary habitat; they display high growth rate, high fecundity and short generation times. *K strategists* are best adapted to stable habitats and have low growth rates and long generation times.

stratification (1) Banding seen in thick cell walls, due to presence of wall layers differing in water content, chemical composition

and physical structure. (2) Grouping of vegetation into two or more fairly well-defined layers of different height, as trees, shrubs and ground vegetation in a wood. (3) Method of breaking dormancy period of seeds by storage in moist sand, often at around 4°C. (4) The vertical structure or layering within a terrestrial or aquatic environment. (5) In statistics, the division of a population to be sampled into subsets, within each of which a sample of observations will be taken.

stratified See **storied.**

stratified epithelium A type of epithelium consisting of several layers of cells, the outer ones flattened and horny, the inner ones polygonal and protoplasmic.

stratum A layer of cells; a tissue layer. adjs. *stratiform, stratose.*

stratum corneum, -granulosum, -lucidum, -Malpighii Layers of the skin in Vertebrates.

stratum germinativum See **Malpighian layer.**

streak An elongated chlorotic or necrotic spot as a symptom of virus infection.

streaming Flowing of protoplasm in the cytoplasm, either unidirectionally (e.g. in a growing fungal hypha) or in a circulation (e.g. cyclosis).

streptococcus A Gram-positive coccus of which the individuals tend to be grouped in chains. Many forms possess species-specific capsular polysaccharides by which they can be divided into groups, which include the causative agents of scarlet fever, erysipelas and one form of mastitis. Other streptococci include *Diplococcus pneumoniae* (one cause of pneumonia). Some types occur normally in the mouth, throat and intestine.

streptomycin An antibiotic which inhibits protein synthesis in prokaryotes. Much used in the laboratory to inhibit growth of microorganisms in cell culture media.

streptostyly In Vertebrates, the condition of having the quadrate movably articulated with the squamosal. Cf. **monimostyly.**

stress Excessive and aversive environmental factors which produce physiological responses in the individual.

stress fibres Bundles of actin filaments (*microfilaments*) and associated proteins found in the cytoplasm.

stria medullaris See **habenula.**

stria, striation A faint ridge or furrow; a streak; a linear mark.

striated muscle Contractile tissue in which the **sarcomeres** are aligned, e.g. skeletal and cardiac muscle. Cf. **unstriated muscle.**

stridulating organs Those parts of the body concerned in sound production by stridulation.

stridulation Sound production by friction of one part of the body against another, as in some Insects.

Strigiformes An order of Birds containing nocturnal birds of prey with hawk-like beaks and claws. Plumage adapted for silent flight, retina containing mainly rods, eyes large and immovable, probably hunt mainly by sound. Owls.

strigose With stiff, appressed hairs or bristles.

stripe See **streak**.

striped muscle See **striated muscle**.

strobila In Scyphozoa, a scyphistoma in process of production of medusoids by transverse fission; in Cestoda, a chain of proglottides. Also called *strobile*. adjs. *strobilate*, *strobilaceous*, *strobiliferous*, *strobiloid*.

strobilate Bearing or pertaining to a strobilus or cone.

strobilization Production of strobilae; in Scyphozoa, transverse fission of a scyphistoma to form medusoids; in Cestoda, production of proglottides by budding from the back of the scolex; in some Polychaeta, reproduction by gemmation.

strobilus *Cone*. (1) The reproductive structure of most gymnosperms and some pteridophytes, consisting of a well-defined group of packed sporophores or sporophylls bearing sporangia and arranged around a central axis. (2) An angiosperm inflorescence of similar appearance to (1).

stroma (1) The matrix of the chloroplast, in which the dark reactions of photosynthesis take place. Cf. **granum**. (2) A mass of fungal tissue formed from intertwined adherent hyphae (plectenchyma), e.g. the major part of a mushroom. Cf. **sclerotium**. (3) A supporting framework, as the connective tissue framework of the ovary or testis in Mammals. pl. *stromata*. adjs. *stromate*, *stromatic*, *stromatiform*, *stromatoid*, *stromoid*, *stromatous*.

stroma lamellae Thylakoids that cross the stroma of a chloroplast, interconnecting the grana.

stromatolites Rounded, multilayered structures up to say 1 m across, found in rocks back to at least 2800 million years ago; what are, apparently, the present-day equivalents result from the growth, under special conditions, of blue-green algae.

strophiole See **caruncle**.

strophism, strophic movement Growth movement in which an organ or its stalk twists in response to a directional stimulus, e.g. the twisting of leaf bases and petioles on many horizontal branches in response to light and/or gravity resulting in the horizontal orientation of the leaf blades.

structural Of changes, aberrations, etc. in the number or arrangement of chromosomes.

See **structural gene**.

structural colours Colour effects produced by some structural modification of the surface of the integument, as the iridescent colours of some Beetles. Cf. **pigmentary colours**.

structural gene The stretch of DNA specifying the *amino acid* sequence of a *polypeptide*, as distinct from the interspersed and associated DNA, some of which is concerned with control of gene expression.

Struthioniformes An order of Birds retaining only two toes and whose feathers lack an aftershaft. They extend their rudimentary wings when running. Known from the Pliocene onwards. Ostriches.

student's t-test See **t-distribution**.

style The part of the carpel between the ovary and the stigma, often relatively long and thin.

stylet A small pointed bristlelike process.

styliform Bristle-shaped.

stylo- Prefix from Gk. *stylos*, pillar.

stylopodium (1) The swollen base of a style, as in some Umbelliferae. (2) The proximal segment of a typical pentadactyl limb; brachium or femur; upper arm or thigh.

sub- Prefix from L. *sub*, under, used in the following senses: (1) deviating slightly from, e.g. *subtypical*, not quite typical; (2) below, e.g. *subvertebral*, below the vertebral column; (3) somewhat, e.g. *subspatulate*, somewhat spatulate; (4) almost, e.g. *subthoracic*, almost thoracic in position.

subchelate In Arthropoda, having the distal joint of an appendage modified so that it will bend back and oppose the penultimate joint, like the blade and handle of a penknife, to form a prehensile weapon. Cf. **chelate**.

subclavian Passing beneath or situated under the clavicle, as the *subclavian artery*.

subclimax Vegetation held more or less permanently at some stage of a succession before the **climax**.

subconscious Syn. for **unconscious**.

subcortical Below the cortex or cortical layer; as certain cavities in Sponges.

subculture A culture of a micro-organism, tissue or organ prepared from a pre-existing culture.

subcutaneous Situated just below the skin.

subdorsal Situated just below the dorsal surface.

suberin A mixture of fatty substances, esp. of cross-linked polyesters of long chain (C_{20}) ω-hydroxy and ω-dicarboxylic aliphatic acids, in some plant cell walls, esp. cork.

suberin lamella A layer of wall material impregnated with suberin.

suberization The deposition of suberin on or in a cell wall.

subgenital Below the genital organs, as the

subgenital pouches of *Aurelia*.

subgenual organ In many Insects, chordotonal organs in the fibia adapted for perceiving vibrations of the substrate.

subimago In Ephemeroptera (Mayflies) the stage in the life history emerging from the last aquatic nymph. It has wings and moults to give the true imago. This **ecdysis** is unique among Insects, involving the casting of a delicate pellicle from the whole body, including the wings. adj. *subimaginal*.

sublimation In psychoanalytic theory, the developmental process by which forbidden impulses are gratified in socially acceptable ways; instinctual energy is displaced from its primary object to a more acceptable one, and tension associated with the gratification of repressed impulses is reduced (e.g. intellectual curiosity as a sublimation of childhood voyeurism).

subliminal perception Refers to the phenomena whereby stimuli presented below the threshold of conscious awareness may influence behaviour. Also *subception*.

sublingua In Marsupials and Lemurs, a fleshy fold beneath the tongue.

sublittoral plant A plant which grows near the sea, but not on the shore.

sublittoral zone In a lake, the lake bottom below the paralimnion, extending from the lakeward limit of rooted vegetation to the upper limit of the hypolimnion.

submaxillary Situated beneath the lower jaw.

subset Term used to classify functionally or structurally different populations of cells within a single cell type. Used esp. of T lymphocytes (helper, suppressor, cytotoxic).

subsidiary cell, accessory cell An epidermal cell associated with the guard cells of a stoma and morphologically different from the other epidermal cells.

subspecies Taxonomic subdivision of a species, with some morphological differences from the other subspecies and often with a different geographical distribution or ecology.

substantia Substance; matter.

substantive variation Variation in the constitution of an organ or organism, as opposed to variation in the number of parts; Also *qualitative variation*.

substrate (1) Reactant in a reaction catalysed by an enzyme. (2) Surface or medium on or in which an organism lives and from which it may derive nourishment.

substrate level phosphorylation The conversion of ADP to ATP which is brought about by the concomitant hydrolysis of some other **high energy phosphate compound**.

substratum See **substrate (2)**.

subtectal Lying beneath the roof, as of the skull; in some Fish, a cranial bone.

subtend To be situated immediately below, e.g. as a leaf is situated immediately below the bud in its axil.

subulate Awl-shaped, tapering from base to apex.

succession The sequence of communities (i.e. a sere) which replace one another in a given area, until a relatively stable community (i.e. the climax) is reached, which is in equilibrium with local conditions.

succise Ending below abruptly, as if cut off.

succulent (1) Juicy, having a high water content. (2) A plant with succulent stems or leaves; most are xerophytes and CAM plants, e.g. cacti or halophytes, e.g. *Salicornia*.

succus entericus A collective name for the enzymes secreted by glandular cells (*Brunner's glands* and *Lieberkühn's crypts*) in the walls of the duodenum, these including erepsin (itself a mixture), invertase, maltase, lactase, nucleotidase, nuclease, lipase and enterokinase.

sucker (1) An upward-growing shoot arising from the base of a stem or adventitiously from a root. (2) In animals, a suctorial organ adapted for adhesion or imbibition, as one of the muscular sucking disks on the tentacles of Cephalopoda; the suctorial mouth of animals like the Leech and the Lamprey; a newly born Whale; one of a large number of Fishes having a suctorial mouth or other suctorial structure, as the Remora (*Echeneis*), members of the genus *Lepadogaster* etc.

sucrose *Saccharobiose*; $C_{12}H_{22}O_{11}$, a disaccharide carbohydrate; mp 160°C; it crystallizes in large monoclinic crystals, is optically active and occurs in beet, sugar cane and many other plants. Hydrolyses to glucose and fructose. Colloq. *sugar, cane sugar*.

sucrose gradient Used in centrifugation to separate molecules on the basis of their sedimentation velocity.

suction pressure Obsolete term equivalent to minus the water potential of a cell.

suctorial Drawing in; imbibing; tending to adhere by producing a vacuum; pertaining to a sucker.

suctorial mouthparts Tubular mouth parts adapted for the imbibition of fluid nourishment; found in some Insects and many ectoparasites.

sudoriferous, sudoriparous Sweat producing; sweat-carrying.

suffructescent, suffruticose Somewhat woody; diminutively shrubby; woody at base with herbaceous stems, e.g. alpine willows.

sugar plant Plant in which carbohydrate is stored as sugar. Cf. **starch plant**.

sulcus A groove or furrow, as one of the grooves on the surface of the cerebrum in

Mammals; in Dinoflagellata, a longitudinal groove in which a flagellum lies; in Anthozoa, the 'ventral' siphonoglyph.

sulphur bacteria Bacteria which live in situations where oxygen is scarce or absent, and which act upon compounds containing sulphur, liberating the element. Occur in two families of true bacteria, the *Thiorhodaceae* (purple sulphur bacteria) and the *Thiobacteriaceae* (colourless sulphur bacteria).

summer annual A plant which completes its life cycle over a few weeks in the summer, surviving the winter as seed. Cf. **winter annual**.

summer egg In many freshwater animals, a thin-shelled, rapidly developing egg laid during the warm season. Cf. **winter egg**.

summer wood See **late wood**.

sun plant A plant adapted to living at high light intensities. Cf. **shade plant**.

superciliary Pertaining to, or situated near, the eyebrows; above the orbit.

superego In psychoanalytic theory, that part of the ego which has incorporated the moral standards of one's parents; the function of self observation and self criticism are associated with it. It differs from the notion of *conscience* in that its activities are often unconscious and often at odds with the individual's conscious values.

superior Placed above something else; higher, upper (as the *superior* rectus muscle of the eyeball). Specifically, an ovary in a flower that is **hypogynous** or **perigynous**.

superior vena cava See **precaval vein**.

supernormal stimulus A stimulus that surpasses a natural stimulus in its ability to evoke a response.

supernumary chromosomes Same as **B-chromosomes**.

superovulation Hormone-induced excess **ovulation**. See **insemination**.

superoxide anion O'_2. Oxygen molecule that carries an extra unpaired electron, and is therefore a free radical. Generated in neutrophil leucocytes and mononuclear phagocytes when activated, e.g. by ingestion of particles or immune complexes. O'_2 is highly reactive and toxic. It may be further reduced to H_2O_2 or, when two radicals interact, one is oxidized and one reduced in a dismutation reaction to form O_2 and H_2O_2. This reaction is catalysed by *superoxide dismutase*, an enzyme present in phagocytic cells. These substances are important in the microbicidal activity of the cells.

superstitious behaviour in animals Refers to behaviour that is produced by the joint action of **reinforcement** and accident; certain acts which happen to coincide with reinforcement will tend to increase; these are often of a bizarre and fixed nature.

super-, supra Prefixes from L. *super*, over, above.

supination In some higher Vertebrates, movement of the hand and forearm by which the palm of the hand is turned upwards and the radius and ulna are brought parallel to one another. Cf. **pronation**. adj. *supinate*.

supinator A muscle effecting supination.

supplemental, supplementary Additional; extra; supernumerary, as (in some Foraminifera) *supplemental* skeleton, a deposit of calcium carbonate outside the primary shell.

suppression The process by which a *mutant* phenotype is restored to normal by a mutation at another locus. Cf. **reversion**.

suppression Absence of some organ or structure normally present. adj. *suppressed*.

suppressor cell A lymphoid cell capable of suppressing antibody production or a specific cell mediated response made by other cells. Suppression may be antigen-specific or non-specific.

suppressor mutation A base change which suppresses the effect of mutations elsewhere. Thus a base change at the anticodon site of a tRNA can suppress lethal mutations in genes, which would otherwise result in chain termination or the insertion of an unacceptable peptide into a protein.

suppressor T cell factor Soluble product of a suppressor T lymphocyte responsible for suppressing other lymphocyte functions.

supradorsal On the back; above the dorsal surface; a dorsal intercalary element of the vertebral column.

supra-occipital A median dorsal cartilage bone of the Vertebrate skull forming the roof of the brain case posteriorly.

suprarenal Situated above the kidneys.

suprarenal body, gland In higher Vertebrates it is one of the endocrine glands lying close to the kidney and releasing into the blood, secretions having important effects on the metabolism of the body; *adrenal gland*. See **adrenal cortex**, **adrenal medulla**.

survival curve One showing the percentage of organisms surviving at different times after they have been subjected to large radiation dose. Less often, one showing percentage of survivals at given time against size of dose.

survivorship curve The number or percentage of an original population surviving, plotted against time, giving an indication of the mortality rate at different ages.

suspension culture A method of culturing large quantities of cells which are kept in vessels continuously stirred and aerated. Sterile nutrient media can be added and spent media removed. Used, e.g. for producing cell products like *interferon* or *anti-*

bodies. See **plant cell culture**.

suspensor A file or files of cells that develop from the proembryo of a seed plant and anchors the embryo in the embryo sac and pushes it into the endosperm.

suspensorium In Vertebrates, the apparatus by which the jaws are attached to the cranium.

suspensory Pertaining to the suspensorium; serving for support or suspension.

suture The line at the junction of fused parts or a line of weakness along which splitting may occur, as in a dehiscent fruit; as the line of junction of adjacent chambers of a Nautiloid shell; a synarthrosis or immovable articulation between bones, e.g. the bones of the cranium; junctions of exoskeletal cuticular plates in Insects. adj. *sutural*.

Sv Abbrev for *Sievert*.

SV 40 A small virus normally infecting monkey cells, either causing a lytic infection or being integrated into the host chromosome. It has a circular chromosome which has been fully sequenced, and vectors derived from it are used to transfer inserted DNA into mammalian cells.

swarm A large number of small animals in movement together; esp. a number of Bees emigrating from one colony to establish another under the guidance of a queen.

swarm cell (1) Flagellated naked cell in Myxomycetes, interconvertible with myxamoeba, capable of encysting and of acting as an isogamete. (2) See **swarmer**.

swarmer (1) Flagellated reproductive cell, esp. a zoospore. (2) See **swarm cell**.

swim bladder See **air bladder**.

swimmerets In some Crustacea, paired biramous abdominal appendages used in part for swimming.

switch plant Plant with small, scale-like or fugacious leaves and long, thin, photosynthetic stems, e.g. Ephedra, Cytisus.

switch region Sequences of DNA 5′ to each segment of DNA which encodes an **immunoglobulin heavy chain** constant region (except the delta segment). Recombination between switch regions repositions the variable (V-D-J) portion of an H-chain gene next to a different constant region segment, resulting in a switch in synthesis from one **isotype** to another. The commonest switches are from IgM to an IgG or IgA isotype.

Sylvian aqueduct In Vertebrates, the cavity of the mesencephalon.

Sylvian fissure In Mammals, a deep lateral fissure of the cerebrum.

sym- See **syn-**.

symbiosis An intimate partnership between two organisms (*symbionts*), in which the mutual advantages normally outweigh the disadvantages. See **mutualism, social symbiosis**.

symbol Words, objects, events which represent or refer to something else, established by convention. In psychoanalytic theory, a symbol's referent is unconscious, and the meaning is hidden from the individual's unconscious awareness (e.g. in dreams).

symmetrical See **actinomorphic**.

symmetry (1) The method of arrangement of the constituent parts of the animal body. (2) In higher animals, the disposition of such organs as show bilateral or radial symmetry.

sympathetic nervous system In some Invertebrates (Crustaceans and Insects), a part of the nervous system supplying the alimentary system, heart, and reproductive organs and spiracles. In Vertebrates, a subdivision of the autonomic nervous system, also known as the *thoracicolumbar system*. The action of these nerves tends to increase activity, speed the heart and circulation, and slow digestive processes. Cf. **parasympathetic nervous system**.

sympathetic ophthalmia An injury which perforates one eye may be followed by inflammatory disease in the sound eye, characterized by lymphocyte infiltration and granuloma formation, esp. in the uveal tract. This is postulated as due to development of cell mediated immunity against uveal pigment or other antigen liberated from the damaged eye, which is normally sequestered away from the immune system.

sympathomimetics A class of drug which mimics the stimulation of the sympathetic nervous system to produce **tachycardia** and increase output from the heart. Some members can increase heart output without tachycardia. Others mimic the β_2-sympathetic stimulation to produce bronchodilatation and vasodilatation.

sympatric Said of two species or populations having common or overlapping geographical distribution(s). Cf. **allopatric**.

sympetalous Gamopetalous. See **gamopetaly**.

symphysis Union of bones in the middle line of the body, by fusion, ligament or cartilage, as the mandibular *symphysis*, the pubic *symphysis*; growing together or coalescence of parts, as acrodont teeth with the jaw; the point of junction of two structures; chiasma; commissure. adj. *symphysial*.

symplast The continuum of protoplasts, linked by plasmodesmata and bounded by the plasmalemma. adj. *symplastic*, as in **symplastic growth**.

symplastic growth *Co-ordinated growth*. Type of growth of a tissue, in which touching walls grow equally so that cell contacts persist. Cf. **intrusive growth, sliding growth**.

sympodial growth *Definite growth, determinate growth*. Pattern of growth in which, after a period of extension, a shoot ceases to grow and one or more of the lateral buds next to the apical bud grow out and repeat the pattern. Cf. **monopodial growth**. See **cymose inflorescence**.

sympodium A branch system that shows **sympodial growth**.

synandrium A group of united anthers or microsporangia.

synandrous Having the stamens united to one another.

synangium A number of sporangia fused into a single structure.

synapsid In the skull of Reptiles, the condition when there is one temporal vacuity, this being low behind the eye, with the post-orbital and squamosal meeting above. Found in Pelycosauria and mammal-like Reptiles. Cf. **diapsid**.

Synapsida A subclass of Reptiles, often mammal-like, with a single lateral temporal vacuity, primitively lying below the post-orbital and the squamosal. The brain case was high, and the inner ear low down. Teeth in more advanced forms were heterodont, the lower jaw was flattened from side to side, and the dentary was relatively large. They had both a coracoid and pre-coracoid. Permian to Jurassic.

synapsis Pairing of strictly homologous regions of homologous chromosomes during meiosis.

synaptic vesicles Structures about 50 nm in diameter found in presynaptic nerve terminals and concerned with the storage of the chemical transmitters.

synaptonemal complex Ladder-like structure of DNA and protein observed to lie between the synapsed homologues of a **pachytene** bivalent in first meiotic prophase. Essential for crossing over and chiasma formation to occur.

synarthrosis An immovable articulation, esp. an immovable junction between bones. Cf. **amphiarthrosis**, **diarthrosis**.

syncarpous Said of a **gynoecium** consisting of 2 or more fused carpels. Cf. **apocarpous**.

syncaryon See **synkaryon**.

synchondrosis Connection of two bones by cartilage, usually with little possibility of relative movement.

syncytium Tissue containing many nuclei, which is not divided into separate compartments by cell membranes. adj. *syncytial*.

syndactyl Showing fusion of two or more digits, as some Birds. n. *syndactylism*.

syndesmochorial placenta A chorioallantoic placenta in which the uterine epithelium disappears so that the chorion is in contact with the endometrium or glandular epithelium of the uterus. Usually cotyledonary, e.g. sheep.

syndesmosis Connection of two bones by a ligament, usually with little possibility of relative movement.

syndrome A concurrence of several symptoms or signs in a disease which are characteristic of it, but do not in themselves constitute a disease and may be associated with several conditions; a set of concurrent symptoms or signs.

synecology The study of relationships between communities and their environment. Cf. **autecology**.

synergic, synergetic Working together; said of muscles which co-operate to produce a particular kind of movement.

synergid Either of the two nuclei that with the egg nucleus constitute the egg apparatus at the micropylar end of the embryo sac of an angiosperm.

synergism (1) The condition in which the result of the combined action of two or more agents, e.g. two growth substances, is greater than the sum of their separate, individual actions. (2) A type of **social facilitation** in which the nearby presence of another organism enhances the efficiency or intensity of a physiological process or behaviour pattern in an individual.

syngamy Sexual reproduction; fusion of gametes.

syngeneic Genetically identical. Usually applied to grafts made or cells transferred within an inbred strain.

syngenesis Lateral fusion of plant members, as the anthers, which unite laterally to form a hollow tube round the style in the Compositae.

syngnathous Of certain Fish, having the jaws fused to form a tubular structure.

synkaryon A pair of nuclei in close association in a fungal hypha, dividing together to give the same close association. In animals a zygote nucleus resulting from the fusion of two pronuclei.

synosteosis See **ankylosis**.

synovia In Vertebrates *glair*-like lubricating fluid, occurring typically within tendon sheaths and the capsular ligaments surrounding movable joints.

synovial membrane The delicate connective tissue layer which lines a tendon sheath or a capsular ligament, and is responsible for the secretion of the **synovia**.

synsacrum In Birds, part of the pelvic girdle formed by the fusion of some of the dorsal and caudal vertebrae with the sacral vertebrae. See **sacrum**.

syn-, sym- Prefixes from Gk. *syn.* with, generally signifying fusion or combination.

syntechnic Said of unrelated forms showing resemblance due to environmental factors; convergent.

syntenosis Union of bones by means of

tendons, as in the phalanges of the digits.

synthetic oligonucleotide Defined DNA sequence chemically polymerized *in vitro*.

synusia A group of plants with similar life form and of the same or unrelated species, occupying a similar habitat, e.g. woodland herbs.

syrinx The vocal organs in Birds, situated at the posterior end of the trachea. pl. *syringes*. adj. *syringeal*.

systaltic Alternately contracting and dilating; pulsatory, as the movements of the heart. Cf. **peristaltic**. n. *systalsis*.

system (1) Tissues of the same histological structure, e.g. the osseous system. (2) Tissues and organs uniting in the performance of the same function, e.g. the digestive system. (3) A method or scheme of classification as found in the *Linnaean system*. (4) A systematic treatise on the animal or plant kingdom, or on any part of either. adj. *systematic*.

systematic Affecting the whole organism; (1) an infection in which the pathogen has spread throughout the host; (2) an insecticide, fungicide etc. which following local application, spreads throughout a plant.

systematics The branch of biology which deals with classification and nomenclature.

systemic Pertaining to the body as a whole, not localized, as the *systemic* circulation.

systemic arch In Vertebrates, the main vessel or vessels carrying blood from the heart to the body as a whole.

systemic lupus erythematosus A disease of humans which is characterized by widespread focal degeneration of connective tissue and disseminated lesions in many tissues including skin, joints, kidneys, pleura, peripheral vessels, peripheral nervous system and transient abnormalities of the central nervous system. It may follow the administration of drugs or other antigenic substances, but the initiating factor is usually unknown. Numerous autoantibodies are present in the blood, of which the most constant are anti-nuclear antibodies. The lesions are mainly the result of the deposition of immune complexes.

systole Rhythmical contraction, as of the heart, or of a contractile vacuole. Cf. **diastole**.

T

2,4,5,-T *2,4,5,-trichlorophenoxyethanoic acid.* Widely used selective herbicide. Thought to have undesirable human side effects in part caused by impurities.

T6 marker chromosome A mouse chromosome originally derived from an irradiated male. It is about half the length of the shortest pair of **autosomes** and has a clearly detectable constriction near the centromere, so being easily recognizable in stained preparation of cells arrested in metaphase. The T6 chromosome has been introduced into an inbred strain of mice (see **recombinant inbred strain**) so as to replace the normal strain. The two strains are histocompatible so that the cells can be grafted from one to another without causing graft rejection. This has made it possible to follow the fate of particular cells or cell lines after transfer, and was the first technique to be used for this purpose.

tabescent Shrivelling.

taboo An anthropological term for the prohibition of some class of people, objects or acts, because they violate the fundamental beliefs and values of a culture.

tabular Having the form of a tablet or slab.

TAB vaccine Vaccine used in the prophylaxis of enteric fevers. Contains heat-killed *Salmonella typhi* and *S. paratyphi A* and *B* in the smooth specific phase and possessing their normal complement of O antigens. Usually preserved with phenol, though other methods (e.g. acetone) are also used. Repeated doses are needed and the protection produced is not complete. Because the O antigens contain lipopolysaccharides some fever and local inflammation commonly follow injection.

tachitoscope A mechanical instrument capable of flashing visual displays on a screen for very short periods of time; used in perceptual research.

tachygenesis Accelerated development with elimination of certain embryonic stages, as in some Caecilians, in which the free-living tadpole stage is suppressed. adj. *tachygenetic.*

tactic movement See **taxis.**

tactile Pertaining to the sense of touch.

tactile bristle A stiff hair which transmits a contact stimulus.

taenia A ribbon-shaped structure, such as the *taenia pontis,* a bundle of nerve fibres in the hind brain of Mammals.

taeniasis The state of infestation of the human body with tapeworms (*Taenia*), which as adults may inhabit the intestine and as larvae, the muscles and other parts of the body.

T-agglutinin Antibody present in the blood of normal persons which agglutinates erythrocytes which have been incubated with **neuraminidase** or acted on by bacteria which produce the enzyme. This reveals the T-antigen present on erythrocytes but not normally available to react with antibody. The stimulus to production of the antibody may be effete erythrocytes.

tagma A distinct region of the body of a metameric animal, formed by the grouping or fusion of somites, as the thorax of an Insect. pl. *tagmata.*

tagmosis In a metameric animal, the grouping or fusion of somites to form definite regions (*tagmata*).

tail See **cauda.**

talon A sharp-hooked claw, as that of a bird of prey.

talus Syn. for *astragalus.* pl. *tali.*

tangential longitudinal section A section cut longitudinally along a more or less cylindrical organ parallel to a tangent at its surface. Abbrev. *TLS.*

tanning In newly-formed cuticle of terrestrial Arthropods, the process in which the spaces between the chitin micelles are filled with sclerotin, which consists of protein molecules linked together or tanned by quinones. This makes the cuticle tougher and darker.

tannin sac A cell containing much tannin.

T-antigens A group of surface antigens defining subpopulations of human T lymphocytes. Many of these have been defined using monoclonal antibodies and are classified in the **CD system.** Corresponding antigens have been identified on mouse, rat and bovine T lymphocytes, and will presumably be found in any species studied sufficiently intensively.

tapetum (1) A layer of cells in a sporangium of a vascular plant surrounding the spore mother cells, becoming absorbed as the spores mature. (2) In the eyes of certain nightflying Insects, a reflecting structure; in some Vertebrates, a reflecting layer of the retinal side of the choroid; in the Vertebrate brain, a tract of fibres in the corpus callosum.

tapeworm Parasitic worms of the class Cestoda, generally taking the form of a scolex with hooks and/or suckers for attachment to the host and a chain of individual proglottids in successive stages of development. Species infecting Man are *Hymenolepis nana* or Dwarf Tapeworm, *Taenia solium* from infected pork, *Diphyllobothrium latum* from infected Fish, and *Echinococcus granulosus* which spends the larval stage only in Man, causing hydatid cysts; the adult worm inhabiting Dogs.

taproot The first (primary) root of a plant developed directly from the radicle. Sometimes develops as a fleshy storage organ, e.g. carrot.

taproot system Root system, characteristic of dicotyledons and conifers, based on a tap root with laterals of various orders. Cf. **fibrous root system**.

Tardigrada A subphylum of minute Arthropods with suctorial mouthparts and four pairs of stumpy clawed legs; common forms, of wide distribution, are found among moss and debris in ditches and gutters, and on tree trunks, and can survive desiccation.

target cell (1) Antigen-bearing cell which is the target of attack by lymphocytes or by specific antibody. (2) In haematology, used to describe an abnormally shaped and unusually thin red cell with central stained area seen in blood films, esp. in certain disorders of haemoglobin formation.

tarsus (1) In Vertebrates, an elongate plate of dense connective tissue which supports the eyelid. (2) In Insects, Myriapods and some Arachnida (as Mites), the terminal part of the leg, consisting typically of five joints. (3) In land Vertebrates, the basal podial region of the hind limb; the ankle. adj. *tarsal.*

taste bud In Vertebrates, an aggregation of superficial sensory cells subserving the sense of taste; in higher forms, usually on the tongue.

TAT Abbrev. for *Thematic Apperception Test.*

taungya Form of land use in the humid tropics, in which villagers are given the right to farm on good forest soils in exchange for their services in tending young trees on the same land. A practical form of **agroforestry**.

taxis Orientation with respect to environmental stimuli; often combined with locomotion, so that the animal moves towards, or away, or at a fixed angle to source. There are various classifications of types of taxes, e.g. *positive taxis* when the organism moves towards the stimulus and the reverse in *negative taxis.* See **kinesis**. pl. *taxes.*

taxon Any group of organisms to which any rank of taxonomic name is applied.

taxonomic series The range of extant living organisms, ranging from the simplest to the most complex forms.

taxonomy The science of classification as applied to living organisms, including study of means of formation of species etc.

T-bands See **banding techniques**.

T cell Abbrev. for *T lymphocyte.*

T cell growth factor See **interleukin-2**. Abbrev. *TCGF.*

T cell leukaemia viruses Group of **retroviruses** which infect T lymphocytes and cause leukaemias. Some cause malignant transformation of T lymphocytes (e.g. HTLV-1 or feline leukaemia virus) whereas others cause **acquired immunodeficiency syndrome** (HTLV-3, now renamed HIV for human immunodeficiency virus). Similar viruses are present in monkeys.

T cell replacing factor A soluble factor derived from helper T lymphocytes which can replace the presence of T lymphocytes in stimulating antibody production by B lymphocytes which have been activated by antigen. Synonymous with B cell differentiation factors of which one is identified as interleukin-4.

TCGF Abbrev. for *T Cell Growth Factor.*

t-distribution The sampling distribution of the mean of a set of observations from a normal distribution with unknown variance. The (central) t-distribution has one parameter, the degrees of freedom, and describes the sampling distribution of the deviation of the sample mean from the population mean.

TdT Abbrev. for *Terminal desoxynucleotidyl Transferase.*

tear gland See **lacrimal gland**.

teats In female Mammals, paired projections from the skin on which the lactiferous tubules of the mammary glands open.

tectorial Covering as the tectorial membrane (*membrana tectoria*) of Corti's organ.

tectrices In Birds, small feathers covering the bases of the remiges and filling up the gaps between them. Also *auriculars.*

tectum A covering or roofing structure, as the *tectum synoticum*, part of the roof of the cartilaginous skull which connects the two auditory capsules.

tegulated Composed of or covered by plates overlapping like tiles.

tele-. Prefix from Gk. *tele*, afar, at a distance.

teleceptor, telereceptor A sense organ which responds to stimuli of remote origin.

telegony The supposed influence of a male with which a female has previously been mated, as evinced in offspring subsequently borne by that female to another mate.

telencephalon One of the two divisions of the Vertebrate forebrain or prosencephalon (the other being the *diencephalon*), comprising the cerebral hemispheres (with the cerebral cortex or pallium and the corpus striatum), the olfactory lobes and the olfactory bulbs.

teleo- Prefix from Gk. *teleios*, perfect.

teleology The interpretation of animal or plant structures in terms of purpose and utility. adj. *teleological.*

teleonomy Impression of purpose arising from adaptation through natural selection.

Teleostei An infraorder of Osteichthyes, including Fish with a wide diversity of form and physiological adaptation. Gills fully filamentous, tail externally (and in many cases internally) homocercal, endoskeleton completely ossified, fins completely fanlike with no trace of an axis. Bony Fishes.

telepathy Communication between individuals that takes place independently of all known sensory channels.

telo- Prefix from Gk. *telos*, end.

telocentric Having the **centromere** at one end of the chromosome.

telolecithal A type of egg which is large in size, with yolk constituting most of the volume of the cell, and with the relatively small amount of cytoplasm concentrated at one pole. Found in Sharks, Skates, Reptiles and Birds. Cf. **mesolecithal, oligolecithal**.

telome Hypothetical morphological unit of primitive vascular plants; an ultimate branch of an axis that repeatedly branches dichotomously.

telomere The structure which terminates the arm of a chromosome.

telome theory Proposal that the shoots of modern land plants have evolved from repeatedly dichotomously branched axes.

telomorph The sexual or perfect stage of a fungus. Cf. **anamorph**.

telophase Final phase of mitosis with cytoplasmic division; the period of reconstruction of nuclei which follows the separation of the daughter chromosomes in mitosis.

telson The post-segmental region of the abdomen in some Crustacea and Chelicerata.

TEM See **transmission electron microscope**.

temperature coefficient The ratio of the rate of progress of any reaction or process, at a given temperature, to the rate at a temperature 10°C lower. Also Q_{10}.

temperature sensitive mutant A mutant organism able to grow at one temperature, the *permissive temperature*, but unable to do so at another, which may be higher or lower than the permissive. This class of mutant has been particularly important in the analysis of mutants affecting vital functions like cell division.

temporal A cartilage bone of the Mammalian skull formed by the fusion of the petrosal with the squamosal.

temporal summation The phenomenon in which the summational effect of a sub-threshold stimulus, which is presented over an extended period of time, may produce a response.

temporal vacuities, openings In Reptiles, openings in the skull, varying in number (none, one or two) and position, and used in classification. The various conditions found in different groups are known as *anapsid, synapsid, parapsid, euryapsid* and *diapsid*.

tendency A general term referring to some measure of the probability that a behaviour will occur, without specifying the nature of the underlying causal factors.

tendinous See **tendon**.

tendon A cord, band or sheet of fibrous tissue by which a muscle is attached to a skeletal structure, or to another muscle. adj. *tendinous*.

tendril A slender, simple or branched, elongated organ used in climbing, at first soft and flexible, later becoming stiff and hard. May be a modified stem, leaf, leaflet or inflorescence.

tenia, teniasis See **taenia, taeniasis**.

tension wood The **reaction wood** of dicotyledons, formed e.g. on the upper side of horizontal branches and characterized by a lower than normal lignin content.

tensor A muscle which stretches or tightens a part of the body without changing the relative position or direction of the axis of the part.

tentacle An elongate, slender, flexible organ, usually anterior, fulfilling a variety of functions in different forms, e.g. feeling, grasping, holding and sometimes locomotion. Also *tentaculum*. adjs. *tentacular, tentaculiferous, tentaculiform*.

tentorium (1) In the Mammalian brain, a strong transverse fold of the dura mater, lying between the cerebrum and the cerebellum. (2) In Insects, the endoskeleton of the head.

tepal One of the members of a perianth which is not clearly differentiated into a calyx and a corolla.

teratogen An agent which raises the incidence of congenital malformations.

teratology The study of monstrosities, as an aid to the understanding of normal development. (Gk. *teras*, gen. *teratos*, a wonder.) See **teratoma**

teratoma A tumour in the body consisting of tissues believed to be derived from the three germ layers (ectoderm, mesoderm and endoderm). May occur in a variety of locations but commonly in the testis and mediastinum. If it contains a predominance of ectodermal elements it is called *dermoid cysts*.

terebrate Possessing a boring organ or a sting.

terete Rounded, more or less cylindrical and neither ridged, grooved nor angled.

tergum The dorsal part of a somite in Arthropoda; one of the plates of the carapace in Cirripedia. adj. *tergal*.

terminal deoxynucleotidyl transferase An enzyme found in pre-T and pre-B lymphocytes and cortical thymocytes but absent from their progeny. Function uncertain, but perhaps involved in the initial arrangement

---

tetanus, toxin and antitoxin

Lockjaw. A disease due to infection with the tetanus bacillus, *Clostridium tetani,* which secretes a toxin that causes the symptoms and signs of the disease, viz., painful tonic spasms of the muscles. These usually begin in the jaw and then spread to other parts.

The *toxin* is produced by *Clostridium tetani* and is a neurotoxin that blocks synaptic transmission in the spinal cord and also blocks neuromuscular transmission. Tetanus toxin binds to a glycolipid, *disialosyl ganglioside* which is particularly rich in the membranes of nerve cells. After inactivation by treatment with formaldehyde to form tetanus *toxoid,* the latter is used for prophylactic immunization against tetanus.

Tetanus antitoxin is an antibody to the toxin, usually prepared in horses which have been hyper-immunized against the exotoxin of *Clostridium tetani.* Used for the prevention of tetanus in humans and animals following possible contamination of wounds. Repeated use is dangerous because of the likelihood that the subject will have been primed against horse serum proteins and may develop **anaphylactic shock** or **serum sickness.** Even a single dose may cause serum sickness. These dangers are avoided by using antiserum or immunoglobulin from humans immunized against tetanus, available in some countries. Since the production of tetanus toxin at the site of infection takes place only slowly, if a person has already been immunized against tetanus a *booster injection* of tetanus toxoid will elicit a secondary response rapidly enough to prevent tetanus from developing.

Tetanus is also the name for the state of prolonged contraction which can be induced in a muscle by a rapid succession of stimuli.

---

of genes for the antigen receptors. The presence of the enzyme is used as a marker for cells which contain it. Abbrev. *TdT.*

termitarium A mound of earth built and inhabited by termites and containing an elaborate system of passages and chambers.

ternate Arranged in threes; esp. a compound leaf with three leaflets.

terpenoids A group of plant **secondary metabolites** based on one to four or more isoprene (C_5) units, includes many essential oils, the gibberellins, carotenoids, plastoquinone, rubber.

territory Areas defended against other individuals, usually of the same species. Territoriality exists among many different species and may involve aggressive encounters, but it is often regulated by less overt behaviour. Serves a wide range of functions: feeding, mating etc. Not all species are territorial and many occupy a **home range** which is not actively defended.

tertiary structure The 3-dimensional configuration of polymers which is a stable folding of the sequence of units (bases or peptides) along the polymer, i.e. of their secondary structure.

tertiary wall, tertiary thickening A deposit of wall material on the inner surface, next to the lumen, of a secondary wall, often in the form of helical strips, as in the

tracheids of the yew.

testa, test (1) The seed coat, several layers of cells in thickness, derived from the integuments of the ovule. (2) A hard external covering, usually calcareous, siliceous, chitinous, fibrous or membranous; an exoskeleton; a shell; a lorica. adjs. *testaceous, testacean.*

testis, testicle A male gonad or reproductive gland responsible for the production of male germ cells or sperms. adj. *testicular.* pl. *testes.*

tetanic contraction See **tetanus**.

tetanus See panel.

tetanus antitoxin See panel.

tetanus toxin See panel.

tetra Prefix from Gk. *tetra-,* four, e.g. *tetracyte.*

tetrad The four haploid cells formed at the end of **meiosis**. The term was formerly used for the four chromatids making up a chromosome pair at the first division of meiosis.

tetradactyl Having four digits.

tetrad analysis The genetic analysis of tetrads in studies of mapping, recombination etc.

tetragonous A stem etc. having four angles and four convex faces.

tetramerous Having 4 parts; arranged in 4s; arranged in multiples of 4.

tetramethyl rhodamine isothiocyanate

A red fluorescent dye used in immuno-fluorescence techniques. In conjunction with **FITC** it allows two colours to be used together. Abbrev. *TRITC*.

tetraparental chimera A chimera (usually mouse) resulting from the artificially induced fusion of two blastocysts at the 4 or 8 cell stage. The resulting animal contains cells from both parents in all its tissues, and is an example of the maintenance of mutual immunological tolerance.

tetraploid Possessing four sets of chromosomes, each chromosome of a set being represented 4 times. Cf. **diploid**, **haploid**.

tetrapod Having four feet.

tetrapterous Having four wings.

tetrarch Said of a stele having 4 strands of protoxylem.

tetrasomic A tetraploid nucleus (or organism) having 1 chromosome 4 times over, the others in duplicate.

tetrasporophyte The typically diploid phase of the red algal life cycle, developing from carpospores and producing haploid tetraspores.

thalamus (1) Same as the **receptacle** of a flower. (2) In the Vertebrate brain, the larger, more ventral part of the dorsal zone of the thalamencephalon.

thalassaemia A group of inherited anaemias in which there is a defect in the alpha or beta chains of haemoglobin, *alpha-, beta-thalassaemia*. Thalassaemia *minor* is used to describe heterozygotes and thalassaemia *major* for the homozygotes.

thalasso- Prefix from Gk. *thalassa*, sea.

thalassophyte A seaweed.

thallus Plant body not differentiated into leaves, stems and roots but consisting of a single cell, a colony, a filament of cells, a mycelium or a large branching multicellular structure. The plant body of the algae, fungi and thalloid liverworts. adj. *thalloid*.

thanatoid Poisonous, deadly, lethal, as some venomous animals.

thanatosis Same as **sham death**.

Thebesian valve An auricular valve of the Mammalian heart.

theca A case or sheath covering or enclosing an organ, as the *theca vertebralis* or dura mater enclosing the spinal cord; a tendon sheath; the wall of a coral cup. adjs. *thecal, thecate*.

thecodont Having the teeth implanted in sockets in the bone which bears them.

thelytoky Parthenogenesis resulting in the production of females only.

thematic apperception test A **projective technique** in which persons are shown a set of pictures and asked to write a story about each. Abbrev. *TAT*.

Theria A subclass of the Mammals, containing the extinct Patriotheria, the Metatheria

(marsupials) and the Eutheria (placentals).

thermal death point The temperature at which an organism is killed or a virus inactivated.

thermo- Prefix from Gk. *therme*, heat.

thermocline In lakes, a region of rapidly changing temperature, found between the epilimnion and the hypolimnion.

thermogenesis Production of heat within the body.

thermolysis Loss of body heat.

thermonasty A nastic movement in response to a change in temperature, e.g. the opening and closing of Crocus flowers.

thermoperiodism The response of a plant to daily (or other) cycles of temperature.

thermophile, thermophilous, thermophilic Requiring, adapted to, or sometimes tolerating high temperatures. Cf. **thermotolerant**.

thermophyllous Having leaves only in the warmer part of the year; deciduous.

thermotolerant Able to endure high temperatures, but not growing well under such conditions.

therophyte Plant which passes the unfavourable season as seeds, and thus has no perennating vegetative buds. See **Raunkaier system**.

thiamin See **vitamin B**.

thigmo- Prefix from Gk. *thigma*, touch.

thigmocyte See **thrombocyte**.

thigmotropism Turning of an organism (or of part of it) towards or away from an object providing touch stimulus. See **haptotropism**.

third ventricle In Vertebrates, the cavity of the diencephalon, joining the two lateral ventricles of the cerebral hemispheres via the foramen of Monro, and the fourth ventricle in the medulla oblongata by the cerebral aqueduct.

thirst A state of motivation which arises primarily as a result of dehydration of body tissues.

thorax In Crustacea and Arachnida, a region of the body lying between the head and the abdomen and usually fused with the former; in Insects, one of the three primary regions of the body, lying between the head and the abdomen, and bearing in the adult three pairs of legs and the wings (if present); in some tubicolous Polychaeta, a region of the body behind the head, distinguished by the form of its segments and the nature of its appendages; in land Vertebrates, the region of the trunk between the head or neck and the abdomen which contains heart and lungs and bears the fore limbs, esp. in the higher forms, in which it is enclosed by ribs and separated from the abdomen by the diaphragm. adj. *thoracic*.

thorn A woody sharp-pointed structure;

usually restricted to those representing modified branches, as in the hawthorn. Cf. **prickle, spine**.

Thr Symbol for **threonine**.

threat behaviour A form of communication usually occurring in situations of conflict between fear and aggression; used to repel conspecifics or members of other species without undue risk or injury. Threat displays are very varied, often involving ritualized postures and expressions, as well as specialized morphological features (e.g. the rattle of the rattlesnake).

threonine *2-amino-3-hydroxybutanoic acid*. The L-isomer is a polar amino acid and constituent of proteins. Symbol Thr, short form T.

Side chain: CH₃

CH→CH₂—
HO

See **amino acid**.

thrombin The proteolytic enzyme which converts fibrinogen to fibrin resulting in blood clotting.

thrombocyte A minute greyish circular or oval body found in the blood of higher Vertebrates; it plays an important role in coagulation; a blood platelet.

thrombosis Coagulation; clotting.

thromb-, thrombo- Prefixes from Gk. *thrombos*, lump, clot.

thrombus A clot formed in a blood vessel during life and composed of thrombocytes (platelets), fibrin and blood cells.

thrum The short-styled form of such heterostyled flowers as the primrose, with the anthers visible at the top of the corolla tube. See **heterostyly**. Cf. **pin**.

Thy 1 antigen A cell surface iso-antigen of mice present on thymus derived lymphocytes in the thymus and peripheral tissues. Antibodies specific for Thy 1 are used to identify such lymphocytes. A similar antigen (Thy 2) is present on rat lymphocytes. Thy antigens are also present in central nervous tissue. Their general structure is similar to immunoglobulin peptide chains, but their function is unknown.

thylakoid A flattened, membrane-bounded sac in a chloroplast. Thylakoids may be single or associated in pairs or threes or more. See **granum, stroma lamella**.

thymic epithelial cells Those which ramify throughout the cortex and medulla of the thymus. Although similar in most respects, differences between cortical and medullary epithelial cells are detected by monoclonal antibodies. They are believed to control the maturation of lymphocyte precursors by secretion of peptide hormones (*thymosins, thymopoietin*).

thymic hypoplasia A congenital cell-mediated immunodeficiency syndrome in human infants. In Di George's syndrome there is also parathyroid hypoplasia, causing additional problems due to hypocalcaemia. Thymic hypoplasia is characterized by recurrent infections of the skin and respiratory tract, marked depletion of lymphocytes in the thymus dependent area of peripheral lymphoid tissues, and failure to express cell-mediated immunity. Antibody production and blood immunoglobulin levels are normal.

thymine *5-methyl-2,6-dioxytetrahydropyrimidine*. One of the two pyrimidine bases in DNA in which it pairs with adenine. It does not occur in RNA which has uracil instead. Its radioactive nucleotide derivative is therefore an important radiolabel for DNA synthesis.

Structure: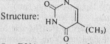

See **DNA, genetic code**.

thymocyte Any lymphocyte found within the thymus.

thymoma Rare thymic tumour. About half are associated with *myasthenia gravis*, and a few others with red cell aplasia, immunoglobulin deficiency, rheumatoid arthritis or polymyositis. There appears to be some abnormality of immune regulation.

thymopoietin Factor derived from the thymus believed to influence the maturation of T lymphocytes. It induces the appearance of T lymphocyte markers on resting lymphocytes both in vitro and after administration in vivo.

thymosins Group of peptides derived from thymus epithelial cells (not the same as **thymopoietin**) which are present in blood. They can partially restore T lymphocyte function in thymectomized animals, and induce differentiation and maturation of immature T lymphocytes.

thymus See panel.

thymus dependent antigen See panel.

thymus dependent area See panel.

thymus derived cells See panel.

thymus independent antigen See panel.

thyroid antibodies Organ specific autoantibodies found in a variety of thyroid diseases, esp. Hashimoto thyroiditis and thyrotoxicosis. The major antibodies are against thyroglobulin, against another antigen in thyroid colloid, against a microsomal antigen of thyroid acinar cells, and against the receptors for thyroid stimulating hormone (TSH). The latter may mimic TSH and cause thyroid overactivity.

thyroid gland In Vertebrates, a ductless gland originating as a median ventral outgrowth from a point well forward on the

thymus, antigens and cells

The *thymus* is an organ of major importance for the development of the immune response. It develops in the embryo from the 3rd and 4th branchial pouches, and in most mammals it consists of two lobes situated in the anterior part of the thorax lying in front of the great vessels of the heart. It consists mainly of lymphocytes arranged between a network of epithelial cells, clearly separated into a cortical and medullary area. **Dendritic interdigitating cells** and nests of macrophage-like cells containing apparently breakdown products of other cells (Hassall's corpuscles) are also present in the medulla.

It is the organ in which prothymocytes derived from bone marrow mature into functional T lymphocytes, beginning in the cortex where prothymocytes differentiate and divide several times. It is here that the genes controlling the T lymphocyte antigen receptor become rearranged from their germ line configuration, and the diversity of receptors is generated, with the elimination of most of the T lymphocytes capable of recognizing 'self' antigenic determinants. They then undergo further differentiation in the medulla, becoming separated into mature helper T lymphocytes, which recognize antigenic determinants associated with Class II MHC molecules and cytotoxic/suppressor T lymphocytes, recognizing antigenic determinants associated with Class I MHC molecules.

The great majority of the thymocytes so generated die *in situ*, leaving only an estimated 5% to become mature T lymphocytes. These leave the thymus and enter peripheral tissues, where they live an independent existence, unaffected e.g. by thymus removal. The thymus is largest during early post-natal life, and gradually declines after puberty to become largely atrophic in old age.

Thymus derived cells are those lymphocytes derived from the thymus. They are usually identified by the presence of thymus-specific surface antigens.

T-dependent antigens are those which fail to stimulate an antibody response if T lymphocytes are absent. Co-operation between B lymphocytes and helper T lymphocytes is required for the B lymphocytes responding to such antigens to differentiate into antibody secreting cells. Most proteins and complex antigens are thymus dependent.

T-independent antigens are those able to stimulate B lymphocytes to produce antibody without the co-operation of T lymphocytes, e.g. in animals lacking a thymus. Such antigens are usually polymers which are poorly digestible by macrophages and which carry a repeated array of antigenic determinants, enabling them to bind firmly to Ig receptors on B lymphocytes. They probably stimulate a subset of B lymphocytes, and do not stimulate memory cells or cell mediated immunity. Some thymus dependent antigens (including lipopolysaccharides) can also stimulate non-specific T cell help, and elicit larger responses in normal than in thymus-deprived animals. These are designated TI-1, whereas those which do not are TI-2.

Thymus dependent areas are areas in *peripheral* lymphoid organs which are predominantly occupied by T lymphocytes which circulate through them, and are depleted if the thymus fails to liberate T thymocytes into the circulation. These areas are anatomically segregated, and contain dendritic interdigitating cells with which the T lymphocytes come into close contact and which may be involved in antigen presentation to them.

floor of the pharynx. It may be a single structure, bilobed or paired, and there may be small accessory masses of thyroid tissue in other places. The gland consists of spherical follicles composed of an outer layer of cuboidal secretory cells surrounding and discharging into a central cavity. In this are found the hormones thyroxine and 3,5,3,-triiodothyronine, which are concerned with the rate of tissue metabolism, and the development of the nervous system and behaviour (deficiency causing *cretinism*), and in Amphibia, with the control of metamorphosis. Evolutionarily the thyroid originates from the endostyle of amphioxus (of the Cephalochorda) and Tunicata, and the ammocoete larva of Lampreys.

Thysanoptera An order of minute insects with asymmetrical piercing mouthparts, a large and free prothorax and tarsi having a protrusible adhesive terminal vesicle. Some are serious pests causing malformation of plants and sometimes inhibiting the development of fruit. Thrips.

tibia In land Vertebrates, the pre-axial bone of the crus; in Insects, Myriapoda and some Arachnida, the fourth joint of the leg.

ticks Blood sucking Arachnids related to mites which are vectors for a number of diseases in man and animals, e.g. **typhus, relapsing fever** and **virus encephalitis**.

tidal volume The volume of air moving in and out of the lungs of Vertebrates (and of the tracheal system of Insects) during normal (unforced) breathing; in Man about 500 ml.

tight junction Junction between epithelial cells where the membranes are in close contact, with no intervening intercellular space. Tight junctions can bind epithelial cells into sheets which permit no leakage of solutes across the sheets between the cells.

tiller A shoot that develops from an axillary or adventitious bud at the base of a stem; characteristic of grasses, including cereals.

timber line A line or zone on a mountain or at high latitudes beyond which trees fail to grow to their normal size or form. Cf. **krummholz**.

Tinamiformes An order of Birds, containing small superficially partridge-like, almost tailless birds which are essentially cursorial, but can fly clumsily for short distances, and have a keeled sternum. Tinamus.

Ti plasmid *Tumour inducing principle*. Plasmid, carried by virulent strains of the crown gall bacterium, *Agrobacterium tumefaciens*, part of which (the T-DNA) may, when the bacterium infects a plant, become transferred and incorporated into the nuclear genome of the host cells, inducing them to grow and form the characteristic galls. Ti plasmids are possible vectors for the 'genetic engineering' of dicotyledonous plants. See **crown gall**.

tissue An aggregate of similar cells forming a definite and continuous fabric, and usually having a comparable and definable function; as *epithelial tissue, nervous tissue, vascular tissue*.

tissue culture The growth of cells, including tissues and organs, outside the organism in artificial media of salts and nutrients.

Depending on the cell type, the cells may be capable of a limited number of divisions or may divide indefinitely. In plants and under appropriate conditions, cultured tissues can often be made to regenerate new plants. See **transformation (2), plant cell culture**.

tissue specific antigen Cell antigen present in a given tissue but not found in other tissues, e.g. thyroglobulin is specific to the thyroid. Auto-immune diseases have different pathological consequences according to whether auto-antibodies and/or cell mediated immunity are directed against tissue specific antigens, or whether auto-antibodies are reactive with non-tissue specific antigens such as nucleic acid or nucleoproteins common to many tissues.

tissue tensions The mutual compression and stretching, of deeper and more superficial tissues respectively, exerted by the tissues of a living plant.

tissue typing The identification of histocompatibility antigens, usually done on blood leucocytes from prospective donor and recipient prior to tissue or organ transplantation.

titre In serological reactions involving the use of serial dilutions of antiserum, describes the highest dilution at which the measured effect is detected.

T lymphocyte antigen receptor   See panel.

T lymphocyte repertoire See panel.

Tm The temperature at which DNA is half denatured. syn. *melting temperature*.

T-maze A T-shaped pathway with a starting box at the base of the T and goal boxes containing reinforcing stimuli at the ends of either or both arms. Discriminative stimuli may be placed in the arms of the T near the **choice point**.

tocopherol See **vitamin E**.

token economy A behaviour modification procedure, based on operant conditioning principles, in which patients are given artificial rewards for socially desirable behaviour; the rewards or tokens can be exchanged for desirable items (e.g. cigarettes).

tolerance See **immunological tolerance**.

tolerogen A substance capable of inducing immunological tolerance.

tomentum Covering of felted cotton hairs. adj. *tomentose*.

-tomy Suffix from Gk. *tome*, a cut.

tone The resting level of muscle contraction due to background neuromuscular activity.

tongue In Vertebrates, the movable muscular organ lying on, and attached to, the floor of the buccal cavity; it has important functions in connection with tasting, mastication, swallowing, and (in higher forms) sound production; in Invertebrates, esp. Insects,

T lymphocyte antigen receptor

T-cell receptor. Molecule present at the surface membrane of T lymphocytes capable of specifically binding antigen in association with **MHC** antigen at the surface of an antigen presenting cell. Such molecules have been isolated from cloned T lymphocyte **hybridomas** and from antigen specific T lymphocyte cell lines, and found to consist of two polypeptide chains α and β. A similar but distinct pair of polypeptide chains, γ and δ, has been found to be expressed in early thymocytes but not in mature T lymphocytes. The chains each contain a variable and a constant region, whose structure is controlled by genes with several **exons** in much the same way as those controlling **immunoglobulins**, to which they are structurally similar with considerable sequence homology, but they are clearly distinct. Each antigen molecule is tightly associated with another molecule (CD3) which is required if activation of the T lymphocyte by the antigen is to take place.

Yet another molecule (CD4 in helper cells and CD6 in cytotoxic/suppressor cells) determines whether the lymphocyte will recognize the antigen in association with MHC Class II or MHC Class I determinants respectively. See also **thymus**. Unlike B lymphocytes, T lymphocytes recognize relatively short linear peptide sequences rather than the shape of the tertiary structure of a protein molecule.

The *T lymphocyte repertoire* is the number of different antigenic determinants to which the T lymphocytes of an individual animal are capable of responding, thought to be comparable in size to the B lymphocyte repertoire. There are however certain peptide sequences which are not recognized, either because they are not recognizably different from 'self' or because they are unable to associate with a particular MHC determinant.

any conformation of the mouthparts which resembles the tongue in structure, appearance or function: proboscis; antlia; haustellum; radula; ligula; any structure which resembles the tongue.

tonicity See **tone**.

tonofilament Filaments composed of keratin found in epithelial cells.

tonoplast The membrane round a vacuole in a plant cell.

tonsils In Vertebrates, lymphoid bodies of disputed function situated at the junction of the buccal cavity and the pharynx.

tonus A state of prolonged tension in a muscle without change in length.

tooth (1) A hard projecting body with a masticatory function. In Vertebrates, a hard calcareous or horny body attached to the skeletal framework of the mouth or pharynx, used for trituration or fragmentation of food; in Invertebrates, any similar projection of chitinous or calcareous material used for mastication or trituration. (2) Any small pointed projection as on the margin of a leaf.

topotype A specimen collected in the same locality as the original type specimen of the same species.

top yeast Sorts of brewer's yeast, *Saccharomyces cerevisiae*, which accumulate at the top of the medium during fermentation and are used in brewing traditional British ales. Cf. **bottom yeast**.

torsion The preliminary twisting of the visceral hump in Gastropod larvae which results in the transfer of the pallial cavity from the posterior to the anterior face, as distinct from the secondary or spiral twisting of the hump exemplified by the spiral form of the shell.

torus (1) The thickened central part of the **pit membrane** of the bordered pits of many gymnosperm tracheids. It seems to act as a valve closing a pit to prevent the threatened spread of an embolism. (2) A ridge or fold, as in Polychaeta; a ridge bearing uncini.

toti- Prefix from L. *totus*, all, whole.

totipotency The ability, possessed by most living plant cells, differentiated or not, to regenerate a plant when isolated and cultured in a suitable medium. See **plant cell culture**. In animals it is confined to the capability of some embryo cells to develop into any organ or a complete embryo. adj. *totipotent*.

touchwood Wood much decayed as a result of fungal attack; it crumbles readily, and when dry is easily ignited by a spark.

toxicology The branch of medical science which deals with the nature and effects of poisons.

toxin A poisonous substance of biological origin.

toxoid Bacterial exotoxin which has been treated (usually with formaldehyde) so that it has lost its toxic properties but retains its ability to stimulate an immune response against the toxin.

trabecula A rod-like structure, e.g. of cell wall material across the lumen of a cell, or of a cell or cells across some larger cavity.

trace element See **micronutrient**.

trachea An air tube of the respiratory system in certain Arthropoda, as Insects; in air-breathing Vertebrates, the windpipe leading from the glottis to the lungs.

tracheal gills In some aquatic Insect larvae, filiform or lamellate respiratory outgrowths of the abdomen richly supplied with tracheae and tracheoles.

tracheal system In certain Arthropoda, as Insects and Myriapods, a system of respiratory tubules containing air and passing to all parts of the body.

tracheary elements Water-conducting cells of the xylem of a vascular plant. After the death and loss of their protoplasmic contents, the cell walls become thickened with lignin so that they can act as tubes for the conduction of water under tension. Cf. **hydroid**.

tracheid(e) An elongated element with pointed ends, occurring in wood. It is derived from a single cell, which lengthens and develops thickened pitted walls, losing its living contents, Tracheides conduct water.

trachelate Necklike, from Gk. *trachelos*, neck.

tracheole The ultimate branches of the tracheal system.

tracheophyte A vascular plant (i.e. pteridophyte or seed plant). Sometimes made a division, Tracheophyta.

tract The extent of an organ or system, as the *alimentary tract*; an area or expanse, as the *ciliated tracts* of some Ctenophora; a band of nerve fibres, as the *optic tract*.

tragus In the ear of some Mammals, including the Microchiroptera (bats), an inner lobe to the pinna.

training (1) *General*, the application of learning principles to improve social or technical skills through some systematic activity. (2) *Operant conditioning*, the procedure of conditioning an animal through the use of reinforcements in order to establish a desired behaviour.

trait A stable and enduring attribute of a person or animal which varies from one individual to another; traits may be physical (eye colour) or psychological (spatial intelligence) and are often used in the study of individual differences in personality.

trance Occurs under hypnosis, and in various conditions such as sleepwalking, a state of *disassociation* in which the individual's will is suspended and he or she acts on wishes or fantasies that are otherwise kept under control.

transaminase An enzyme which catalyses the transfer of amino groups from amino acids to keto acids, thus converting a keto acid into an amino acid, e.g. the conversion of α-ketoglutarate to glutamate, *transamination*.

transcribing genes Genes which are in the process of being actively transcribed into RNA. See **coding sequences**.

transcription The process by which an RNA polymerase produces single-stranded RNA complementary to one strand of the DNA or, rarely, RNA.

transcriptionally-active chromatin That chromatin which is being transcribed into RNA.

transcription complex Functional association of DNA, nascent RNA, protein and ribonucleoprotein actively transcribing and processing RNA.

transduction During phage infection and consequent bacterial lysis, the integration of segments of host DNA into that of the phage. It can then be transferred to another host.

transect A line or belt of vegetation marked off for study.

transfection The alteration of the host genome after infection by phage.

transferase An enzyme which catalyses the transfer of chemical groups between compounds, e.g. glycosyl transferase transfers a sugar residue onto the growing oligosaccharide complex of a glycoprotein.

transfer cell Parenchymatous cell with elaborate ingrowths of the cell wall, greatly increasing the area of the plasmalemma. It occurs where there is substantial movement of solutes between symplast and apoplast in scattered families of bryophytes and vascular plants.

transference In psychoanalytic theory, the tendency of the client to displace onto the analyst feelings and ideas derived from previous experience with other figures (e.g. one's parents).

transfer factor A dialysable substance, obtained from extracts of individuals with delayed type hypersensitivity, that is claimed to be able to transfer to another individual the ability to give a delayed hypersensitivity response.

transfer of training The facilitation (positive) or hindrance (negative) of performance on a learning or training task as a result of previous activity. The positive and negative transfer effects appear to be a function of the similarity of the tasks.

transfer RNA See **tRNA**.

transformation (1) The alteration of the

bacterial or eukaryotic cell genotype following the uptake of purified DNA. (2) The alteration of cells in tissue culture by various agencies so that they behave in many ways like cancer cells, e.g. their lack of growth control and the ability to divide indefinitely. See **contact inhibition**, **growth in soft agar**, **nude** mice.

transfusion reaction Disturbance following transfusion of blood, due to antibodies in the recipient reactive with donor blood cells or, more rarely, to antibodies in the transfused blood reactive with the recipient's blood cells. The antigens involved are usually those on erythrocytes, although after multiple transfusions those on leucocytes can sometimes be important. The most severe and common form of such reactions is due to **ABO blood group system** incompatibility when natural antibodies are present against ABO antigens absent from the recipient's own red cells.

transfusion tissue A tissue of short tracheids and parenchyma cells surrounding or associated with the vascular bundle(s) in the leaves of many gymnosperms, presumably functioning in the distribution of water and collection of photosynthate.

transgenic Used to describe animals, e.g. *transgenic mice*, which are derived from embryos into which isolated genomic DNA from another species has been introduced at an early stage of development. Such foreign genes may be incorporated into the nucleus and chromosomes so that the animal can express the foreign gene product by e.g. micro-injecting purified DNA into the nucleus of a fertilized egg. Also applied to plants. See **plant genetic manipulation**.

transitional epithelium A stratified epithelium consisting of only three or four layers of cells, esp. that found lining the ureters, the bladder and the pelvis of the kidney in Vertebrates.

transitional object Winnicott's concept of objects (e.g. a doll or piece of cloth) which act as comforters during the child's initial development from total dependence to self reliance.

transition region The region of the axis of a plant in which the change from root to shoot structure occurs.

translation The process by which ribosomes and tRNA decipher the **genetic code** in a messenger RNA in order to synthesize a specific polypeptide.

translocated herbicide A plant poison which, if absorbed in one region, will be conducted to all parts of the plant and e.g. kill the roots as well. Cf. **contact herbicide**.

translocated injury Injury occurring in an area remote from the original directly affected part of an animal or plant, but associated with it in type and extent.

translocation (1) An exchange between non-homologous chromosomes whereby a part of one becomes attached to the other, or a re-arrangement within one chromosome. (2) The transport of solutes about a plant, including the upward movement of inorganic salts in the transpiration stream in the xylem, and the movement of sugars in the phloem. See **mass flow hypothesis**.

transmission electron microscope See panel.

transpiration The loss of water by evaporation, mainly through the stomata in vascular plants.

transpiration stream The flow of water from the soil through the tissues of the plant to the evaporating surfaces, all driven by transpiration. Cf. **cohesion theory**.

transplant (1) In surgery and experimental zoology, the process of transferring a part or organ from its normal position to another position in the same individual or to a position in another individual. Also *transplantation*. (2) The part or organ transferred in this way.

transposon A sequence of DNA which is capable of inserting itself into many different sites in the host's chromosome. Also *transposable element*

trans-sexualism Gender identification with the opposite sex.

transverse, transversal Broader than long; lying across or perpendicular to the long axis of the body or of an organ; lying crosswise between two structures; connecting two structures in crosswise fashion.

transvestism Sexual gratification through dressing in the clothes of the opposite sex.

trapezium In the Mammalian brain, a part of the medulla oblongata consisting of transverse fibres running behind the pyramid bundles of the pons varolii.

trauma (1) *Medically*, structural damage to the body caused by the impact of some object or substance (e.g. a burn). (2) In *psychiatry*, any totally unexpected experience which the person cannot assimilate. See **shock**.

traumatic Relating to wounds.

traumatic neurosis A psychiatric illness resulting from severe and unexpected experience; characterized by periods of trance when the events are re-experienced, and often by traumatic dreams. Differs from other neuroses in that the symptoms have no unconscious meaning, but are an attempt to assimilate the experience by repeating it.

tree A tall, woody perennial plant having a well-marked trunk and few or no branches persisting above the base. A form known as *excurrent*.

tree ferns Ferns (Cyathea and Dicksonia and several extinct genera) which form a

transmission electron microscope, TEM

A form of **electron microscope** in which the specimen is usually either a thin (<70nm) section of fixed, embedded material, often stained with heavy metals, or virus particles or macromolecules, often negatively stained or shadowed with heavy metals and supported on a thin film. The specimen is evenly illuminated by a broad beam of electrons at 40–100 kV and the image is formed directly by focusing those electrons which pass more or less unscattered through the specimen (the *transmitted electrons*) on a fluorescent screen for direct viewing or on a photographic film for recording (see diagram). The screen image is darker where the specimen is denser.

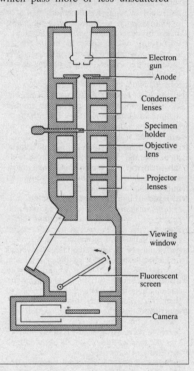

The TEM has the merit of much finer resolution (0.3 nm or less) than the light microscope but is usually not suitable for living specimens. In passing through the specimen some electrons are scattered and lose amounts of energy characteristic of the atoms with which they have interacted. Some more sophisticated instruments can select electrons of a particular energy band and form images showing e.g. the distribution of a chemical element. Energy absorption by the specimen can also damage the specimen, particularly at high magnifications.

Some TEMs are equipped for **X-ray analysis** and in others, *high voltage TEMs,* the electron beam is accelerated through 500–1500 kV; such beams can penetrate relatively thick, even hydrated and living, specimens.

Diagram labels: Electron gun; Anode; Condenser lenses; Specimen holder; Objective lens; Projector lenses; Viewing window; Fluorescent screen; Camera.

trunk up to 20 m high, typically unbranched and with a relatively slender stem surrounded and supported by matted adventitious roots and persistent leaf bases.

trematic Pertaining to the gill clefts.

Trematoda A class of Platyhelminthes, all the members of which are either ectoparasites or endoparasites, and have a tough cuticle, a muscular non-protrusible pharynx, and a forked intestine; a ventral sucker for attachment is usually present, and a sucker surrounding the mouth. Sometimes divided into three classes: Digenea, Aspidogastrea, Monogenea. Liver Flukes.

Treponemataceae A family of mainly parasitic, small spirochaetes, many of which are pathogenic, e.g. *Treponema pallidum*

(syphilis), *Treponema pertenue* (yaws).

trial and error learning In learning theory, refers to an essentially passive type of learning in which behaviour changes occur as a result of their association with positive or negative consequences (**reinforcements**). Originally the term was used to compare it with **insight learning** and is now more frequently referred to as operant conditioning.

triandrous Having three stamens.

triarch Said of a stele having three strands of protoxylem, e.g. the roots of some dicotyledons.

tribe A section of a family consisting of a number of related genera.

tricarboxylic cycle A cyclical series of metabolic interconversions of di- and tricarboxylic acids (including the sometimes eponymous citric acid) which brings about the oxidative degradation of acetyl-CoA. The electrons released generate reductive power which is exploited during their passage along the electron transfer chain to generate ATP.

tricarpellary Consisting of three carpels.

triceps A muscle with three insertions.

trichocyst In some Ciliophora, a minute hairlike body lying in the sub cuticular layer of protoplasm; it is capable of being shot out, and is an organ of attachment.

trichogyne An outgrowth from the female sex organ of the red algae, some fungi and lichens and a few green algae for the reception of the male gamete.

trichoid Hairlike.

trichome Any outgrowth of the epidermis of a plant, composed of one or more cells but without vascular tissue.

trichosis Arrangement or distribution of hair.

trichotomous Branching into three. Cf. **dichotomous**.

trich-, tricho-. Prefixes from Gk. *thrix*, gen. *trichos*, hair.

tricipital Adj. from **triceps**.

tricuspid Having 3 points, as the tight auriculo-ventricular valve of the Mammalian heart.

trifacial The 5th cranial or trigeminal nerve of Vertebrates.

trifid Split into three parts but not to the base.

trifoliate Having three leaves or, sometimes, three leaflets.

trifoliolate Said of a compound leaf having three leaflets, e.g. clover.

trifurcate Having 3 branches.

trigeminal Having 3 branches; the 5th cranial nerve of Vertebrates, dividing into the ophthalmic, maxillary and mandibular nerves.

trigonous A stem, triangular in cross section but obtusely angled and with convex faces. Cf. **triquetrous**.

trimerous Arranged in 3s or in multiples of 3.

trimonoecious Said of a species in which the plants bear male, female and hermaphrodite flowers.

trimorphic Having 3 forms. See **heterostyly**.

trioecious Said of a species in which some individuals bear male flowers only, others female only and the rest hermaphrodite.

tripinnate Three times **pinnate**.

triple fusion The fusion of the 2 **polar nuclei** with the second male gamete in angiosperms.

triplet The sequence of three bases in an mRNA which specify a particular amino acid. See **genetic code**.

triplets In Mammals, 3 individuals produced at the same birth.

triple vaccine Vaccine containing a mixture of diphtheria toxoid, tetanus toxoid and pertussis vaccine. Routinely used to produce active immunity against diphtheria, tetanus and whooping cough in infants.

triploblastic Having three types of tissue in the body, there being mesoderm between the ectoderm and endoderm, which gives rise to connective, skeletal and muscular tissues etc. Cf. **diploblastic**.

triploid Having 3 times the haploid number of chromosomes for the species.

triquetrous A stem, triangular in cross section but with acute angles and concave faces. Cf. **trigonous**.

trisomic Said of an otherwise normal diploid organism in which one chromosome type is represented thrice instead of twice.

trisomy 21 Also *Down's syndrome*. Condition in which an individual has three copies of chromosome 21, either in all cells or in a proportion of their cells. Affected individuals show many abnormalities to varying degrees of severity, including the characteristic *mongoloid* eye fold, and usually subnormal intelligence. Incidence increases with mother's age.

TRITC Abbrev. for *Tetramethyl Rhodamine IsoThioCyanate*. Used for making antibodies fluorescent in fluorescent antibody techniques.

triton X-100 *iso-Octylphenoxypolyethoxyethanol*. A non-ionic detergent which is commonly used to solubilize membrane proteins in their biologically active state.

tritor The masticatory surface of a tooth.

triturate To grind to a fine powder, esp. beneath the surface of a liquid.

trivalent Said of association of 3 chromosomes at meiosis.

tRNA An RNA molecule about 80 nucleotides long, with complementary sequences which result in several short *hairpin-like* structures. The loop at the end of one of these carries the anticodon triplet, which binds to the codon of the mRNA. The corresponding amino acid is bound to the 3′ end of the molecule. See **adaptor hypothesis**.

trochal Wheel-shaped.

trochanter The second joint of the leg in Insects; a prominence for the attachment of muscles near the head of the femur in Vertebrates.

trochlea Any structure shaped like a pulley, esp. any foramen through which a tendon passes. adj. *trochlear*.

trochophore, trochosphere The free-swimming pelagic larval form of Annelida, Mollusca and Bryozoa, possessing a prominent pre-oral ring of cilia and an apical tuft of cilia.

trophallaxis　Mutual exchange of food between *imagines* (sing. *imago*) and their larvae, as in some social Insects.

trophic　Pertaining to nutrition.

trophic level　Broad class of organisms within an *ecosystem* characterized by mode of food supply. The first trophic level comprises the green plants, the second is the herbivores and the third is the carnivores which eat the herbivores.

trophic structure　A characteristic feature of any ecosystem, measured and described either in terms of the standing crop per unit area, or energy fixed per unit area per unit time, at successive trophic levels. It can be shown graphically by the various **ecological pyramids**.

trophoblast　The differentiated outer layer of epiblast in a segmenting Mammalian ovum.

trophozoite　In Protozoa, the trophic phase of the adult, which generally reproduces by schizogony.

troph-, tropho-　Prefixes from Gk, *trophe*, nourishment.

tropism　A reflex response of a cell or organism to an external stimulus; movement that orients an organism to achieve a certain distribution of stimulation. See **geotropism**, **phototropism**. Cf. **taxis**.

tropomyosin　A filamentous protein aligned along the actin fibres of muscle. Under the influence of troponin it controls the interaction of actin and myosin.

troponin　A complex of 3 polypeptide chains which mediates the effect of calcium on muscle contraction.

truncate　Square-ended; appearing as if cut off.

truncus　A main blood vessel; as the *truncus transversus* or **Cuvierian duct** and the *truncus arteriosus* or great vessel, through which blood passes from the ventricle. Also *trunk*.

trunk　(1) Upright, massive main stem of a tree. (2) The body, apart from the limbs. (3) The proboscis of an elephant. See **truncus**.

Try　Symbol for **tryptophan**.

trypanosomes　A group of flagellate Protozoa including many causing disease of man (see **trypanosomiasis**) and animals.

trypanosomiasis　*Chagas' disease*. A disease, occurring in parts of South America and Africa, due to infection of the muscles, heart, and brain of Man with the protozoal parasite, *Trypanosoma cruzi*, the infection being conveyed by the bite of an insect.

trypsin　A protease secreted by the pancreas which is specific for peptide bonds adjacent to lysine and arginine residues.

tryptophan　*2-amino-3-indolepropanoic acid*. The L-isomer is a polarizable amino acid and constituent of proteins. Symbol Try, short form W.

Side chain: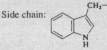

See **amino acid**.

tube　The cylindrical proximal part of a gamosepalous calyx or gamopetalous corolla.

tubefeet　See **podium** (2).

tuber　Swollen underground stem acting as a storage and perennating organ, e.g. the potato.

tubercle　(1) Any small rounded projection or swelling on a bone or other part of the body; a cusp of a tooth. (2) A solid elevation of the skin larger than a papule. (3) A small mass or nodule of cells resulting from infection with the bacillus of tuberculosis. (4) Loosely, tuberculosis; the tubercle bacillus. (5) In plants a **nodule**. Also called *tuberculum*. adjs. *tubercled, tubercular, tuberculate, tuberculose*.

tuberculate　Covered with small wart-like projections.

tuberculin　A protein or mixture of proteins derived from *Mycobacterium tuberculosis*, which is employed in the tuberculin test as a diagnostic reagent for detecting sensitization by, or infection with *M. tuberculosis*. Old tuberculin (OT) is a heat concentrated filtrate from the medium in which the organism has been grown. Purified protein derivative (tuberculin PPD) is a soluble protein fraction, precipitated by trichloroacetic acid from a synthetic medium in which *M. tuberculosis* has been grown. Tuberculins can be derived from human, bovine or avian strains of the bacillus but show extensive antigenic cross reactivity.

tuberculin test　Test for delayed hypersensitivity to **tuberculin** in human or other animals. Positive reactions are presumptive evidence of cell-mediated immunity to, and therefore of past or present exposure to *Mycobacterium tuberculosis*, but does not necessarily indicate active disease. See **Mantoux test**.

tuberculosis　Infection by *Mycobacterium tuberculosis*, esp. of the lungs; characterized by the development of tubercles in the bodily tissues and by fever, anorexia and loss of weight. Spread by air droplets and raw milk.

tuberosity　A prominence on a bone, generally from muscle attachment, esp. those near the head of the humerus.

tuberous　Of or like a tuber; having tubers.

tubicolous　Living in a tube.

tubifacient　Tube building, as certain Polychaeta.

tubule, tubulus　A fine tube; any small tubular structure. adjs. *tubulate, tubuliferous, tubuliform, tubulose*. See **microtubule**.

tubulin A globular protein of two closely related variants, α- and β-tubulin which forms, as an α/β dimer, the basic unit for the contraction of microtubules.

tufted Grass shoots, clustered or clumped rather than scattered. *Caespitose.*

tumid Swollen; inflated.

tumour-inducing principle   See **Ti plasmid**.

tumour necrosis factor Abbrev. *TNF*. Name given to the substance, secreted by macrophages which have been activated by bacterial endotoxin or by mycobacteria, which causes necrosis of a number of tumour cell lines. It also causes muscle necrosis and general wasting of the body, perhaps by inhibiting lipoprotein lipase (normally required for fat transport in cells). Its primary structure is similar (about 30% homology) to that of lymphotoxin and the mechanism by which it kills cells is presently unknown. There is evidence that it enhances resistance to parasitic infection. Also *TNFα* and *LTβ*.

tumour specific antigen Antigen present in tumour cells which is not expressed (or only very minimally) by their normal counterparts. Sometimes these are foetal antigens not normally expressed in the adult, and sometimes coded for by viral material incorporated into the host's genome. Significant in respect of the existence of specific immune responses to tumours which are present in tumour-bearing animals (though insufficiently effective), and in respect of the use of radiolabelled or potentially immunotoxic antibodies targetted on tumour cells *in vivo*.

tundra A biome which is essentially an Arctic grassland, where the vegetation consists of lichens, grasses, sedges and dwarf woody plants. It covers two large areas, one in the Palearctic, and one in the Nearctic region.

tunic An investing layer. adj. *tunicate*.

tunica Outer layer(s) of cells in the shoot apical meristem of many angiosperms, which give rise to the epidermis and which divide anticlinally and thus do not displace the underlying cells from the meristem. See **tunica-corpus concept**.

tunica-corpus concept The concept that the shoot apex in many angiosperms is organized into a **tunica** and a **corpus** the distinctness of which is maintained more or less indefinitely. The concept accounts for the existence of **periclinal chimeras**.

Tunicata See **Urochordata**.

tunicate bulb A bulb composed of a number of swollen leaf bases each of which completely encloses the next younger, as in the onion.

tunicate, tunicated Enclosed by a non-living test or mantle. Generally, having a coat or covering.

Turbellaria A class of Platyhelminthes comprising forms of free-living habit, marine, freshwater or terrestrial; with a ciliated ectoderm; usually with a muscular protrusible pharynx and a pair of eyespots; rarely have suckers. Planarians.

turbinal Coiled in a spiral; one of certain bones of the nose in Vertebrates which support the folds of the olfactory mucous membrane.

turbinate In the form of a whorl or an inverted cone, as certain Gastropod shells.

turgid (1) Said of a cell which is distended and stiff as a result of the osmotic uptake of water, having a positive turgor pressure. (2) Said of a non-woody tissue which is stiff as a result of the cells being turgid. Cf. **flaccid**.

turgor movement Movement of a plant part resulting from changes in the turgor of its cells or the cells of its support. See **pulvinus**. Cf. **growth movement**.

turgor potential *Pressure potential*, ψp. That component of the **water potential** due to the hydrostatic pressure; equal to the **turgor pressure**. An important component in turgid cells and in the xylem.

turgor pressure The hydrostatic pressure of the contents of a cell; normally positive in most plant cells; normally negative in the conducting cells of the xylem of transpiring plants.

Turner's syndrome A condition in Man in which a person looks superficially like a female but has only one X-chromosome.

turtle shell The horny plates of the hawk's bill turtle. Commonly *tortoise shell*.

twenty-four hour rhythm See **circadian rhythm**.

twiner A plant that climbs by winding around a support.

twins (1) *Identical twins* arise from the same fertilized egg which has subsequently divided into two, each half developing into a separate individual. (2) In Mammals, *non-identical twins* are produced from separate eggs fertilized at the same time.

twisted aestivation Same as *contorted aestivation*.

tylosis, tylose The bladder-like expansion of the wall of a living parenchyma cell through a pit into the lumen of a xylem tracheid or vessel. Tyloses apparently form in non-functional conduits after spontaneous or wound-induced **embolism**, and may restrict the spread of pathogens.

tympanic bulla In some Mammals, a bony vesicle surrounding the outer part of the tympanic cavity and external auditory meatus formed by the expansion of the tympanic bone.

tympanum A drumlike structure; in some Insects, the external vibratory membrane of a chordotonal organ; in some Birds an inflatable air sac of the neck region; in Vertebrates, the middle ear, or the resonating membrane of the middle ear; in Birds, the resonating sac of the syrinx. adjs. *tympanic*, *tympanal*.

type The individual specimen on which the description of a new species or genus is based; the sum total of the characteristics of a group. adj. *typical*.

type specimen The actual specimen from which a given species was first described.

typhlosole In some Invertebrates, a longitudinal dorsal inwardly projecting fold of the wall of the intestine, by which the absorptive surface is increased.

typhoid, typhoid fever The most serious form of enteric fever (which includes *paratyphoid*) due to infection with the bacillus

Salmonella typhi. Prolonged fever, a rose rash and inflammation of the small intestine with ulceration occur. Infection is by faecal contamination of food or water.

typical intensity The high degree of stereotyping observed in many patterns of behaviour that have a communicative function.

Tyr Symbol for tyrosine.

tyrosine *2-amino-3-(4-hydroxyphenyl)propanoic acid*. The L-isomer is an 'acidic' amino acid and constituent of proteins. Symbol Tyr, short form Y.

Side chain: HO—⬡—CH₂—

See **amino acid**.

U

ubiquinone Small highly mobile electron carrier mediating the transfer of electrons from flavoprotein to cytochrome in the *electron transfer chain*.

ubiquitin Polypeptide of wide distribution in both prokaryotes and eukaryotes. It is attached to proteins prior to their degradation in the course of cellular protein turnover.

UCR Abbrev. for *UnConditioned Reflex*. See **classical conditioning**.

UCS Abbrev. for *UnConditioned Stimulus*.

uliginose, uliginous Growing in places which are wet.

ulna The post-axial bone of the antebrachium in land Vertebrates. adj. *ulnar*.

ulotrichous Having woolly or curly hair.

ultra- Prefix from L. *ultra*, beyond.

ultracentrifuge A high-speed centrifuge much used for molecular separation and capable of creating forces up to 500 000 times gravity.

ultramicroscope An instrument for viewing particles too small to be seen by an ordinary microscope, e.g. fog or smoke particles. An intense light projected from the side shows, against a dark background, the light scattered by the particles. See **dark-ground illumination**.

ultramicrotome A modified microtome developed for cutting ultra-thin sections for examination with the electron microscope. The cutting surface may be of steel, but is more usually glass or diamond, and the movement of the specimen block towards the knife is very delicately controlled, e.g. by the thermal expansion of a rod.

ultrastructure The submicroscopic structure of a cell, particularly as shown by the electron microscope.

ultraviolet microscope Instrument using ultraviolet light for illuminating the object. Its resolving power is therefore about doubled as the resolution varies inversely with the wavelength of the radiation, but more usefully the nucleic acids absorb strongly in this region and can therefore be localized and measured.

umbel Inflorescence or simple umbel of many flowers borne on stalks arising together from the top of a main stalk. Often this sort of branching is repeated in a compound umbel with several main stalklets arising together from the top of a larger stalk.

umbellate Having the characters of an umbel; producing umbels.

umbellifer A plant which has its flowers in umbels, esp. a member of the family Umbelliferae.

Umbelliferae *Apiaceae*. The carrot family, ca 3000 spp. of dicotyledonous flowering plants (superorder Rosidae). Mostly herbs, more or less cosmopolitan, but esp. in temperate and upland regions. The flowers are in simple or compound umbels, have five free petals, five stamens and an inferior ovary of two carpels. Includes several vegetables and flavouring plants, e.g. carrot, parsnip, celery, parsley, coriander, cumin. Several are poisonous including hemlock.

umbilical cord In eutherian Mammals, the vascular cord connecting the foetus with the placenta.

umbilicus In Gastropod shells, the cavity of a hollow columella; in Birds, a groove or slit in the quill of a feather; in Mammals, an abdominal depression marking the position of former attachment of the umbilical cord. pl. *umbilici*.

umbo A boss or protuberance; the beak-like prominence which represents the oldest part of a Bivalve shell. pl. *umbones*. adj. *umbonate*.

umbrella A flat cone-shaped structure, esp. the contractile disk of a medusa.

umwelt The relevant aspects of the environment which constitute the subjectively significant, or meaningful, surroundings for an animal or individual, i.e. that class of environmental variables capable of influencing behaviour.

unarmed Without spines, thorns, prickles, sharp teeth etc.

unavailable Said of that fraction of e.g. water or mineral nutrient in soil or mineral fertilizer which cannot be used by plants. Cf. **available**.

uncate, unciform, uncinate Hooked; hook-like.

uncinus A hook, or hook-like structure, e.g. a hook-like chaeta of Annelida; in Gastropoda, one of the marginal radula teeth.

unconscious mind Refers to mental processes of which the subject is unaware, but which influence thought and action.

underleaf One of a row of leaves on the underside of the stem of a liverwort.

undulating membrane An extension of the flagella membrane in some Protozoa (e.g. *Trypanosoma*) by which it is attached to the cell for part of its length.

ungual Pertaining to or affecting the nails. See **unguis**.

unguiculate Provided with claws. Applied specifically in plants to a petal with an expanded limb supported on a long narrow stalklike base.

unguis In Insects, one of the tarsal claws; in Vertebrates, the dorsal scale contributing to a nail or claw; more generally, a nail or claw.

pl. *ungues.* adjs. *ungual, unguinal.*

ungula A hoof. adj. *ungulate.*

ungulate The term applied to several groups of superficially similar hoofed animals which are not necessarily closely related taxonomically. Horses, Cows, Deer, Tapirs.

unguligrade Walking on the tips of enlarged nails of one or more toes, i.e. hoofs, as in horses etc. Cf. **digitigrade, plantigrade.**

uniaxial Having a main axis consisting of a single row of large cells with only clearly subordinate branches. Cf. **multiaxial.**

unicellular Consisting of a single cell.

unilocular Having a single compartment. Cf. **multilocular, bilocular.**

uninemy hypothesis Each chromatid contains a single, DNA, double-helical molecule organized linearly with respect to the chromosomal axis.

uninucleate Containing one nucleus.

uniparous Giving birth to one offspring at a time.

unipolar Said of nerve cells having only one process. Cf. **bipolar, multipolar.**

unipotent Of embryonic cells, capable of forming a single cell type only. Cf. **totipotent.**

unique sequence DNA Those sequences which are only represented once in the **haploid** genome. Most genes are in this category.

uniramous Of appendages having only one branch, like those of some Crustaceans. Cf. **biramous.**

uniseriate Arranged in a single row, series or layer.

unisexual Showing the characters of one sex or the other; distinctly male or female. Cf. **hermaphrodite.**

unit character A character that can be classified into two distinct types, usually the normal and the mutant, and displaying *Mendelian inheritance.*

unit leaf rate See **net assimilation rate.**

univalent One of the single chromosomes which separate in the first meiotic division.

universal veil Membrane which encloses the developing fruiting body of some agarics, rupturing, as the stalk grows, to leave the **volva.**

univoltine Producing only one set of offspring during the breeding season or year. Cf. **multivoltine.**

unstirred layer Same as **boundary layer.**

unstriated muscle A form of contractile tissue composed of spindle-shaped fibrillar uninucleate cells, occurring principally in the walls of the hollow viscera. Also *smooth muscle.* Cf. **striated muscle.** See **voluntary muscle.**

upper quartile The argument of the cumulative distribution function corresponding to a probability of 0.75; (of a sample) the value below which occur three quarters of the observations in the ordered set of observations.

uracil *2,6-dioxypyrimidine*, one of the four bases in RNA and the only one that does not occur in DNA. Its radiolabelled derivative is therefore a specific label for RNA synthesis. It pairs with adenine.

Structure:

See **genetic code.**

urea cycle The cyclic interconversion of 4 amino acids which converts carbamyl phosphate into urea. It represents the major pathway for the excretion of nitrogenous waste in terrestrial vertebrates.

uredosorus A pustule consisting of uredospores, with their supporting hyphae, and some sterile hyphae.

uredospore, urediniospore, urediospore A binucleate spore which rapidly propagates the dikaryotic phase of a rust fungus.

ured-, uredo-. Prefixes from L. *uredo*, blight.

ureotelic Excreting nitrogen in the form of urea.

ureter The duct by which the urine is conveyed from the kidney to the bladder or cloaca.

urethra The duct by which the urine is conveyed from the bladder to the exterior, and which in male Vertebrates serves also for the passage of semen. adj. *urethral.*

uricotelic Excreting nitrogen in the form of uric acid.

urine In Vertebrates, the excretory product elaborated by the kidneys, usually of a more or less fluid nature. adj. *urinary.*

uriniferous, uriniparous Urine-secreting, urine-producing, as the glandular tubules of the kidney.

urinogenital Pertaining to the urinary and genital systems.

urinogenital system The organs of the urinary and genital systems when there is a direct functional connection between them, as in male Vertebrates.

uro-. Prefix from (1) Gk. *ouron*, urine, (2) Gk. *oura*, tail.

urochord Having the notochord confined to the tail region.

Urochordata A subphylum of Chordata, in which only the larvae have a hollow dorsal nerve cord and a notochord, the adults being without coelom, segmentation and bony tissue, and having a dorsal atrium, a reduced nervous system, and a test composed of tunicin, a substance closely related to cellulose. Sea Squirts. Also *Tunicata*.

Urodela An order of Amphibians, the adults having four similar pentadactyl limbs and a prominent tail. The larvae have external gills which persist in the adults of neotenous forms, and in some others gill slits persist. Newts and Salamanders. Also *Caudata*.

urodelous Having a persistent tail, as Salamanders.

uropod In Malacostraca, an appendage of the abdominal somite preceding the telson.

uropygial gland See **oil gland**

uropygium In Birds, the short caudal stump into which the body is prolonged posteriorly.

urosome In aquatic Vertebrates, the tail region; in Crustacea, the hinder part of the abdomen.

urostyle In Fish, the hypural bone; in Anura, a rodlike bone formed by the fusion of the caudal vertebrae.

urticant, urticating Irritating; stinging.

urticaria Skin rash typically with localized, elevated erythematous, itchy weals due to local release of histamine and other vasoactive substances. Frequently associated with immediate hypersensitivity (type 1 reaction) on contact of the skin with antigens, or more generalized after absorption of allergens from the gut. Can result from taking drugs or certain foods (e.g. shellfish), or as a reaction to the injection of serum, insect bites or plant stings (*nettle rash*). Also *hives*.

Ustilaginales An order of the Basidiomycotina containing the **smut** fungi.

uterus In female Mammals, the muscular posterior part of the oviduct in which the foetus is lodged during the prenatal period; in lower Vertebrates and Invertebrates, a term loosely used to indicate the lower part of the female genital duct, or in certain cases (as in Platyhelminthes) a special duct in which eggs are stored or young developed. adj. *uterine*.

utricle A small sac; in Vertebrates, the upper chamber of the inner ear from which arise the semicircular canals. In plants, any one of a variety of inflated bladder-like structures. Also *utriculus*. adjs. *utricular, utriculiform*.

uvea (1) In Vertebrates, the posterior pigment-bearing layer of the iris of the eye. (2) The iris, the ciliary body and the choroid, considered as one structure. Also *uveal tract*.

V

vaccination Production of active immunity by administration of a vaccine.

vaccine Therapeutic material, treated to lose its virulence and containing antigens derived from one or more pathogenic organisms, which on administration to humans or other animals, will stimulate active immunity and protect against infection with these or related organisms.

vaccinia Syn. for the virus used in vaccine procedures to produce immunity to smallpox. Differs from cowpox and smallpox virus in minor antigens only and was probably derived originally from cowpox virus. The virus has proved very stable (i.e. not liable to mutation) and safe to use in humans, and produces long-lasting cell mediated immunity. Genes for other viral antigens have been introduced into vaccinia virus, by **recombinant DNA**, techniques with the intention that immunization with the recombinant strain will produce protective immunity against other viruses also.

vaccinial Of, pertaining to, or caused by, vaccinia.

vacuolar membrane The membrane surrounding an intracellular **vesicle**.

vacuole See **vesicle**.

vacuole A cavity, containing sap and separated from the cytoplasm by a membrane, the *tonoplast*.

vacuum activity Behaviour manifested in the apparent absence of the external stimuli that normally elicit the activity, presumably because of internal factors governing the motivation to perform the behaviour.

vagina (1) Any sheathlike structure, esp. the leaf sheath of grasses. (2) The terminal portion of the female genital duct leading from the uterus to the external genital opening. adjs. *vaginal, vaginant, vaginate, vaginiferous*.

vaginal plug In female Rodents and Insectivores, the coagulated secretion of Cowper's glands which blocks the vagina and prevents premature escape of seminal fluid and further mating.

vagus (1) The tenth cranial nerve of Vertebrates, supplying the viscera and heart and, in lower forms, the gills and lateral line system. (2) In Mammals, supplying the larynx.

Val Symbol for **valine**.

valency In immunology it refers to the number of antigen binding sites on an antibody molecule. Those belonging to most Ig classes have two, but IgM has 10 combining sites and IgA, which can exist as monomer, dimer and higher polymers has multiples of 2. The valency of an *antigen* can likewise be expressed in terms of the number of antigen

combining sites with which it can combine. Most large antigen molecules are multivalent. See **lattice hypothesis**.

valine *2-amino-3-methylbutanoic acid*. The L-isomer is a nonpolarizable amino acid and constituent of proteins. Symbol Val, short form V.

$$\text{Side chain:} \quad \begin{array}{c} CH_3 \\ \diagdown \\ CH_3 \end{array} CH-$$

See **amino acid**.

valley bog Type of *Sphagnum* bog forming where water draining from relatively acid rocks stagnates in a flat-bottomed valley so as to keep the soil constantly wet.

valvate (1) Arranged so that the edges touch but not overlap. See **aestivation, vernation**. Cf. **imbricate**. (2) Opening by valves.

valve (1) The flattened part of a theca of a diatom frustule. (2) That part of a fruit wall which separates at dehiscence. (3) Any structure which controls the passage of material through a tube, duct or aperture, usually in the form of membranous folds, as the auriculo-ventricular valves of the heart. (4) In Mollusca, Cirripedia and Brachiopoda, one of several separate pieces composing the shell. (5) In Insects, a covering plate or sheath, esp. one of a pair which can be opposed to form a tubular structure, as the valves of the ovipositor. Also *valva*.

VAM Abbrev. for *Vesicular-Arbuscular Mycorrhiza*.

vane The web of a feather, composed of the barbs and barbules. Also *vexillum*.

variable-interval schedule See **interval schedule of reinforcement**.

variable-ratio schedule See **ratio schedule of reinforcement**.

variable region The N-terminal half of light chains (V_L) and the N-terminal half of the Fab portion of the heavy chains (V_H) of immunoglobulin molecules. The amino acid sequences in these regions are variable within a single immunoglobulin class and light chain type. This variability is controlled by the **immunoglobulin genes**. Similarly the N-terminal region of the α, β and γ chains of the T lymphocyte antigen receptors have variable regions, controlled in a similar manner.

variance The mean of the sum of squared deviations of a set of observations from the corresponding mean; the second moment about the mean of a random variable; the dispersion parameter of a probability distribution.

variant A specimen which differs in its characteristics from the type and which is produced either by changed environmental conditions and/or by mutation.

variate A quantity, measurement or attribute which is the subject of statistical analysis.

variation The differences between the offspring of a single mating; the differences between the individuals of a race, subspecies, or species; the differences between analogous groups of higher rank.

variegation The occurrence of differently coloured areas on leaves or petals due to (1) virus infection (streaks, spots, mottles), (2) mineral deficiency (veinal or interveinal chlorosis), (3) genetically determined patterns as in leaves of *Coleus* cultivars, (4) chimerical structure (light coloured borders on leaves), (5) a **transposon**, transposable element.

variety A race; a stock or strain; a sport or mutant; a breed; a subspecies; a category of individuals within a species which differ in constant transmissible characteristics from the type but which can be traced back to the type, by a complete series of gradations; a geographical or biological race.

variola See **smallpox**.

vas A vessel, duct or tube carrying fluid. pl. *vasa.* adj. *vasal.* See **vas deferens**.

vasa efferentia A series of small ducts by which the semen is conveyed from the testis to the vas deferens.

vasa vasorum In Vertebrates, small blood vessels ramifying in the external coats of the larger arteries and veins.

vascular Relating to vessels which convey fluids or provide for the circulation of fluids, e.g. xylem and phloem; provided with vessels for the circulation of fluids.

vascular area See **area vasculosa**.

vascular bundle Strand of conducting tissue composed of xylem and phloem and, usually in dicotyledons, cambium. See **eustele**.

vascular cylinder Same as stele.

vascular plant *Tracheophyte.* Member of those plant groups that have a vascular system of xylem and phloem, the pteridophytes and seed plants.

vascular ray A *ray* in secondary xylem or phloem.

vascular system (1) In a vascular plant, all the conducting tissues, both xylem and phloem. (2) In animals, the organs responsible for the circulation of blood and lymph, collectively.

vasculum A receptacle for collecting botanical specimens.

vas deferens A duct leading from the testis to the ejaculatory organ, the urino-genital canal, the cloaca or the exterior. pl. *vasa.*

vasifactive See **vasoformative**.

vaso- Prefix from L. *vas*, vessel.

vasochorial placenta A chorioallantoic placenta in which the epithelium and the endometrium of the uterus disappear, and the chorion is in intimate contact with the endothelial wall of the maternal capillaries as in some Carnivora. Also *endotheliochorial placenta.*

vasoconstrictor Of certain autonomic nerves, causing constriction of the arteries. Also *vasohypertonic.*

vasodilator Of certain autonomic nerves, causing expansion of the arteries. Also *vasohypotonic, vasoinhibitory.*

vasoformative Pertaining to the formation of blood or blood vessels.

vasomotor Causing constriction or expansion of the arteries; as certain nerves of the autonomic nervous system.

vasopressin A nonopeptide secreted by the posterior pituitary gland. It elevates blood pressure by the contraction of small blood vessels and aids water resorption by the kidney. Used in the diagnosis and treatment of **diabetes insipidus**.

vasopressor Substance which causes a rise of blood pressure.

VDRL test A rapid screening test for syphilis which depends on the flocculation of a cardiolipin-cholesterol-lecithin preparation, if the serum being tested contains anticardiolipin antibody. Used in venereal disease reference laboratories, hence *VDRL.*

vector (1) An agent (usually an insect) which transmits a disease caused by a parasite or micro-organism from one host to another. (2) Insect, bird, wind etc. carrying pollen from stamen to stigma. (3) A DNA molecule derived from a self-replicating phage, virus, plasmid or bacterium which can accept inserted DNA sequences. Used to transfer DNA from one organism to another to make recombinant DNA.

vegan A very strict vegetarian who abstains from all food of animal origin.

vegetable pole The lower portion or pole of an ovum in which cleavage is slow owing to the presence of yolk. Cf. **animal pole**.

vegetative Not reproducing sexually; not carrying flowers or other sexually reproducing structures.

vegetative functions The autonomic or involuntary functions, as digestion, circulation.

vegetative propagation The natural and esp. the horticultural production of new plants from bulbs, offsets, stolons, rhizomes etc. and by layering, taking cuttings, grafting etc. In the absence of mutation, the offspring will be genetically identical to the parent plant. See **asexual reproduction**, **plant cell culture**.

vegetative reproduction Usually natural rather than horticultural **vegetative propagation**. In animals, propagation by budding.

veil See **partial veil**, **universal veil**, **velum**.

veiled cell Cell characterized by large veil-like processes found in the lymph draining skin, esp. after local antigenic stimulation. They express large amounts of surface MHC Class II antigens on their surface, and represent Langerhans cells in transit from the skin to the draining lymph node, where they take the form of interdigitating cells, which are very effective in antigen presentation to T lymphocytes.

vein (1) A vascular bundle and supporting tissues in a leaf. (2) A vessel conveying blood back to the heart from the various organs of the body. (3) In Insects, a wing nervure. adj. *venous*.

vein islet See **areole**.

veliger The secondary larval stage of most Mollusca, developing from the trochophore and characterized by the possession of a velum.

vellus In Man, the widespread short downy hair which replaces the fine lanugo which almost covers the foetus from the fifth or sixth month until shortly before birth.

velum A veil-like structure, as the *velum pendulum* or posterior part of the soft palate in higher Mammals; in some Ciliophora, a delicate membrane bordering the oral cavity; in Porifera, a membrane constricting the lumen of an incurrent or excurrent canal; in hydrozoan medusae, an annular shelf projecting inwards from the margin of the umbrella; in Rotifera, the trochal disk; in Mollusca, the ciliated locomotor organ of the veliger larva; in Cephalochorda, the perforated membrane separating the buccal cavity from the pharynx.

velvet The tissue layers covering a growing antler, consisting of periosteum, skin and hair.

venae cavae The caval veins; in higher Vertebrates, three large main veins conveying blood to the right auricle of the heart.

venation The pattern formed by the veins of a leaf or the veins of an animal's circulation; by extension, the veins themselves considered as a whole.

venomous Having poison-secreting glands.

venous system That part of the circulatory system responsible for the conveyance of blood from the organs of the body to the heart.

vent The aperture of the anus or cloaca in Vertebrates.

venter A protuberance; a median swelling; the abdomen in Vertebrates; the ventral surface of the abdomen.

ventral (1) Pertaining to that surface of a flattened thalloid plant which faces the substrate. (2) See **adaxial**, but see note on the use of **dorsal** in reference to leaves.

ventral suture The presumed line of junction of the edges of the infolded carpel.

ventricle, ventriculous A chamber or cavity, esp. the cavities of the Vertebrate brain and the main contractile chamber or chambers of the heart (in Vertebrates or Invertebrates). adj. *ventricular*.

ventricose (1) Swollen in the middle. (2) Having an inflated bulge to one side.

venule In Chordata, small blood vessels which receive blood from the capillaries and unite to form veins.

verbal test Mental test consisting primarily of items measuring vocabulary, verbal reasoning, comprehension etc. Cf. **performance test**.

Verbenaceae Family of ca 3000 spp. of dicotyledonous flowering plants (superorder Asteridae). Trees, shrubs and herbs, almost all tropical and subtropical. Flowers gamopetalous, usually zygomorphic, and with a superior ovary. Includes teak and other important timber trees.

vermicule A small worm-like structure or organism, as the motile phase of certain Sporozoa.

vermiform Worm-like, as the *vermiform appendix*.

vermis In lower Vertebrates, the main portion of the cerebellum; in Mammals, the central lobe of the cerebellum.

vernal Of, or belonging to, spring.

vernalization The natural or artificial promotion of flowering by a period of low temperature, around 4°C.

vernation (1) Arrangement of unexpanded leaves in the vegetative bud. See **aestivation**. (2) Same as **ptyxis**.

verruca A wart-like process; esp. one of a number of wart-like processes situated around the base of certain kinds of alcyonarian polyp.

verrucose Warty; covered with wart-like outgrowths.

versatile Capable of free movement as an anther which is attached to the tip of the filament by a small area on its dorsal side, so that it turns freely in the wind, helping the pollen dispersal, or like the toes of birds when they may be turned forwards or backwards.

versicolorous Not all of the same colour, changing in colour with age.

vertebra One of the bony or cartilaginous skeletal elements of mesodermal origin which arise around the notochord and compose the backbone. pl. *vertebrae*. adjs. *vertebral, vertebrate*.

Vertebrata A subphylum of Chordata in which the notochord stops beneath the forebrain and a skull is always present. There are usually paired limbs. The brain is complex and associated with specialized sense organs, and there are at least ten pairs of cranial nerves. The pharynx is small and there are rarely more than seven gill slits. The heart has at least three chambers and the blood has corpuscles containing haemoglobin.

vertebraterial canals In Vertebrates, small canals found one on each side of all or most of the cervical vertebrae. They are formed by the articulation or fusion of the two heads of the small or vestigial cervical ribs to the centra and transverse processes, and the vertebral arteries run through them.

vertex In higher Vertebrates, the top of the head, the highest point of the skull; in Insects, the dorsal area of the head behind the epicranial suture.

verticil A whorl.

verticillaster A kind of inflorescence found in dead nettles and related plants. It looks like a dense whorl of flowers, but is really a combination of two crowded dichasial cymes, one on each side of the stem.

verticillate Arranged in whorls.

vesica The urinary bladder.

vesicant Causing blisters; any agent which does this. See **war gas**.

vesicle (1) A structure like a lysosome, which is surrounded by a membrane and situated in the cellular cytoplasm. (2) A small cavity containing fluid. (3) A small saclike space containing gas. (4) One of the three primary cavities of the Vertebrate brain. (5) A small bladder-like sac. Also *vesicula*. adj. *vesicular*.

vesicular-arbuscular mycorrhiza Endotrophic **mycorrhiza** in which the fungus invades the cortical cells to form vesicles and arbuscules (finely branched structures). VAMs are very common among herbaceous plants, including many crop plants, and may significantly improve the mineral nutrition of the host. Abbrev. *VAM*.

vesicula seminalis In many animals, including Man, a sac in which spermatozoa are stored during the completion of their development.

vesiculate, vacuolate Having vesicles or vacuoles.

vessel In plants, a long, from 1 cm to 10 m, unbranched, water-conducting tube of the xylem, formed from a longitudinal file of cells by the perforation of their common end walls. Water moves through perforations within a vessel but through pits into and out of vessels and from one vessel to the next. Vessels are found in very few pteridophytes, a few gymnosperms and most angiosperms. In animals, a channel or duct with definitive walls, as one of the principal vessels through which blood flows.

vessel element, -member, -segment A tracheary element of the xylem which with others in a file forms a vessel. Cf. **tracheid**.

vestibule A passage leading from one cavity to another or leading into a cavity from the exterior; in Protozoa, a depression in the ectoplasm at the base of which is the mouth; in a female Mammal, the space between the vulva and the junction of the vagina and the urethra (urinogenital sinus); in Birds, the posterior chamber of the cloaca; in Vertebrates generally, the cavity of the internal ear. adjs. *vestibular, vestibulate*.

vestigial Of small or reduced structure; of a functionless structure representing a useful organ of a lower form. n. *vestige*.

vestiture A covering, e.g. of hairs, feathers, fur or scales.

V-gene That coding for the variable region of immunoglobulin heavy or light chain. It is widely separated from the C-gene (constant region) in germ line DNA (as present in lymphocyte precursors and other body cells such as in liver), but during maturation of B lymphocytes it is rearranged by translocation to a position close to the 5′ end of the C-gene, but still separated from it by an **intron**. A similar process takes place during the maturation of T lymphocytes.

viable Capable of living and developing normally.

Vi antigen Surface somatic antigen present in freshly isolated strains of *Salmonella typhi* and *S. paratyphi*, which masks the O antigen and renders the organisms relatively unable to combine with antibody against the O antigen. Vi antigen is associated with virulence, possibly for this reason.

vibrissa (1) In Mammals, one of the stiff tactile hairs borne on the sides of the snout and about the eyes. (2) One of the vaneless rictal feathers of certain Birds, e.g. Flycatchers. pl. *vibrissae*.

vigilance A term for the state of readiness to detect changes in the environment.

villose, villous Shaggy.

villus A hair-like or finger-shaped process, such as the absorptive processes of the Vertebrate intestine; one of the vascular processes of the Mammalian placenta which fit into the crypts of the uterine wall. pl. *villi*. adjs. *villous, villiform*.

vimentin Intermediate filament protein characteristic of fibroblasts.

vinca alkaloid Alkaloids used in the treatment of cancer, **leukaemia** and **lymphoma**, which interfere with cell division by causing metaphase arrest. *Vincristine* and *vinblastine* are common examples.

virescence Abnormal, usually pathogenic, condition in which flowers remain green.

virology The study of viruses.

virulence Capacity of a pathogen to cause disease.

virulent phage One which always kills its host. Ant. *lysogenic phage*. See **lysogeny**.

virus A particulate infective agent smaller

vitamins

Organic substances required in relatively small amounts in the diet for the proper functioning of the organism. Lack causes *deficiency diseases* curable by administration of the appropriate vitamin. There are two main groups: the *fat soluble*, vitamins A, D, E and K, and the *water soluble*, vitamin C and the vitamins of the B complex.

Vitamin A or *retinol*. A precursor of the prosthetic group of the light sensitive protein, rhodopsin. Deficiency of vitamin A causes night blindness. It is also required by young animals for growth. Fish liver oils and dairy products are rich sources of vitamin A.

Vitamin B complex. (1) B1 (*thiamin*). As its pyrophosphate it functions as a coenzyme of various enzymes. Deficiency results in the disease beri-beri. Present in yeast and cereal germs. (2) B2 (*riboflavin*). Forms part of the prosthetic group of flavoproteins. Deficiency causes skin and corneal lesions. (3) Niacin (*nicotinic acid*). Component of the coenzyme nicotinamide adenine dinucleotide, NAD. Deficiency results in the disease pellagra. (4) B6 (*pyridoxal*). As its phosphate it acts as a coenzyme for transaminases. (5) Pantothenic acid. A component of coenzyme A. (6) Biotin. The prosthetic group of the enzyme carboxylase. (7) Folic acid (*tetrahydrofolate*). Serves as a donor of 1-carbon fragments for several biosyntheses. Deficiency inhibits these reactions which include the synthesis of purines. (8) B12 (*cobalamin*). Component of the coenzyme cobalamin which takes part in enzymic interconversions of acyl CoAs and methylations. Used in the treatment of pernicious anaemia. Liver is a rich source.

Vitamin C or *ascorbic acid*. Important in the hydroxylation of collagen which in its absence is inadequately hydroxylated. The defective collagen produces the skin lesions and blood vessel weaknesses which are characteristic of scurvy, the deficiency disease of this vitamin. Fresh fruit and green vegetables are important sources.

Vitamin D or *calciferol*. The vitamin involved in calcium and phosphorus metabolism. Deficiency impairs bone growth and causes the disease ricketts. Fish liver oils are a rich source and the vitamin can, in sunlight, be synthesized in the skin from cholesterol.

Vitamin E or *α-tocopherol*. A vitamin involved in reproduction. Its absence leads to sterility in both sexes.

Vitamin K. A necessary requirement for the production of prothrombin, the precursor of thrombin, and consequently essential for normal blood coagulation.

than accepted bacterial forms, invisible by light microscopy, incapable of propagation in inanimate media and multiplying only in susceptible living cells, in which specific cytopathogenic changes frequently occur. Causative agent of many important diseases of man, lower animals and plants, e.g. poliomyelitis, foot and mouth disease, tobacco mosaic. See **bacteriophage**.

virus neutralization tests Tests used to identify antibody response to a virus or, using a known antibody, to identify a virus. Depends on specific antibody neutralizing the infectivity of a virus by preventing it from binding to the target cell. They may be carried out *in vivo* in susceptible animals or chick embryos or, more usually, in tissue culture.

visceral arch See **gill arch**.

visceral clefts The gill clefts, esp. the abortive gill clefts of higher Vertebrates.

viscus Any one of the organs situated within the chest and the abdomen; heart, lungs, liver, spleen, intestines etc. pl. *viscera*. adj. *visceral*.

visual cliff An experimental set up in which there is a vertical drop, over which an animal is prevented from falling by a sheet of glass. Some animals avoid this area as a result of the visual perception of the drop.

vital stain A stain which can be used on living cells without killing them.

vitamins See panel.

vitellarium A yolk-forming gland.

vitelligenous Yolk-secreting or yolk-producing.

vitelline Egg yellow; pertaining to yolk.

vitelline membrane A protective membrane formed around a fertilized ovum to prevent the entry of further sperms.

vitellus Yolk of egg.

vitreous humour The jelly-like substance filling the posterior chamber of the Vertebrate eye, between the lens and the retina.

viviparous Giving birth to living young which have already reached an advanced stage of development. Cf. **oviparous**. n. *viviparity*.

vivipary (1) The production of bulbils or small plants in place of flowers, as in e.g. *Festuca vivipara*. (2) The premature germination of seeds or spores before they are shed from the parent plant, as in many mangrove trees.

vocal cords In air-breathing Vertebrates, folds of the lining membrane of the larynx by the vibration of the edges of which, under the influence of the breath, the voice is produced.

vocal sac In many male Frogs, loose folds of skin at each angle of the mouth which can be inflated from within the mouth into a globular form, and act as resonators.

volant Flying; pertaining to flight.

voltinism Breeding rhythm; brood frequency. See **univoltine**, **bivoltine**, **multivoltine**.

voluntary muscle Any muscle controlled by the motor centres in the brain, the skeletal muscles. All such muscles are striated. Cf. **involuntary muscles**.

volva Cup-like structure at the base of the fruiting bodies of many basidiomycetes, e.g. mushroom, representing the remains of the universal veil.

vomer A paired membrane bone forming part of the cranial floor in the nasal region of the Vertebrate skull; believed not to be homologous in all groups. adj. *vomerine*.

vomerine teeth In most Fish and Amphibia, teeth, sometimes atypical, borne on the vomers.

voyeurism Sexual gratification through the clandestine observation of other people's sexual activities or anatomy.

vulva The external genital opening of a female Mammal. adj. *vulviform*.

W

waggle dance Semicircular movements of a hive bee, including an *abdominal waggle*, on returning from a foraging trip of more than 150 yd from the hive; conveys information about the direction and distance of a food source to worker bees, and stimulates them to visit the site. Cf. **round dance**.

Waldenstrom's macroglobulinaemia Disease occurring mainly in elderly males, characterized by the presence of large amounts of monoclonal IgM in the blood, lymphoid tissue enlargement, splenomegaly, a haemorrhagic tendency and depression. The IgM is occasionally found to have detectable antibody activity, e.g. rheumatoid factor. The disease is probably a relatively benign and slowly progressing form of myelomatosis.

waldsterben Symptoms of tree decline in Central Europe from the 1970s, not attributable to known diseases, and widely held to be caused by atmospheric pollution.

wall See **cell wall**.

Wallace's line An imaginary line passing through the Malay Archipelago and dividing the oriental faunal region from the Australasian region.

wandering cells Migratory amoeboid cells; may be leucocytes or phagocytes.

warm-blooded Said of animals which have the bodily temperature constantly maintained at a point usually above that of the environment, of which it is independent. Also called *idiothermous* or *homoiothermous*.

war neurosis A preferable syn. for *shellshock*. The term was originally used (World War I) for all types of nervous conditions resulting from war experiences, esp. those caused by a bursting shell, which might result in (*a*) a condition of physical shock or concussion to the nervous system, (*b*) the precipitation of a psychoneurosis in a predisposed individual, (*c*) a combination of these conditions.

warning coloration See **aposematic coloration**.

Wassermann reaction Complement fixation test formerly used in the diagnosis of syphilis. Cardiolipin derived from ox heart is used as antigen because the blood of persons with syphilis regularly contains antibody which reacts with this substance. However, false results may be obtained in cases where there are increased immunoglobulin levels, e.g. due to parasitic infection or in autoimmune conditions. Confirmatory tests using *Treponema pallidum* itself as the antigen may need to be made. The latter procedure is now generally used.

WAT Abbrev. for *Word Association Test*.

water culture See **hydroponics**.

water-in-oil emulsion adjuvant An adjuvant in which the antigen, dissolved or suspended in water, is enclosed in tiny droplets within a continuous phase of mineral oil. The antigen solution constitutes the dispersed phase, stabilized by an emulsifying agent such as mannitol mono oleate. Cf. **Freund's adjuvant**.

water pore, water stoma Opening in the epidermis, associated with a hydathode, through which water exudes. It is often a modified stoma.

water potential ψ_w. A measure of the free energy of water in a solution, as in a cell or soil sample, and hence of its tendency to move by diffusion, osmosis or as vapour. It is the chemical potential of water in solution minus that of pure water (0 at standard temperature) divided by the partial molar volume of water; expressed as units of pressure, MPa or bar. Water diffusing or osmosing always moves down a water potential gradient. The components of water potential are **osmotic potential**, **turgor potential** and **matric potential**.

water-storage tissue Tissue composed of large, highly vacuolate cells with relatively extensible walls, which can buffer the water supply. Water can also be stored in tree trunks, and in tracheids which can be emptied and refilled.

water-vascular system A system of coelomic canals in the Echinodermata, associated with the tubefeet and in which water circulates; in Platyhelminthes, the excretory system.

W-chromosome See **sex determination**.

web The mesh of silk threads produced by Spiders, some Insects and other forms; the vexillum of a feather; the membrane connecting the toes in aquatic Vertebrates, such as the Otter.

webbed Having the toes connected by membrane, as in Frogs, Penguins, Otters.

Weberian ossicles *Weberian apparatus*. In some Teleostei, e.g. Carp, Catfish and others, a chain of small bones, derived from processes of the anterior vertebrae, which connect the air bladder to the ear, transmitting vibrations from the former to a perilymphatic sac from which they pass to the endolymph of the inner ear. They correspond functionally to the middle ear ossicles of higher Vertebrates.

weed A plant growing where it is not wanted by Man. Weeds of cultivated land are often natural plants of disturbed habitats and are often apomictic or self-pollinating, or spread vegetatively. See **R-strategist**.

Weil-Felix reaction An agglutination test used in the diagnosis of rickettsial infections (typhus, etc.) which depends upon a carbohydrate cross-reacting antigen shared by *Ricckettsiae* and certain strains of *Proteus*. The agglutination pattern of patients with rickettsial disease against O-agglutinable strains of *Proteus* OX19, OX2 and OXK is diagnostic of the various rickettsial diseases.

Weltanschauung German word for *world outlook, philosophy of life*.

western blotting Technique for the analysis and identification of protein antigens. The proteins are separated by polyacrylamide gel electrophoresis and then transferred electrophoretically, *blotted*, to a nitrocellulose membrane or chemically treated paper to which the proteins bind in a pattern identical to that in the gel. Bands of antigen bound to the membrane or paper are detected by overlaying with antibody, followed by anti-immunoglobulin or protein-A labelled with a radioisotope, fluorescent dye, enzyme or colloidal gold. Termed *western* because it is similar to the Southern and northern blotting methods used for DNA and RNA.

wet deposition Deposition of materials from the atmosphere in rain.

wet rot (1) A rot in which the tissue is rapidly broken down with the release of water from the lysed cells, as in the brown rots of stored fruits. (2) The rot of timber that is often wet, caused by the fungus, *Coniophora puteara*.

whalebone See **baleen**.

whiplash flagellum *Acronematic flagellum*. One without hairs on its surface. Ant. *tinsel flagellum*, a decorated flagellum.

white cell See **leucocyte**.

white fibres Unbranched, inelastic fibres of connective tissue occurring in wavy bundles. Cf. **yellow fibres**.

white fibrocartilage A form of fibrocartilage in which white fibres predominate.

white matter An area of the central nervous system, mainly composed of cell processes, and therefore light in colour.

whorl (1) A group of 3 or more plant structures arising at the same level on a stem and forming a ring around it. (2) A ring of floral organs round the receptacle of a flower. (3) A single turn of a spirally-coiled shell or other spiral structure.

Widal reaction Bacterial agglutination test used in the diagnosis of enteric fevers. The patient's serum is titrated against reference strains of each of the likely organisms. The organisms must be motile, smooth and in the specific phase. Formalinized and alcoholized suspensions are used respectively for testing for H (flagellar) and O agglutinins. The Widal test is not likely to be positive be-

fore the 10th day of the disease and prior immunization with TAB vaccine may cause false positive results. In enteric infection however the titre against the infecting organism will continue to rise.

wide spectrum Of antibiotics etc. effective against a wide range of micro-organisms. Also *broad spectrum*.

wild type The normal phenotype with respect to a specified gene locus; usually symbolized by +.

wilt A type of plant disease characterized by wilting as an early symptom and usually caused by the infection of the vascular system by a fungus or bacterium, e.g. Dutch elm disease, caused by the fungus *Ceratocystis ulmi*.

wilting Loss of stiffness due to shortage of water and the loss of turgor by the cells. See **permanent wilting point**.

wind dispersal The dispersal of spores, seed and fruits by the wind.

wind pollination The conveyance of pollen from anthers to stigmas by means of the wind. Also *anemophily*.

wing (1) Any broad flat expansion. (2) An organ used for flight, as the forelimb in Birds and Bats. (3) The membranous expansions of the mesothorax and metathorax in Insects. (4) A longitudinal flange on a stem or stalk; the downwardly continued lamina of a decurrent leaf. (5) A flattened outgrowth of a seed or fruit aiding in wind dispersal.

wing coverts See **tectrices**.

winter annual A plant which completes its life cycle over a few months in the coldest part of the year, surviving the summer as seeds. See **annual**.

winter egg In some freshwater animals, a thick shelled egg laid at the onset of the cold season which does not develop until the following warm season. Cf. **summer egg**.

Wirsung's duct The ventral or main pancreatic duct of Mammals.

Wiskott-Aldrich syndrome Sex-linked recessive disease of infants characterized by haemorrhagic diathesis, eczema and recurrent infections. Delayed hypersensitivity reactions are absent and there is a defective antibody response to polysaccharide antigens, with low blood IgM levels. A protein normally present in platelet membranes is missing. This combined defect of cell-mediated and humoral immunity probably involves failure to recognize or process antigens. Patients die early of infection or bleeding due to increased destruction of platelets.

witches' broom A dense tuft of twigs formed on a woody plant as a response to infection.

withdrawal symptoms Temporary psychological and physiological disturbances resulting from the body's attempt to readjust to the absence of a drug.

Wolffian body The **mesonephros**.

Wolffian duct The kidney duct of Vertebrates. In adult anamniotes (Agnatha, Fish, and Amphibia) it serves as a kidney duct and a sperm duct in males, while in adult amniotes (Reptiles, Birds and Mammals) whose metanephric kidney has a separate duct, it is present only in males, forming the *vas deferens*.

wood See **xylem**. Wood, the constructional material, is the secondary xylem of conifers (softwood) and dicotyledons (hardwood). See **heartwood**, **sapwood**.

wood fibre, -parenchyma, -ray See **fibre, xylem parenchyma, ray**

woodland Natural or semi-natural vegetation containing trees, but not forming a continuous canopy. See **forest**.

woody tissues Tissues which are hard because of the presence of lignin in the cell walls.

wool A modification of hair in which the fibres are shorter, curled, and possess an imbricated surface. Specifically, the covering of a sheep. The fibres are covered with small scales and are composed of keratin. Also *fleece wool*.

word association test A psychological test in which the subject is presented with a stimulus word and asked to produce the first word that comes to mind; latency to response, and the nature of the association word are interpreted as revealing verbal habits, thought processes, personality characteristics and emotional state. Abbrev. *WAT*.

word salad A schizophrenic speech pattern in which words and phrases are combined in a disorganized fashion, apparently devoid of logic and meaning.

worker In social Insects, one of a caste of sterile individuals which do all the work of the colony.

worm An imprecise term applied to elongated Invertebrates with no appendages, as in Flatworm (Platyhelminthes), Roundworm, Eelworm (Nematoda), Earthworm (Lumbricus spp.) etc. Also applied to immature forms of some Insects, as in Mealworm, (Tenebrio, Coleoptera), Cutworm (some Lepidoptera), Wireworm (Elateridae, Coleoptera), Click beetles and Millipedes (Diplopoda).

wound tissue Those formed in response to wounding as the wound vessels which reconnect severed xylem strands. See **callus**.

XYZ

x, 2x, 3x.... Symbols for the number of copies of the *haploid chromosome number* or *basic chromosome set*.

xanthochroism A condition in which all skin pigments other than golden and yellow ones disappear, as in the Goldfish.

xanthophore A cell occurring in the integument and containing a yellow pigment, as in Goldfish. Also *guanophore, ochrophore*.

Xanthophyceae The yellow-green algae, a class of eukaryotic algae in the division Heterokontophyta. Without fucoxanthin. Naked or walled. Flagellated and amoeboid unicellular, palmelloid, coccoid, dendroid, simply filamentous and siphoneous types, many remarkably similar to analogous green algae. Mostly phototrophs. Mostly fresh water and in soil.

xanthophylls Yellowish, oxygenated carotenoids acting as minor or, in a few cases, major **accessory pigments** in photosynthesis. Each major algal and plant group has its characteristic set of xanthophylls. Xanthophyll itself, $C_{40}H_{56}O_2$, is one of the two yellow pigments present in the normal chlorophyll mixture of green plants; also a yellow pigment occurring in some Phytomastigina.

X-chromosome See **sex determination**.

xenia The influence of the pollen on the seed through its effect, by double fertilization, on the nature of the endosperm. Cf. **metaxenia**.

xeno- Prefix from Gk. *xenos*, strange, foreign.

xenogamy Fertilization involving pollen and ovules from flowers on genetically non-identical plants of the same species (i.e. different *genets*) See **cross pollination, cross fertilization, outbreeding**.

xenogeneic Grafted tissue that has been derived from a species different from the recipient. Hence *xenograft*.

xeric Dry enough to restrict plant growth.

xeroderma pigmentosum A heritable disease of young children in which prolonged exposure to sunlight on the skin causes erythematous patches which later become pigmented, scaly, wartlike and finally cancerous.

xeromorphic Morphologically characteristic of xerophytes.

xerophyte A plant adapted to a dry habitat, where growth may be limited by water shortage.

xerosere A succession beginning on dry land, as opposed to starting under water. Cf. **hydrosere**.

X-inactivation Permanent condensation of one or other X-chromosome which occurs in the early development of female mammals, with accompanying repression of most of the genes on that chromosome.

xiphisternum A posterior element of the sternum, usually cartilaginous.

xiphi-, xipho- Prefixes from Gk. *xiphos*, sword.

xiphoid Sword-shaped.

X-linkage See **sex linked**.

X-ray analysis The collection and analysis of the X-rays emitted when a beam of electrons in e.g. an **electron microscope** interacts with matter. Such X-rays have wavelengths characteristic of the interacting atoms and can thus give information about both elemental composition and spatial distribution in a specimen.

X-ray crystallography See panel.

xylem The **vascular tissue** with the prime function of water transport; it consists of tracheids and vessels and associated parenchyma and fibres. Secondary xylem (wood) may also be important for support (tracheids and fibres) and storage (xylem parenchyma). See **primary xylem, secondary xylem, ray, apoplast, axial** system, **conduit, cohesion theory**.

xylem parenchyma Parenchyma cells chiefly within the secondary xylem, mostly with lignified walls and with living contents in the sap wood. Has storage and defensive functions.

xylogenous, xylophilous Growing on wood; living on or in wood.

xylophagous Wood-eating.

xylophilous See **xylogenous**.

xylose Xylose is a pentose found in many plants. It is a stereoisomer of arabinose. Also *wood sugar*.

xylotomous Wood-boring; wood-cutting.

XYY syndrome Condition in which the human male has an extra Y chromosome. They are normal males, except for slight growth and sometimes minor behavioural abnormalities.

Y-chromosome See **sex determination**.

yeast Unicellular fungus reproducing asexually by budding or division, esp. the genus, Saccharomyces including *S. cerevisiae* (baker's and brewer's yeast) and *S. ellipsoides* (used in wine making).

yeast genetics See panel.

yellow cells In Oligochaeta, yellowish cells forming a layer investing the intestine and playing a role in connection with nitrogenous excretion; chloragogen cells.

yellow fever *Yellow jack.* An acute infectious disease caused by a virus, conveyed to

X-ray crystallography

This is the main technique providing images of biological molecules in which the arrangement of atoms can be seen. It has had a dramatic impact in explaining, in particular, enzyme function, antibody-antigen recognition, light harvesting in chloroplasts, self-assembly of biological fibres and virus activity in terms of physics and chemistry.

However, all biological mechanisms have not yet been studied at the atomic level because certain technical criteria must be fulfilled before X-ray crystallography can be applied. The most obvious is that the molecule or complex structure to be imaged must be crystallized. This involves the isolation in a highly pure form of adequate quantities of molecules, often extremely rare. Even then crystallization is by no means inevitable and is still the rate-limiting step in this kind of structural biochemistry.

In an ordinary light microscope, image formation takes place in two stages. An incident light beam is scattered by the object and these scattered rays are collected by a lens and recombined to form an image. Unfortunately there is, at present, no lens able to refocus X-rays. This is because X-ray physics determines that their refractive index at most interfaces is very close to unity. The scattered rays can however be collected from the object, giving the *Fraunhofer diffraction pattern* which contains the **amplitudes** of the scattered rays but not their **phases**. Thus only half the information needed by a computer to calculate the image is available. The phase problem has been solved in a number of ways of which **multiple isomorphous replacement** (*MIR*) has been most commonly used for proteins.

The structures of about 200 biological molecules (mostly proteins) have been determined together with a number of complexes and assemblies. There have also been a number of important technical advances. These include: (1) the ability to analyze the thermal dependence of diffraction patterns, providing information about both structure and dynamics, which has shown that the mobility as well as the geometry of proteins is important for their function; (2) the development of *synchroton radiation sources* which have provided X-ray beams which are much easier to use. They have beams of smaller cross-section and greatly enhanced brightness with easier tunability and lower wavelength than conventional laboratory sources, but are so expensive that they can only be operated as central facilities. Better sources have been matched by better X-ray detectors, allowing rapid and time-resolved data collection and facilitating the study of protein structure during the time course of a process like enzyme catalysis or muscle contraction. They have also made possible the study of smaller crystals at higher resolution, even of delicate molecules enclosed in containers.

These developments are now being applied to proteins produced as the result of **site specific mutagenesis**, whereby specific substitutions of one amino acid by another with different properties at e.g. the active site of an enzyme, can be made as a result of a rational choice involving knowledge of the structure of the site. Such methods in *protein engineering* should provide novel substances for medical and industrial purposes.

Man by the bite of the Mosquito *Aëdes aegypti* (*Stegomyia fasciata*); characterized by high fever, acute hepatitis, jaundice, and haemorrhages in the skin and from the stomach and bowels; it occurs in tropical America and West Africa.

yellow fibres   Straight, branched, elastic fibres occurring singly in areolar connective tissue. Cf. **white fibres**.

yellow fibrocartilage A form of fibrocartilage in which yellow fibres predominate.

yellows A disease of plants in which there is considerable yellowing (chlorosis) of normally green tissue, caused by viruses or mycoplasmata.

yellow spot *Macula lutea*; the small area at

yeast genetics

Yeasts are simple microbial **eukaryotes** with several features making them highly suitable for genetical and biochemical studies, esp. of fundamental cellular processes. Two species, *Saccharomyces cerevisiae* and *Schizosaccharomyces pombe* have been used most. In these and some others, both haploid and diploid cells can proliferate by vegetative growth but a sexual system is also present, with mating between haploid cells leading to nuclear fusion and the formation of a diploid nucleus. The resulting diploid cell can enter meiosis under suitable conditions, leading to the formation of four haploid progeny spores. Each spore can germinate and form a colony of cells with identical genotypes, allowing the progeny of a cross to be analysed. Thus yeasts behave genetically liker higher eukaryotes, classical genetic procedures being readily carried out with the major advantage that all four spores from a single meiosis can be micro-dissected apart, allowing sophisticated analysis.

Some mutants have simple biochemical defects, resulting in a failure to grow in the absence of particular components in the growth medium; others are unable to use certain carbon or nitrogen sources for growth. **Temperature sensitive mutants**, defective in central cellular processes, have also been obtained.

Additionally the methods of **genetic manipulation** or *recombinant DNA technology* can be applied to yeasts. DNA molecules carrying particular genes can be introduced into yeasts by simple procedures in the process called *transformation*. If a yeast mutant is transformed with a gene library of wild-type DNA, constructed in a suitable plasmid vector, then a small proportion of the transformants will acquire a copy of the functional gene and lose their mutant character. The plasmid responsible can be recovered from the yeast transformant and the structure and properties of the gene itself then investigated. Other types of transformation allow particular genes, present in the yeast chromosomes to be altered in a controlled manner, allowing mutations to be directed towards specific effects in the cell. Such mutations, generated *in vitro*, can then be studied at the molecular level.

the centre of the retina in Vertebrates at which day vision is most distinct.

Y-maze A maze similar to a **T-maze**, but in which the arms are not at right angles to the stem.

yolk The nutritive nonliving material contained by an ovum.

yolk duct Vitelline duct.

yolk epithelium The epithelium surrounding the yolk sac.

yolk gland See **vitellarium**.

yolk plug A mass of yolk-containing cells which partially occludes the blastopore in some Amphibians.

yolk sac The yolk-containing sac which is attached to the embryo by the yolk stalk in certain forms.

Z-chromosome See **sex determination**.

Z-DNA A form of duplex DNA, in which purines and pyrimidines alternate in a strand and which results in a left-handed helix.

zeitgeber Literally, a *time-giver* that synchronizes various rhythmic behaviours with external events.

zeugopodium The second segment of a typical pentadactyl limb, lying between the stylopodium and the autopodium; ante-brachium or crus; forearm or shank.

Z-helix A helix winding in the sense of a conventional, right-handed, screw.

Z-line These limit the sarcomeres of striated muscle and contain α-actin.

zoidiophilous Pollinated by animals.

zona An area, patch, strip or band; a zone. adjs. *zonal, zonary, zonate*.

zona granulosa The mass of membrana granulosa cells of the Graafian follicle around the ovum; *discus proligerus* or *cumulus oophorus*.

zona pellucida A thick transparent membrane surrounding the fully formed ovum in a Graafian follicle.

zona radiata The envelope of the Mammalian egg outside the vitelline membrane.

zonary placentation The condition in which the villi are on a partial or complete girdle around the embryo, as in Carnivora and Proboscidea.

zonation The occurrence, in an area, of distinct bands of vegetation each with its own characteristic dominant and other species, as the seaweeds on a shore or the vegetation on a mountain side. Cf. **succession**.

zonula ciliaris In the Vertebrate eye, a double fenestrated membrane connecting the ciliary process of the choroid with the capsule surrounding the lens.

zonule A small belt or zone, such as the zonula ciliaris of the Vertebrate eye.

zoo- Prefix from Gk. *zoon*, animal.

zoobiotic Parasitic on, or living in association with, an animal.

zooblast An animal cell.

zoochlorellae Symbiotic green algae found in various animals.

zoochorous Spores or seeds dispersed by animals.

zoocyst See **sporocyst**.

zoogamete A motile gamete.

zoogamy Sexual reproduction of animals.

zoogeography The study of animal distribution.

zooid An individual forming part of a colony in Protozoa (Volvocina), Coelenterata, Hemichordata, Urochorda and Bryozoa; in Polychaeta, a posterior sexual region formed by asexual reproduction; polyp; a polypide.

zooidogamy Fertilization by motile spermatozoids. Cf. **siphonogamy**.

zooplankton Floating and drifting animal life.

zoosperm A spermatozoid.

zoosporangium A sporangium in which zoospores are formed.

zoospore A motile, usually naked, asexual (i.e. not a gamete) reproductive cell found in some algae and fungi, swimming by means of one to several flagella.

z scheme Scheme linking **photosystem II** and **photosystem I** such that oxygen is produced by the former, NADP is reduced by the latter and ATP is generated by (noncyclic) photophosphorylation by *electron transport* from PS II to PS I.

zygapophyses Articular processes of the vertebrae of higher Vertebrates, arising from the anterior and posterior sides of the neurapophyses.

zygodactylous Said of Birds which have

the first and fourth toes directed backwards, as Parrots.

zygogenetic A product of fertilization.

zygoma The bony arch of the side of the head in Mammals which bounds the lower side of the orbit. adj. *zygomatic*. See **jugal.**

zygomatic arch See **zygoma.**

zygomatic bone See **jugal.**

zygomorphic Bilaterally symmetrical; divisible into 2 by one (vertical) plane only. Also *irregular*. Cf. **actinomorphic**.

Zygomycotina, Zygomycetes A subdivision or class of the Eumycota or true fungi. No motile stages. Usually mycelial, aseptate. Asexual spores formed in sporangia; sexual reproduction by formation of a zygospore. Mostly saprophytic, e.g. *Mucor*, the pin-mould; some insect parasites. Also *Glomus* spp. which form **vesicular arbuscular mycorrhizas** with very many plants.

zygonema The zygotene phase of meiosis.

zygospore (1) Any thick-walled resting spore formed directly from a zygote, as in many algae and in some fungi. See **oöspore**. (2) A thick-walled resting spore formed from the zygote resulting from the union of isogametes, i.e. in Zygnemaphyceae and Zygomycotina.

zygote The cell that results from the fusion of 2 gametes.

zygotene The second stage of meiotic prophase, intervening between leptotene and pachytene, in which the chromatin threads approximate in pairs and become looped. adj. *zygotic*.

zyg-, zygo- Prefixes from Gk. *zygon*, yoke.

zymogen Inert precursor of many active proteins and degradative enzymes. It is converted into the active form at the required site of activity. Thus trypsin is formed in the intestinal lumen from the inactive trypsinogen fibrin generated at the site of blood clotting from the inactive fibrinogen.

zymosan Cell wall fraction of yeast which activates the alternative **complement** pathway, and thus binds C3b. Frequently used for study of the capacity of cells to phagocytose opsonized materials.

zym-, zymo- Obsolescent prefixes relating to fermentation by enzymes.